Guide to
Electroporation and Electrofusion

Guide to Electroporation and Electrofusion

Edited by

Donald C. Chang

Department of Molecular Physiology and Biophysics
Baylor College of Medicine
Houston, Texas

Department of Biology
Hong Kong University of Science and Technology
Kowloon, Hong Kong

Bruce M. Chassy

Department of Food Science
University of Illinois at Urbana-Champaign
Urbana, Illinois

James A. Saunders

Plant Sciences Institute
Beltsville Agricultural Research Center
United States Department of Agriculture
Beltsville, Maryland

Arthur E. Sowers

Department of Biophysics
School of Medicine
University of Maryland at Baltimore
Baltimore, Maryland

Academic Press, Inc.
Harcourt Brace Jovanovich, Publishers
San Diego New York Boston London Sydney Tokyo Toronto

In Memorium

The editors would like to recognize the untimely passing of Eve Morris, co-founder and President of BTX, whose death on January 3, 1991 struck us all with a great sense of professional and personal loss. Pioneering work by BTX, one of several commercial companies, resulted in the introduction and widespread use of commercial pulse generators for electroporation and electrofusion, which contributed greatly to the development of this field of research.

This book is printed on acid-free paper. ∞

Copyright © 1992 by ACADEMIC PRESS, INC.
All Rights Reserved.
No part of this publication may be reproduced or transmitted in any form or by any means, electronic or mechanical, including photocopy, recording, or any information storage and retrieval system, without permission in writing from the publisher.

Academic Press, Inc.
San Diego, California 92101

United Kingdom Edition published by
Academic Press Limited
24–28 Oval Road, London NW1 7DX

Library of Congress Cataloging-in-Publication Data

Guide to electroporation and electrofusion / Donald C. Chang ...
    [et. al.].
        p.   cm.
    Includes index.
    ISBN 0-12-168040-1 (Hardcover)
    ISBN 0-12-168041-X (Paperback)
    1. Electroporation.   2. Electrofusion.   I. Chang Donald C.
GH585.5E48H36   1992
574.8--dc20   91-18976
                                                                           CIP

PRINTED IN THE UNITED STATES OF AMERICA
91 92 93 94   9 8 7 6 5 4 3 2 1

Contents

Part II
Applications of Electroporation and Electrofusion in Current Research

Part III
Practical Protocols for Electroporation and Electrofusion

Part IV
Instrumentation for Electroporation and Electrofusion

Preface

In the last decade, there has been a widespread use of electroporation and electrofusion in many different fields, including molecular biology, cell biology, plant genetics, hybridoma technology, agricultural research, and others. The number of published papers that involved the use of electroporation has grown in an exponential manner since the mid-1980s. Consequently, we feel there is a strong need for a comprehensive treatise on the subjects of electroporation and electrofusion.

This book includes four major parts. Part I deals with the basic principles and the fundamental processes of electroporation and electrofusion. The chapters in this part review the recent progress of studies aimed at providing a better understanding of the molecular mechanisms by which the externally applied electric field is able to induce membrane permeation and cell fusion. These studies include both theoretical and experimental works, and use a wide range of approaches including physical, chemical, and biological methods.

In Part II, important applications of electroporation and electrofusion in many different areas of biological research are reviewed. These applications include gene transfer in mammalian cells, genetic manipulation in plant cells, genetic transformation of bacteria and yeast, electroinjection of exogenous molecules for studying cellular functions, and production of hybridoma and human monoclonal antibodies. Other innovative applications such as embryo cloning, electroinsertion of proteins into cell membranes, and electrofusion of cells to tissues are also included.

The chapters of Part III are a collection of practical protocols. This section provides useful information to assist readers in designing experiments using electroporation and electrofusion. Step-by-step procedures are provided for the use of electroporation and electrofusion to introduce genes (and other molecules) into various biological cells, including mammalian cells, plant cells, bacteria, and yeasts. Methods to improve hybridoma production by electrofusion are also provided in some of these chapters.

Finally, in Part IV, the instrumentation required for electroporation and electrofusion is discussed. We have summarized the features of most of the commercially available equipment designed for electroporation and electrofusion. Some useful tips for optimizing the operation of this equipment are also provided.

This book is written to satisfy the needs of scientists in many different fields and at different levels of experience. For the beginners, this book can serve as a practical laboratory manual; for the research scientist, it can serve as a guide for experimental design, particularly in the development of research proposals; and for experienced scientists in this field, the book will serve as a current overview of the ongoing research in leading laboratories around the world.

Chapters in the book are written by top experts in the studies of electroporation and electrofusion. The authors include not only most of the well-known pioneers in these fields, but also many international authorities whose work has provided the most recent understanding of the basic mechanisms of electroporation and electrofusion. Chapters in Part II and Part III are contributed by leading scientists who are largely responsible for the development of many of the applications of electroporation and electrofusion in various areas of biological research. Because of their collective efforts, this book is able to provide an extraordinary breadth and depth of coverage.

One important reason we were able to attract so many top experts to contribute to this book was that the editors were involved in organizing the first International Conference on Electroporation and Electrofusion held at Woods Hole, Massachusetts, in late 1990. Without the impetus provided by the conference, it would have been nearly impossible to recruit all the experts to contribute to this book. We would like to gratefully acknowledge Life Technologies, Bio-Rad Laboratories, BTX, Inc., Hoefer Scientific Instruments, IBI/Kodak, the U. S. Department of Agriculture, and the Marine Biological Laboratory of Woods Hole. We also want to thank Huntington Potter and T. Y. Tsong, who made valuable contributions. Finally, the editors appreciate the support by Lorraine Lica of Academic Press.

1

Overview of Electroporation and Electrofusion

Donald C. Chang,[1] James A. Saunders,[2] Bruce M. Chassy,[3] and Arthur E. Sowers[4]

[1]Department of Molecular Physiology and Biophysics, Baylor College of Medicine, Houston, Texas 77030

[2]Plant Sciences Institute, Beltsville Agricultural Research Center, United States Department of Agriculture, Beltsville, Maryland 20705

[3]Department of Food Science, University of Illinois, Urbana, Illinois 61801

[4]Department of Biophysics, School of Medicine, University of Maryland at Baltimore, Baltimore, Maryland 21201

I. Introduction
   A. Electroporation
   B. Electrofusion
II. Advantages of Electroporation and Electrofusion
III. Mechanisms of Electroporation and Electrofusion
IV. Applications of Electroporation
V. Application of Electrofusion
   References

I. Introduction

A. Electroporation

Electroporation is a phenomenon in which the membrane of a cell exposed to high-intensity electric field pulses can be temporarily destabilized in specific regions of the cell. During the destabilization period, the cell membrane is highly permeable to exogenous molecules present in the surrounding media.

[1]Present address: Department of Biology, Hong Kong University of Science and Technology, Kowloon, Hong Kong.

Guide to Electroporation and Electrofusion

Electroporation thus can be regarded as a massive microinjection technique that can be used to inject a single cell or millions of cells with specific components in the culture medium.

Several publications appeared in the 1950s and 1960s that showed that an externally applied electric field can induce a large membrane potential at the two poles of the cell (Cole, 1968). It was known that an excessively high field could also cause cell lysis (Sale and Hamilton, 1967; Sale and Hamilton, 1968). By the early 1970s, several laboratories had found that when the induced membrane potential reaches a critical value, it can cause a dielectric breakdown of the membrane. Such breakdown was demonstrated in red blood cells (Crowley, 1973; Zimmermann *et al.*, 1974) and in model membranes (Neumann and Rosenheck, 1972; Neumann and Rosenheck, 1973).

By the late 1970s, the concept of "membrane pore" formation or membrane destabilization, as a result of dielectric breakdown of the cell membrane, was formally discussed (Kinosita and Tsong, 1977). At about the same time, it was found that if the electric field was applied as a very short duration pulse, the cells could recover from the electrical treatment. This implied that these electric field-mediated "pores" were resealable and could be induced without permanent damage to the cells (Baker and Knight, 1978a; 1978b; Gauger and Bentrup, 1979; Zimmermann *et al.*, 1980).

By the early 1980s, reports began appearing that showed that many small molecules, such as sucrose, dyes, or monovalent or divalent ions, could pass through these electric field-induced "membrane pores" to a broad array of cell types. Many laboratories started to use pulsed electric fields to introduce a variety of molecules into the cells, including drugs (Zimmermann *et al.*, 1980), catacholamine and Ca-EGTA (Knight and Baker, 1982), and DNA (Wong and Neumann, 1982; Neumann *et al.*, 1982; Potter *et al.*, 1984; Fromm *et al.*, 1986). In the last decade, there has been an explosion in the number of groups using this "electroporation" technique to incorporate various molecules into many different types of cells. Recently, a new method of electroporation which utilizes a pulsed radio-frequency electric field to break down the cell membrane has been developed (Chang, 1989).

B. Electrofusion

When neighboring cells are brought into contact during the electrically mediated membrane destabilization process outlined above, these cells can be induced to fuse. The number of cells that can be fused by the application of a pulse (or pulses) of this high-intensity electric field is dependent on the size and type of cell, as well as the field intensity of the electrical pulse. The experimental procedures are very similar to those of electroporation, except that the cells to be fused must be brought into contact first. This cell contact can be accomplished by (1) mechanical manipulation, (2) chemical treatment, or (3) dielectrophoresis (in which the cells are lined up in chains by applying a low-intensity, high-frequency, oscillating electric field).

The phenomenon of electrofusion is closely related to that of electroporation. Late in 1979 and early in the 1980s several laboratories had reported success in using electrical pulses to induce fusion in various systems, including plant cells (Senda *et al.*, 1979; Zimmermann and Scheurich, 1981) and red blood cells (Scheurich *et al.*, 1980). One significant contribution of Zimmermann's group was their utilization of the phenomenon of dielectrophoresis (Pohl, 1978) to facilitate cell contact, thus making the electrofusion method more widely useful. Since the beginning of the 1980s, "electrofusion" has been applied to fuse many different cell types, and has become the method of choice for cell hybridization.

II. Advantages of Electroporation and Electrofusion

The pulsed electric field method has a number of advantages over the conventional methods of cell permeabilization or cell fusion. It is a noninvasive, nonchemical method that does not seem to alter the biological structure or function of the target cells. Electrofusion is relatively easy to perform and is much more time efficient than the traditional chemical or biological fusion techniques. Also, unlike the other chemical or biological methods, the electric field method can be relatively nontoxic. The efficiency of the electric field method is generally significantly better than most alternative methods, and finally, because the electric field method is a physical method, it can be applied to a much wider selection of cell types.

III. Mechanisms of Electroporation and Electrofusion

The basic phenomenology of electroporation and electrofusion are reasonably well known, although the molecular mechanisms by which the electric field interacts with the cell membrane are still under active investigation. Basically, a membrane potential is induced by the externally applied electric field. The electrical field is usually induced by a relatively short DC pulse. The pulse can be either a square-wave pulse, usually with a duration of less than 100 μs, or it can be an exponentially decaying capacitive discharge pulse with a duration in the millisecond range (Saunders *et al.*, 1989).

When the induced potential reaches a critical value, it causes an electrical breakdown of the cell membrane. The value of this critical potential is about 1 V, but can vary depending on the pulse width, composition of membrane, etc. Multiple membrane pores are formed as a result of breakdown. Many studies have been done to characterize the structure and properties of these electropores (see Part I of this book). Very recently, porelike structures have been visualized for the first time in red blood cells using a rapid-freezing microscopy technique (see Chapter 2 of this book). The dynamics of pore formation and resealing are also under active investigation at this time (See Part I). The mechanisms by which membranes of neighboring cells are induced to fuse by the electric field is not yet clearly understood, but several theories have been proposed (See Chapters 6, 7, 8, 10, and 11).

Issues to be resolved include: Does the applied field cause a reversible or irreversible breakdown of the cell membrane? Does electroporation or electrofusion occur exclusively at the lipid bilayer region of the cell membrane? In other words, what is the role of membrane proteins?

IV. Applications of Electroporation

The applications of electroporation or electrically mediated gene transfer techniques are responsible for the major part of the popularity of this rapidly expanding field. The ability of a high-voltage pulse to reversibly change the permeability of the cell membrane leaves the tantalizing possibility of incorporating specific genes into relatively large numbers of isolated cells. Although it is not 100% effective, transformation yields as high as 60–70% have been obtained with some regularity (Saunders, *et al.,* 1989). Different researchers have used a variety of names to describe the electrically mediated gene transfer processes, including electroinjection, electrotransfection, and electrical microinjection, as well as electroporation, but the basic process is similar in all cases.

Specific applications for electroporation have involved the introduction of both DNA and RNA to a variety of plant, animal, bacterial, and yeast cells. Although marker genes were originally the most popular type of DNA to be incorporated into the recipient cells, recent trends have used functional genes that are important to biotechnology. Other major applications are injection of drugs, proteins, metabolites, molecular probes, and antibodies for studies of cellular structure and function.

V. Applications of Electrofusion

The applications of electrofusion extend into many different areas using a wide variety of cell types. In plants, where individual cells have the potential to regenerate into mature differentiated tissue, somatic hybridization of isolated protoplasts by electrofusion has been a popular method of genomic gene transfer. This is an extremely efficient method of cell fusion, which results in relatively high yields of multinucleate cells containing the entire combined genomes of each parental cell type. Unfortunately, fusions among each of the parental cell types are as common, if not more so, than fusions between the different parental cell types. Thus, the selection of the hybrid cell of choice is an integral part of any plant fusion protocol.

A second area that has gained considerable interest in electrofusion research has been that of hybridoma/monoclonal antibody production. The selection system for the proper fusion partners, that is, antibody production, is already built into the system. Harvested cells producing the desired antibody can be collected in culture, processed, and relatively large amounts of antibodies recovered. The use of electrofusion techniques in this application has sometimes improved the yields and recoverability of hybridoma cells by 100-fold in comparison to chemical fusion methods.

Another exciting area of electrofusion that is just emerging is the area of cell/tissue electrofusion. Experimental protocols in which isolated cells are electrofused to various tissue either *in vitro* or in some cases *in vivo,* are being used to effect genetic transformations that were previously not possible (Heller and Gilbert, Chapter 24 of this book).

In summary, the electroporation and electrofusion techniques are highly versatile and widely useful physical methods that have tremendous potential applications in cell biology, molecular biology, biotechnology, and other branches of biological research.

References

Baker, P. F., and Knight, D. E. (1978a). A high voltage technique for gaining rapid access to the interior of secretory cells. *J. Physiol.* **284,** 30.

Baker, P. F., and Knight, D. E. (1978b). Influence of anions on exocytosis in leaky bovine adrenal medullary cells. *J. Physiol.* **296,** 106.

Chang, D. C. (1989). Cell poration and cell fusion using an oscillating electric field. *Biophys. J.* **56,** 641–652.

Cole, K. S. (1968). A chapter of classical biophysics. *In* "Membranes, Ions, and Impulses." University of California Press, Berkeley, pp. 12–18.

Crowley, J. M. (1973). Electrical breakdown of bimolecular lipid membranes as an electromechanical instability. *Biophys. J.* **13,** 711–724.

Fromm, M. L., Taylor, P., and Walbot, V. (1986). Stable transformation of maize after gene transfer by electroporation. *Nature* **319,** 791–793.

Gauger, B., and Bentrup, F. W. (1979). A study of dielectric membrane breakdown in the *Fucus* egg. *J. Membrane Biol.* **48,** 249–264.

Kinosita, K., Jr., and Tsong, T. Y. (1977). Hemolysis of human erythrocytes by a transient electric field. *Proc. Natl. Acad. Sci. USA* **74,** 1923–1927.

Knight, D. E., and Baker, P. F. (1982). Calcium dependence of catecholamine release from bovine adrenal medullary cells after exposure to intense electric fields. *J. Membrane Biol.* **68,** 107–140.

Neumann, E., and Rosenheck, K. (1972). Permeability changes induced by electric impulses in vesicular membranes. *J. Membrane Biol.* **10,** 279–290.

Neumann, E., and Rosenheck, K. (1973). Potential difference across vesicular membranes. *J. Membrane Biol.* **14,** 194–196.

Neumann, E., Schaefer-Ridder, M., Wang, Y., and Hofschneider, P. H. (1982). Gene transfer into mouse myloma cells by electroporation in high electric fields. *EMBO J.* **1,** 841–845.

Pohl, H. A. (1978). "Dielectrophoresis." Cambridge University Press, London.

Potter H., Weir, L., and Leder, P. (1984). Enhancer-dependent expression of human k immunoglobulin genes introduced into mouse pre-B lymphocytes by electroporation. *Proc. Natl. Acad. Sci. USA* **81**, 7161–7165.

Sale, A. J. H., and Hamilton, W. A. (1967). Effects of high electric fields on microorganisms. I. Killing of bacteria and yeast. II. Mechanism of action of the lethal effect. *Biochim. Biophys. Acta* **148**, 781–800.

Sale, A. J. H., and Hamilton, W. A. (1968). Effects of high electric fields on microorganisms. III. Lysis of erythrocytes and protoplasts. *Biochim. Biophys. Acta* **163**, 37–43.

Saunders, J. A., Smith, C. R. and Kaper, J. M. (1989). Effects of electroporation pulse wave on the incorporation of viral RNA into tobacco protoplasts. *BioTechniques* **7**, 1124–1131.

Scheurich, P., Zimmermann, U., Mischel, M., and Lamprecht, I. (1980). Membrane fusion and deformation of red blood cells by electric fields. *Z. Naturforsch* **35c**, 1081–1085.

Senda M., Takeda, J., Abe, S., and Nakamura, T. (1979). Induction of cell fusion of plant protoplasts by electrical stimulation. *Plant Cell Physiol.* **20**, 1441–1443.

Wong T. K., Neumann, E. (1982). Electric field mediated gene transfer. *Biochem. Biophys. Res. Commun.* **107**, 584–587.

Zimmermann, U., and Scheurich, P. (1981). High frequency fusion of plant protoplasts by electric fields. *Planta* **151**: 26–32.

Zimmermann, U., Pilwat, G., and Riemann, F. (1974). Dielectric breakdown of cell membranes. *Biophys. J.* **14**, 881–889.

Zimmermann U., Vienken, J., and Pilwat, G. (1980). Development of drug carrier systems: electric field induced effects in cell membranes. *J. Electroanal. Chem.* **116**, 553–574.

Part I

Mechanisms and Fundamental Processes in Electroporation and Electrofusion

2

Structure and Dynamics of Electric Field-Induced Membrane Pores as Revealed by Rapid-Freezing Electron Microscopy

Donald C. Chang[1]

Department of Molecular Physiology and Biophysics, Baylor College of Medicine,
Houston, Texas 77030

[1]Present address: Department of Biology, Hong Kong University of Science and Technology, Kowloon, Hong Kong.

I. Introduction

The cell membrane can be transiently permeabilized by exposing the cell to a high-intensity electric field pulse (Kinosita and Tsong, 1977; Benz and Zimmermann, 1981; Neumann et al., 1982; Knight and Baker, 1982; Sowers and Lieber, 1986). Molecules can enter or leave the cell during this permeabilized state. Because this transient permeability is thought to result from the creation of membrane pores by the applied electric field, this process is called electroporation.

In recent years, electroporation has become the most promising method of gene transfer (Chu et al., 1987; Potter, 1988; Chang et al., 1991). It has also been used to introduce proteins (Hashimoto et al., 1989; Winegar, 1989; Tsongalis et al., 1990), metabolites (Swezey and Epel, 1988; Sokolowski et al., 1986), and antibodies (Chakrabarti et al., 1989) into living cells (see also Chapter 19 by Chang et al. in this book). In spite of these successes, the basic mechanisms of electroporation still remain largely unknown. The concept that membrane pores may be created by the electric field is mainly a theoretical hypothesis. There have been conflicting estimates about the structure and properties of these hypothetical "electropores" (Kinosita and Tsong, 1977; Chernomordik et al., 1983; Powell and Weaver, 1986; Sowers and Lieber, 1986); the so-called "pores" could be "craters," "cracks" or other forms of defects of the membrane structure (Stenger and Hui, 1986; Forster and Neumann, 1989). In fact, even if pores exist as a result of exposure to the applied field, it is still not clear whether cells could take up exogenous DNA via these pores. Early attempts to characterize the transport properties of the membranes in electropermeabilized red blood cells (Kinosita and Tsong, 1977) suggested that electropores were only large enough to allow sucrose molecules to pass through (about 1 nm in diameter). Thus, it is difficult to explain how DNA molecules (which may be several micrometers long and more than 6 nm wide) could enter the cells through these small electropores (Wong and Neumann, 1982).

In order to reveal the ultrastructure of the electropermeabilized membrane and to understand the dynamics of pore formation and resealing, we collaborated with Dr. Thomas S. Reese of the National Institutes of Health (NIH) to examine the electroporated cells using a rapid-freezing electron microscopy (EM) technique. Before our investigation, no electropores in the cell membrane had been directly visualized. With a combination of a specially developed cryofixation method and freeze-fracture EM, we were able to capture the changes in membrane structure with a time resolution of less than 1 ms. Preliminary findings of this study have provided the first evidence and characterization of membrane pores created by the electric field (Chang and Reese, 1990).

In order to simplify the interpretation of the morphological data, it is desirable to choose a proper cell model in which the cell membrane structure is relatively smooth. Thus, we have used scanning EM to examine the surface structures of a number of cell types. Some of the eukaryotic cells (such as COS-M6 cells) were

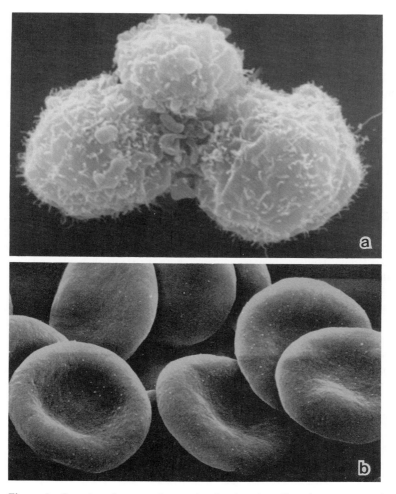

Figure 1   Scanning electron micrographs showing the cell surface structure of (a) COS M-6 cells and (b) human red blood cells. (Panel b was adapted from Fujita *et al.*, 1981.)

found to be covered with numerous microvilli structures (Fig. 1A), which would make interpretation of the freeze-fracture results very complicated. (For example, fractured microvilli could be easily confused with membrane openings.) The cell model we decided to use was human red blood cell (RBC). The membrane structure of the normal RBC is very well known (Pinto da Silva, 1972; Weinstein, 1974). Since the structure of RBC is relatively simple and uniform (Fig. 1B), it would be easier to detect the electric field-induced structural changes in such a cell.

II. Methods

A. Sample Preparation

Red blood cells were collected from human blood. They were washed twice and then resuspended in a low ionic strength medium, composed of 14 mM Na phosphate, 13 mM NaCl, and 150 mM sucrose (pH 7.4). The suspended cells were then ready for use in the freeze-fracture EM experiments. In the freeze-substitution experiment, red cells were prepared in a similar manner, except that 3% bovine serum albumin was added to the medium. The purpose of adding the albumin was to keep the suspended cells from falling apart during the substitution.

B. Cryofixation and Electroporation

Suspended human red cells (0.2 μl) were sandwiched between two thin copper plates (Balzers Union, Hudson, NH) which served both as the sample holder and electrodes for the applied electric field (see Fig. 2). These copper plates (about 5 mm long and 3 mm wide) were separated by a thin layer of insulator made of Parafilm. The cell sample sandwich was mounted on the tip of a spring-driven vertical plunger on which two electronic sensors were installed to monitor the position of the specimen during the plunging action (Chang and Reese, 1990). Upon the release of a solenoid, the sample sandwich was plunged into a liquid propane/ethane mixture cooled by liquid nitrogen. Using trigger signals from the electronic sensors and a delay circuit, we could apply an electrical pulse to porate the cells at a specific interval before the

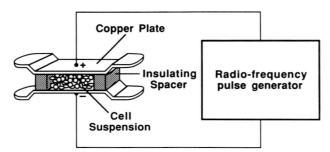

Figure 2   Schematic showing the arrangement of the cell sample for rapid freezing. Cells suspended in poration medium were sandwiched between two thin copper plates, which served as both the sample holder and electrodes for the applied electric field.

specimen reached the coolant. This time interval could be adjusted from 1 ms to many minutes.

By measuring the specimen capacitance during freezing, we estimated that the first layer of cells next to the copper plate surface was frozen within 0.1 ms after the sample sandwich was immersed by the liquid coolant. Such a rapid freezing rate has two advantages: First, it insures that the observed membrane structure is not distorted by formation of large ice crystals; second, it allows us to capture transient morphological changes in the cell membrane.

Unlike conventional electroporation methods, which utilize a DC (direct current) pulse to porate cells, we used a DC-shifted radio-frequency (RF) pulse (oscillating frequency 100 kHz, 0.3 ms wide, field strength 4–5 kV/cm). It has been shown in our earlier studies that the RF field was more effective than the DC field in electroporating cells, and, the RF field also resulted in better cell viability (Chang, 1989a, 1989b; also, see Chapter 19 by Chang et al., in this book).

C. Freeze-Fracture and Freeze-Substitution Electron Microscopy

Following the rapid-freezing procedure, the frozen samples were stored in liquid nitrogen for later processing. Most of the cell samples were examined by standard complementary-replica freeze-fracture electron microscopy techniques using a Balzers 301 freeze-fracture apparatus. Basically, frozen samples were fractured at −112°C under high vacuum, etched slightly for 2 min, and rotary-shadowed with platinum and carbon. The replicas were cleaned in nitric acid and sodium hypochlorite and rinsed twice in distilled water. For samples processed by freeze-substitution, we followed the procedures of Ornberg and Reese (1981). Rapid-frozen samples were submerged in a freshly prepared solution of 4% osmium tetraoxide in acetone, precooled in liquid nitrogen. The vials containing the samples were placed in a freezer (−80°C) to incubate for 1 day and then were allowed to warm slowly to room temperature over a 9- to 12-h period. After 1–2 h at room temperature, the samples were washed 3 times for 20 min in acetone, and stained with 0.1% hafnium tetrachloride in acetone for 3–4 h. The samples were then embedded in Araldite. After the Araldite had polymerized, then sections were cut from well-frozen areas of each sample and stained with uranyl acetate and lead citrate.

Thin sections from freeze-substitution samples and the processed freeze-fracture replicas were examined using a transmission electron microscope (JEOL model CX 200). Images were recorded on photographic plates. To simplify the interpretation of the freeze-fracture micrographs, we used a reversed-print technique to make the shadows appear as dark areas in the printed micrographs.

III. Results

A. Membrane Structure of Electropermeabilized Cells Revealed by Freeze-Fracture EM

The red cell membrane is a classical model of plasmalemma, which was studied extensively during the early development of freeze-fracture electron microscopy. The membrane structure of the normal red cell thus is well known (Pinto da Silva, 1972; Weinstein, 1974). Figure 3 shows the typical membrane structure of the control red cell prepared by our method. In this freeze-fracture micrograph, the outer leaflet of the cell membrane (E-face) is shown to have a relatively simple and uniform structure. When the red cell was permeabilized by an electric pulse, the structure of the cell membrane differed markedly. Figure 4 shows the E-face of a red cell membrane frozen 40 ms after being permeabilized by an RF electrical pulse, where numerous circular membrane openings can be seen. The membrane face curved into these openings, suggesting that these structures might be shaped like volcanos with their apices pointing away from the viewer, toward the outside of the cell. Most of

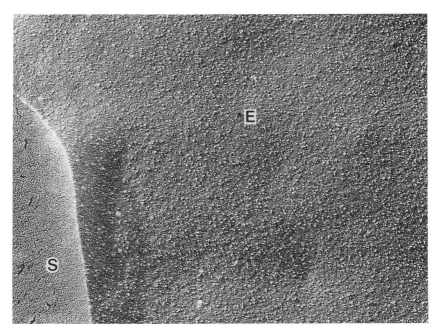

Figure 3   E-face membrane (labeled E) of a control human red blood cell. The frozen extracellular medium is labeled S. 60,000 ×.

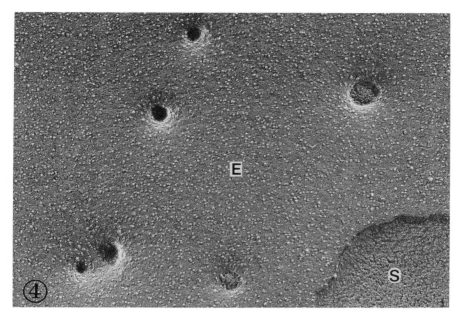

Figure 4 E-face membrane of an electropermeabilized human red blood cell frozen at 40 ms following the application of a pulse of an RF electric field. 60,000×.

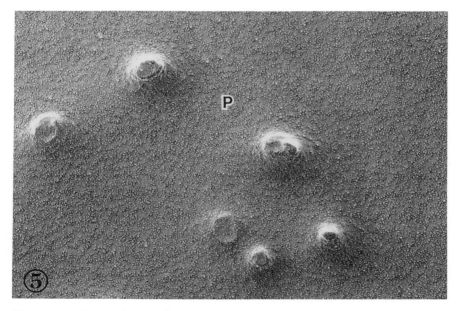

Figure 5 P-face membrane (labeled P) of an electropermeabilized human red blood cell frozen at 220 ms following the application of the electrical pulse. 60,000×.

these openings ended as a planar disk of granular material, which resembled the appearance of fracture faces running through a frozen aqueous medium.

The P-face of the electropermeabilized red cell membrane also showed many porelike structures complementary to those found in the E-face membranes. Figure 5 shows a typical P-face membrane of a red cell frozen 220 ms after the electrical pulsation. Volcano-shaped membrane evaginations with their apices pointing outward can be clearly seen. These volcano-shaped membrane openings were observed only in red-cell membranes that were porated by a high-intensity RF electric field; they were not observed in any of the control cell membranes. Furthermore, these

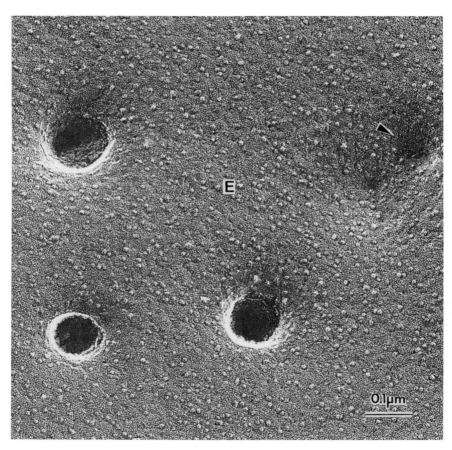

Figure 6 Magnified view of porelike structures found in the E-face membranes of electropermeabilized red cells. Three deep pores and one partially open (arrowhead) can be seen here.

circular openings were the only major discernible changes in the structure of the cell membrane after electroporation, in both E-face and P-face fracture views. Therefore, we think that these circular openings must represent the poration sites of the electropermeabilized cell membrane.

From freeze-fracture micrographs alone, it is difficult to determine conclusively whether the observed opening is the true opening of the electropore, or whether it represents the cross-fractured neck of an elongated membrane evagination. However, our observations, as a whole, suggest that most membrane openings are pore-like structures rather than cross-fractured evaginations. First, in the magnified view of membrane openings (Fig. 6), one can see that the E-face membrane ended directly in the granular material (which is typical of frozen medium). If the opening was a fractured face of evagination, one should see a ring of P-face membrane between the E-face and the granular material. Second, partially opened pores can occasionally be observed (see arrow in Fig. 6). The structure of such partially developed membrane openings clearly indicates a discontinuity of the plasma membrane, which cannot be interpreted as fractured faces of membrane evaginations.

At this point, the detailed structures of the apices of the volcano-shape openings is not clear. These structures could be relatively diverse in nature. In some P-face membranes, intramembrane particles (IMP) can occasionally be observed at the apices of volcano-shape structures (Fig. 5). The density of these particles, however, was significantly less than that of the IMP in the P-face of the normal membrane. Thus, some of the apices may consist of a patch of permeabilized membrane rather than an aqueous pathway.

B. Membrane Structure of Electropermeabilized Cells Revealed by Freeze-Substitution EM

The freeze-fracture examinations can only provide a panoramic view of the structures in the plane of the cell membrane. In order to obtain morphological information in the dimension perpendicular to the membrane plane, we examined thin sections cut from freeze-substituted specimens that were rapidly frozen under conditions identical to the freeze-fracture study. For red cells that were not exposed to an applied electric field, the cell membrane normally appeared as a typical double railroad track pattern (Weinstein, 1974). In freeze-substituted red cells that had been electropermeabilized by an applied RF electrical pulse, discontinuities in their cell membranes could occasionally be observed (Fig. 7). We believe that such discontinuities are related to the poration sites represented by the larger circular membrane openings observed in the freeze-fracture experiments. Indeed, we frequently observed depletion of cytoplasmic proteins at the region near the membrane discontinuity. This observation suggests that the membrane discontinuities are openings through which cellular contents (mainly hemoglobin molecules) can escape.

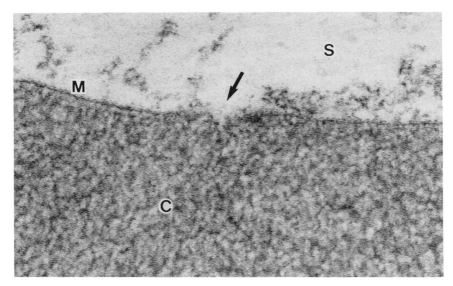

Figure 7   Micrograph showing a thin section of a freeze-substituted red cell, which was rapidly frozen at 220 ms after being permeabilized by an electrical pulse. The porelike structure is marked by an arrow. Symbols: C, cytoplasm; M, cell membrane; S, extracellular solution. 60,000 × .

C. Pore Size and Density

From the freeze-fracture views of the electropermeabilized red cell membrane (Figs. 4–6), it is apparent that the size of the membrane openings is not uniform. The pore size also appears to change at different stages of pore development. We have made a survey of 10 membrane fracture planes from cells frozen at 40 ms after the electric pulse. The diameters of the membrane openings were found to vary from 20 to 120 nm (Fig. 8). At later times, the smaller pores tended to widen slightly. For example, most of the membrane openings observed at 220 ms after electric pulsing were between 50 and 120 nm.

The density of poration sites also varies in a relatively wide range between membranes. Some pieces of membrane might have only one or two porelike structures, while others might have many dense poration sites. Figure 9 shows the freeze-fracture view of a piece of cell membrane that had the highest density of poration sites found in this study (7 pores μm^2). The variation of pore density is probably related to the orientation of the membrane in the fracture plane. Calculations based on a spherical cell show that the induced membrane potential at a given point of the cell surface is proportional to the cell diameter and the cosine of an angle between

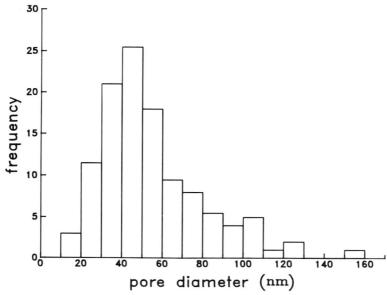

Figure 8 Histogram showing the distribution of pore sizes at 40 ms after the cells were permeabilized by the electrical pulse.

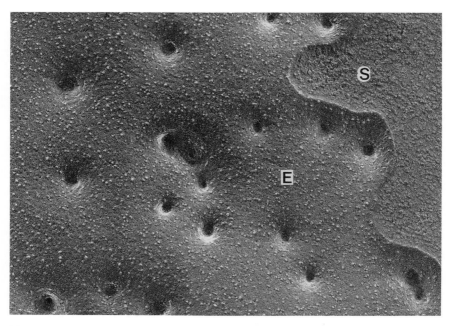

Figure 9 E-face membrane of an electropermeabilized red cell that had a high density of membrane pores. 60,000 ×.

the electric field and the normal vector of the membrane (Cole, 1968). Hence, the membrane perpendicular to the electric field would have the highest magnitude of induced membrane potential and thus is likely to have the maximum number of pores created. Furthermore, since the red cell is not spherical, its effective diameter (measured along the direction of the field) also varies with the orientation of the cell. The induced membrane potential (and thus the density of electropores) could vary depending on the cell orientation.

D. Dynamics of Pore Formation and Resealing

With our rapid-freezing technique, we could control the time delay (t) between the application of the electric pulse and the time of freezing. Thus, we can examine the dynamic changes of membrane structure after the cell was exposed to the field. Figure 10a shows an E-face membrane from a cell frozen at $t = 0.5$ ms. The membrane appears to be smooth; it appeared practically identical to the membrane of the control red cell. Cell membranes frozen at slightly later times (i.e., $t = 1$ ms) also had a similar appearance. These observations suggest that in the first millisecond following application of the electric field, the diameter of pores must be very small (less than the resolution of the freeze-fracture EM, about 2 nm). The earliest time that we could detect any significant structural changes in the membrane was at $t = 3$ ms (Fig. 10b), where deep porelike membrane openings (20–40 nm in diameter) were found in a few E-face membranes of the electropermeabilized red cells. At $t = 40$ ms, porelike membrane openings could be observed in most cell membranes, and their diameters had expanded to the range of 20–120 nm (Fig. 10c). These membrane openings generally were shaped like volcanos.

The volcano-shaped openings could be observed in cell membranes of most electropermeabilized red cells frozen at $t = 54$ ms, 220 ms, 1.0 s, and 1.7 s. The shapes of the openings were very similar, but the sizes of the smaller pores may have increased slightly with time. A few seconds after the electric pulse, the electric field-induced pores appeared to have begun to reseal. At $t = 5$ s, most of the openings deviated from the volcano shape and became more shallow. The diameters also appeared to be smaller (Fig. 10d). At $t = 10$ s, the porelike structures had almost disappeared and were replaced by numerous pitlike indentations in the membrane (Fig. 10e). These pits may represent partially resealed membrane pores.

The pitlike indentations gradually disappeared with time. At about half a minute after the electrical pulsing, the membrane in most red cells appeared almost like that of control red cells. However, occasionally, some small volcano-shape membrane openings could still be found in a few red cells. Apparently, there are significant variations in the resealing process of the electroporated cells.

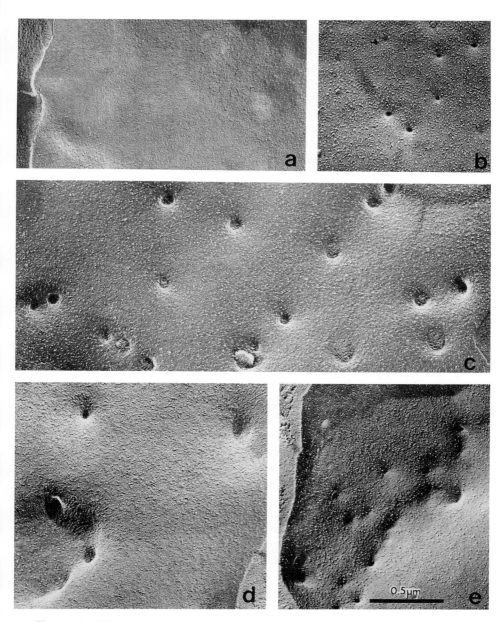

Figure 10   Micrographs showing the structure of the cell membranes of red cells frozen at different times (t) following the application of the porating electrical pulse. (a) $t = 0.5$ ms, (b) 2.5 ms, (c) 40 ms, (d) 5 s, and (e) 10 s. (From Chang and Reese, 1990.)

IV. Discussion

A. Formation of Electropores and the Pathways for Gene Transfer

The results of this study provide morphological evidence that the electric field-induced permeabilization of cell membrane is related to the formation of transient membrane pores. Our observations show that the evolution of electropores is a dynamic process, which may be divided roughly into three stages: In the first stage (consisting of the first few milliseconds after the electrical pulsing), pores were created. Initially, the newly formed pores were very small (<2 nm); within a few milliseconds, they expanded rapidly to a diameter of 20–40 nm. In the second stage (from a few milliseconds to several seconds), the pore structures became relatively stable. After pores had expanded to 20–100 nm in the first 20 ms of this stage, they remained more or less unchanged during the next few seconds. In the third stage (from seconds to minutes), pores underwent a resealing process. The pore diameters became reduced with time. However, some of the partially resealed pores might have a long lifetime.

This finding solves a major puzzle about how genes may enter the cell by electroporation. Previously, the typical diameter of electropores was estimated to be on the order of 1 nm, based on transport studies of carbohydrate molecules in electropermeabilized red cells (Kinosita and Tsong, 1977). It was difficult to explain how large macromolecules like circular DNA could enter the cell through such small pores. This puzzle can now be solved in view of the findings of this study. The discrepancy is related to the time resolution of the measurement. In the study based on measurements of carbohydrate transport, the pore size was estimated from data collected many minutes after the cells were porated. The membrane pores would have been partially resealed by then. Naturally, the estimated diameter of pores was very small. Our study, on the other hand, examined the structure of cell membrane from milliseconds to many minutes after the cells were permeabilized by the applied electrical pulse. Our findings suggest that electropores formed at the first few seconds following the electrical pulsing were significantly larger than the size estimated by the transport study. The diameters of the transient electropores could be as large as 20–120 nm, which should be sufficient to allow most macromolecules to pass through. The lifetime of these membrane pores (in the range of seconds) was also long enough to allow elongated molecules (such as DNA) to diffuse into the cell.

B. Influence of Material Flow on the Electroporation Process

A fundamental question about electroporation is whether the membrane pores are shaped mainly by the primary effects of interactions between the applied electric field and the cell membrane, or by a secondary effect of material flow following the initial permeabilization of the cell membrane induced by the electric field. The diameters of our observed membrane openings are larger than those predicted in many theories that consider principally the primary effects (Chernomordik *et al.*, 1983; Powell and Weaver, 1986). This difference suggests that material flow could be important in shaping the pores. Furthermore, we found in this study that the membrane opening always appeared as a volcano pointing outward from the cell, regardless of whether the membrane was facing the anode or the cathode electrodes. This observation is consistent with the view that formation of pores may be influenced by an ejection of cellular contents.

The shape and direction of opening cannot be explained by water movement because the same phenomenon was observed when the poration medium has hypo-osmotic. One probable cause of such outward-pointing pores is the flow of hemoglobin molecules. The RBC is much like a bag full of hemoglobin. Upon the permeabilizing of the cell membrane by the applied electric field, hemoglobin molecules might move out of the cell quickly, and this material movement could influence the shape of the openings of the permeabilized membrane. Indeed, the results of our freeze-substitution study (see Fig. 7) suggest a partial loss of hemoglobin near the membrane pores.

C. Implications of the Mechanisms of Pore Formation

In most earlier studies of electroporation, it was assumed that membrane pores were formed by a mechanism called "reversible breakdown," which was based on observations in lipid bilayers (Zimmermann and Vienken, 1982; Chernomordik *et al.*, 1983; Glaser *et al.*, 1988). Recently we suggested an alternative hypothesis, that the electropore may be created by an "irreversible breakdown" of the membrane in a localized region (Chang, 1989a). The findings in this EM study are consistent with this alternative view. The red cell membrane is known to attach to a network of cytoskeletal (or membrane-skeletal) proteins consisting mainly of spectrin, membrane actin, and ankyrin (Bennett, 1985; Cohen and Branton, 1979). The dimension of the holes of this network is on the order of 40–100 nm (Steck, 1989; Liu *et al.*, 1987). Due to the support of this protein network, the plasmalemma is far more difficult to rupture than a lipid bilayer. Thus, an electropore could be formed by an irreversible breakdown of a localized patch of cell membrane, which is enclosed within the hole of the cytoskeletal network. If such a hypothesis is correct, one would expect that (a) the peak size of the membrane opening should be similar to the size of the holes of the cytoskeletal network, that is, on the order of 40–100 nm, and (b) the pore may expand very quickly initially when its diameter is very small, but once it reaches a size comparable to the holes of the cytoskeletal network, the pore will stop expanding and its structure will become more stable. Both of these predictions are in good agreement with our experimental findings. Not only does the dimension of the volcano-shape membrane openings match the size of the holes in the cytoskeletal network, the dynamics of pore formation also indicate that the membrane openings became stabilized when they reach a size comparable to that of the network holes. Hence, results of this study suggest that membrane–cytoskeletal interactions may play an important role in the shaping of electropores.

D. Comparison with Results of Electrical Measurements

The process of electroporation has been studied previously using measurements of electrical properties of the electroporated cells. Some of these electrical measurements suggested that certain electropores could be formed (and reseal) within microseconds (Kinosita et al., 1988). Since the electrical properties of a membrane are mainly determined by its permeability to ions, such fast-forming pores could be very small. [For instance, a pore with a diameter of 0.5 nm is sufficient to pass most ions (Hille, 1984).] Thus, it is not always possible to correlate information obtained from electrical measurements to those obtained from structural measurements. The typical resolution of freeze-fracture EM is about 2 nm and thus would not reveal pores of ion size. On the other hand, it is very difficult to deduce from the electrical measurements the structure and size of the membrane pathways that are responsible to changes of membrane permeability. The difference in cell models used in different studies also causes another difficulty of interpretation. Nevertheless, if one combines the information obtained from the electrical measurements and those obtained from this study, one may conclude that the process of electroporation may consist of three major stages: (1) There appears to be an active process of formation and resealing of tiny pores during the first few microseconds following the application of the electrical pulse. (2) Some of these pores may rapidly expand in the next few milliseconds. It is due to the existence of these expanded electropores that large molecules (such as hemoglobin or DNA) are allowed to pass through the cell membrane. (3) In a period of about 10 s, many of the pores may begin to reseal. According to previous studies of membrane transport properties, the partially resealed pores could have a long lifetime, depending on temperature or other physical conditions (Kinosita and Tsong, 1977; Zimermann et al., 1980).

In summary, we have presented structural evidence that electropermeabilization of red blood cell is related to the creation of volcano-shape membrane pores. These pores are dynamic structures, the development of which involves at least three major stages. The transient pores observed in stage 2 have large openings (20–120 nm in diameter), which are most likely to be the pathways for the passage of macromolecules. Results of this study suggest that the electroporation process could be influenced by two factors, namely, the membrane–cytoskeletal interactions and the material flow immediately following the initial permeabilization of the cell membrane. Thus, it is reasonable to expect that the structure and dynamics of electropores may vary from cell type to cell type.

Acknowledgment

I am grateful to Dr. Tom S. Reese for his invaluable support and contributions to this project. I thank P. Q. Gao, Q. Zheng, and J. R. Hunt for their technical assistance, J. Chludzinski for his photography work, and Dr. Julian Heath for his comments. This work was supported in part by a grant from the Texas Advanced Technology Program.

References

Bennett, V. (1985). The membrane skeleton of human erythrocytes and its implications for more complex cells. *Annu. Rev. Biochem.* **54**, 273–304.

Benz, R., and Zimmermann, U. (1981). The resealing process of lipid bilayers after reversible electric breakdown. *Biochim. Biophys. Acta* **640**, 169–178.

Chakrabarti, R., Wylie, D. E., and Schuster, S. M. (1989). Transfer of monoclonal antibodies into mammalian cells by electroporation. *J. Biol. Chem.* **264**, 15494–15500.

Chang, D. C. (1989a). Cell poration and cell fusion using an oscillating electric field. *Biophys. J.* **56**, 641–652.

Chang, D. C. (1989b). Cell fusion and cell poration by pulsed radio-frequency electric fields. *In* "Electroporation and Electrofusion in Cell Biology," Neumann *et al.*, eds. Plenum Press, New York, pp. 215–227.

Chang, D. C. and Reese, T. S. (1990). Changes in membrane structure induced by electroporation as revealed by rapid-freezing electron microscopy. *Biophys. J.* **58**, 1–12.

Chang, D. C., Gao, P. Q., and Maxwell, B. L. (1991). High efficiency gene transfection by electroporation using a radio-frequency electric field. *Biochim. Biophys. Acta* **1992**, 153–160.

Chernomordik, L. V., Sukharev, S. I., Abidor, I. G., and Chizmadzher, Y. A. (1983). Breakdown of lipid bilayer membranes in an electric field. *Biochim. Biophys. Acta* **736**, 203–213.

Chu, G., Hayakawa, H., and Berg, P. (1987). Electroporation for the efficient transfer of mammalian cells with DNA. *Nucleic Acids Res.* **15**, 1311–1326.

Cohen, C. M., and Branton, D. (1979). The role of spectrin in erythrocyte membrane stimulated actin polymerization. *Nature (Lond.)* **279**, 163–165.

Cole, K. S. (1968) "Membranes, Ions, and Impulses: A Chapter of Classical Biophysics." University of California Press, Berkeley, pp. 12–18.

Fujita, T., Tanaka, K., and Tokunaga, J. (eds.) (1981). "SEM Atlas of Cells and Tissue." Igaku-Shoin, Tokyo.

Glaser, R. W., Leikin, S. L., Chernomordik, L. V., Pastushenko, V. L., and Sokirko, A. I. (1988). Reversible electrical breakdown of lipid bilayers: Formation and evolution of pores. *Biochim. Biophys. Acta* **940**, 275–287.

Hashimoto, K., Tatsumi, N., and Okuda, K. (1989). Introduction of phalloidin labelled with fluorescein isothiocyanate into living polymorphonuclear leukocytes by electroporation. *J. Biochem. Biophys. Methods* **19**, 143–153.

26   Donald C. Chang

Hille, B. (1984). *In* "Ionic Channels in Excitable Membranes." Sinauer Associates, Inc., New York, pp. 181–190.

Kinosita, K., and Tsong, T. Y. (1977). Formation and resealing of pores of controlled sizes in human erythrocyte membranes. *Nature (Lond.)* **268**, 438–441.

Kinosita, K., Ashikawa, I., Saita, N., Yoshimura, H., Itoh, H., Nagayama, K., and Ikegami, A. (1988). Electropore formation of cell membrane visualized under a pulsed-laser fluorescence microscope. *Biophys. J.* **53**, 1015–1019.

Knight, D. E., and Baker, P. F. (1982). Calcium-dependence of catecholamine release from bovine adrenal medullary cells after exposure to intense electric fields. *J. Membrane Biol.* **68**, 107–140.

Liu, S. C., Derick, L. H., and Palek, J. (1987). Visualization of the hexagonal lattice in the erythrocyte membrane skeleton. *J. Cell Biol.* **104**, 527–536.

Neumann, E. (1989). The relaxation hysteresis of membrane electroporation. *In* "Electroporation and Electrofusion in Cell Biology," eds. E. Neumann, A. E. Sowers, and C. A. Jordan, Plenum Press, New York, pp. 61–82.

Neumann, E., Schaefer-Ridder, M., Wang, Y., and Hofschneider, P. N. (1982). Gene transfer into mouse lyoma cells by electroporation in high electric fields. *EMBO J* **1**, 841–845.

Ornberg, R. L., and Reese, T. S. (1981). Beginning of exocytosis captured by rapid freezing of *Limulus* amebocytes. *J. Cell Biol.* **90**, 40–54.

Pinto Da Silva, P. (1972). Translational mobility of membrane intercalated particles of human erythrocyte ghosts—pH dependent, reversible aggregation. *J. Cell Biol.* **53**, 777–787.

Potter, H. (1988). Electroporation in biology—Methods, applications and instrumentation. *Anal. Biochem.* **174**, 361–373.

Powell, K. T., and Weaver, J. C. (1986). Transient aqueous pores in bilayer membranes: A statistical theory. *Bioelectrochem. Bioelectroenerg.* **15**, 211–227.

Sokolowski, J. A., Jastreboff, M. M., Bertino, J. R., Sartorelli, A. C., and Narayanan, R. (1986). Introduction of deoxynucleoside triphosphates into intact cells by electroporation. *Anal. Biochem.* **158**, 272–277.

Sowers, A. E, and Lieber, M. R. (1986). Electropore diameters, lifetimes, numbers and locations in individual erythrocyte ghosts. *FEBS Lett.* **205**, 179–184.

Steck, T. L. (1989). Red cell shape, *In* "Cell Shape: Determinants, Regulation and Regulatory Role," eds. W. D. Stein and F. Bonner, Academic Press, New York, pp. 205–246.

Stenger, D. A., and Hui, S. W. (1986). Kinetics of ultrastructural changes during electrically induced fusion of human erythrocytes. *J. Membrane Biol.* **93**, 43–53.

Swezey, R. R., and Epel, D. (1988). Enzyme stimulation upon fertilization is revealed in electrically permeabilized sea urchin eggs. *Proc. Natl. Acad. Sci. USA* **85**, 812–816.

Tsongalis, G. J., Lambert, W. C., and Lambert, M. W. (1990). Electroporation of normal human DNA endonucleases into xeroderma pigmentosum cells corrects their DNA repair defect. *Carcinogenesis* **11**, 499–503.

Weinstien, R.S. (1974). The morphology of adult red cells. *In* "The Red Blood Cell," Vol. 1, D. M. Surgenor, ed., Academic Press, New York, pp. 213–268.

Wong, T. K., and Neumann, E. (1982). Electric field mediated gene transfer. *Biochem. Biophys. Res. Commun.* **107**, 584–587.

Winegar, R. A., Phillips, J. W., Youngblom, J. H., and Morgan, W. F. (1989). Cell electroporation is a highly efficient method for introducing restriction endonucleases into cells. *Mutat. Res.* **225**, 49–53.

Zimmermann, U., and Vienken, J. (1982). Electric field-induced cell-to-cell fusion. *J. Membrane Biol.* **67**, 165–182.

Zimmermann, U., Vienken, J., and Pilwat, G. (1980). Development of drug carrier systems: Electric field induced effects in cell membranes. *J. Electroanal. Chem.* **116**, 553–574.

3

Events of Membrane Electroporation Visualized on a Time Scale from Microsecond to Seconds

Kazuhiko Kinosita, Jr.,[1] Masahiro Hibino,[1] Hiroyasu Itoh,[2]
Masaya Shigemori,[3] Ken'ichi Hirano,[2] Yutaka Kirino,[4] and
Tsuyoshi Hayakawa[2]

[1]Department of Physics, Faculty of Science and Technology, Keio University,
Kohoku-ku, Yokohama 223, Japan

[2]Tsukuba Research Laboratory, Hamamatsu Photonics, K. K.,
Tokodai, Tsukuba 300-26, Japan

[3]Department of Physics, Faculty of Science, Gakushuin University, Toshima-ku,
Tokyo 171, Japan

[4]Faculty of Pharmaceutical Sciences, Kyushu University, Higashi-ku, Fukuoka 812, Japan

I. Introduction

Electroporation is the phenomenon in which a cell exposed to an electric pulse is permeabilized as though aqueous pores are introduced in the cell membrane (Zimmermann, 1982; Tsong, 1983; Neumann *et al.*, 1989). In 1967, Sale and Hamilton observed that treatment of cells with intense electric pulses led to cell lysis (Sale and Hamilton, 1967), and suggested that the cell membranes of the cells were damaged by the transmembrane potential induced by the applied electric field (Sale and Hamilton, 1968). Subsequent studies (Neumann and Rosenheck, 1972; Zimmermann *et al.*, 1976; Kinosita and Tsong, 1977a, 1977b) showed that the cell membranes of pulse-treated cells were permeable to molecules with a size less than a certain limit, suggesting the creation of a porous structure in the membrane. A visual demonstration of the pores had to wait for the recent electron microscopic observations by Chang and Reese (1990), presumably because of the difficulty in detecting the pores which are transient structures. Though electroporation is a widely used tool for artificially altering cellular contents, not much is known about the molecular events in the membrane.

   In this chapter, we describe our recent work toward understanding the mechanisms of electroporation. All the work is based on observations under an optical microscope, which allowed us to resolve spatially the events in a single cell. The microscope system we used had a time resolution of a submicrosecond, which far exceeds the resolution of a subsecond obtained in ordinary microscopy. Achieving the submicrosecond resolution was important, since the pore formation is thought to be a microsecond process. We present below several aspects of the electroporation process, all resolved in space and time.

II. Visualization of Transmembrane Potential

The primary event that leads to pore formation is believed to be the induction of a large transmembrane potential by the applied electric field (Fig. 1). For a spherical cell of radius a in a uniform electric field E_0, the potential difference $\Delta\Psi$ between the extracellular and intracellular surfaces of the cell membrane is given, in the absence of pores, by

$$\Delta\Psi = 1.5\, faE_0 \cos \theta[1 - \exp(-t/\tau)] \tag{1}$$

$$\tau = faC_m\, (r_i + r_e/2) \tag{2}$$

$$f = 1/[1 + aG_m(r_i + r_e)] \tag{3}$$

where t is the time after the constant field is turned on, C_m is the membrane capacitance per unit area, r_i and r_e are specific resistances of the intra- and extracellular media, and G_m is the membrane conductance per unit area (here assumed to be

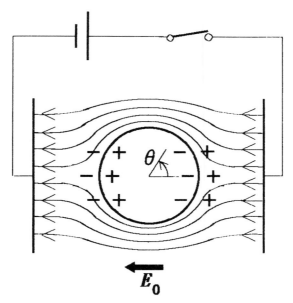

Figure 1   Induction of transmembrane potential in a cell exposed to an external electric field.

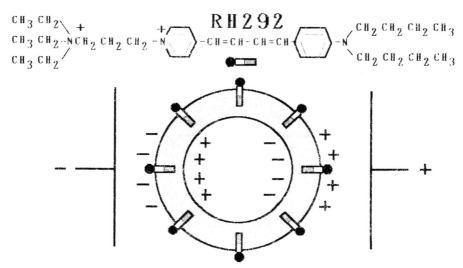

Figure 2   Structure of the voltage-sensitive dye RH292 (top) and its orientation in the membrane (bottom).

uniform over the entire cell surface) (Cole, 1972). Note that $\Delta\Psi$ in our notation is the extracellular potential minus intracellular potential. For normal cells, G_m is sufficiently small such that $f = 1$.

The transmembrane potential $\Delta\Psi$ can be visualized by staining the cell membrane with a voltage-sensitive fluorescent dye. We used the styryl dye RH292 (Grinvald *et al.*, 1982), the structure of which is shown in Fig. 2, top. When added to the extracellular medium, the dye, with its hydrophilic head and two hydrophobic tails, is expected to partition in the cell membrane as depicted in Fig. 2, bottom. The dye fluoresces strongly when bound to the membrane. The intensity of the dye fluorescence is sensitive to $\Delta\Psi$: the intensity increases for positive $\Delta\Psi$, and decreases for negative $\Delta\Psi$. The image of the fluorescence therefore indicates the spatial variation of $\Delta\Psi$. Imaging at a temporal resolution of 0.3 μs has been achieved

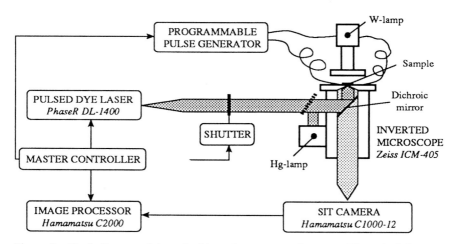

Figure 3   Block diagram of the pulsed-laser fluorescence microscope. The pulsed dye laser delivers a light pulse of duration 0.3 μs into the inverted epifluorescence microscope. A fluorescent sample on the specimen stage receives this light pulse and emits fluorescence. The fluorescence image is captured by the SIT (silicon intensified target) camera. In the camera the incoming light, the flashing image lasting for only 0.3 μs, is converted into, and held in the form of, a charge distribution on a target plate. The electrical image is then read, and at the same time erased, by raster scanning over the plate at the video rate of 33 ms per frame. The output signal from the camera is digitized and recorded in a frame memory in the image processor. The pulse generator shown at the top of the figure is for electrical stimulation of the sample. To record a microsecond response of the sample in the form of sequential images, the electrical stimulus is repeated at 0.5 Hz; each time the laser is triggered at a different time in the stimulus. For details, see Hibino *et al.* (1991). A later version, multishot laser microscope (Itoh *et al.*, 1990) enables sequential imaging in a single event.

under a pulsed laser fluorescence microscope (Kinosita *et al.*, 1988a; Hibino *et al.*, 1991) shown in Fig. 3.

Figure 4 shows snapshots, taken under the pulsed laser fluorescence microscope, of the fluorescence of the dye adsorbed on the cell membrane of a sea urchin egg. Before the external electric field was turned on at time 0, the fluorescence was uniform along the cell periphery. When the field was applied, the fluorescence intensity on the positive-electrode side increased whereas it decreased on the negative side. The change was almost complete by 2 μs. The change in fluorescence intensity, which is expected to be proportional to $\Delta\Psi$, was in accord with Eq. (1): spatially the intensity change was approximately cosinelike, and temporally the rise time was consistent with the prediction from Eq. (2) of ~1 μs ($a = 50$ μm, $C_m \approx 1$ μF/cm^2, $r_i \approx 200$ Ω · cm, and r_e of the Ca^{2+}-free sea water of 20 Ω · cm). In low-salt media the change in fluorescence intensity was slower in accordance with Eq. (2) (Kinosita *et al.*, 1988a, 1988b).

The intensity of applied field in Fig. 4 was 100 V/cm, for which the maximal $\Delta\Psi$, $\Delta\Psi_{max} \equiv \Delta\Psi(\theta = 0°, t = \infty)$, is calculated to be 0.75 V. Figure 4 therefore

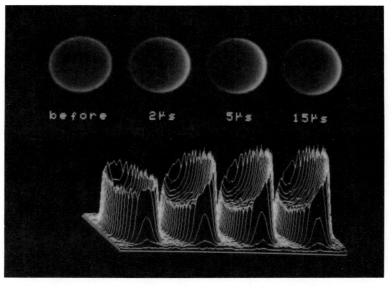

Figure 4   Induction of transmembrane potential in a sea urchin egg. Snapshots, at indicated times after the onset of an external field, were taken with an exposure time of 0.3 μs under the pulsed laser fluorescence microscope (Kinosita *et al.*, 1988a, 1988b). The egg was stained with RH292 and exposed to an electric field of 100 V/cm in Ca^{2+}-free seawater. Positive electrode was to the right of the cell and negative electrode to the left. Intensity profiles are shown at the bottom.

indicates that the cell membrane can withstand such a high $\Delta\Psi$ at least for tens of microseconds. For $|\Delta\Psi|$ of at least up to 0.75 V, the cell behaved as predicted in Eq. (1), or as a spherical conductor surrounded by a thin dielectric membrane of negligible conductance.

III. Electroporation Revealed in the Behavior of Transmembrane Potential

A. Saturation of Transmembrane Potential

Under an external electric field for which the theoretical $\Delta\Psi_{max}$ is greater than about 1 V, the behavior of fluorescence response was quite different from that in Fig. 4. An example is shown in Fig. 5, where the egg was exposed to a field of 400 V/cm for which $\Delta\Psi_{max}$ is 3 V and for which a fluorescence change 4 times greater than that in Fig. 4 would have been expected. The actual change was much smaller and indicated saturation, particularly in the two regions directly opposing the electrodes. In Fig. 6, the fluorescence change at 2 μs is plotted against θ for field intensities of 100 V/cm and 400 V/cm. The profile at 400 V/cm (solid line) clearly deviates

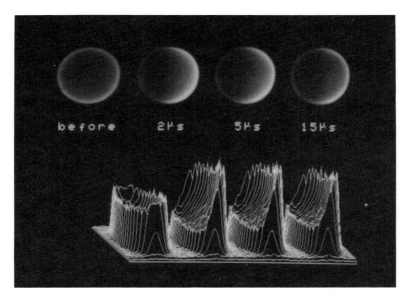

Figure 5 Saturation of transmembrane potential. The experiment was the same as in Fig. 4 except the field intensity was 400 V/cm.

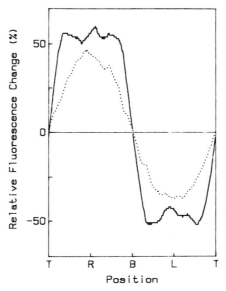

Figure 6   Comparison of the fluorescence signals at 2 μs under electric fields of 100 V/cm (dashed line) and 400 V/cm (solid line). From the images in Figs. 4 and 5, fluorescence intensities relative to the intensities before the application of the field were calculated and plotted against θ defined in Fig. 1. T, top of the image (θ = 90°); R, right (θ = 0°); B, bottom (θ = 90°); L, left (θ = 180°).

from a cosine curve and shows a flat top and bottom (saturation). Comparison of this profile with the dashed profile suggests that the saturation level corresponds to $|\Delta\Psi|$ of about 1 V, or slightly less, whereas Eq. (1) would predict $\Delta\Psi(\theta = 0°, t = 2 \mu s) = 2.6$ V. After 2 μs the fluorescence intensity slowly returned toward the original level on the microsecond time scale (Fig. 5).

Our interpretation of these observations is the following (see also Hibino *et al.*, 1991). When $|\Delta\Psi|$ reaches a critical value of about 1 V, the membrane is porated and starts to conduct current. The magnitude of the current, or the extent of poration, is large enough to counteract the further rise of $|\Delta\Psi|$ and to keep $|\Delta\Psi|$ at or below the critical value. The poration is fast enough to compete with the rise of $\Delta\Psi$, which was 1 μs in Fig. 5. In other words, the cell membrane cannot sustain a voltage difference greater than the critical value of about 1 V even for a period as short as 2 μs. The slow return of the fluorescence intensity after 2 μs is interpreted as the gradual increase in the number of pores and/or the pore size.

The above results and interpretations are consistent with earlier studies on different samples. In red blood cells, for example, the poration takes place within 1 μs after $\Delta\Psi$ reaches the critical value of about 1 V (Kinosita and Tsong, 1977c). A longer electric pulse produces larger pores (Kinosita and Tsong, 1977b). A large current (measured indirectly) starts to flow across the cell membrane as soon as $\Delta\Psi$ exceeds 1 V (Kinosita and Tsong, 1979). Coster and Zimmerman (1975) directly injected current in a cell of *Valonia utricularis* and measured the transmembrane potential. When the potential augmented by the injected current reached a critical value of about 1V, any further increase was counteracted by a dramatic increase in the membrane conductance, which was just large enough to keep the potential at the critical value. When the current was held constant after the breakdown, the potential gradually decreased on the microsecond time scale.

B. Estimation of Membrane Conductance

Our spatially resolved observations outlined above allowed a semiquantitative estimation of the distribution of membrane conductance in an electroporated cell (Kinosita *et al.*, 1988a, 1988b; Hibino *et al.*, 1991). If electroporation uniformly increases the membrane conductance over the entire cell surface, $\Delta\Psi$ in the porated cell should still be proportional to $\cos\theta$ [Eq. (1) with $f < 1$]. This is clearly not the case, as can be seen in Figs. 5 and 6. The finite membrane conductance introduced by the pores should be concentrated in the two regions facing the electrodes. The observed $\Delta\Psi$ can in fact be approximated by a theoretical $\Delta\Psi$ in which the membrane conductance $G(\theta)$ is given by

$$G(\theta) = \begin{cases} G_0 \dfrac{(|\cos\theta| - \cos\theta_c)}{(1 - \cos\theta_c)} & (0° \leqq \theta \leqq \theta_c, \ 180° - \theta_c \leqq \theta \leqq 180°) \\ 0 & (\theta_c < \theta < 180° - \theta_c) \end{cases} \tag{4}$$

This equation implies that only those regions of the membrane where the induced potential $\Delta\Psi$ reached a critical value are porated, and that the conductance introduced is proportional to the excess potential, that is, $|\Delta\Psi|$ that would be obtained in the absence of pores minus the critical potential for poration; θ_c is the angle at which $|\Delta\Psi|$ in the absence of pores equals the critical potential.

The value of G_0, the maximal membrane conductance, depended on the intensity and duration of the applied field. For field intensities of a few hundred V/cm (theoretical $\Delta\Psi_{max}$, in the absence of pores, a few V) and for field duration of tens of microseconds, G_0 was in the range 1–10 S/cm² both at high and low ionic strengths. This conductance is equivalent to the replacement of the order of 0.1% of the membrane area by aqueous openings.

Equation (4) also implies that the two regions opposing the positive and negative electrodes are equally conductive. Experimental profiles of $\Delta\Psi$ in fact appeared more or less symmetric, indicating a symmetric $G(\theta)$. This is expected from a theoretical point of view: if only one side of the cell is porated, the next moment $\Delta\Psi$ on the other side almost doubles (Hibino *et al.*, 1991), and therefore that side would also be porated. The potential is always induced in such a way as to balance the membrane conductance on the two sides.

The symmetric $G(\theta)$, however, does not necessarily mean that structural changes in the membrane are also strictly symmetric. For example, different combinations of the pore density and pore size can yield an apparently same conductance. As will be discussed below, many pieces of evidence exist that point to asymmetric structural changes. In the presence of the applied field, the tendency toward asymmetry may be masked by the equalizing action of $\Delta\Psi$ above. In this regard, we noticed that $\Delta\Psi$ in a rapid transient, within 2 μs after the application of the field, was often clearly asymmetric (M. Hibino, unpublished). The asymmetry rapidly disappeared after 2 μs, presumably due to the equalizing action above, but the underlying structural changes in the membrane may well remain asymmetric.

C. Kinetics of Electroporation

As discussed above, pores are created in those parts of the membrane where $\Delta\Psi$ reaches the critical value of approximately 1 V. The poration takes place well within 2 μs. Similarly, in a low-salt medium where the rise of the potential is slow, the poration takes place at the moment the potential reaches the critical value. After the initial poration, if the external field persists, the size and/or number of pores gradually increases, which results in a slow decrease of $|\Delta\Psi|$ as seen in Fig. 5 (also Kinosita *et al.*, 1988b; Hibino *et al.*, 1991). This is in accord with previous results with red blood cells (Kinosita and Tsong, 1977b, 1977c). However, with our limited measurements of up to a few milliseconds and up to 1 kV/cm in Ca^{2+}-free sea water, we have never seen the pores grow to the extent of total abolishment of the induced $\Delta\Psi$: the fluorescence of RH292 always indicated the presence of residual $\Delta\Psi$ even in the regions directly opposing the electrodes. This implies that, within the experimental conditions given above, the membrane conductance does not grow beyond 10^2 S/cm^2. [The combination of Eqs. (1) and (3), although only applicable to the case of a uniform poration, is nevertheless useful for an order-of-magnitude estimation.]

The critical potential of about 1 V for poration applies only to short electric pulses. For example, when we applied an electric field of 100 V/cm ($\Delta\Psi_{max} = 0.75$ V) for 1 ms, $\Delta\Psi$ clearly deviated from cos θ and showed saturation (M. Hibino, unpublished). The critical potential is a very slowly decreasing function of the duration of the applied field, as in the case in red blood cells (Kinosita and Tsong, 1977c).

D. Recovery after Poration

The recovery of the cell membrane after poration can also be inferred from measurements of the transmembrane potential. For this, a small, subcritical electric pulse was applied to the cell and the response to this monitor pulse was examined. If the membrane had recovered by the time the monitor pulse was given, the response should be identical to that of an intact cell.

The sea urchin egg treated with a 186 V/cm, 20 μs pulse recovered within 2 s by the above criterion (Kinosita *et al.*, 1988a). On the other hand, immediately after the poration the membrane remained highly conductive. At 20 μs after the pulse, for example, G_0 measured under a monitor pulse of 67 V/cm was still as high as 0.3 S/cm^2. This conductance value, however, should be interpreted with caution, since it was estimated under a relatively high field; the monitor pulse may have reopened once-closed pores. This measurement is useful nevertheless in estimating the time course of membrane recovery. By 1 ms the G_0 value dropped by an order of magnitude, indicating that the fast recovery took place on the microsecond time scale. Though the potential measurement was not sensitive enough to follow further recovery, the complete recovery must take at least seconds, as will be discussed below.

E. Molecular Orientation in Porated Membranes

Figure 2 suggests that RH292 penetrates the cell membrane in a more or less perpendicular fashion. This was confirmed to be the case by polarized fluorescence microscopy (Kinosita *et al.*, 1990). The polarization of fluorescence also depends on the degree of orientational order of the dye, which is a lipid analog, in the membrane. We therefore compared the polarization between an intact egg and the egg 5 μs after the application of a 400-V/cm, 20-μs pulse. No appreciable difference was found, indicating that the poration did not induce a gross membrane disorder at least for an electric pulse of the above magnitude. Structural changes in the membrane appear to be confined in the immediate vicinity of the pores.

IV. Large-Hole Formation and Deformation in Giant Liposomes

A. Large-Hole Formation

In sea urchin eggs, we have never observed a pore that is large enough to allow direct detection under an optical microscope. In contrast, in asolectin liposomes with a diameter several tens of micrometers (Mueller *et al.*, 1983), we found a micrometer-sized hole(s), as shown in Fig. 7 (note the sharp edge of the hole in the image; H. Itoh and M. Hibino, unpublished). The hole appeared a few hundred microseconds after the onset of an external electric field for which $\Delta \Psi_{max} > 1$ V. The hole(s) was most often found on the positive-electrode side, but there were cases in which it appeared only on the negative side or where several holes were created on both sides. The number of holes observed was also variable.

When the applied field was turned off, the hole progressively shrank and eventually closed. The closure time of typical holes was in the millisecond range; closure of a large hole took hundreds of milliseconds. When the applied field was too intense or too long, the liposome punctured in many ways. In some cases, for example, the lipid membrane rolled up from the edge of the hole, leading to the rupture of the

Figure 7   A micrometer-sized hole created in a giant liposome. A snapshot 300 μs after the onset of an electric field of 600 V/cm (direction from right to left). The diameter of the liposome was 50 μm. The liposome was stained with RH292.

liposome; this process did not always proceed to completion and sometimes half a liposome (hemisphere) remained that was stable for many seconds. When the applied pulse was too strong, the liposome simply broke into pieces.

We are not yet certain whether the holes we observe in giant liposomes are in the same category as the electropores that we discussed in Section III. The image of RH292 fluorescence indicates that phenomena similar to those in Section III, that is, the induction of $\Delta\Psi$ and its saturation under a high field, also take place in liposomes before a large hole becomes visible. The initial event thus appears to be the creation of electropores on the nanometer scale. The large hole may be formed by the growth of a small pore over hundreds of microseconds, or by coalescence of many small pores. Deformation of the liposome, described below, is also a possible cause.

B. Deformation under Intense Electric Fields

Helfrich (1974) has shown that when a spherical insulating membrane, such as a liposome, is exposed to an electric field, the sphere experiences a compressing force in the direction perpendicular to the field and becomes prolate. Under an intense field that causes electroporation, however, the membrane can no longer be regarded as an insulator. For a conductive membrane, theory (Hyuga *et al.*, 1991a, 1991b) predicts that the force on the membrane depends on the ratio of the specific resistance of the external medium to that of the internal medium; a spherical membrane will become prolate when this ratio is greater than one, and oblate when the ratio is less than one.

For the giant liposomes, the above prediction has been confirmed experimentally as shown in Fig. 8. Sea urchin eggs also underwent the two types of deformation under an intense (e.g., 800 V/cm) and long (e.g., 1 ms) electric pulse, depending on the specific resistance of the extracellular medium. Since both types of deformation tend to expand the membrane area, the formation of large holes in liposomes could be the result of, or accelerated by, the deformation.

Harbich and Helfrich (1979) have already shown that giant lecithin vesicles under a relatively weak electric field (a few tens of V/cm) are elongated along the field and large openings are formed at the ends facing the electrodes. They estimated $|\Delta\Psi|$ at the ends to be about 0.1 V, which, when sustained for a long time, may be sufficient to cause electroporation. Time dependence was not described, but presumably the process took seconds. Thus the large holes that we observed in the microsecond range, under a much higher field, may be an accelerated version of the same phenomenon. The possible role of deformation in making large holes has yet to be studied in both systems.

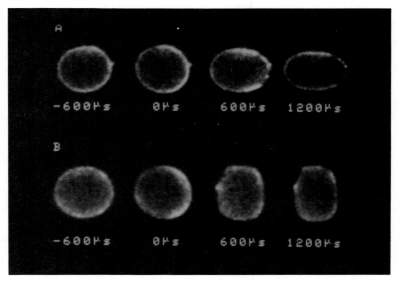

Figure 8   Deformation of giant liposomes under intense electric field. Snapshots, at indicated times after the onset of an external field, taken with an exposure time of 100 μs (starting at indicated time) under the multishot laser microscope (Itoh *et al.*, 1990). A, field = 700 V/cm (from right to left); internal medium, 0.5 m*M* KCl, 0.5 m*M* MgSO$_4$, 0.1 *M* sucrose; external medium, 0.1 *M* glucose. B, field = 1000 V/cm (from right to left); internal medium, 0.5 m*M* KCl, 0.5 m*M* MgSO$_4$, 0.1 *M* sucrose; external medium 0.7 m*M* KCl, 0.7 m*M* MgSO$_4$, 0.1 *M* glucose. Liposomes were stained with RH292.

V. Permeation across Porated Membranes

A. Calcium Influx into Porated Eggs

When a sea urchin egg is electroporated in normal sea water containing Ca^{2+}, the partial fertilization envelope rises as a result of Ca^{2+} influx into the cell (Baker *et al.*, 1980). Rossignol *et al.* (1983) observed this process only on the positive-electrode side. We, in contrast, noticed that when experiments described in Section III were performed in normal seawater, the envelope was formed on the negative-electrode side. In some cases the envelopes were formed on both sides but the envelope on the negative side was much larger (Hibino *et al.*, 1991).

To investigate the reason for this apparent discrepancy, we injected the fluorescent Ca^{2+} probe indo-1 (Grynkiewicz *et al.*, 1985) into an egg and directly imaged the

Ca^{2+} level in the egg (Kinosita *et al.,* 1991). After a 500-V/cm, 400-μs pulse was given to the cell, rapid Ca^{2+} influx was seen on the positive side, followed by a slower but larger influx from the negative side (both on the subsecond to second time scales). The influxes did not last long, and the original low Ca^{2+} level was restored within a few minutes. A large fertilization envelope was formed on the positive-electrode side, together with a small one on the negative side. The Ca^{2+} influxes after a 500-V/cm, 200-μs pulse were smaller but their time courses were similar to the 400-μs case above. In this case, however, the envelope was larger on the negative side than on the positive side. When the external medium was Ca^{2+}-free seawater, neither Ca^{2+} influx nor envelope elevation were observed.

The above results show, first, that the pores that admit Ca^{2+} remain open for seconds (under the above conditions) and then close (or shrink). Second, poration is somehow asymmetric; the influxes from the positive and negative sides differ not only in magnitude but also in the time course. Third, elevation of the fertilization envelope requires a certain amount of Ca^{2+}. When, after a very intense and/or long electric pulse, the fast Ca^{2+} influx from the positive side fulfills this requirement, the envelope rises on the positive side. If the influx from the positive side is not large enough, then the envelope rises on the negative side. This third point resolves the discrepancy between Rossignol *et al.* and Hibino *et al.,* since the former used 1-kV/cm, 100-μs pulse whereas the pulse of the latter was several hundred V/cm and its duration tens of microseconds.

B. Influx of Fluorescent Dyes into Porated Eggs

After treatment with an intense and long pulse (e.g., 800 V/cm for 1 ms), relatively large molecules such as RH292 or its zwitterionic relative RH160 (Grinvald *et al.,* 1982) were seen to penetrate into the cell interior. The penetration was slow, requiring observation over seconds (in a low-salt medium) to minutes (in Ca^{2+}-free seawater). For both RH292 and RH160, penetration was mainly from the negative-electrode side (Fig. 9) (Hibino, unpublished). [The symmetric uptake of RH237 shown in Kinosita *et al.* (1988b) was due to the use of a pulse with reversed polarities.]

C. Efflux of Fluorescent Dyes from Porated Liposomes

Giant liposomes loaded with the fluorescent dye tetramethylrhodamine, or fluorescein-dextran, showed various efflux patterns after electroporation (M. Hibino, unpublished). Two basic patterns were observed: a rapid burst in the subsecond time range and a slow leakage in seconds. The rapid burst was usually seen on the positive-electrode side, but there were some cases where the burst was observed on the negative side or on both sides. The slow leakage could not be located.

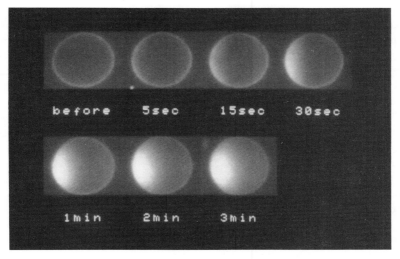

Figure 9   Penetration of the dye RH160 into an electroporated sea urchin egg. At time 0, the egg was exposed to an electric pulse of 800 V/cm with 1 ms duration (positive electrode to the right of the cell). Medium, Ca^{2+}-free seawater.

When the membrane of a liposome was simultaneously visualized by RH292, the rapid efflux was seen to emerge through a micrometer-sized hole(s) discussed in Section IV,A above. The burst decayed as the hole closed. In many, but not all, cases the burst was followed by a slow leakage, which indicated that the lipid membrane remained in an unstable state for many seconds after closure of the large hole.

D. Asymmetric Poration

The magnitude of the field-induced potential $\Delta\Psi$, which is thought to be the primary cause of electroporation, is symmetric, except for sign, with respect to the field direction [Eq. (1)]. Yet the final effects, permeation of ions or molecules through the porated membrane or the secondary effects thereof, have often been found to be asymmetric (Rossignol *et al.*, 1983; Mehrle *et al.*, 1985; Sowers, 1988). Our results presented above also show asymmetric behaviors: in sea urchin eggs the dye and Ca^{2+} permeation was much higher on the negative-electrode side, whereas the dye burst was seen mainly on the positive side in liposomes. In liposomes, in particular, the hole formation in the presence of the field was already asymmetric.

The physiological membrane potential, to which the field-induced potential is added, has been suggested to be the cause of the asymmetry (Mehrle *et al.*, 1985).

This explanation cannot be applied to the sea urchin egg in a straightforward manner, since the total potential is larger on the positive-electrode side whereas higher permeation was observed on the negative side. Sowers (1988) postulated electroosmosis as a possible cause of the asymmetric fluxes. The electroosmosis, however, does not operate in the absence of field and cannot explain the asymmetry of Ca^{2+} and dye fluxes observed long after the termination of the applied pulse. The cause of the asymmetry clearly calls for further studies.

The membrane conductance inferred from the measured transmembrane potential was more or less symmetric, except for an initial transient lasting for a microsecond or two. This is not necessarily contradictory to the marked asymmetry observed in later stages. As already discussed in Section III,B, the membrane conductance in the presence of an external field is forced to be balanced, but for the same membrane conductance the pore size and pore density can be different. Even if these parameters are common to both hemispheres, structural changes on the pore site can be different. For example, the membrane surrounding the pore may protrude, in a volcano-like shape, toward the extracellular medium or toward the cytoplasmic side, depending on whether the pore is on the positive or negative side. Electroosmosis might play a role in inducing such an asymmetry. Slight differences in the pore structure may well be amplified in the subsequent resealing stage.

VI. Summary

1. Induction of the transmembrane potential in sea urchin eggs exposed to an external electric field was visualized. Spatial distribution of the potential and the time course of induction were in accord with theory.
2. Those parts of the cell membrane where the absolute magnitude of the induced potential reached a critical value of about 1 V were porated within 2 μs. The extent of poration was such as to keep the membrane potential at or below the critical value. Thereafter, the membrane conductance gradually increased on the microsecond time scale.
3. The magnitude of induced membrane conductance was of the order of 1–10 S/cm^2 under an external electric field of a few hundred V/cm. Except for an initial transient of ~1 μs, the conductance of the positive and negative hemispheres were similar.
4. After the external field was removed, the membrane conductance returned toward the original level. The conductance level decreased by an order of magnitude by 1 ms.
5. When a liposome with a diameter of tens of micrometers was exposed to an electric field of several hundred V/cm for a few hundred microseconds, a micrometer-sized hole(s) was formed in the membrane. In the majority of cases, the hole was observed on the positive side.
6. The hole created under a mild electric pulse spontaneously closed in subseconds.

7. Both sea urchin eggs and liposomes were deformed under intense electric fields. The ratio of the specific resistances of the external and internal media determined the direction of deformation.
8. Permeability of porated eggs and liposomes was markedly asymmetric. In sea urchin eggs, higher permeability was observed on the negative-electrode side, while the positive-electrode side was more permeable in liposomes.

Acknowledgments

We thank Dr. A. Ikegami, Dr. K. Nagayama, and Dr. Y. Inoue for their continuous support on which this work is based. We also thank Mr. M. Hosoda for extensive support in developing the image-analysis system and Dr. G. Marriott for critically reading the manuscript.

This work was supported by Grants-in-Aid from Ministry of Education, Science and Culture of Japan, and by Special Coordination Funds for Promoting Science and Technology from the Agency of Science and Technology.

References

Baker, P. F., Knight, D. E., and Whitaker, M. J. (1980). The relation between ionized calcium and cortical granule exocytosis in eggs of the sea urchin *Echinus esculentus*. *Proc. R. Soc. Lond. B.* **207**, 149–161.

Chang, D. C., and Reese, T. S. (1990). Changes in membrane structure induced by electroporation as revealed by rapid-freezing electron microscopy. *Biophys. J.* **58**, 1–12.

Cole, K. S. (1972). "Membranes, Ions and Impulses." University of California Press, Berkeley.

Coster, H. G. L., and Zimmermann, U. (1975). The mechanism of electrical breakdown in the membranes of *Valonia utricularis*. *J. Membrane Biol.* **22**, 73–90.

Grinvald, A., Hildesheim, R., Farber, I. C., and Anglister, L. (1982). Improved fluorescent probes for the measurement of rapid changes in membrane potential. *Biophys. J.* **39**, 301–308.

Grynkiewicz, G., Poenie, M., and Tsien, R. Y. (1985). A new generation of Ca^{2+} indicators with greatly improved fluorescence properties. *J. Biol. Chem.* **260**, 3440–3450.

Harbich, W., and Helfrich, W.(1979). Alignment and opening of giant lecithin vesicles by electric fields. *Z. Naturforsch.* **34a**, 1063–1065.

Helfrich, W. (1974). Deformation of lipid bilayer spheres by electric fields. *Z. Naturforsch.* **29c**, 182–183.

Hibino, M., Shigemori, M., Itoh, H., Nagayama, K., and Kinosita, K., Jr. (1991). Membrane conductance of an electroporated cell analyzed by submicrosecond imaging of transmembrane conductance. *Biophys. J.* **59**, 209–220.

Hyuga, H., Kinosita, K., Jr., and Wakabayashi, N. (1991a). Deformation of vesicles under the influence of strong electric fields. *Jpn. J. Appl. Phys.*, **30**, 1141–1148.

Hyuga H., Kinosita, K., Jr., and Wakabayashi, N. (1991b). Deformation of vesicles under the influence of strong electric fields II. *Jpn. J. Appl. Phys.*, **30**, 1333–1335.

Itoh, H., Hibino, M., Shigemori, M., Koishi, M., Takahashi, A., Hayakawa, T., and Kinosita, K., Jr. (1990). Multi-shot pulsed laser fluorescence microscope system. *Proc. SPIE (Society of Photo Optical Instrumentation Engineers)* **1204**, 49–53.

Kinosita, K., Jr., and Tsong, T. Y. (1977a). Hemolysis of human erythrocytes by a transient electric field. *Proc. Natl. Acad. Sci. USA* **74**, 1923–1927.

Kinosita, K., Jr., and Tsong, T. Y. (1977b). Formation and resealing of pores of controlled sizes in human erythrocyte membrane. *Nature (Lond.)* **268**, 438–441.

Kinosita, K., Jr., and Tsong, T. Y. (1977c). Voltage-induced pore formation and hemolysis of human erythrocytes. *Biochim. Biophys. Acta* **471**, 227–242.

Kinosita, K., Jr., and Tsong, T. Y. (1979). Voltage-induced conductance in human erythrocyte membranes. *Biochim. Biophys. Acta* **554**, 479–497.

Kinosita, K., Jr., Ashikawa, I., Saita, N., Yoshimura, H., Itoh, H., Nagayama, K., and Ikegami, A. (1988a). Electroporation of cell membrane visualized under a pulsed-laser fluorescence microscope. *Biophys. J.* **53**, 1015–1019.

Kinosita, K., Jr., Ashikawa, I., Hibino, M., Shigemori, M., Yoshimura, H., Itoh, H., Nagayama, K., and Ikegami, A. (1988b). Submicrosecond imaging under a pulsed-laser fluorescence microscope. *Proc. SPIE (Society of Photo-Optical Instrumentation Engineering)* **909**, 271–277.

Kinosita, K., Jr., Hibino, M., Shigemori, M., Ashikawa, I., Itoh, H., Nagayama, K., and Ikegami, A. (1990). Submicrosecond imaging under a pulsed-laser fluorescence microscope. Electroporation of cell membrane time- and space-resolved. *In* "Science on Form," S. Ishizaka, ed., KTK Scientific Publishers, Tokyo, pp. 97–104.

Kinosita, K., Jr., Itoh, H., Ishiwata, S., Hirano, K., Nishizaka, T., and Hayakawa, T. (1991). Dual-view microscopy with a single camera. Real-time imaging of molecular orientations and calcium. *J. Cell Biol.,* in press.

Mehrle, W., Zimmermann, U., and Hampp, R. (1985). Evidence for asymmetrical uptake of fluorescent dyes through electro-permeabilized membranes of *Avena* mesophyll protoplasts. *FEBS Lett.* **185**, 89–94.

Mueller, P., Chien, T. F., and Rudy, B. (1983). Formation and properties of cell-sized lipid bilayer vesicles. *Biophys. J.* **44**, 375–381.

Neumann, E., and Rosenheck, K. (1972). Permeability changes induced by electric impulses in vesicular membranes. *J. Membrane Biol.* **10**, 279–290.

Neumann, E., Sowers, A. E., and Jordan, C. A. (eds.) (1989). "Electroporation and Electrofusion in Cell Biology." Plenum Publishing, New York.

Rossignol, D. P., Decker,G. L., Lennarz, W. J., Tsong, T. Y., and Teissie, J. (1983). Introduction of calcium-dependent, localized cortical granule breakdown in sea-urchin eggs by voltage pulsation. *Biophys. Biochim. Acta* **763**, 346–355.

Sale, A. J. H., and Hamilton, W. A. (1967). Effects of high electric fields on microorganisms. I. Killing of bacteria and yeasts. *Biochim. Biophys. Acta* **148**, 781–788.

Sale, A. J. H., and Hamilton, W. A. (1968). Effects of high electric fields on microorganisms. III. Lysis of erythrocytes and protoplasts. *Biochim. Biophys. Acta* **163**, 37–43.

Sowers, A. E. (1988). Fusion events and nonfusion contents mixing events induced in erythrocyte ghosts by an electric pulse. *Biophys. J.* **54**, 619–626.

Tsong, T. Y. (1983). Voltage modulation of membrane permeability and energy utilization in cells. *Biosci. Rep.* **3**, 487–505.

Zimmermann, U. (1982). Electric field-mediated fusion and related electrical phenomena. *Biochim. Biophys. Acta* **694**, 227–277.

Zimmermann, U., Pilwat, G., Holzapfel, C., and Rosenheck, K. (1976). Electrical hemolysis of human and bovine red blood cells. *J. Membrane Biol.* **30**, 135–152.

4

Time Sequence of Molecular Events in Electroporation

Tian Y. Tsong

Department of Biochemistry, University of Minnesota College of Biological Sciences,
St. Paul, Minnesota 55108

I. Introduction
II. Primary Events of Electroporation
    A. Electric Field-Induced Transmembrane Potential
    B. Kinetics of Pore Formation in Lipid Bilayer
    C. Kinetics of Electroporation in Cell Membranes
III. Other Effects of Pulsed Electric Fields
    A. Colloid Osmotic Lysis
    B. Membrane Fusion
    C. Some Related Phenomena
IV. Conclusion
    References

I. Introduction

To understand the phenomenon of electroporation of cell membranes and to broaden the application of electroporation in biotechnology, several questions have to be addressed. [1] How does an electric field interact with a cell membrane? [2] What are the subsequent events leading to the breakdown of the membrane permeation barrier? [3] What would happen to a cell when the permeation barrier of its plasma membrane is impaired? And [4] how would a cell then repair the damaged membrane and restore its normal functions? The answer to the first question seems straightforward; the interaction is electrostatic. Any other explanation would violate our current understanding of thermodynamics. The second question deals with the primary event of electroporation and is a mechanistic one. To answer it, we need

Guide to Electroporation and Electrofusion
Copyright © 1992 by Academic Press, Inc. All rights of reproduction in any form reserved.

information on the physical properties and the structural dynamics of the constituent molecules of the cell membrane. The third question appraises the consequences of electroporation to a cell and the fourth question examines the ability of a cell to react to and to recover from an electric insult. The main goal of this chapter is to summarize what we have learned in the past two decades concerning the molecular dynamics of the lipid bilayer and the cell membrane and to formulate our understanding of the primary and the secondary events of electroporation. Such analysis should provide us a sense of direction in our future experimentation.

Historically, study of electrical properties of cells paralleled the development of the membrane hypothesis [1, 2]. Reversible electrical modification of cell membrane permeability was noticed and has been studied in great detail since the 1940s [1, 3]. Coster called it "reversible electric punch through" [4]. Irreversible electric breakdown of the bilayer lipid membrane (BLM) and cell membranes have also been reported [5–7]. It was clear from these early studies that electric breakdown of cell membranes requires a transmembrane potential drop of approximately 1 V when microsecond to millisecond electric pulses are used [5–7]. The study of electric breakdown of cell membranes took a quantum jump with the discovery that such techniques could be used to load living cells with drugs [8, 9] and DNA [10, 11], and to induce cell fusions for the preparation of hybridomas [12] and other heterokaryons of agricultural interest [13–16].

II. Primary Events of Electroporation

A. Electric Field-Induced Transmembrane Potential

We now know that when a cell (radius $= R_{cell}$) suspended in a medium of physiological salt concentrations (~ 0.15 M NaCl or KCl) is exposed to an applied electric field (direct current of strength E_{appl}), there is a rapid redistribution of cations in the vicinity of the plasma membrane thus, generating a transmembrane potential $\Delta\psi_{membr}$, with a rise time, τ_{membr}.

$$\Delta\psi_{membr} = 1.5 \, R_{cell} \, E_{appl} \cos\theta \, [1 - \exp(-t/\tau_{membr})] \qquad (1)$$

$$\tau_{membr} = R_{cell} \, C_{membr} \, (r_{int} + r_{ext}/2) \qquad (2)$$

in which θ is the angle between the field line and the normal from the center of the spherical cell to a point of interest on the membrane surface, and C_{membr}, r_{int}, and r_{ext} are the capacitance per unit area of the membrane and the resistivities of the cytoplasmic fluid and the external medium, respectively [1, 15, 17]. For biological cells of diameters in micrometers, $\tau_{membr} < 1$ μs and the exponential term in Eq. (1) approaches zero within 1 μs. Cells of larger diameters have τ_{membr} greater than 1 μs [18]. The maximum transmembrane potential generated in a cell with a DC electric pulse of duration a few times longer than τ_{membr} is

$$\Delta\psi_{membr,max} = 1.5 \, R_{cell} \, E_{appl} \tag{3}$$

When an alternating field (AC), $E_{appl} = E^{\circ}_{ac} \sin(2\pi f)$, E°_{ac} and f being respectively the amplitude and the frequency of the field, is used, the Schwann equation should apply [19, 20]

$$\Delta\psi_{membr} = 1.5 \, R_{cell} \, E_{appl} \, \cos\theta/[1 + (2\,\pi\,f\,\tau_{membr})^2]^{1/2} \tag{4}$$

and

$$\Delta\psi_{membr,max} = 1.5 \, R_{cell} \, E^{\circ}_{ac}/[1 + (2\,\pi\,f\,\tau_{membr})^2]^{1/2} \tag{5}$$

For AC with $f < 100$ Hz, Eqs. (1) and (3) may replace Eqs. (4) and (5) [20, 21].

Eqs. (1)–(3) have been verified to be applicable to cell membranes by two types of experiments. The first type uses electric potential sensitive fluorescence dyes to image, in real time, $\Delta\psi_{membr}$ when cells are exposed to an electric pulse [18, 22–24]. A time resolution to submicroseconds has been achieved [18, 24]. The second type measures the critical breakdown potential of uniform-sized lipid vesicles [25, 26]. The critical $\Delta\psi_{membr}$ terms calculated by Eq. (3) are then compared with the breakdown potential of BLM. The agreement indicates that Eq. (3) is applicable to lipid vesicles.

Equation (5) has also been tested by using a fluorescence probe for DNA, propidium iodide [20]. When myeloma cells are electroporated, the dye permeates into the cells and binds to DNA, giving rise to two bright fluorescence bands at the two ends of cells facing the electrodes (Fig. 1A). The critical breakdown potential of myeloma cell membranes depends on the frequency of the AC field as is predicted by Eq. (5) (Fig. 1B).

B. Kinetics of Pore Formation in Lipid Bilayer

A BLM formed from the lipid extract of cell membranes is a good permeation barrier for ions and hydrophilic molecules. The membrane specific conductance for Na^+ or K^+ is usually smaller than 10^{-8} S cm^{-2} (3). Since lipid molecules are either charged or polarizable, an electric potential imposed across the membrane will disturb the highly ordered bilayer arrangement of lipids. In a BLM of single lipid component, the conformation of lipid molecules is synergetically controlled. At a low temperature, they typically assume a solidlike state (S-state) in which all the methylene to methylene bonds (CH_2—CH_2) of the hydrocarbon chains are in the trans conformation. When the temperature is raised above a critical point, T_c, many of these carbon/carbon bonds convert to the cis conformation [27]. Cis bonds or kinks in the hydrocarbon chain can cause the bilayer to become disorganized. Thus,

A

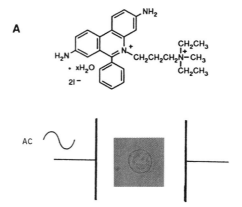

AC

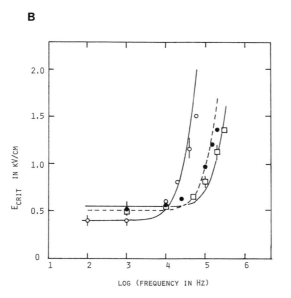

B

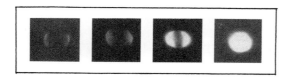

for temperatures beyond T_c, the bilayer exists in the fluidlike phase (F-state). A bilayer in the F-state generally has a higher conductivity for ions than when it is in the S-state. However, experiments show that the permeability of a lipid bilayer to ions is maximal at T_c, a temperature at which lipid exists half in the S-state and half in the F-state [26, 28]. Analysis based on a cluster model of lipid phase transition indicates that the ionic permeability of lipid bilayer is proportional to a parameter that specifies the boundary between the S- and the F-states of lipid crystalline phases [29]. In other words, ions or hydrophilic molecules permeate the bilayer through lattice defects. Lattice defects are highly fluctuating structures, and the compressibility of the bilayer is highest in these boundary regions, at the T_c [30].

An electric field can also cause lipid phase transitions. The molecular events induced by an electric field should be similar to those induced by heat, both reflecting different modes of motion of molecules in the bilayer. Kinetics of lipid phase transitions have been studied by various physicochemical methods [31–33]. The trans/cis isomerization of a CH_2—CH_2 bond takes place in 5 ns [33]. A lattice defect develops only if the kink formation propagates to its neighboring molecules. This process takes place in microseconds. After these rapid reactions there are some cooperative structure relaxations, which last for minutes [31, 34]. Other modes of motion for lipid include lateral diffusion ($\sim 1 \times 10^{-8}$ cm^2 s^{-1}) and the flip-flop from one monolayer to another in hours or days [2, 27]. Rotation or wobbling of lipid molecules occurs in the nanosecond to microsecond time ranges [2]. The ranges of rates given here are in the absence of an applied electric potential. An electric field can enhance a rate by a factor of $\exp(zq\,\Delta\psi_{membr}/kTd)$, where z, q, k, and d are the net charge in the molecule, the elementary charge, the Boltzman constant, and the thickness of the membrane [35]. An electric field is a vector quantity and it can drive the motion of molecules anisotropically [35].

Figure 1   (A) Observation of electroporation of a myeloma cell (Tib9) by the fluorescence changes of propidium iodide. The chemical formula of propidium iodide is given in the upper figure. Propidium iodide is weakly fluorescent in solution but becomes strongly fluorescent when bound to DNA. The middle figure shows a typical myeloma cell under the light microscope. The relative positions of the platinum electrodes are indicated. The lower figure gives four fluorescence photographs taken at different times after the cell was electroporated by an AC field. Within 1–3 s, two narrow, bright bands appeared at the two loci facing the electrodes (the leftmost photo), indicating that the dye enters the cells at these two loci. The next three photos, from left to right, were taken at 20 s, 1 min, 3 min, respectively, after electroporation. (From reference 20, with permission.) (B) Critical AC field strength (E_{crit}) for electroporation of myeloma cells (Tib9) as a function of the AC frequency. Data were obtained in three media of different resistivities, 52,600 Ω cm (○), 7050 Ω (●), and 2380 Ω cm (□). The curves drawn through these data points were by calculation (○) or by optimization (● and □) according to the Schwan equation [Eq. (5)]. (From reference 20, with permission.)

As mentioned, the rise time of a $\Delta\psi_{membr}$, τ_{membr}, is about 1 μs for a lipid vesicle of micrometer diameter or a sea urchin egg. For cells with diameters of a few micrometers, it is in the submicrosecond range. The pore initiation step is as fast as τ_{membr}, suggesting that electroporation occurs at the existing lattice defects. Once a pore is initiated, there would be a strong current across the pore, which can produce local heating. The pore expands in the 50 μs time range. Once an electric pulse is terminated, a rapid shrinking of pores, in the millisecond time, is detected by the conductivity and the fluorescence measurements. The physical meaning of the slower reactions is till not clear [34]. If the electric pulse is prolonged, pores will expand, and beyond a certain limit, these pores will cause a BLM to rupture irreversibly. For liposomes, intense or long electric pulses cause fragmentations and produce small vesicles.

These complex molecular events may be represented by Schemes (6) and (7) [37].

$$
\begin{array}{ccc}
A \rightleftharpoons B & \rightarrow & C \\
\nwarrow & & \downarrow \\
& B' \leftarrow C' &
\end{array} \tag{6}
$$

$$
\begin{array}{ccc}
& C \rightarrow A_1 & \\
& \nearrow & \\
A \rightleftharpoons B \rightarrow & D \rightarrow A_2 & \\
& \searrow & \\
& E \rightarrow A_3 &
\end{array} \tag{7}
$$

The A to B transition in both schemes denotes the pore initiation step. Lattice defects of the bilayer are converted into hydrophobic pores. At this stage, the normal orientations of lipid molecules are not drastically altered and the process is considered reversible. The B to C transition in Scheme 6 denotes a pore expansion step. Here some changes in lipid orientations will occur in order to form relatively stable hydrophilic pores. When the electric pulse is terminated, the system will relax back to the initial state A. The backward pathway is necessarily different from that of the forward reaction because in the backward pathway the reaction happens in the absence of an electric field while in forward pathway the reaction happens in the presence of an electric field. In Scheme (7), the B to C, D, and E transitions represent fragmentation mainly due to the repulsion of polarized lipid bilayer. The end products are small vesicles A_1, A_2, and A_3.

C. Kinetics of Electroporation in Cell Membranes

The plasma membrane of a cell is the first site of the electric interaction. Beside lipids, there are proteins, carbohydrates, and other types of molecules, most of which are either charged or polarizable. Channel proteins are especially sensitive to the $\Delta\psi_{membr}$, and each type of channel has a range of $\Delta\psi_{membr}$ in which it becomes conductive. The range of $\Delta\psi_{membr}$ for opening protein channels is approximately 50 mV, considerably smaller than the dielectric strength of the lipid bilayer, which is in the range 150–400 mV. Like a lattice defect of the lipid bilayer, once a protein channel is forced to open, a strong current much exceeding the normal conductance of the channel will generate local heat sufficient to denature the protein. This denaturation could be reversible or irreversible, depending on the extent of temperature change and the properties of the channel. The opening/closing of a protein channel is in the submicrosecond time range [38]. Thermal denaturation of a protein takes milliseconds to seconds [39]. Renaturation of a protein occurs in seconds [39]. If a protein is irreversibly denatured, it will be removed from the membrane by the cellular recycling process, which is relatively slow. Since the opening of protein channels may not be sufficient to prevent the rise of $\Delta\psi_{membr}$ beyond the breakdown potential of the lipid bilayer, electroporation will also occur in the lipid domain. Electroporations through protein channels and lipid domains can take place independently of each other. Kinetic schemes representing these parallel reactions are [37, 40, 41].

$$A \rightleftharpoons B \rightarrow C \rightarrow C' \rightarrow B' \rightarrow A \qquad (8)$$

$$P_{close} \rightleftharpoons P_{open} \rightleftharpoons P_{denatr} \nearrow^{P_{irrev}}_{\searrow [O]} \qquad (9)$$

Scheme (8) is identical to Scheme (6) and represents electroporation of lipid domains, and Scheme (9) represents electroporation of protein channels. A channel protein is shown to change from closed, open, reversibly denatured, and irreversibly denatured states. The boxed circle denotes a protein, which after being irreversibly denatured is excised from the membrane for recycling. Teissie and Tsong [42] and Serpersu and Tsong [43, 44] have presented evidence that the Na,K-ATPase of human erythrocyte is punctured with an intense pulsated electric field (PEF). Membrane conductance sensitive to the specific inhibitor of the enzyme, ouabain, has been detected to occur in the microsecond time range.

   Electropores in lipid domains will reseal within seconds [24, 25]. Closing of PEF-perforated protein channels should transpire in milliseconds. However, repairing of a PEF-damaged cell membrane will take minutes to hours [41, 45]. The best monitor of a complete recovery of electroperforated cell membranes is by measuring the recovery of membrane permeability to K^+ or Rb^+ tracer. As was shown by Serpersu et al. for human erythrocytes, even small pores that allowed only the permeation of Rb^+ but not the sucrose tracer took more than 15 min to reseal. Larger pores that allowed permeation of sucrose tracer took more than 10 h to reseal against leakage of Rb^+ tracer [41].

III. Other Effects of Pulsed Electric Fields

A. Colloid Osmotic Lysis

A major difference between electroporation of lipid vesicles and that of cells is the colloid osmotic lysis of cells [17, 45, 46]. A PEF-perforated cell membrane loses its permeation barrier to ions and small molecules but not necessarily to proteins. The electroperforated membrane becomes semipermeable to cytoplasmic macro-molecules. The osmotic pressure of these macromolecules will cause the cell to swell. This process, known as colloidal swelling, eventually leads to rupture of the plasma membrane because of the excessive osmotic pressure imposed on the cells (Fig. 2).

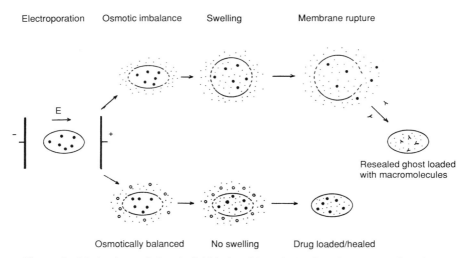

Figure 2 Mechanisms of electric field-induced hemolysis of erythrocytes, and loading of enzymes or antibodies into ghosts, or drug loading into hemoglobin-retained erythrocytes. In the figure, small dots denote drugs; small dark circles, cytoplasmic macromolecules; small open circles, inulin or oligosaccharide; and Y-shaped molecules, antibodies. In the upper path, an electroporated erythrocyte swells due to the colloidal osmotic pressure of hemoglobin. When the membrane is ruptured, hemoglobin leaks out and in the process of membrane resealing, antibodies are loaded into the ghost. In the lower path, the colloidal osmotic pressure of hemoglobin is counterbalanced by inulin or oligosaccharide added to the external medium. Swelling of the electroporated cells is prevented. A drug of small molar mass (smaller than 2000) can be loaded, and the cell membrane resealed completely. The final product is the drug-loaded intact erythrocytes. For most cells with stronger membranes than erythrocytes, or with cell walls, the upper path may be used to load macromolecules without membrane rupture or cell lysis. (From reference 37, with permission.)

Colloidal osmotic pressure in the PEF-treated red blood cells was identified as the main cause of the electric field-stimulated hemolysis. Colloidal swelling depends on the osmotic imbalance of the cytoplasmic and the suspending media. When the difference is large, PEF-treated cells will swell in the minute time range. During this swelling phase, electropores in cell membranes also begin to reseal. If the resealing takes place faster than the swelling, cells will shrink again and recover their original volume, thus averting membrane rupture. If, on the other hand, the resealing is slower than the swelling, the plasma membrane of cells will be ruptured. The colloidal osmotic lysis may be prevented by balancing the osmotic pressure of the cytoplasm and the suspending medium. Molecules impermeant to electropores, such as cytochrome c, myoglobin, or oligosaccharides, have been added to the suspension of the electroporated erythrocytes to thwart hemolysis [8, 9, 41, 45]. In an osmotic pressure balanced medium, cell swelling is completely prevented and the membrane resealing can be accomplished without much loss of the cytoplasmic macromolecules [8, 9, 41, 45]. Electroporated, drug-loaded and resealed erythrocytes, sea urchin eggs, and dictyostelium discoideum [47] have been shown to behave like normal control cells with respect to their survivability, mobility, or the ability to differentiate.

B. Membrane Fusion

Small lipid vesicles exposed to a PEF will fuse to become larger vesicles [36]. The effect has been attributed to temperature-induced lipid phase transitions, although current understanding favors electroporation as the basis for membrane fusion [13–15]. Scheme (10) characterizes such reactions.

$$n\,A \rightleftharpoons n\,B \rightarrow n\,C \rightarrow D \qquad (10)$$

The n small vesicles undergo A to B pore initiation followed by pore expansion to C. If these vesicles are in contact at the time of, or collide after, electroporation, they will fuse into a large vesicle D. The arrangement of lipids at the two foci facing the electrodes will be perturbed by the applied field. When the field is terminated, the disorganized lipid molecules will reform the original bilayer. During the restoration of the bilayer arrangement of lipid, an intervesicle bilayer may also form if two vesicles are in contact. Since lipid reorientation involves slow flip-flop reaction, the "fusogenic" state of the bilayer may persist for minutes [48, 49]. Thus, intervesicle bilayer formation will also occur for vesicles brought into contact after electroporation. The end result of the intervesicle bilayer formation is the formation of large vesicles. In cells, the initial fusion process may be similar to that of lipid vesicles. But after fusion of the lipid bilayer, intercellular weaving of cytoskeletal networks must also take place in order to form a stable cell doublet. This latter reaction may take minutes to hours.

C. Some Related Phenomena

Infection of cells by plasmid DNA is greatly facilitated by the electroporation of cell membranes. Until now the mechanisms of PEF-induced DNA transfection was not known. Because the size of a plasmid DNA is in micrometers (molecular weight several million), any electropore that allows DNA passage must have a large diameter. Experiments measuring ion conductance or molecular permeability of electroperforated cells give estimates of stable pores to be in the 1–10 nm range [17, 24,

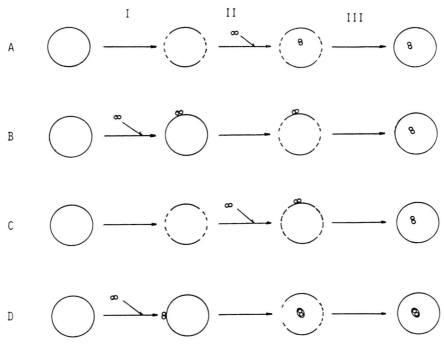

Figure 3   Comparison of different mechanisms of DNA transfection by the electroporation method. (A) After the electroporation of the cell (Step I), circular and superhelical plasmid DNA (indicated by ∞) diffuses into the cell through the bulk solution (Step II). Cell membranes reseal (Step III). (B) DNA binds to the cell surface and then enters the cell through surface diffusion. (C) Electroporation is done before the addition of DNA. DNA binds to the surface and diffuses into the cell. Because the size of the pores is reduced during the incubation time (the time between electroporation and addition of DNA), the transfection efficiency is much lower than that of (B). (D) DNA is assumed to enter the cell by the electrophoretic force of the applied field. When DNA enters, it carries with it a piece of cell membrane and is protected by a shell of lipid. Other mechanisms such as electric field-stimulated endocytosis for DNA uptake are possible. (From reference 51, with permission.)

50]. Our study of the electrotransfection of *Escherichia coli* indicates that only DNA that binds to cell surface has a high probability of transfection [51]. Furthermore, an AC field as low as 100 V/cm is found also to facilitate DNA entry into *E. coli* [52]. Since this intensity is too low to cause electroporation of cell membranes, some other mechanisms have to be considered. Chernomordik *et al.* suggest that DNA entry may occur by the endocytotic-like mechanisms [53]. Possible mechanisms of high electric field-stimulated DNA transfection are compared in Fig. 3. Although our results favor Scheme B over other schemes, further investigation is warranted. For the low-field experiments, we have used the electroconformational coupling concept to interpret the result [35, 52].

Reversible electroconformational change of a membrane protein may be coupled to a biochemical reaction to provide an efficient mechanisms for energy and signal transduction. Such phenomena are currently under intense investigations from several laboratories [35].

Table 1
Some Effects of an Electric Field on Chemical Reaction

	Effects common to all systems	Effects specific to cells in suspension
Electric field effects	Electrophoresis	Field-induced transmembrane potential
	Orientation of molecules	Amplification of common effects within cell membrane by 1.5 R_{cell}/d for a spherical cell
	Electroconformational change $\Delta K/K$ $= \Delta M (\Delta E/RT)$	Electrocompression of membrane Large transmembrane current
For alternating field only	Dielectrophoresis	Amplification of common effects within cell membrane
	Enforced conformational oscillation for energy and signal transductions	
Thermal effects $\Delta T = i^2 r$ $\Delta t/(4.18 C_p)$	Enthalpy effects $\Delta K/K =$ $(\Delta H/RT^2)\Delta T$	Colligative effects of $\Delta \pi = cR \Delta T$
	Solvent expansion or shock wave $\Delta P = (\alpha/\kappa) \Delta T$	Thermal osmosis effects $\Delta p = - (Q/vT)\Delta T$

Symbols used: R_{cell}, radius of cell; ΔT, temperature change; r, specific resistivity of solution; C_p, specific heat capacity of solution; c, concentration of solute; K, equilibrium constant; ΔM, change in the molar electric moment; ΔH, enthalphy of reaction; ΔP, pressure generated by shock wave; α, thermal expansion coefficient of solvent; κ, compressibility of solvent; Q, heat of transfer of water across cell membrane; d, membrane thickness; i, current; Δt, field exposure time; $\Delta \pi$, change in osmotic pressure; R, gas constant; ΔK, change in equilibrium; ΔE, effective electric field; T, Kelvin temperature; and v, partial molar volume of solvent. See reference 34 for details.

IV. Conclusion

A pulsed electric field can have many different effects upon a chemical reaction. Some originate from coulombic interaction and some from Joule heating. For a reaction that occurs in a cell membrane, many effects are enhanced because the intensity of the electric field is amplified in the membrane according to Eqs. (3) and (5). The maximal field intensity across the cell membrane is $\Delta\psi_{membr,max}/d$, with d being the thickness of the membrane. When the field is oscillatory, some constituents of the membrane may absorb energy from the applied field [35]. This phenomenon may provide yet another mechanism for modifying the permeability of a cell membrane. Different effects of a PEF to a chemical reaction are summarized in Table 1.

Acknowledgments

I thank my former and current colleagues, K. Kinosita, Jr., J. Teissie, E. H. Serpersu, E. M. El-Mashak, T.-D. Xie, and P. Marszalek, for their contributions to this project, and C. J. Gross for help with the manuscript.

References

1. Cole, K. S. (1972). "Membrane, Ions and Impulses." University of California Press, Berkeley.
2. Jain, M. K., and Wagner, R. C. (1980). "Introduction to Biological Membranes." Wiley & Sons, New York.
3. Tien, H. T. (1974). "Bilayer Lipid Membranes (BLM): Theory and Practice." Marcel Dekker, New York.
4. Coster, H. G. L. (1965). A quantitative analysis of the voltage-current relationships of fixed charge membranes and the associated property of "punch-through." *Biophys. J.* **5**, 669–686.
5. Huang, C.-H., Wheeldon, L., and Thompson, T. E. (1964). The properties of lipid bilayer membranes separating two aqueous phases: Formation of a membrane of simple composition. *J. Mol. Biol.* **8**, 148–160.
6. Sale, A. J. H., and Hamilton, W. A. (1968). Effects of high electric fields in microorganisms. III. Lysis of erythrocytes and protoplasts. *Biochim. Biophys. Acta* **163**, 37–43.
7. Coster, H. G. L., and Zimmermann, U. (1975). The mechanism of electrical breakdown in the membranes of *Valonia utricularis*. *J. Membrane Biol.* **22**, 73–90.
8. Kinosita, K., and Tsong, T. Y. (1978). Survival of sucrose-loaded erythrocytes in circulation. *Nature* **272**, 258–260.
9. Tsong, T. Y. (1987). Electric modification of membrane permeability for drug loading into living cells. *Methods Enzymol.* **149**, 248–259.
10. Wong, T.-K., and Neumann, E. (1982). Electric field medicated gene transfer. *Biochem. Biophys. Res. Commun.* **107**, 584–587.
11. Neumann, E., Schaefer-Ridder, M., Wang, Y., and Hofschneider, P. H. (1982). Gene transfer into mouse lyoma cells by electroporation in high fields. *EMBO J.* **1**, 841–845.

12. Lo, M. M. S., Tsong, T. Y., Conrad, M. K., Strittmater, S. M., Hester, L. H., and Snyder, S. H. (1984). Monoclonal antibody production by receptor-mediated electrically induced cell fusion. *Nature* **310**, 792–794.
13. Sowers, A. E. (ed.) (1987). "Cell Fusion." Plenum Press, New York.
14. Zimmermann, U. (1982). Electric field-mediated fusion and related electrical phenomena. *Biochim. Biophys. Acta* **694**, 227–277.
15. Neumann, E., Sowers, A. E., and Jordan, C. A. (ed.) (1989). "Electroporation and Electrofusion in Cell Biology." Plenum Press, New York.
16. Teissie, J., Knutson, V. P., Tsong, T. Y., and Lane, M. D. (1982). Electric pulse-induced fusion of 3T3 cells in monolayer culture. *Science* **216**, 537–538.
17. Kinosita, K., and Tsong, T. Y. (1977). Voltage induced pore formation and hemolysis of human erythrocyte membranes. *Biochim. Biophys. Acta* **471**, 227–242.
18. Kinosita, K ., Ashikava, I., Saita, N., Yosimura, H., Itoh, H., Nagayama, K., and Ikegami, A. (1988). Electroporation of cell membrane visualized under pulsed laser fluorescence microscope. *Biophys. J.* **53**, 1015–1019.
19. Schwan, H. P. (1983). Biophysics of the interaction of electromagnetic energy with cells and membranes. *In* "Biological Effects and Dosimetry of Nonionizing Radiation," M. Grandolfo, S. M. Michaelson, and A. Rindi (eds), Plenum Press, New York, pp. 213–231.
20. Marszalek, P., Liu, D.-S., and Tsong, T. Y. (1990). Schwan equation and transmembrane potential induced by alternating electric field. *Biophys. J.* **58**, 1053–1058.
21. Pliquett, V. F. (1968). Das Verhalten von oxytrichiden unter einfluss does elektrischen felds. *Z. Biol.* **116**, 10–22.
22. Gross, D., Loew, L. M., and Webb, W. W. (1986). Optical imaging of cell membrane potential changes induced by applied electric fields. *Biophys. J.* **50**, 339–348.
23. Ehrenberg, B., Farkas, D. L., Fluhler, E. N., Lojewska, Z., and Loew, L. M. (1987). Membrane potential induced by external electric field pulses can be followed with a potentiometric dye. *Biophys. J.* **51**, 833–837.
24. Hibino, M., Shigemori, M., Itoh, H., Nagayama, K., and Kinosita, K. (1991). Membrane conductance of an electroporated cell analyzed by submicrosecond imaging of transmembrane potential. *Biophys. J.* **59**, 209–220.
25. Teissie, J., and Tsong, T. Y. (1981). Electric field-induced transient pores in phospholipid bilayer vesicles. *Biochemistry* **20**, 1548–1554.
26. El-Mashak, E. M., and Tsong, T. Y. (1985). Ion selectivity of temperature-induced and electric field induced pores in dipalmitoylphosphatidylcholine vesicles. *Biochemistry* **24**, 2884–2888.
27. Thompson, T. E., and Huang, C.-H. (1980). Dynamics of lipids in biomembranes. *In* "Membrane Physiology," T. E. Andreoli, J. F. Hoffman, and D. D. Fanestil (eds.), Plenum, New York, pp. 27–48.
28. Tsong, T. Y., Greenberg, M., and Kanehisa, M. I. (1977). Anesthetic action on membrane lipids. *Biochemistry* **16**, 3115–3121.
29. Kanehisa, M. I., and Tsong, T. Y. (1978). Cluster model of lipid phase transitions with application to passive permeation of molecules and structure relaxation in lipid bilayers. *J. Am. Chem. Soc.* **100**, 424–432.
30. Ipsen, J. H., Jorgensen, K., and Mouristsen, O. G. (1990). Density fluctuations in saturated phospholipid bilayers increase as the actyl-chain length increases. *Biophys. J.* **58**, 1099–1107.

31. Tsong, T. Y., and Kanehisa, M. I. (1977). Relaxation phenomena in aqueous dispersions of synthetic lecithins. *Biochemistry* 16, 2674–2680.
32. Genz, A., Holzwarth, J. F., and Tsong, T. Y. (1986). The influence of cholesterol on the main phase transition of unilamellar dipalmitoylphosphatidylcholine vesicles. *Biophys. J.* 50, 1043–1051.
33. Gruenewald, B., Frisch, W., and Holzwarth, J. F. (1981). The kinetics of the formation of rotational isomers in the hydrophobic tail region of phospholipid bilayers. *Biochim. Biophys. Acta* 641, 311–319.
34. Caffrey, M. (1989). The study of lipid phase transition kinetics by time-resolved x-ray diffraction. *Annu. Rev. Biophys. Biophys. Chem.* 18, 159–186.
35. Tsong, T. Y. (1990). Electrical modulation of membrane proteins: Enforced conformational oscillations and biological energy and signal transductions. *Annu. Rev. Biophys. Biophys. Chem.* 19, 83–106.
36. Tsong, T. Y. (1974). Temperature jump relaxation kinetics of aqueous suspensions of phospholipids and B. subtlis membranes. *Fed. Proc.* 33, 1342.
37. Tsong, T. Y. (1990). On electroporation of cell membranes and some related phenomena. *Bioelectrochem. Bioenerg.* 24, 271–295.
38. Tsien, R. W., Hess, P., McClesky, E. W. and Rosenberg, R. L. (1987). Calcium channels: Mechanisms of selectivity, permeation, and block. *Annu. Rev. Biophys. Biophys. Chem.* 16, 265–290.
39. Kim, P. S., and Baldwin, R. L. (1982). Specific intermediates in the folding reactions of small proteins an the mechanisms of protein folding. *Annu. Rev. Biochem.* 51, 459–489.
40. Kinosita, K., and Tsong, T. Y. (1979). Voltage-induced conductance in human erythrocyte membranes. *Biochim. Biophys. Acta* 554, 479–497.
41. Serpersu, E. H., Kinosita, K., and Tsong, T. Y. (1985). Reversible and irreversible modification of membrane permeability by electric field. *Biochim. Biophys. Acta* 812, 779–785.
42. Teissie, J., and Tsong, T. Y. (1980). Evidence of voltage induced channel opening in Na,K-ATPase of human erythrocyte membranes. *J. Membrane Biol.* 55, 133–140.
43. Serpersu, E. H., and Tsong, T. Y. (1983). Stimulation of Rb^+ pumping activity of Na,K-ATP-ase in human erythrocytes with an external electric field. *J. Membrane Biol.* 74, 191–201.
44. Liu, D.-S., Astumian, R. D., and Tsong, T. Y. (1990). Activation of Na_+ and K^+ pumping modes of Na,K-ATPase by an oscillating electric field. *J. Biol. Chem.* 265, 7260–7267.
45. Kinosita, K., and Tsong, T. Y. (1977). Formation and resealing of pores of controlled sizes in human erythrocyte membrane. *Nature* 268, 438–441.
46. Kinosita, K., and Tsong, T. Y. (1977). Hemolysis of human erythrocytes by a transient electric field. *Proc. Natl. Acad. Sci. USA* 74, 1923–1927.
47. Tsong, T. Y. (1983). Voltage modulation of membrane permeability and energy utilization in cells. *Biosci. Rep.* 3, 487–505.
48. Sowers, A. E. (1986). A long-lived fusogenic state is induced in erythrocyte ghosts by electric pulses. *J. Cell Biol.* 102, 1358–1362.
49. Sowers, A. E. (1988). Fusion events and nonfusion content mixing events induced in erythrocyte ghosts by an electric pulse. *Biophys. J.* 54, 619–626.

50. Sowers, A. E., and Lieber, M. L. (1986). Electropores in individual erythrocyte ghost: Diameter, lifetimes, numbers, and locations. *FEB Lett.* **205,** 179–184.
51. Xie, T.-D., Sun, L., and Tsong, T. Y. (1990). Study of mechanisms of electric field-induced DNA transfection I. DNA entry by surface binding and diffusion through membrane pores. *Biophys. J.* **58,** 13–19.
52. Xie, T.-D., and Tsong, T. Y. (1990). Study of mechanisms of electric field-induced DNA transfection II. Transfection by low-amplitude, low frequency alternating electric fields. *Biophys. J.* **58,** 897–903.
53. Chernomordik, L. A., Sokolov, A. V., and Budker, V. G. (1990). Electrostimulated uptake of DNA by liposomes. *Biochim. Biophys. Acta* **1024,** 179–183.

5

Electropores in Lipid Bilayers and Cell Membranes

Leonid V. Chernomordik[1]

A. N. Frumkin Institute of Electrochemistry, Academy of Sciences of the USSR, Moscow 117071, U.S.S.R.

I. Introduction

The application of a high electric field pulse to membranes results in a drastic increase in permeability due to the appearance of pores in the membranes. Thanks to the universality of this phenomenon in a variety of biological membranes, as well as for simple model systems based on lipid bilayers and the physical character of its trigger (i.e., electrostimulation), electroporation can be regarded as a relatively simple subject for the investigation of the mechanism. A variety of data has been accumulated on the phenomenology and mechanism of the early stages of electroporation related to the pore arising and evolving in the lipid matrix (lipid bilayer) of cell membranes. However, the later stages of the pore evolution process and the role of electropores in the related phenomena of electrofusion and electrotransfection are still unclear.

In the present chapter I shall consider briefly what is known and not known about the mechanism of electroporation and the properties of electropores.

[1]Present address: NICHD, National Institutes of Health, Bethesda, Maryland 20892.

II. Experimental Approaches

The majority of experimental techniques used in studies of electroporation are based mainly on membrane permeability [1–3] or electrical conductance [4–7] measurements. Both approaches have their own advantages and limitations. The permeability measurement is a simple and fundamental approach used to reveal the increase in permeability for test molecules of different sizes caused by electroporation. The measurement is usually carried out after the electric field application. In contrast, the electrical conductance measurements make it possible to observe the arising and evolution of pores just during the electric field pulse. This approach is limited because of the non-ohmic behavior of small pores [5, 8]. Lastly, the elegant approach based on the analysis of the transbilayer movement of a lipid marker was suggested in reference 9.

III. The Common Features of Electroporation

The reversible (i.e., the initial low level of permeability is restored with time after the electric field pulse) or irreversible disturbance of the barrier function of membranes due to electroporation appears to be a universal phenomenon. It can be initiated by the electrical treatment with a broad range of the following parameters: (1) induced transmembrane voltage in the range of 0.2–1.5 V applied for (2) durations of tens of nanoseconds to milliseconds and even seconds [10]. The phenomenon of electroporation has been observed for the membranes of a number of different species of procaryotes and plant and animal cells (see the other chapters in this monograph). However, in spite of the universal qualitative character of phenomenon, the efficiency of electroporation and properties of electropores depend dramatically on different parameters. The longer the duration of the pulse, the lower the value of the pulse amplitude required to induce electroporation and, for the same pulse amplitude, the larger the mean diameter and number of the pores [5, 8]. The characteristic relationships between amplitude and duration of the electrical treatment required to develop the electroporation of model (curves 2, 3) and biological (curve 1) membranes are presented in Fig. 1. These results [see also refs. 2, 11–13] argue against any interpretations of the phenomenon described based on the instability of the membrane at certain critical induced transmembrane voltage [14, 15].

The electroporation process is modulated considerably by the structure of membranes. There is a significant difference between the values of the pulse amplitude required to induce electroporation for membranes of erythrocytes of different mammalian species [16]. The lipid composition of model membranes can change even the qualitative character of electroporation process. For example, in planar lipid bilayers of oxidized cholesterol there is a five order of magnitude reversible increase

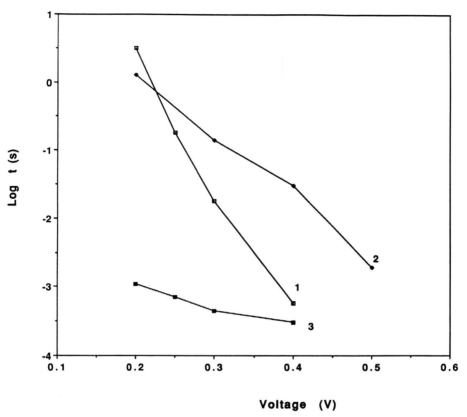

Figure 1   The relationships between the amplitudes and durations of square electric field pulses induced electroporation for erythrocyte membrane (curve 1) [5], planar lipid bilayer (curve 2) [24], and liposomes of egg phosphatydilcholine (curve 3). The parameters of square electric field pulse inducing two order of magnitude increase of conductance (curves 1 and 2) or permeability for fructose (curve 3) are shown in the figure. For the liposomes (curve 3) the membrane voltage U generated across the membrane was calculated from the electric field intensity E using the equation $U = 1.5Er$, where r is the mean radius of the liposomes.

of conductance induced by the electric field pulse attributed to arising of a number of pores [17]. In contrast, for bilayers of phosphatidylcholine the formation of pores of above the critical radius will lead to the rupture of the membrane. This always precedes the development of a large population of small pores. Thus, in this case a more than two order of magnitude increase in the membrane conductance was irreversible [17, 18].

Electroporation can be divided into two phases: (1) induction, and (2) resealing of pores. The loss of the barrier function of membranes during electroporation can develop in less than microseconds for sufficiently high intensities of electrical treatments [5, 17, 19]. As shown in references 8 and 20, the increase of the radii of electropores during the electric field pulse as well as the decrease of the pore radii during the postpulse resealing process takes milliseconds. Finally, the time for complete resealing of the electropores in lipid bilayer (i.e., the time required the pores to disappear) varies over the range of seconds to minutes [5]. It is necessary to mention that though the close range of the resealing times was observed in many papers for the electropores in cell membranes (see other chapters in this book), much longer times (more than 30 min) were also reported in some cases [5, 21, 22].

IV. Mechanisms of Lipid Bilayer Electroporation

The striking similarity of the phenomenology of the initial stages of the electroporation for lipid and different biological membranes [5] points out that at least the main part of initial pores arises in the lipid matrix of cell membranes. The experimental and theoretical investigation [8] of the mechanism of this stage of the electroporation addressed two main questions: (1) why does the electric field promote the pore arising and expansion, and (2) what is (are) the structure(s) of the pores in lipid bilayers? According to references 18 and 23 the drawing of water into the cylindrical pore is caused by the decrease of the free energy of the membrane with a pore of radius r due to electrical polarization in the presence of a voltage U across the membrane:

$$\pi r^2 C_m \left(\frac{\varepsilon_w}{\varepsilon_m} - 1 \right) \frac{U^2}{2}$$

which is determined by the difference of the dielectric constant of water in pore ε_w, and that of membrane, ε_m. Here r is the pore radius, and C_m the specific capacitance of the membrane. The effect of the electric field is counteracted by the additional energy of the pore edge. For large enough pore radii it can be expressed as $2\pi r \gamma$, where γ is linear tension in the pore (i.e., the work of the formation of the unit of pore perimeter). The walls of pores can, in principle, be formed by hydrocarbon lipid tails (hydrophobic pores) (Fig. 2b) or by the polar heads of reoriented lipid molecules (hydrophilic pores) (Fig. 2c) [8, 18]. For relatively large pore radii (more than 0.5 nm) the energy of the aqueous pore will be significantly lower for the pore with a hydrophilic edge [8, 18]. In contrast, for small pores the mechanical limitations of the lipid packing in the pore edge as well as the hydration repulsion between the polar heads of lipid molecules that form the pore edge must significantly increase the linear tension of the pores. In fact, large electropores, which can lead to the rupture of planar lipid bilayers in an electric field, were experimentally shown

PORE DEVELOPMENT AND RESEALING

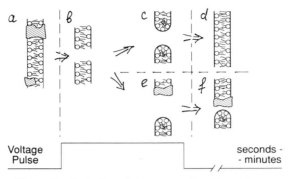

Figure 2   Evolution of electropores in cell membrane.

to be hydrophilic. This conclusion is based on (1) the determination that the values of the linear tension of pores ($0.3–2 \times 10^{-11}$ N) [24] for membranes made of different lipids were characteristic for a hydrophilic surface, as well as on (2) the observation of cation–anion selectivity of the pores in membranes made from negatively charged lipids [25]. The exchange of lipids between outer and inner monolayers of erythrocyte membranes as result of electroporation, as was observed in reference 9, suggests the existence of continuous bridges between the monolayers in hydrophilic pores. Such an assumption is also supported by an increase in the rate of accumulation of electropores in a planar lipid bilayer containing lysolecithin [8]. This lipid is known to facilitate the packing of lipids in hydrophilic pores, which decreases the linear tension of the pores [24]. The relation between the effective molecular shape of lipid molecules forming the bilayer and the linear tension was experimentally and theoretically studied in reference 24. It was shown that the larger the ratio of the effective area of the polar head of lipids to the area of its hydrophobic part, the less are the values of the linear tension of pores, which "predicts" the higher number of the electropores.

Thus the presence of hydrophilic pores in lipid bilayer can be regarded as the structural basis of the increase of membrane conductance and permeability at some stage of the reversible electroporation. On the other hand, the initial stage of the process is apparently based on the evolution of small hydrophobic pores ($r < 0.5$ nm) formed by the lateral thermal fluctuations of lipid molecules in a bilayer [8] (Fig. 2b). If the radius of the pore exceeds the value where the energies of the hydrophobic and hydrophilic pores coincide, then hydrophobic pores transform into hydrophilic pores by lipid molecules reorientating (Fig. 2c). This model was developed and experimentally substantiated in references 8 and 18.

V. The Specificity of Electroporation for Biomembranes

The application of a strong electric field to the cell membrane results in the appearance of hydrophilic pores in the lipid matrix of membrane. The investigation of the mechanism of this stage of the electroporation by using simple model systems allowed us to obtain a physical description of the process [8, 18]. However, it is quite clear that the structure of any biological membrane is much more complex than pure lipid bilayers. The processes of electropore formation and evolution in the lipid matrix of cell membranes may significantly differ from those for model lipid bilayers. The most remarkable distinguishing feature observed for cell membrane electroporation is related to the time of pore resealing. If the amplitude of electric treatment used is too high or its duration is too long, then the pores will be extremely long-lived (lifetime > 30 min) or even nonresealing [5, 21, 22]. These electropores may be similar to the nonresealing (residual) pores developed in erythrocytes by hemolysis. It was suggested in reference 5 that the stabilization of pores occurs if a given small pore induced in membrane lipid matrix by a short but very intense electric field pulse is sufficiently close to some membrane protein for the pore edge to reach it during pore expansion (Fig. 2e and f). A similar situation can occur if a few but very large pores are formed in membranes by a long-duration electric pulse of even low amplitude or by osmotic hemolysis. The presence of proteins in the pore edges may prevent or inhibit the complete resealing of pores after treatment.

The presence of different membrane proteins has significant effects on the behavior of the lipid bilayers in a high electric field [26]. The observed changes in the field strength needed to induce electroporation after introduction of proteins into the experimental chamber may be due to the local (in the immediate vicinity of the protein molecules inserted into bilayer) or macroscopic changes of lipid bilayer proteins. Also, the spontaneous curvature of the membrane monolayers, reflecting the tendency to bend in different directions [24], can increase or decrease depending on the structure of the proteins and the depth of their insertion into membrane. Furthermore, under certain conditions some part of the permeability increase during electroporation was traced to activation of the membrane pump proteins [27]. In addition, glycophorins present on the cell surface may modulate the membrane response to high electric field. The positive spontaneous curvature created by the electrostatic interactions in the erythrocyte glycocalix [28] has to increase the probability of pore formation in lipid matrix [24]. Thus, the distrubance of the glycocalix by a proteolytic treatment may cause the stabilization of the erythrocyte membranes against electroporation described in references 5 and 6 after incubation of cells with pronase. Lastly, the electroporation of cell membranes can be influenced by the the state of mammalian erythrocyte spectrin–actin network located on the inner surface of the membrane (a rudimentary cytoskeleton) [29]. If the pressure of the lipid matrix created by tension from the protein skeleton [30, 31] can condense the bilayer, then it may decrease the probability of pore formation by electric field.

The influence of the membrane skeleton on the process of electropore expansion was studied recently by Chernomordik and Sowers [32; also see Sowers, Chapter 8, this book]. The modification of the spectrin–actin skeleton of ghosts of rabbit and human erythrocytes by heating, low pH, or low ionic strength treatments results in a dramatic transformation of the electrofusion character. The application of a fusogenic electric field pulse to untreated (control) ghosts brought into contact by dielectrophoretic alignment in pearl chains results in the formation of area contact zones. The redistribution of membrane marker, DiI, from initially labeled to initially unlabeled membranes in these structures was observed by fluorescence microscopy. The observed morphology was stable. Neither giant ghost formation (i.e., a completely spherical fusion end product) nor dissociation of aggregates of ghosts into separate ghosts was observed up to 48 hr after the application of the electric field pulse. Thin-section electron microscopy showed that the fusogenic pulse caused the contact zone in these irreversible ghost aggregates to develop a fusion zone composed of a septum perforated with numerous fusion pores. The expansion of the pores to form a spherical fusion product (i.e., disappearance of the septum between fused cells) was prevented by the membrane skeleton. The modification of skeleton by heating of ghosts in isotonic sodium phosphate buffer at 46°C for 10 min led to rapid formation of giant sphere fusion products. Thus, the fate of pores in erythrocyte membrane may be determined by the integrity of the membrane skeleton.

The role of cytoskeleton for electrofusion of nucleated animal cells was discussed earlier in references 33 and 34. No qualitative effects of the drugs affecting cytoskeleton organization on the fusion process were found in both studies. However, it is difficult to compare these results with the data presented since the agents used (cytochalasin B, colchicine, and colcemid) do not disturb the membrane cortex [35].

Thus the structures of cytoskeleton can modify the evolution of the pores formed by electric field application. However, the existence of the opposite effect, that is, the modification of protein skeleton under conditions of electroporation, was also reported in literature on the effects on the cytoskeleton of animal cells after electric treatment reported by Teissie et al. [36].

Further, the formation of blebs on the membranes of different animal cells was shown to occur after electric pulse treatment [37, 38]. The blebs appeared on the surface of viable cells mainly in the region of membranes where the transmembrane voltage generated was the highest. These relatively large deformations of the membrane (0.2–1 μm) developed in seconds to minutes after electric field application in osmotic-dependent process. Since membrane blebs are considered in the literature as protein-deficient regions of membrane [39], their appearance after pulse treatments was interpreted in reference 38 as a consequence of the local disturbances of the membrane cortex at electroporation. The local change of ionic strength, pH, or

Ca^{2+} ion concentration in the cytoplasm near the entry point (the pores) during electric treatment may induce the local rupture of protein skeleton network under the membrane. The process of rupture and redistribution of the erythrocyte membrane skeleton was analyzed theoretically in reference 40. According to this analysis the spectrin–actin network caused the redistribution of the emerged rupture in the cytoplasmic surface of the membrane bilayer. Due to the interaction of the membrane skeleton and integral proteins, this leads to the formation of protein deficient patches. The probable sizes and the directions of the evolution of such patches are determined by the local shape of the membrane, electrical interactions of the membrane proteins, and the elasticity of the membrane skeleton.

VI. Pores and Electrotransfection of Cells

It is well known now that the treatment of cells by short-duration high electric field pulses causes not only electropore formation but also electrofusion and electrotransfection of cells. This phenomenon is extensively utilized in modern biotechnology (see the other chapters in this monograph). It is widely assumed that the relationships between these phenomena are quite simple. To a first approximation it can be suggested that electrofusion reflects just the occasional joining of two electropores formed in the contacting membranes of cells coming together [16]. However, the mechanism of electrofusion and the role of electropores in it are not yet clear. The possible participation of different secondary effects such as colloid-osmotic swelling of cells [2, 16] and the attraction of pore edges created by the current going through the coaxial pores in contacting membranes during pulse application have been also discussed [41]. It should also be mentioned that the relatively macroscale disturbances of membrane cortex caused by electrical treatment may be a possible electrofusion intermediate. This suggestion may explain the long-lived fusogenic state of membranes observed in references 42–44 as well as the delay between fusogenic pulse and transfer of membrane dye between fusing cells [45]. Finally, the formation of local perforations in stable septums of fusion zones at junctions between compartments of fusing cells can be insufficient for the completion of the fusion process [32].

A first look at the role of electropores in the electrotransfection as pathways for DNA molecules to diffuse through the membranes into cells seem to be rather clear. However, the role of electropores in these phenomena may not be direct but rather may mediate some secondary effect. In fact, the electrotransfection technique was used to introduce huge DNA molecules of 65 kb [46] to 150 kb [47] into cells. The rough estimation of the sizes of these molecules as random coils gives the values in the range of micrometers. On the other hand, the typical sizes of the electropores estimated by different experimental approaches are equal to or less than 10 nm effective diameter, even during the pulse [2, 16, 37, 48]. A recent rapid-freezing electron microscopy study revealed "volcano-shaped" membrane openings with 20–120 nm diameters in electropermeabilized human erythrocytes [49]. However, the pores were found only 3 ms after the 0.3-ms electric pulse treatment. Thus, the development of huge pores observed in this study is related to the further (i.e., secondary) expansion of electropores *after* the pulse, which may be due to a secondary process such as the colloid-osmotic swelling of cells accelerated by the hypotonicity of the poration medium used in reference 49. The loading of erythrocytes by huge macromolecules and particles under conditions of electrically induced colloid-osmotic hemolysis was described [16].

Thus, the sizes of initial electropores in membranes may be insufficient to pass DNA molecules during electrotransfection. There are several secondary processes that may mediate DNA translocation.

First, there are some indications against a significant role of the relatively macroscale changes of membrane structure in the electrotransfection. Hence, the introduction of DNA into cells and unilamellar liposomes requires quite specific field strengths to be applied to the membranes (Fig. 3). The efficiencies of genetic transformation for *Escherichia coli* and animal cells (curves 2 and 3) and uptake of labeled DNA into liposomes (curve 1) are plotted in Fig. 3. against the field strength of electric field pulse used. The rough estimation of the membrane voltages generated during a pulse across the vesicles of different radii [16] shows the closeness of the transmembrane voltage ranges (0.4–0.8 V) needed for DNA translocation through the membrane of quite different structures. It suggests that the general properties of the lipid matrix of the membranes rather than the changes of specific organization of cell membranes after electroporation are what determine primarily the efficiency of electrotransfection for different cells.

The role of colloid-osmotic processes in DNA translocation through the membrane appears improbable because, in fact, the electrotransfection of DNA takes place even in hyperosmotic poration medium [50]. Furthermore, the uptake of DNA by cells takes place during, or at least not later than a few seconds after, the electrical pulse [50]. The measurements of the transient expression in Cos-1 cells of the plasmid pCH110 carrying galactosidase gene was used to study the kinetics of the electrotransfection [50]. A 20-fold drop in the transfection efficiency was observed in case of the addition of DNA 5 s *after* instead of 5 s before the electric pulse. It was shown also that the treatment of suspension containing cells and DNA 3 s after pulse with DNase had practically no influence on the transfection efficiency. These results are consistent with the data reported in [51].

The data suggest that DNA translocation occurs during the pulse. Since both electroosmosis [52]. and electrophoresis phenomena could cause the translocation, experiments were carried out [50, 53] to differentiate between these two mechanisms. The transient expression of the bacterial galactosidase gene was measured after application of electric pulses of different polarity to Cos-1 cells in monolayer on porous films. Transfection in a monolayer allows the creation of the spatial asymmetry

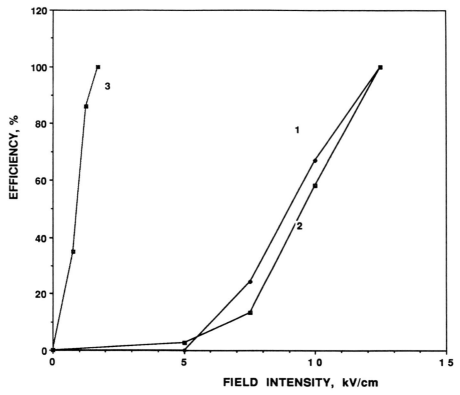

Figure 3 The efficiency of electrotransfection of *Escherichia coli* cells (curve 2), of animal cells of Rat 1 line (curve 3), and that of DNA uptake by liposomes (curve 1) [54]; 100% is the maximal level of efficiency reached.

of cell–plasmid interaction along the direction of applied electric field. It was observed in this experimental system that the efficiency of transfection was 10-fold higher for the pulse polarity inducing DNA electrophoresis toward cells than for the pulse of reverse polarity that had to result in electroosmotic flow into the cells. It has been shown also that increase of medium viscosity by Ficoll 400 or decrease of DNA effective charge by Mg^{2+} both reduced transfection efficiency [50]. These data indicate an involvement of DNA electrophoresis in electrotransfection of cells. We suggest that the electrotransfection process includes the following stages. After electroporation of cell membrane takes place, the lines of electric field are concentrated in the pores. The movement of DNA molecules is focused by electrophoretic force directed toward the pores. Then the DNA crosses the cell membrane via the via the pores in some process requiring the DNA interaction with pores. DNA uptake by cells in an endocytosislike process, described for the case of electrostimulated DNA uptake by liposomes [54], can be considered as a possible alternative.

Only further experiments will clarify the role of electropores in the electro-stimulated fusion and transfection of cells.

Acknowledgments

I express my gratitude to my colleagues Drs. Yuri Chizmadzhev, Dimitry Klenchin, Sergey Leikin, Arthur Sowers, and Sergey Sukharev for fruitful discussions.

References

1. Neumann, E., and Rosenheck, K. (1972). Permeability changes induced by electric impulses in vesicular membranes. *J. Membrane Biol.* **10**, 279–290.
2. Kinosita, K., and Tsong, T. (1977). Hemolysis of human erythrocytes by a transient electric field. *Proc. Natl. Acad. Sci. USA* **74**, 1923–1927.
3. Zimmermann, U., Vienken, J., and Pilwat, G. (1980). Development of drug carrier systems: electrical field induced effects in cell membranes. *Bioelectrochem. Bioenerg.* **7**, 553–574.
4. Kinosita, K., and Tsong, T. (1979). Voltage induced conductances in human erythrocyte membranes. *Biochim. Biophys. Acta* **554**, 479–497.
5. Pastushenko, V. F., Sokirko, A. V., Abidor, I. G., and Chizmadzhev, Y. A. (1987). The electrical breakdown of cell and lipid membranes: The similarity of phenomenology. *Biochim. Biophys. Acta* **902**, 360–373.
6. Abidor, I. G., Barbul, A. I., Zhelev, D., Sukharev, S. I., Kuzmin, P. I., Pastushenko, V. F., and Zelenin, A. V. (1989). Electrofusion and electrical properties of cell pellets in centrifuge. *Biol. Membrany* **6**, 527–531.
7. Powell, K. T., Morgenthaler, A. W., and Weaver, J. C. (1989). Tissue electroporation: Observation of reversible electrical breakdown in viable frog skin. *Biophys. J.* **56**, 1163–1171.
8. Glaser, R., Leikin, S. L., Chernomordik, L. V., Sokirko, A. V., and Pastushenko, V. F. (1988). Pore arising and development in course of reversible electrical breakdown. *Biochim. Biophys. Acta* **940**, 275–287.
9. Dressler, V., Schwister, K., Haest, C. W. M., and Deuticke, B. (1983). Dielectric breakdown of the erythrocyte membrane enhances transbilayer mobility of phospholipids. *Biochim. Biophys. Acta* **732**, 304–307.
10. Neumann, E., Sowers, A. E., and Jordan, C. A. (eds.) (1989). "Electroporation and Electrofusion in Cell Biology." Plenum Press, New York.
11. Benz, R., and Zimmermann, U. (1980). Pulse-length dependence of the electrical breakdown in lipid bilayer membranes. *Biochim. Biophys. Acta* **597**, 637–642.
12. Sowers, A. E. (1989). The mechanism of electroporation and electrofusion in erythrocyte membranes. In "Electroporation and Electrofusion in Cell Biology," E. Neumann, A. E. Sowers, and C. A. Jordan (eds.), Plenum Press, New York, pp. 229–256.

13. Zhelev, D. V., Dimitrov, D. S., and Doinov, P. (1988). Correlation between physical parameters in electrofusion and electroporation of protoplasts. *Bioelectrochem. Bioenerg.* 20, 155–167.

14. Crowley, J. M. (1972). Electrical breakdown of bimolecular lipid membranes as an electromechanical instability. *Biophys. J.* 13, 711–718.

15. Needham, D., and Hochmuth, R. M. (1989). Electro-mechanical permeabilization of lipid vesicles: Role of membrane tension and compressibility. *Biophys. J.* 55, 1001–1009.

16. Zimmermann, U. (1982). Electric field mediated fusion and related electrical phenomena. *Biochim. Biophys. Acta* 694, 227–277.

17. Benz, R., Beckers, F., and Zimmermann, U. (1979). Reversible electrical breakdown of lipid bilayer membranes: A charge-pulse relaxation study. *J. Membrane Biol.* 48, 181–204.

18. Abidor, I. G., Arakelyan, V. B., Chernomordik, L. V., Chizmadzhev, Y. A., Pastushenko, V. F., and Tarasevich, M. R. (1979). Electrical breakdown of BLM: Main experimental facts and their qualitative discussion. *Bioelectrochem. Bioenerg.* 6, 37–52.

19. Benz, R., and Conti, F. (1981). Reversible electrical breakdown of squid giant axon membrane. *Biochim. Biophys. Acta* 645, 115–123.

20. Chernomordik, L. V., Sukharev, S. I., Abidor, I. G., and Chizmadzhev, Y. A. (1983). Breakdown of lipid bilayer membranes in an electric field. *Biochim. Biophys. Acta* 736, 203–213.

21. Knight, D. E., and Baker, P. F. (1982). Calcium-dependence of catecholamine release from bovine adrenal medullary cells after exposure to intense electric fields. *J. Membrane Biol.* 68, 107–140.

22. Tsong, T. Y. Chapter 4, this book.

23. Weaver, J. C., and Powell, K. (1989). Theory of electroporation. In "Electroporation and Electrofusion in Cell Biology," E. Neumann, A. E. Sowers, and C. A. Jordan, (eds.), Plenum Press, New York, pp. 111–126.

24. Chernomordik, L. V., Kozlov, M. M., Melikyan, G. B., Abidor, I. G., Markin, V. S., and Chizmadzhev, Y. A. (1985). The shape of lipid molecules and monolayer membrane fusion. *Biochim. Biophys. Acta* 812, 643–655.

25. Chernomordik, L. V., and Abidor, I.G. (1980). The voltage induced local defects in unmodified BLM. *Bioelectrochem. Bioenerg.* 7,617–623.

26. Chizmadzhev, Y. A., Chernomordik, L. V., Pastushenko, V. F., and Abidor, I. G. (1981). Electrical breakdown of BLM. In "Itogy Nauky y Techniky," VINITY, vol. 2, pp. 161–266.

27. Teissie, J., and Tsong, T. (1980). Evidence of voltage induced channel opening in Na/K ATPase of human erythrocyte membrane. *J. Membrane Biol.* 55, 133–140.

28. Lerche, D., Kozlov, M. M., and Markin, V. S. (1987). Electrostatic free energy and spontaneous curvature of spherical charged layered membrane. *Biorheology* 24, 23–34.

29. Marchesi, V. T. (1985). Stabilizing infrastructure of cell membranes. *Annu. Rev. Cell Biol.* 1, 531–561.

30. Vertessy, B. G., and Steck, T. L. (1989). Elasticity of the human red cell membrane skeleton: effects of temperature and denaturants. *Biophys. J.* 55, 255–262.

31. Kozlov, M. M., and Markin, V. S. (1986). A model for skeleton of a red blood cell. *Biol. Membrany* 3, 404–422.

32. Chernomordik, L. V., and Sowers, A. E. (1990). Evidence that the spectrin network and a non-osmotic force control the fusion product morphology in electrofused erythrocyte ghosts. *Biophys. J.* (in press).

33. Sukharev, S. I., Bandrina, I. N., Barbul, A. I., Fedorova, L. I., Abidor, I. G., and Zelenin, A. V. (1990). Electrofusion of fibroblasts on the porous membrane. *Biochim. Biophys. Acta* **1034,** 125–131.

34. Blangero, C., Rols, M. P., and Teissie, J. (1989). Cytoskeletal reorganization during electric-field-induced fusion of Chinese hamster ovary cells grown in monolayers. *Biochim. Biophys. Acta* **981,** 295–302.

35. Vasiliev, Y. M., and Gelfand, I. M. (1981). "Neoplastic and Normal Cells in Culture." University Press, Cambridge.

36. Teissie, J., Rols, M. P., and Blangero, C. (1989). Electrofusion of mammalian cells and giant unilamellar vesicles. In "Electroporation and Electrofusion in Cell Biology," E. Neumann, A. E. Sowers, and C. A. Jordan, (eds.), Plenum Press, New York, pp. 203–214.

37. Escande-Geraud, M. L., Rols, M. P., Dupont, M. A., Gas, N., and Teissie, J. (1988). Reversible plasma membrane ultrastructural changes correlated with electropermeabilization in Chinese hamster ovary cells. *Biochim. Biophys. Acta* **247,** 247–259.

38. Gass, G. V., and Chernomordik, L. V. (1990). Reversible large-scale deformations in the membranes of electrically treated cells: Electroinduced bleb formation. *Biochim. Biophys. Acta* **1023,** 1–11.

39. Godman, J., Mirranda, A. F., Deitch, A. D., and Tanenbaum, S. W. (1975). *J. Cell Biol.* **64,** 644–667.

40. Kozlov, M., Chernomordik, L., and Markin, V. (1990). A mechanism of formation of protein-free regions in the red blood cell membrane: The rupture of the membrane skeleton. *J. Theoret. Biol.* **144,** 347–365.

41. Kuzmin, P. I., Pastushenko, V. F., Abidor, I. G., Sukharev, S. I., Barbul, A. I., and Chizmadzhev, Y. A. (1988). Theoretical analysis of a cell electrofusion mechanism. *Biol. Membrany* **5,** 600–612.

42. Sowers, A. E. (1986). A long-lived fusogenic state is induced in erythrocyte ghosts by electric pulses. *J. Cell Biol.* **102,** 1358–1362.

43. Sowers, A. E. (1987). The long-lived fusogenic state induced in erythrocyte ghosts by electric pulses is not laterally mobile. *Biophys. J.* **52,** 1015–1020.

44. Teissie, J., and Rols, M. P. (1986). Fusion of mammalian cells in culture is obtained by creating the contact between cells after their electropermeabilization. *Biochim. Biophys. Res. Commun.* **140,** 258–266.

45. Dimitrov, D. S., and Sowers, A. E. (1990). A delay in membrane fusion: lag times observed by fluorescence microscropy of individual fusion events induced by an electric field pulse. *Biochemistry* **29,** 8337–8344.

46. Jastreboff, M. M., Ito, E., Bertino, J. R., and Narayanan, R. (1987). Use of electroporation for high-molecular-weight DNA-mediated gene transfer. *Exp. Cell Res.* **171,** 513–517.

47. Knutson, J. C. and Yee, D. (1987). Electroporation: parameters affecting transport of DNA into mammalian cells. *Anal. Biochem.* **164,** 44–52.

48. Sowers, A. E., and Lieber, M. L. (1986). Electropores in individual erythrocyte ghosts: Diameters, lifetimes, numbers and locations. *FEBS Lett.* **205,** 179–184.

49. Chang, D. C. (1990). Changes in membrane structure induced by electroporation as revealed by rapid-freezing electron microscopy. *Biophys. J.* **58**, 1–12.
50. Klenchin, V. A., Sukharev, S. I., Serov, S. M., Chernomordik, L. V., and Chizmadzhev, Yu. A. (1991). Electrically induced DNA uptake by cells is a fast process involving DNA electrophoresis. *Biophys. J.* (in press).
51. Taketo, A. (1989). Properties of electroporation—mediated DNA transfer in *Escherichia coli. J. Biochem.* **105**, 813–817.
52. Dimitrov, D. S., and Sowers, A. E. (1990). Membrane electroporation—fast molecular exchange by electroosmosis. *Biochim. Biophys. Acta* **1022**, 381–392.
53. Klenchin, V. A., Sukharev, S. I., Serov, S. M., Chernomordik, L. V., and Chizmadzhev, Y. A. (1990). Electrophoresis role in electrostimulated uptake of DNA by animal cells. *Biol. Membrany* **7**, 446–447.
54. Chernomordik, L. V., Sokolov, A. V., and Budker, V. G. (1990). Electrostimulated uptake of DNA by liposomes. *Biochim. Biophys. Acta* **1024**, 179–183.

6

Biophysical Considerations of Membrane Electroporation

Eberhard Neumann, Andreas Sprafke, Elvira Boldt, and Hendrik Wolf

Faculty of Chemistry, University of Bielefeld, D-4800 Bielefeld 1, Germany

---

I. Introduction

Membrane electroporation describes the transient, reversible permeabilization of the membranes of cells, organelles, or lipid bilayer vesicles by electric field pulses; for review see Neumann *et al.* (1989). Electroporation not only renders membranes transiently permeable and leads to material exchange across membranes, but also induces and facilitates fusion of membranes in contact; for review see Sowers (1987). There is an increasing number of practical applications of the electroporation technique, for instance:

Guide to Electroporation and Electrofusion

- Direct transfer of genes, other nucleic acids, proteins, and other molecules into all types of cells and microorganisms.
- Electrofusion of cells.
- Electrostimulation of cell growth and proliferation.
- Electroinsertion of membrane proteins.

Contrary to the impressive success of the electroporation techniques in cell biology, biotechnology, and medicine (gene therapy), the molecular membrane processes of electroporation are not yet well understood. Data analysis and optimization of conditions for the electrotransfer of DNA and of proteins, electrofusions, electrostimulation, and electroinsertion are still primarily empirical.

Since it is primarily the lipid part of biological membranes that is electroporated, lipid bilayer vesicles may be used as a model system to study the elementary processes of the electroporation phenomena.

Recent electrooptic and conductometric data are consistent with electric field-induced changes in the membrane structure of the vesicles. The lipids rearrange in such a way that optically anisotropic light scattering centers appear in the lipid phase of the vesicles. These scattering centers may be identified with geometrically anisotropic pores, which conduct alkali salt ions and other substances if, at given field strengths and pulse durations, a critical pore size is reached.

The relaxation kinetic data are the basis for the interpretation of a variety of observations accumulated in previous studies on membrane permeabilization, cell electrotransfection, and electrofusion. Deeper insight is gained on the stimulus/duration curve, on the electrosensitivity of cells, on the dependences of cell transformations on DNA and cell concentrations, and various other parameters.

II. Electrooptic and Conductometric Relaxations of Lipid Bilayer Vesicles

Recent progress in relaxation kinetics in the presence of high electric field strengths has provided new ways to study structural changes in membrane systems. We have developed an electric field-jump relaxation spectrometer that permits the simultaneous measurements of field pulse-induced changes in electric conductance and in optical properties of solutions and suspensions in the nanosecond to millisecond time range with high resolution. Rectangular field pulses up to 150 kV cm^{-1} and of variable pulse duration (0.5 μs to 1 ms) can be applied by cable discharge in a single DC-pulse mode.

A. Relaxation Data

Figure 1 shows that there are up to three conductance modes $\Delta\lambda/\lambda(0)$ and up to two electrooptic scattering–dichroitic relaxation modes.

Mode I of the conductance relaxation clearly has the feature of a displacement current, as expected for the interfacial ionic polarization preceding, as an exponen-

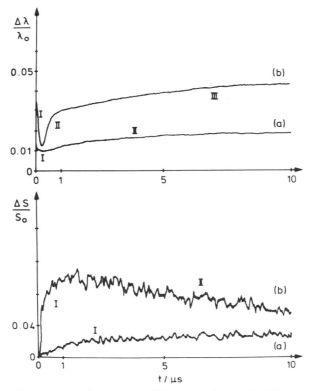

Figure 1   Conductometric ($\Delta\lambda/\lambda_0$) and electrooptic ($\Delta S/S_0$) relaxations of a turbid 0.2 mM NaCl suspension of lipid bilayer vesicles (mean diameter $\theta = 80 + 20$ nm) filled with 0.2 mM NaCl, pH 6.6, 20°C. Vesicle density: 10^{12}/ml ($= 0.5$ mM lipid). Time courses of the relative conductivity $\Delta\lambda/\lambda(0)$ and of the relative scattering dichroism at 332 nm light wavelength in the presence of a constant electric field strength E and of $\Delta t = 10$ μs duration: (a) $E = 20$ kV cm^{-1} and (b) $E = 60$ kV cm^{-1}; $\lambda(0) = 22.6$ μS cm^{-1} at 20°C. The Roman numerals refer to the (exponential) relaxation modes.

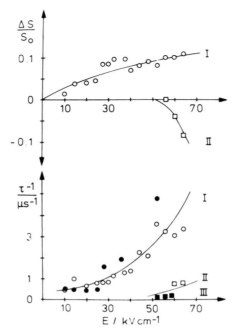

Figure 2   Linear scattering dichroism and relaxation rates
(τ^{-1}) as a function of the field strength: $\Delta S = \Delta S_\parallel - \Delta S_\perp$,
where $\Delta S_\parallel$ is the light intensity change observed with polarized
light parallel to the direction of the applied electric field and
$\Delta S_\perp$ is the signal of the perpendicular light polarization mode.
See legend of Fig. 1. Note the positive value of ΔS of mode
I and the negative value of ΔS of mode II. The relaxation rates
of the conductivity relaxation mode I are the same as those of
the optical model I, and the rates of mode III and the S(II)
mode are roughly the same.

tially decaying forcing function, the subsequent conductance relaxations II
and III.

Mode I of the light-scattering dichroism, defined as $\Delta S = \Delta S_\parallel - \Delta S_\perp$ (see Fig.
1), has the same relaxation rate (τ^{-1}) as the interfacial polarization current, at all
field strengths (Fig. 2). The optical change thus seems to be caused by, and rate-
limited by, the buildup of the electric potential difference across the membrane.

At higher field strengths a second relaxation with opposite amplitude compared
to mode I is visible. This optical mode (II) is slightly faster than the relaxation
mode (III) of the conductance relaxations, but both modes become observable at
about the same field intensity, E_c of the 10-μs pulse.

B. Structural Changes

The conservative scattering dichroism suggests that the field-induced, interfacial ionic polarization of the vesicle causes solvent to enter the membrane. If these aqueous scattering centers are pores, the shape of the pore must be asymmetric as indicated by the light scattering anisotropy ($\Delta S \neq 0$). We may model these pores as narrow hydrophobic (HO) pores $(P_{HO})_r$ of radius r. If the field strength E reaches a critical value E_c at a given pulse duration Δt, a different pore quality appears. This pore type is associated with a negative value of the scattering dichroism and with a larger conductance. We model these pores as broader hydrophilic (HI) pores $(P_{HI})_{r>r_c}$, which are formed when a critical pore radius r_c is reached. The formation of these pores is slower ($\tau^{-1} = k_{HI} \approx 10^5 \text{ s}^{-1}$) compared to the HO pores ($\tau^{-1} = k_{HO} \approx 10^6 \text{ s}^{-1}$) because the transition $P_{HO} \rightleftharpoons P_{HI}$ involves rotation of lipid molecules in the pore wall.

1. Sequence of Membrane Changes

Since the light-scattering signal $\Delta S(I)/S(0)$ increases continuously with increasing field intensity, it is presumed that the number and size of hydrophobic (HO) pores increase continuously up to a saturation value (Fig. 2). The sequence of membrane changes from the poreless state (M) to HO pores $(P_{HO})_r$ and HI pores $(P_{HI})_r$ is modeled in Fig. 3.

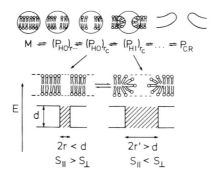

Figure 3   Sequence of events in membrane electroporation, modeled in terms of aqueous pores: M is the closed bilayer state, and $(P_{HO})_r$ is the hydrophobic pore of radius r. The HO pores convert to hydrophilic (HI) pores, $(P_{HI})_r$, if a critical radius r_c is reached; P_{CR} refers to crater-like pores. The geometrically anisotropic pores $2r \neq d$ describe the change of sign of the dichroitic signal $\Delta S = \Delta S_{\parallel} - \Delta S_{\perp} = S_{\parallel} - S_{\perp}$ with respect to the direction of the applied electric field E.

2. Model for the HO ⇌ HI Transition

For geometrical reasons of lipid packing, water will enter to a certain degree between the individual lipids of the pore wall of a hydrophilic pore. The transition

$$(P_{HO})_{r_c} \rightleftharpoons (P_{HI})_{r_c}$$

changes the geometry of the light scattering center. The HI pore $(2r' > d)$ is broadened compared to the HO pore $(2r < d)$; see Fig. 3.

3. Field Strength and Pulse Duration

The scheme of the structural transitions depicted in Fig. 3 is very instructive for the interpretation of the stimulus/duration curves: the higher the field strength, the shorter can be the pulse length in order to achieve the same membrane permeabilization. A low-intensity pulse may need a pulse duration of milliseconds to reach the same extent of structural transitions as a high intensity pulse applied for only microseconds. Thus even very small field intensities may cause DNA transfer, provided the field application is long enough (seconds). It is thus obvious that the specification of a critical field intensity E_c is only meaningful if the pulse duration Δt is also given.

4. Pulse Number

The return to the closed membrane state M (Fig. 3) after switching off the field at the end of the pulse occurs in the absence of the external field. In particular, the return transition $P_{HO} \leftarrow P_{HI}$, involving reorientation of the wall lipids, may face major activation barriers and thus is slow (seconds, minutes).

If now a second pulse hits the membrane patch not in the M state but in the remaining $(P_{HI})_r$ states, the change induced by the second pulse is facilitated, because the transitions $P_{HO} \rightarrow P_{HI}$ have already been caused by the first pulse. Thus in terms of membrane structure and lipid rearrangements the pulses of a pulse train are not equivalent.

III. Electroporation and Membrane Permeability

Among the various practical aspects, the electrotransformation of intact bacteria and other microorganisms requires media with low conductivity to reduce Joule heating, although that by itself can be favorable for efficient gene transfer (Wolf *et al.*, 1989). In any case, the interpretation of the low-conductivity data necessitates an extension of the analytical framework in terms of the external solution conductivity (Neumann and Boldt, 1989).

Using the green algae cells of the species *Chlamydomonas reinhardtii (Ch. reinh.)* as an example, a simple procedure is outlined to determine the critical electroporation voltage ($\Delta\varphi_{M,c}$) and the membrane/envelope conductivity (λ_m). Finally, a strategy is presented that permits the determination of optimum values of electric field strength, pulse length, and pulse number for the transient permeabilization of viable cells, thus providing a useful basis of optimization strategies for cell biological and medical applications.

A. Permeabilization–Resealing Cycle

It was recently recognized that the transient membrane electroporation may be viewed as a cycle of electropore formation and resealing resembling a relaxation hysteresis (Neumann, 1989). In brief, in the presence of the field pulse of limited duration ($\Delta t = 5$ μs to 10 ms), the cell membrane (initial states C, representing the actual state equilibria $M \rightleftharpoons \Sigma_r^{eq} (P_{HO})_r$) becomes more porous (states P representing the state equilibria $\Sigma_r^{eq}(P_{HI})_r$) and thereby also fusogenic. The state transitions $C \rightleftharpoons P$ in the presence of the electric field occur unidirectionally and may lead to irreversible rupture at long pulse durations. If, however, the electric field pulse is switched off before rupture or lysis of the cells occurs, the electropores or electrocracks anneal at $E = 0$, with $C \leftarrow P$ being unidirectional, too. The electroporation–resealing cycle may be represented by the scheme

$$C \underset{k_R}{\overset{k_P(E)}{\rightleftharpoons}} P \tag{1}$$

where $k_P(E)$ and k_R are overall rate coefficients for electroporation and resealing, respectively. The electroporated (and fusogenic) membrane states P usually are long-lived, in particular at low temperatures (4°C).

If the electroporation hysteresis is coupled to other processes (states X), such as material release or uptake, or lysis, Scheme (1) has to be extended to

$$C \rightleftharpoons P \rightarrow \rightarrow X \tag{2}$$

Usually for short pulse times Δt, the reverse transition $P \leftarrow C$ and the transition $P \rightarrow X$ occur at $E = 0$; they are after-field pulse effects.

B. Electrosensitivity

It is known that a biological cell population is inhomogeneous in cell size, in the state of growth, or in metabolic conditions. Nonspherical cells such as bacterium rods have different positions relative to the external electric field direction. Because of the dominantly negative surface charges of the cell wall, the rods will orient with their longest axis into the direction of the electric field. Both oriented rods and the

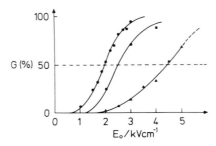

Figure 4 Electrosensitivity $G(\%)$ of a suspension of green algae cells *Chlamydomonas rheinhardtii* (wild type 11-32 c, Göttingen) to quasirectangular electric pulses of the initial field intensity E_0 and of the pulse length $\Delta t = 0.2$ ms at different medium conductivities λ_0: ●, $\lambda_0 = 3.5 \pm 0.1 \times 10^{-4}$ S cm^{-1}; ■, $\lambda_0 = 1.5 \pm 0.1 \times 10^{-4}$ S cm^{-1}; ▲, $\lambda_0 = 5.6 \pm 0.5 \times 10^{-5}$ S cm^{-1}. $G(\%)$ is the percentage of cells that were critically permeabilized such that the (lethal) dye serva blue G ($M_r = 854$, largest dimension 2.5 nm) was taken up, visibly coloring these cells.

oriented polyelectrolyte DNA will migrate with different velocities along the electric field lines. All these factors cause a distribution of the critical electroporation field strength E_c.

Figure 4 shows examples of such a distribution. Here, the state X in Scheme (2) is the colored cell (G), obtained by the uptake of a dye by the electroporated cell. The percentage $G(\%)$ of the G colored cells is given by: $G(\%) = 100 \times G/C_T$, where C_T is the total number of the cells. $G(\%)$ is a measure for the critically permeabilized cells.

For practical purposes we may take the field strength E_0 (50%) where $G(\%) = 50$ as representative for the cell population and define a mean value by

$$\overline{E}_c = E_0(50\%) \tag{3}$$

The width of the electrosensitivity of the cell population may be given in terms of a "variance" ($\pm \Delta\overline{E}_c$). Since E_c depends on the cell size, the mean value $\overline{E}_c$ corresponds to a mean value $\overline{a}$ of the effective radius of spherical cells. Here, too, we have a "variance": $\overline{a} + \Delta\overline{a}$.

C. Ionic Interfacial Polarization

The electroporation data suggest that the membrane electropermeabilization results from an indirect electric field effect. The structural changes of the membrane phase are preceded by the ionic interfacial polarization (Neumann, 1989).

Usually in electroporation experiments the polarization time constant τ_p (Schwan, 1957) is small compared with the pulse duration Δt, the buildup of the stationary value $\Delta\varphi(E)$ of the interfacial potential difference across the membrane is practically instantaneous.

For low-conductivity membranes of thickness d of cells of radius a, the stationary value is given by

$$\Delta\varphi(E) = -1.5f(\lambda)aE|\cos\delta| \tag{4}$$

where δ is the angle between the membrane site considered and the direction of E. The conductivity factor $f(\lambda)$ is a function of the specific conductances or conductivities of the external solution ($\lambda_0 \geqslant 10^{-4}$ S cm^{-1}), of the cell interior ($\lambda_i \approx 10^{-2}$ S cm^{-1}), and of the membrane ($\lambda_m \approx 10^{-7}$ S cm^{-1}), respectively, and of the ratio d/a.

Usually $\lambda_m \ll \lambda_i$, λ_0 and $d \ll a$ such that (Neumann, 1989)

$$f(\lambda) = [1 + \lambda_m(2 + \lambda_i/\lambda_0)/(2\lambda_i\,d/a)]^{-1} \tag{5}$$

If in low-conductivity media $\lambda_0 \ll \lambda_i$, but still $\lambda_0 \gg \lambda_m$, Eq. (5) reduces to a form that is particularly useful for graphical data evaluation:

$$[f(\lambda)]^{-1} = 1 + (\lambda_m a/2d)\,\lambda_0^{-1} \tag{6}$$

Obviously, E_c corresponds to a critical cross-membrane electroporation voltage $\Delta\varphi_{M,c}$; $\Delta\varphi_{M,c} \approx 1$ V (Sale and Hamilton, 1968) for short repetitive pulses of 20 μs. When the contributions of fixed surface charges and the associated ionic atmospheres can be neglected, the total trans-membrane voltage $\Delta\varphi_M(E)$ in the presence of the externally applied field E is a function of the intrinsic membrane potential $\Delta\varphi_m$ (e.g., $\Delta\varphi_m = -70$ mV) and of the interfacial polarization term $\Delta\varphi(E)$:

$$\Delta\varphi_M(E) = \text{fct}[\Delta\varphi_m, \Delta\varphi(E)] \tag{7}$$

The total potential profile in the direction of E is explicitly given by (Neumann, 1989)

$$\Delta\varphi_M(E) = -[1.5f(\lambda)aE + \Delta\varphi_m/\cos\delta]|\cos\delta| \tag{8}$$

Often $|\Delta\varphi(E)| > |\Delta\varphi_m|$, hence $\Delta\varphi_m(E) \approx \Delta\varphi(E)$. The pole caps of spherical membranes are the sites of maximum interfacial polarization. With $|\cos\delta| = 1$, Eq. (8) can be used to relate the mean values of $\bar{E}_c$ and $\bar{a}$ by the expression

$$\bar{E}_c = -\Delta\varphi_{M,c}/[1.5\bar{a}f(\lambda)] \tag{9}$$

Applying Eq. (6) for low-conductivity media we obtain

$$\bar{E}_c = -(\Delta\varphi_{M,c}/3\bar{a})[1 + (\lambda_m\bar{a}/2d)\,\lambda_0^{-1}] \tag{10}$$

When $\bar{a}$ is given by the geometrical radius of the most abundant cell size, the data in Fig. 5 can be used to determine $\Delta\varphi_{M,c}$ and λ_m. For the *Ch. reinh.* cells, $\bar{a} = 3.5$ μm: $\Delta\varphi_{M,c} = -0.8$ V, $\lambda_m \approx 5 \times 10^{-7}$ S cm^{-1}. The critical voltage 0.8 V at $\Delta t = 0.2$ ms compares well with the value of 1 V estimated by Sale and Hamilton

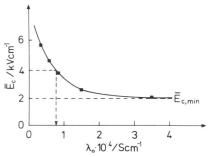

Figure 5 The population mean value $\overline{E}_c = E_0$ (50%) of *Ch. reinhardtii* cells (Fig. 4) decreases with increasing conductivities λ_0 of the pulsing medium. The λ_0 value at which $\overline{E}_c = 2\overline{E}_{c,min}$ [i.e., $f(\lambda) = 0.5$] is given by $(\lambda_0)_{0.5} = \lambda_m \overline{a}/2d$ and permits a simple graphical estimate of the λ_m value. From $\overline{E}_{c,min}$ we obtain $\Delta\varphi_{M,c} = -1.5a\overline{E}_{c,min}$.

(1968) for $\Delta t = 20$ μs. The dependence of E_c on λ_0 is quantitatively consistent with the concept of ionic interfacial polarization, being reduced at low medium conductivity (λ_0), that is, low electrolyte concentration. The data at low λ_0 values suggest that the membrane conductivity λ_m cannot be neglected; instead of $f(\lambda) = 1$, Eq. (6) applies.

D. Lifetime of Membrane Electroporation

The longevity of the electroporated cells is dependent on the temperature and may be measured by the after-plus addition of a dye (which colors the cell) at various times t_{add} after the electroporation pulse. Due to annealing processes more cells lose the permeability property at larger after-pulse addition times (Fig. 6). For the data analysis Scheme (2) is specified to

$$C \underset{k_R}{\overset{}{\rightleftarrows}} P \overset{k_D}{\rightarrow} G \qquad (11)$$

where k_R and k_D are the rate coefficients for resealing (R) and dye uptake (D) and X = G is the (lethally) colored cell state.

The data suggest that $k_D \gg k_R$, such that G is a quantitative measure of the cells in the critically electroporated states P (= G). The rate equation for Scheme (11) under these conditions is given by

$$-d[G]/dt = -d[P]/dt = -k_R[P] \qquad (12)$$

Integration for the boundary conditions $t = 0$: [P] = [P$_0$], and $t \to \infty$: [P] =

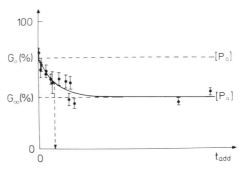

Figure 6   Time course of the recovery of the electroporated cells (resealing), measured by the uptake of the dye serva blue G, which was added at various times t_{add} after the electroporation pulse. For this example $k_R = 2.6 (\pm 0.9) \times 10^{-2}$ s^{-1} at $T = 298$ K (25°C).

[P$_\infty$] yields

$$[P(t)] - [P_\infty] = ([P_0] - [P_\infty]) \exp(-k_R t) \qquad (13)$$

Equation (13) permits the evaluation of the coefficient k_R of cell recovery from the observed exponential decay of $G(\%)$ in Fig. 6. Note that $G(\%)$ is proportional to P such that, for instance, $([P(t)] - [P_\infty])/([P_0] - [P_\infty]) = (G(t) - G_\infty)/(G_0 - G_\infty)$.

The rate coefficient k_R is a measure of the mean lifetime $\bar{t}_e$ of the electroporated cell state: $\bar{t}_e = k_R^{-1}$. In the example given in Fig. 6, $t_e \approx 40$ s; the electroporation state is thus called long-lived compared to the pulse duration of 0.2 ms (Fig. 4).

E. Physical Electroporation Parameters

The dye method of coloring electroporated cells can be used to determine the pulse strength–duration ($\bar{E}_c/\Delta t$) relationship and the dependence of $\bar{E}_c$ on medium conductivity (λ_0), on temperature, and on pulse number. The $\bar{E}_c/\lambda_0$ dependence yields λ_m as a membrane-specific electric parameter. By $G_m = \lambda_m/a$, the membrane conductance, G_m can be determined. For *Ch. reinh.*, $G_m \approx 1.4 \times 10^{-3}$ S cm^{-2}, comparing well with other biomembrane systems.

It is of particular practical importance that the values $G_0(\%)$ and $G_\infty(\%)$ can be used to determine, at a given pulse length and pulse number, the optimum range of E_0 for the electroporation experiments (Fig. 7). The optimum field strength range shifts when the medium conditions (λ_0, temperature, etc.) are changed; ΔE_0 and $\Delta G(\%) = G_0(\%) - G_\infty(\%)$ must be explored for every particular cell type.

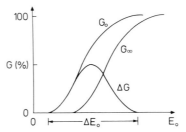

Figure 7   Scheme for the optimization of electroporation conditions; $\Delta G = G_0 - G_\infty$, representing the transiently permeabilized surviving cells. The field strength range ΔE_0 is shifted when the electroporation conditions are changed.

In summary, it is shown that the dye method, originally applied as a qualitative tool to estimate electroporated cells (Neumann *et al.*, 1980, 1982) can be quantified to yield useful information for the optimization of the electroporation technique.

IV. Electroporative DNA Transfer

One of the more recent developments in the technique of the electroporative gene transfer is the electrotransformation of intact bacteria, without pretreatment to remove the cell wall (see, for instance, Wolf *et al.*, 1989).

A. DNA Surface Adsorption

The number T of transformants per milliliter cell suspension increases with increasing DNA concentration [DNA] and with increasing cell density (Wolf *et al.*, 1989). Both correlations, T(DNA) at constant cell density and T(cell density) at constant DNA concentration, are of the Langmuir-type adsorption isotherm. We may model these observations with a scheme in which the electroporation hysteresis $C \rightleftharpoons P$ [see Eq. (1)] is coupled to the adsorptive, (electro)diffusive approach of DNA (D) to the surface of the cell (C), yielding the surface bound states $D \cdot C$ and $D \cdot P$:

$$D + mC \rightleftharpoons D \cdot C$$
$$\updownarrow \qquad \updownarrow \qquad\qquad\qquad (14)$$
$$D + mP \rightleftharpoons D \cdot P \leftrightarrow P \cdot D \rightarrow D_{in} \rightarrow TC$$

In Scheme (14), the process $D \cdot P \leftrightarrow P \cdot D$ denotes the crossmembrane transport of DNA, D_{in} is the DNA that entered the cytoplasm, and TC denotes transformed cell; m is the maximum number of surface binding sites for DNA per cell.

B. Analysis of DNA Surface Binding

It appears reasonable to assume that the probability of cell transformation increases with an increase in the concentration of the state $D \cdot P$. If $[TC] \sim [D \cdot P]$, then

$$T/T_{max} = [TC]/[TC]_{max,} = [DP]/[DP]_{max} \qquad (15)$$

In terms of Scheme (14) the extent of "adsorption" depends on DNA and cell concentration, respectively, according to

$$[DP]/[DP]_{max} = [D]/([D] + \bar{K}_D) = m[C]/(m[C] + \bar{K}_c) \qquad (16)$$

where $\bar{K}_D = \bar{K}_c$ is given by

$$\bar{K}_D = [D](m[C] + m[P])/([DC] + [DP]) \qquad (17)$$
$$= K_1(1 + K_0)/(1 + K'_0)$$

and $K_0 = [P]/[C]$, $K_1 = ([D]m[C])/[DC]$, $K'_0 = [D \cdot P]/[D \cdot C]$, and $K_2 = K_1 K_0/K'_0$ are the equilibrium constants of the individual steps.

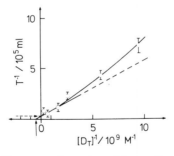

Figure 8   Electrotransformation of intact Corynebacterium glutamicum cells. Cell density 10^{10}/ml at 4°C. Quasirectangular pulse, $E_0 = 10$ kV cm^{-1}, $\Delta t = 5$ ms; medium: 0.272 M sucrose, 0.01 M HEPES buffer, 1 mM MgCl$_2$, KOH adjusted to yield pH 7.4 at 4°C, plasmid DNA pUL-330-plasmid (5.2 kb). The Langmuir adsorption type relationship between the number of transformants T per milliliter and the total concentration of DNA, $[D_T]$, is represented in terms of reciprocal quantities; $[T]^{-1} = T_{max}^{-1}(1 + \bar{K}_D/[D_T])$. The dashed line refers to the linear dependence on $[D]^{-1}$; $[D]$ is the concentration of free DNA, $M = 3.43 \times 10^6$ g/mol. The abscissa and ordinate intercepts yield $\bar{K}_D = 2\,(\pm\,1) \times 10^{-9}\,M$ (DNA) and $T_{max} = 3.3 \times 10^5$/ml.

Equations (15) and (16) can be combined and rewritten as

$$T^{-1} = T_{max}^{-1} (1 + \overline{K}_D/\{D\}) \qquad (18)$$

providing a convenient relationship for a graphical data evaluation. Note that [D] refers to the free, unbound DNA. The data representation in terms of the total DNA concentration $[D_T]$ is nonlinear in $1/[D_T]$, but it is suited to evaluate T_{max} and $\overline{K}_D$; see Fig. 8. For intact *Cornyebacterium glutamicum* cells we obtain $\overline{K}_D = 2(\pm 1) \times 10^{-9}$ M(DNA), $m = 25$, and $T_{max} = 3.3 \times 10^5$ ml^{-1} at $E = 10.5$ kV cm^{-1} and $\Delta t = 5$ ms.

The electroporative cell transformation of intact *Cornyebacterium glutamicum* cells can thus be quantitatively described in terms of preceding surface adsorption of the applied plasmid DNA.

Acknowledgments

We thank Mrs. M. Pohlmann for careful typing of the manuscript, the Deutsche Forschungsgemeinschaft for Grant Ne227/6–1 to E. Neumann, and the Fonds der Chemie.

References

Neumann, E. (1989). The relaxation hysteresis of membrane electroporation. *In* "Electroporation and Electrofusion in Cell Biology" (E. Neumann, A. E. Sowers, and C. A. Jordan, eds.), pp. 61–82, Plenum Press, New York.

Neumann, E., and Boldt, E. (1989). Membrane electroporation: biophysical and biotechnical aspects. *In* "Charge and Field Effects in Biosystems" (M. J. Allen, S. F. Cleary, and F. M. Hawkridge, eds.), Vol. 2, pp. 373–382, Plenum Press, New York.

Neumann, E., Gerisch, G., and Opatz, K. (1980). Cell fusion induced by high electric impulses applied to *Dictyostelium. Naturwissenschaften* 67, 414–415.

Neumann, E., Schaefer-Ridder, M., Wang, Y., and Hofschneider, P. H. (1982). Gene transfer into mouse lyoma cells by electroporation in high electric fields. *EMBO J.* 1, 841–845.

Neumann, E., Sowers, A. E., and Jordan, C. A. (eds.) (1989). "Electroporation and Electrofusion in Cell Biology." Plenum Press, New York.

Sale, A. J. H., and Hamilton, W. A. (1968). Effects of high electric fields on microorganisms. III. Lysis of erythrocytes and protoplasts. *Biochim. Biophys. Acta* 163, 37–43.

Schwan, H. P. (1957). Electrical properties of tissue and cell suspensions. *Adv. Biol. Med. Phys.* 5, 147–209.

Sowers, A. E. (ed.) (1987). "Cell Fusion." Plenum Press, New York.

Wolf, H., Pühler, A., and Neumann, E. (1989). Electrotransformation of intact and osmotically sensitive cells of *Corynebacterium glutamicum. Appl. Microbiol. Biotechnol.* 30, 283–289.

7

Progress toward a Theoretical Model for Electroporation Mechanism: Membrane Electrical Behavior and Molecular Transport

James C. Weaver and Alan Barnett[1]

Harvard-MIT Division of Health Sciences and Technology,
Massachusetts Institute of Technology,
Cambridge, Massachusetts 02139

[1]Present address: Pendergrass Diagnostic Research Laboratory, Department of Radiology, University of Pennsylvania, Philadelphia, Pennsylvania 19104.

I. Introduction

Electroporation is increasingly recognized as a powerful method for transiently altering cell and artificial bilayer membranes, particularly for molecular transport into cells. From the viewpoint of biophysics, dramatic membrane phenomena occur, initially driven by strong electric field interactions, but later probably by mechanical pressures. From the viewpoint of biology, many different kinds of molecules experience greatly enhanced transport into or out of cells, and in many cases a large fraction of the cells survive. With both viewpoints in mind, an understanding of electroporation should combine descriptions of membrane electrical behavior and molecular transport. Experimentally, electroporation caused by short pulses universally occurs at a transmembrane voltage of about 1000 mV for many different types of cell and artificial membranes. This near independence of detailed membrane composition strongly suggests that theoretical modeling based on a small number of key parameters will succeed. We find that a model based on transient aqueous pores can quantitatively describe several key aspects of the striking electrical behavior both in planar bilayer membranes and in cell membranes, can predict contribution of electrical drift to molecular transport, and should be capable of extension to include the contributions of convection and diffusion to molecular transport. As shown in this chapter, a heterogeneous population of transient aqueous pores in the lipid portion of the membrane and a small number of metastable pores may be able to account for much of cell membrane electroporation.

A. Electrical Behavior of Electroporated Cell Membranes

Relatively few electrical measurements have been made on isolated cells, because the transmembrane voltage, $U(t)$, measurement requires the use of either microelectrodes (Benz and Zimmermann, 1980) or voltage-sensitive dyes (Kinosita et al., 1988; Farkas, 1989; Kinosita et al., Chapter 3, this volume). Alternatively, measurement across a monolayer tissue (e.g., frog skin) reflects the average transcellular electrical behavior of a large number of cells (Stämfli, 1958; Fishman and Macey, 1969; Benz and Conti, 1981; Powell et al., 1989). As is the case for artificial planar bilayer membranes, an onset of the high-conductance state of "reversible electrical breakdown" (REB) is an electrical indication of electroporation.

B. Electrical Behavior of Electroporated Artificial Planar Bilayer Membranes

Studies of artificial planar bilayer membranes have been important to the development of understanding of electroporation (Abidor et al., 1979; Benz et al., 1979;

Table 1

Four Observed Outcomes for a Planar Bilayer Membrane[a]

Characteristic electric behavior	Pulse magnitude
"Reversible electrical breakdown" (REB); membrane discharge to $U = 0$	Largest
Incomplete REB (discharge halts at $U > 0$)	Smaller
Rupture (mechanical); slow, sigmoidal electrical discharge	Still smaller
Membrane charging without dramatic behavior of U	Smallest

[a]Benz et al. (1979).

Benz and Zimmermann, 1980; Glaser et al., 1988; Chernomordik and Chizmadzhev, 1989). Charge-pulse experiments have been particularly useful in the development of our theoretical model, because of the short time scale and the resulting less demanding computational requirements. We have initially emphasized comparison to charge-pulse experiments in which a 0.4-μs square pulse was used (Benz et al., 1979). Four different outcomes were observed as the pulse magnitude was varied (Table 1). Unlike cells, planar membranes can rupture directly because of electroporation (Sugar and Neumann, 1984).

II. Theory of Electroporation

A. Theory of Rupture: Bulk Electrocompression versus Explicit Pores

Initial attempts to explain rupture of planar membranes were based on electrocompression of the entire membrane and a resulting instability (Crowley, 1973). However, this approach could not explain the rupture of "solvent-free" and biological membranes, which have small values of compressibility. Related, more complex models were based on the viscoelastic behavior of a membrane, but the instability was identified with reversible permeabilization or REB even though the subsequent behavior of $U(t)$ was not treated (Dimitrov and Jain, 1984; Dimitrov, 1984). It was first suggested by Chizmadzhev and co-workers that local "defects" (pores) are responsible (Fig. 1). Their basic theory was presented in an series of seven back-to-back papers, of which the first is cited here (Abidor et al., 1979). Similar pore approaches to rupture were subsequently developed independently (Weaver and Mintzer, 1981; Sugar, 1981). These theoretical models gave reasonable predictions for a critical transmembrane voltage for rupture, U_c. None, however, simultaneously described the phenomenon of rupture and REB in the "charge injection" experiments.

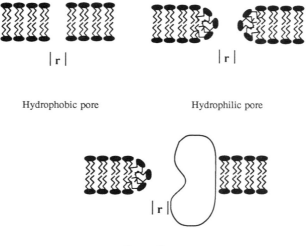

Composite pore

Figure 1 Artist's drawing of hypothetical pore structures. Hydrophobic pores can be regarded as lateral density fluctuations, with entry of water into the lipid interior, and probably are precursors to hydrophilic pores (Abidor *et al.*, 1979). Inverted or hydrophilic pores are believed to allow the passage of ions and molecules, through their larger size (smaller edge energy, γ), and their larger dielectric constant of the pore wall. The present theoretical model is based on transient aqueous pores, which are of this type. The composite pore involves one or more membrane macromolecules or cell structural molecules, with the possibility of both a lower γ and a prolonged lifetime (metastable pore).

B. Theory of "Reversible Electrical Breakdown" due to Transient Pores

The ability of a single membrane to exhibit four different outcomes by changing only the electrical pulse is striking. Particularly dramatic is REB. But this is not a true breakdown (Condon and Odishaw, 1958), because there is not enough energy ($zU \approx 1$ eV) to ionize molecules. Instead, REB can be understood as a rapid electrical discharge due to the high ionic conductance caused by a gentle structural membrane rearrangement, that is, pore formation (Weaver, 1990). Although easy to state qualitatively (Weaver and Mintzer, 1981), a quantitative description of REB required a complex model (Powell *et al.*, 1986), and the first version had several unsatisfying assumptions (Weaver and Powell, 1989). A recent version (Table 3) overcomes these difficulties and provides a plausible, quantitative description of $U(t)$ and $G(t)$ for charge injection conditions (Figs. 2 and 3; Barnett and Weaver, 1991).

1. Basic Approach to the General Electroporation Model

Tables 2, 3 and 4 outline the essential features of both the general approach, and the most recent version. This approach allows individual modules (e.g., a physical model of the pore free energy) to be independently changed, while retaining other features of the general model. For quantitative predictions, these elements are presently formalized by using the elements given in Tables 3 and 4.

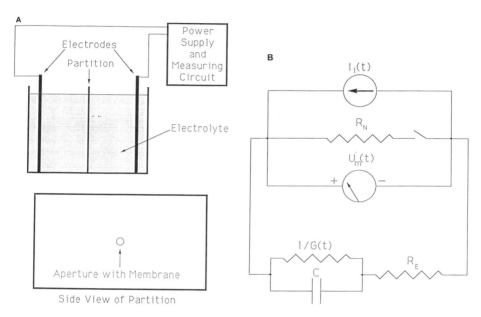

Figure 2   A. Experimental configuration for studying artificial planar bilayer membranes (Tien, 1985). A vessel containing a bathing electrolyte is separated into two compartments by an insulating partition. On each side there is a planar electrode connected to a voltage source (pulse generator) and a measuring circuit. The membrane spans an aperture of area A_m in the partition. B. Equivalent circuit for a charge pulse experiment using the above configuration (Powell *et al.*, 1986; Barnett and Weaver, 1991). The pulse generator is modeled by the current source $I_i(t)$ in parallel with the resistor R_N, which corresponds to a voltage pulse V_0. The electrolyte, electrodes, and wires are modeled by the resistor R_E, and the membrane is modeled by the capacitor C in parallel with the resistor $1/G(t)$. The switch is opened at the end of the current pulse, forcing the discharge current to flow through the membrane. The voltmeter measures the potential difference $U_m(t)$ between the electrodes. The membrane initially charges with a time constant of $\tau_{CHG} = (R_N + R_E)C$ (Barnett and Weaver, 1991), which is typically a few microseconds, that is, a result of the finite resistance of the bathing electrolyte, electrodes, and pulse generator output resistance. This is of the same order of magnitude as for many isolated cells (Kinosita *et al.*, 1988; Lojewska *et al.*, 1989).

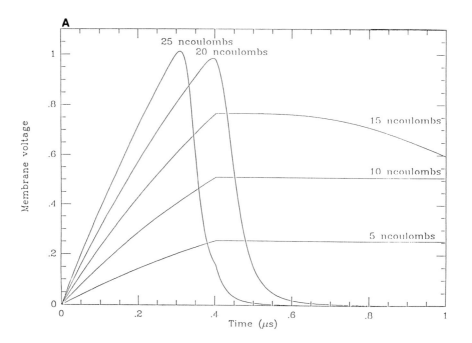

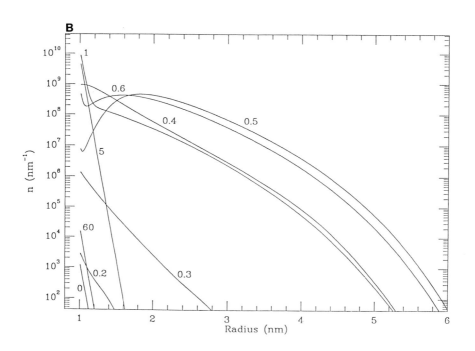

Table 2

Elements of the General Model of Electroporation

---

(1)   A resting membrane contains an equilibrium population of hydrophilic pores.

(2)   The membrane conductance $G(t)$ is due to the pores.

(3)   Changes in the transmembrane voltage $U(t)$ cause changes in the pore population.

(4)   Transient behavior is determined by calculating how the pore population changes because of changes in $U(t)$.

(5)   Feedback between $U(t)$ and $G(t)$ involves both the pore population and external electrical resistances of the experimental system (membrane environment).

(6)   Reversible electrical breakdown (REB) is caused by large ionic conduction through the transient pore population.

(7)   Rupture of a planar membrane is caused by the appearance of one or more large, unstable pores.

---

2. Elements of the Electroporation Model

In order to make quantitative predictions, the general approach is made specific by using the following physical elements in a numerical simulation.

a. Rate of Pore Formation.

Hydrophilic pores are assumed to be created by transitions from hydrophobic pores (Fig. 1) (Abidor et al., 1979). The details of this process are not known. Unless explicitly stated otherwise, the term "pore" here refers to hydrophilic pores, as they are believed to allow most of the ionic and molecular transport. A quantitative description of pore creation and destruction is

---

Figure 3   A. Short-time scale (0–1 μs) behavior of the transmembrane voltage [$U(t)$] predicted by a recent version of the theoretical model for a planar bilayer membrane exposed to a single very short (0.4 μs) pulse, that is, "charge injection" conditions (Benz et al., 1979; Barnett and Weaver, 1991). The key features of reversible electrical breakdown (REB) are predicted by the model, as is the occurrence of incomplete reversible electrical breakdown. In the latter case, the membrane discharge is incomplete, as $U(t)$ does not reach zero after the pulse. Each curve is labeled by the corresponding value of the injected charge Q. The curves for $Q = 25$ nanocoulombs and 20 nanocoulombs show REB while the others do not. B. Corresponding computed pore population distribution (probability density) $n(r,t)$ (Barnett and Weaver, 1991). Each curve is labeled by the corresponding value of the injected charge Q. For $Q = 25$ and 20 nanocoulombs, cases for which REB occurs, N increases to about 10^8 in less than 0.5 μs and then decays exponentially with a time constant of ~ 4.5 μs. For $Q = 15$ nanocoulombs, N increases rapidly to about 10^5 and remains almost constant for about 4 μs before the exponential decrease. For $Q = 10$ nanocoulombs, N increases to about 2×10^3 in about 5 μs and remains almost constant for about 30 μs before the decay phase. The membrane in this case ruptures. For $Q = 5$, nanocoulombs, N increases to about 40 in 80 μs. N will return to its initial value as the membrane discharges with a time constant of about 2 s.

Table 3
Physical Ingredients of a Recent Version of the Model[a]

Ingredient	Significance
$\dot{N}_c$	Rate of pore creation (pores with radius r_{min})
$\dot{N}_d$	Rate of pore destruction (pores with radium r_{min})
$n(r,t)$	Pore probability density function describing a range of pore sizes
$dn(r,t)/dt$	Dynamic behavior of the heterogeneous pore population
σ_p	Born energy-modified conductivity within pores
$H(r,r_i)$	Hindered transport through pores (Renkin, 1954)
$U_p \leq U$	Local transpore potential difference reduced by the spreading resistance
$R_E + R_N$	Membrane charging through external resistances
Circuit equation	Coupling to the pulse generator, electrodes, and electrolyte (Fig. 4)
η_s	Transport of charged molecules (Barnett and Weaver, submitted)

[a]Barnett and Weaver (1991).

Table 4
Simulation Parameters and Values

Parameter		Value
a	coefficient of U^2 in Λ	1.9×10^{-20} farad
A_m	membrane area[a]	1.45×10^{-6} m^2
C	capacitance of membrane	9.61×10^{-9} farad
D_p	diffusion constant in pore radius space[a]	5×10^{-14} m^2 sec^{-1}
dt	time step size [in units of $D_p/(\Delta r)^2$]	0.5
h	membrane thickness[a]	2.8 nm
kT	Boltzmann's constant times temperature	4×10^{-21} joule
R_E	series resistance of electrolyte, electrodes, and wires	30 ohm
R_N	internal resistance of current source (pulse generator)	50 ohm
r_{max}	large pore initial size	40 nm
r_{min}	minimum pore radius[a]	1 nm
r_+	radius of positive ions	0.2 nm
r_-	radius of negative ions	0.2 nm
t_{pulse}	pulse length	0.4 μs
z_+	charge of positive ions (in units of proton charge)	$+1$
z_-	charge of negative ions (in units of proton charge)	-1
γ	pore edge energy density[a]	2×10^{-11} joule m^{-1}
Γ	membrane surface energy[a]	1×10^{-3} joule m^{-2}
δ_c	pore creation energy barrier[a]	2.04×10^{-19} joule
δ_d	pore destruction energy barrier[a]	2.04×10^{-19} joule
Δr	numerical grid spacing of simulation	0.0195 nm
ε_l	dielectric constant of lipid[a]	2.1 ε_0
ε_w	dielectric constant of water	80 ε_0
v	pore creation rate prefactor[a]	10^{28} s^{-1}
κ_e	conductivity of bulk solution	0.98 ohm m^{-1}
χ	pore destruction rate prefactor[a]	5×10^{16} m s^{-1}

[a]Parameters characterizing the membrane (Barnett and Weaver, 1991).

essential, even if it cannot yet be made completely from first principles, such as by molecular dynamics (Karplus and Petsko, 1990). Generally a "pore free energy," $\Delta E(r)$, is used in which the pore is approximated as a circular cylinder of radius r. Both mechanical (Litster, 1975; Taupin *et al.*, 1975) and electrical (Abidor *et al.*, 1979; Weaver and Mintzer, 1981; Sugar, 1981) contributions are considered. This is emphasized by the form $\Delta E(r, U_p)$, where U_p is the local transmembrane voltage at the site of the pore. We use $\Delta E = \Delta E_M + \Delta E_E + \Delta E_0$ where ΔE_M is the mechanical contribution, ΔE_E is the electrical contribution, and ΔE_0 is an arbitrary is constant. The mechanical contribution is

$$\Delta E_M = 2\pi\gamma r - \pi\Gamma r^2 \tag{1}$$

where Γ is the surface energy density of the membrane–water interface and γ is the pore edge energy. In the case of a planar membrane, rupture is believed to occur if one or more pores achieve radii $r > r_c$ where the critical radius for rupture is $r_c = \gamma/\Gamma$ (Litster, 1975; Taupin *et al.*, 1975).

The electrical contribution ΔE_E treats the pore as an electrical capacitor, but with conduction through the pore explicitly included (Pastushenko and Chizmadzhev, 1982; Powell and Weaver, 1986). This gives

$$\Delta E_E = -\frac{\pi(\varepsilon_w - \varepsilon_l)U^2}{h^2} \int_{r_{min}}^{r} \alpha^2 r \, dr, \quad \text{where} \quad \alpha(r) = \left[1 + \frac{\pi r \kappa_p(r)}{2h\kappa_e} \right]^{-1} \tag{2}$$

where $\alpha(r)$ includes the "voltage divider" effect associated with the spreading resistance external to a pore and the internal pore resistance (Section II,A,1,g). We assume that pores have a minimum radius, $r_{min} = 1.0$ nm, because headgroup packing constraints require r_{min} to be somewhat greater than the size of the hydrophilic headgroups ($\sim$0.7 nm) that make up the inner surface of the pore, and the pore wall must contain at least several phospholipid molecules (Sugar and Neumann, 1984). We also assumed that the number of pores changes due to the creation and destruction of pores with radius r_{min}. This yields a boundary condition at $r = r_{min}$ for the flux of pores in radius space

$$J_p = \dot{N}_c - \dot{N}_d \quad \text{at} \quad r = r_{min} \tag{3}$$

Here $\dot{N}_c$ and $\dot{N}_d$ are the pore creation and destruction rates, and the pore flux is

$$J_p = -D_p\left(\frac{\partial n}{\partial r} + \frac{n}{kT}\frac{\partial \, \Delta E}{\partial r}\right) \tag{4}$$

During pore formation the membrane achieves energetically unfavorable configurations—that is, an energy barrier Λ must be overcome. Although the important details are unknown, we assume that Λ depends on the transmembrane voltage U. The contribution of permanent dipoles associated with pore structures is neglected

(Neumann *et al.*, 1982), so that Λ is assumed to depend on U^2, such that $\Lambda = \delta_c - aU^2$. Here δ_c and a are constants. The corresponding absolute rate estimate is

$$\dot{N}_c = \nu \exp\left(-\frac{\delta_c - aU^2}{kT}\right) \tag{5}$$

where ν is an attempt rate (Weaver and Mintzer, 1981).

b. Rate of Pore Destruction.   Pore destruction is also not understood in detail. We assume that the probability that a pore of radius r_{min} is destroyed is independent of U. This gives

$$\dot{N}_d = \chi\, n(r_{min}) \exp\left(-\frac{\delta_d}{kT}\right) \qquad \text{where } \chi = \text{constant (dimensions} \atop \text{of velocity)} \tag{6}$$

c. Pore Probability Density Function.   A combination of diffusion and physical forces governs pore evolution. As a result, pores with a range of sizes appear in the membrane. This distribution of sizes is described by a probability density function $n(r,t)$. At any time t, there are $n(r,t)\,\Delta r$ pores with radii between r and $r + \Delta r$.

d. Dynamic Behavior of the Pore Population.   The pore population rate of change is quantitatively described using Smoluchowski's equation (Pastushenko *et al.*, 1979; Chizmadzhev *et al.*, 1979; Powell and Weaver, 1986).

$$\frac{\partial n}{\partial t} = D_p\left[\frac{\partial^2 n}{\partial r^2} + \frac{\partial}{\partial r}\left(\frac{n}{kT}\frac{\partial \Delta E}{\partial r}\right)\right] \tag{7}$$

Here D_p is the effective diffusion constant for the pore radius (Deryagin and Gutop, 1962), which is independent of radius (Powell and Weaver, 1986). The term ΔE is important because $-(\partial \Delta E/\partial r)_U$ is the effective driving force that acts to change the pore size. This equation is valid for $r_{min} \leqslant r \leqslant r_{max}$ (we use $r_{max} = 2r_c$). For small $U(t)$ relatively few pores are predicted. However, for large $U(t)$ a large number of pores with a wide range of sizes is predicted, and this corresponds to "electroporation."

e. Born Energy-Modified Conduction by Pores.   We assume that the transport of ions across the membrane occurs by passage through pores large enough to accommodate small hydrated ions (e.g., Na^+ and Cl^-). However the presence of ions within small pores requires that the "Born energy" (Parsegian, 1969) and hindered motion both be considered. The bulk electrolyte conductivity σ_e is a function of the concentrations c_i and of the mobilities η_i of its ions:

$$\sigma_e = \sum_i (z_i e)^2 \eta_i c_i \tag{8}$$

Here $e = 1.6 \times 10^{-19}$ coulomb, z_i is the charge of the ith type of ion, and the sum is over the different ion types of the electrolyte. In contrast, the conductivity within a pore, σ_p, is reduced.

$$\sigma_p = \sum_i (z_i e)^2 \eta_i c_i H_i \, \exp\left(\frac{\mu_i^\circ}{kT}\right) \tag{9}$$

Here H_i is a steric hindrance factor (Section II,A,1,f) and μ_i° is the standard chemical potential of an ion of type i inside the pore. We use a previously obtained estimate (Pastushenko and Chizmadzhev, 1982),

$$\mu_i^\circ = \frac{(z_i e)^2}{\varepsilon_1 r} P\left(\frac{\varepsilon_1}{\varepsilon_w}\right) \tag{10}$$

which is based on a point charge on the axis of an infinite cylindrical cavity. Here ε_1 and ε_w are the dielectric constants of the lipid and the water, respectively, and the function P has a maximum value of 0.25 (Parsegian, 1969).

f. Hindered Transport through Pores.   Hindrance of movement within the small space of a pore is described by using Renkin's equation (Renkin, 1954).

$$H(r,r_i) = \left[1 - \left(\frac{r_i}{r}\right)\right]^2 \left[1 - 2.1\left(\frac{r_i}{r}\right) + 2.09\left(\frac{r_i}{r}\right)^3 - 0.95\left(\frac{r_i}{r}\right)^5\right] \tag{11}$$

$$H \to 0 \quad \text{as} \quad r_i \to r$$

g. Local Transpore Potential Difference Reduced.   A heterogeneous electric field exists within the electrolyte near the pore entrances. The associated potential drop in the electrolyte is estimated by considering the "spreading resistance" (Newman, 1966) $R_s(r)$ given by $R_s \approx 1/2\sigma_e r$. The spreading resistance results in a smaller transmembrane voltage U_p at the site of pore. The internal resistance of a pore is estimated by using the reduced conductivity σ_p, so that $R_p = h/\pi r^2 \sigma_p$. These two resistances act in series, creating a "voltage divider" and therefore $U_p \leq U$. This reduces the electrical driving force. The current through a pore is $I_p = U/R_s(r) + R_p(r)$, and the total current I through the membrane is therefore

$$I = U \int_{r_{min}}^{\infty} \left[\frac{n(r,t)}{R_s(r) + R_p(r)}\right] dr \tag{12}$$

$$\text{with the conductance} \quad G(t) = \int_{r_{min}}^{\infty} \left[\frac{n(r,t)}{R_s(r) + R_p(r)}\right] dr$$

as $r_{min} > r_i = r_+ = r_-$. Here $G(t)$ is the strongly time-varying membrane conductance, which is used self-consistently with an equation for the external circuit (Section II,A,1h) in order to determine the membrane electrical behavior.

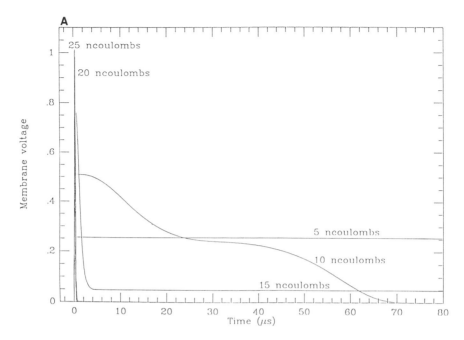

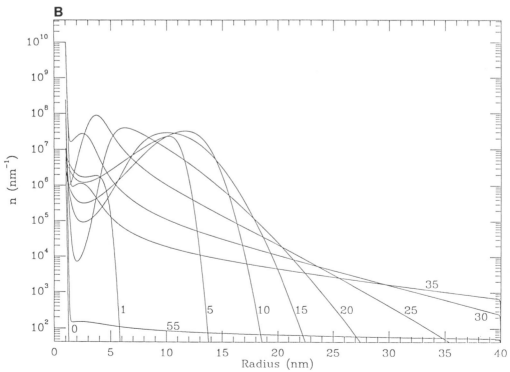

h. Circuit Equation Coupling the Membrane to the Pulse Source.   The circuit of Fig. 4 shows the relationship between the pulse generator, the charging pathway resistance, and the membrane, which is treated as the parallel combination of the membrane capacitance C. The capacitance changes negligibly during electroporation, because only a small fraction of the membrane is occupied by aqueous pores. However, the membrane resistance, $R(t) = 1/G(t)$, changes by orders of magnitude. In a charge pulse experiment a current pulse of amplitude I_i passes through R_N to create a voltage pulse, V_0 (Fig. 4). For $0 < t < t_{\text{pulse}}$ current flows into and/or across the membrane, and at $t = t_{\text{pulse}}$ the pulse is terminated by opening the switch. Because the generator is then disconnected, membrane discharge can occur only through the membrane for a planar membrane (not so for a cell). The appropriate circuit differential equations are

$$C\frac{dU}{dt} = \begin{cases} \dfrac{I_p R_N}{R_E + R_N} - U\left(G(t) + \dfrac{1}{R_E + R_N}\right) & \text{if } t < t_{pulse} \\[2em] -\dfrac{U}{R} & \text{if } t > t_{pulse} \end{cases} \tag{13}$$

subject to the initial condition $U(0) = 0$. A self-consistent numerical solution to these equations is used in order to provide an electroporation simulation.

Figure 4   A. Longer-timescale (0–80 µs) electrical behavior predicted by the same model, showing rupture and simple charging of an artificial planar bilayer membrane (Barnett and Weaver, 1991). The characteristic sigmoidal behavior of $U(t)$ is predicted by the model (Barnett and Weaver, 1991), but the time scale is somewhat shorter than found in experiments (Benz and Zimmermann, 1980). Each curve is labeled by the corresponding value of the injected charge Q. The curves for $Q = 25$ and 20 nanocoulombs are the spikes at $t = 0$. The curve for $Q = 15$ nanocoulombs shows that the membrane underwent REB at $t = 2$ µs, but the membrane recovered before it had time to discharge completely. The curve for $Q = 10$ nanocoulombs shows rupture, while the curve for $Q = 5$ nanocoulombs shows that the membrane conductance did not increase enough to discharge the membrane. B. Corresponding pore population distributions as described by $n(r,t)$ (Barnett and Weaver, 1991). Each curve is labeled by the corresponding value of the injected charge Q. For $Q = 25$ and 20 nanocoulombs, cases for which REB occurs, N increases to about 10^8 in less than 0.5 µs and then decays exponentially with a time constant of ~4.5 µs. For $Q = 15$ nanocoulombs, N increases rapidly to about 10^5 and remains almost constant for about 4 µs before the exponential decrease. For $Q = 10$ nanocoulombs, N increases to about 2×10^3 in about 5 µs and remains almost constant for about 30 µs before the decay phase. The membrane in this case ruptures. For $Q = 5$ nanocoulombs, N increases to about 40 in 80 µs. N will return to its initial value as the membrane discharges with a time constant of about 2 s.

III. Molecular Transport

A. Molecular Transport Mechanisms

Electrical behavior is far less important to biology than the molecular transport across cell membranes (Neumann and Rosenheck, 1972; Potter, 1988; Neumann *et al.*, 1989). As an initial step, we extended the model to predict one contribution (that of electrical drift) to molecular transport across a planar membrane (Table 5).

In electroporative transport we expect that water-soluble molecules are transported through the pores that have $r \geqslant r_s$, where r_s is the effective radius of a molecule. Hindered transport is expected for all three types of transport, because many pores are only slightly larger than r_s. For each type of transport the characteristic time for relaxation into a steady-state transport is important, as a steady-state description is generally much simpler, if justified.

B. Contribution of Electrical Drift

In order to make initial estimates of transport, we consider only the transport of charged molecules due to drift in response to the electric field. For a charged molecule of type s, the drift current across the membrane is larger than the diffusion current as long as U is greater than the Nernst voltage for species s, $U_{Nemst,s} > kT/z_s e$ $\ln(c_{s,1}/c_{s,2})$ where z_s is the charge on a molecule of type s, and the argument of the logarithm is the ratio of the concentrations on the opposite sides of the membrane. We also limit our treatment to those cases in which the associated electric current is small compared to the current carried by the small ions of the bathing electrolyte. This allows us to neglect the effect of molecular transport on the electrical behavior of the membrane, and provides an estimate of the electrical drift flux, $\dot{N}_s$, of charged molecules of type s across the membrane.

$$\dot{N}_s(t) = U \int_{r_s}^{\infty} \left[\frac{n(r,t)}{X_s(r) + X_p(r)} \right] dr \qquad (14)$$

Table 5

General Mechanisms of Molecular Transport through Pores

Mechanism	Molecular basis
Drift	Velocity in response to a local physical (e.g., electrical) field
Diffusion	Microscopic random walk
Convection	Fluid flow carrying dissolved molecules

Note that only pores with $r \geq r_s$ participate, and that the hindrance factor is zero for $r = r_s$ [Eq. (11)]. The net number of molecules transported across the membrane is the time integral of N_s over the duration of the simulation, and is a quantity that should be measurable for different experimental conditions, particularly different pulse magnitudes, duration and shapes. The terms

$$X_s = \frac{1}{(z_s^2 e^2)\eta_s c_s r} \quad \text{and} \quad X_p = \frac{b}{\pi(z_s^2 e^2)\eta_s c_s H_s \, \exp\!\left(\dfrac{\mu_s^\circ}{kT}\right) r^2} \tag{15}$$

are the analogs to the spreading resistance and internal pore resistance for the small ions (Section II,A,1,g). Here η_s is the electrical mobility of a molecule of type s, and c_s is their concentration on the supply side. As is the case for conduction by small ions, μ_s° is the chemical potential of a molecule of type s inside a pore, which is approximated by its Born energy, and $H = H(r,r_s)$ is the steric hindrance factor.

In a recent set of simulations (Barnett and Weaver, 1991), we used the potential energy of a point charge on the axis of an infinite cylinder cavity for the Born energy. However, this is too large, as two significant contributions were not included. First, a hydrophilic pore is lined with phospholipid head groups, whose effective dielectric constant is intermediate between that of lipid and water; this lowers the Born energy. Second, a membrane-spanning pore has finite length, so that the infinite cylinder approximation is too large; this also lowers the Born energy. For this reason, here we have used the results of Jordan, who has treated finite pores lined with material of intermediate dielectric constant (Jordan, 1983, 1984). Thus we have subsequently used a reduced estimate, in which Eq. (10) is multiplied by the factor $0.02(b/r)$. This approximation still treats the charge as a point charge, which is a reasonable approximation for a small ion, but far less so for a larger molecule. For this reason, we have used an effective value for z_s^2 (e.g., $z_{s,\text{eff}}^2 = z^{1.5}$) that is less than z_s^2. In order to test the sensitivity of molecular transport to molecular size and charge we assumed a spherical shape for propidium iodide (PI; Figs. 5 and 6). During a simulation of electrical behavior $U(t)$ and $n(r,t)$ are progressively obtained. Then N_s is simultaneously estimated, and the cumulative flux is computed as the simulation progresses to provide a prediction of the average net transport, $\overline{N}_s$. In short, to date molecular transport was treated using (1) a driving force proportional to z, (2) repulsion due to a Born energy effect, and (3) geometrically reduced transport through hindrance.

C. Diffusion, Convection, and Metastable Pores

1. Diffusion and Convection

Our initial estimates of molecular transport involved only transport due to electrical drift, and should be extended by also including convection [e.g., electroosmosis

(Dimitrov and Sowers, 1990)] and diffusion (Sowers and Lieber, 1986). The same general strategy should be reasonable: a dynamic, heterogeneous pore population should be computed, in which electrical interactions are the dominant source of pore creation and expansion. In the case of a planar membrane with no osmotic or hydrostatic pressure gradient, the final stages of pore population expansion and then

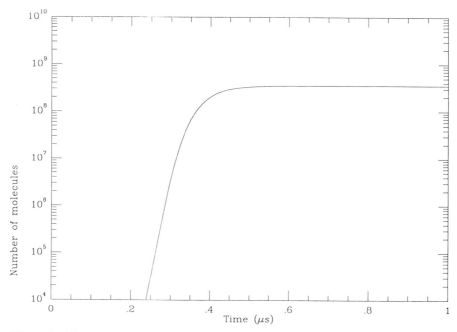

Figure 5   Theoretical estimate of the transport of propidium iodide (PI) across a artificial planar bilayer membrane. The molecule was treated as a sphere with radius $r_s = 0.6$ nm and charge $z_s = +2$. A 0.4-μs square 20 nanocoulomb pulse (Fig. 2) was used. Only transient aqueous pores are used in this version of the model. Future versions should include metastable pores, and also estimates of the contributions of diffusion and convection. The planar bilayer membrane (Figs. 2 and 3) is far larger than a cell membrane (Fig. 6), but can generally not tolerate a long pulse, because the planar membrane can rupture due to electroporation (Fig. 3), whereas as a cell membrane cannot (Sugar and Neumann, 1984). As shown by a comparison with Fig. 6, the model predicts that the ability to use longer pulses with cells results in almost comparable molecular transport, in spite of the much larger area of the artificial planar membrane. (Here $A_{cell}/A_{planar} \approx 10^{-10}$ m^2/1.45$\times$ 10^{-6} m^2 = 7 $\times$ 10^{-5}.) For the 20 nanocoulomb pulse (voltage pulse of 2.5 V), the transmembrane voltage is elevated only briefly (Fig. 2), so that in comparison to the cell, a combination of a significantly elevated transmembrane voltage and the evolution of a large pore population does not occur.

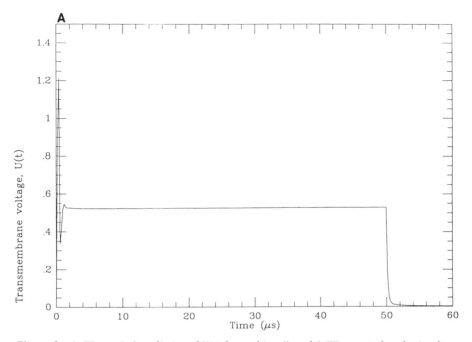

Figure 6 A. Theoretical prediction of $U(t)$ for a cubic cell model (Weaver *et al.*, submitted). This extension of a planar bilayer membrane model assumes that the participating membrane regions are two planar areas on opposite sides of a cell. Unlike the planar membrane, the cubic cell correctly gives a rapid discharge of the externally induced cellular transcellular potential difference even if there is negligible electroporation. Here the behavior was simulated by providing a single 50-μs square pulse of amplitude $V_0 = 4$ V. For a cell with characteristic linear dimension of 8 μm (e.g., the yeast *Saccharomyces cerevisiae*), this corresponds to an electric field pulse of magnitude $E_e \approx 2V_0/L_{cell} = 10^4$ V/cm. This is within the experimental range (about 5–10 $\times$ 10^3 V/cm) where a significant percentage of the cells exhibit large uptake. At the higher field strengths, more and larger pores may occur, with more cells participating and with greater molecular uptake, but (as determined stringently by plating) fewer cells eventually survive (Weaver *et al.*, 1988). In the simulation shown here, the membrane first charges as current flows to the membrane through the external electrolyte resistance, and $U(t)$ initially increases rapidly. At about 0.5 μs, $U(t)$ reaches 1.2 V, and then begins to decrease, because of a tremendous increase in membrane conductance. This discharge is reversible electrical breakdown (REB). A partial discharge then takes place, even though the pulse is still on. Before reaching a plateau, $U(t)$ rises again, as the now highly conductive membrane continues charging, but with a voltage divider (membrane resistance and external resistance) now in effect. Then, for the remainder of the pulse a plateau of $U \approx 525$ mV is reached. At the end of the pulse, $U(t)$ decays to zero in about 2 μs. The maximum transmembrane voltage achieved (1.2 V) is close to the threshold value (about 1 V), at which REB is first discerned in a series of simulations in which V_0 is varied. If somewhat smaller amplitude pulses (e.g., $V_0 = 1.5$ V) are used, REB barely occurs, and only a slight discharge occurs during the 50-μs pulse. This pulse corresponds to $E_e \approx 3.8 \times 10^3$ V/cm (8-μm cell), which is close to the experimentally observed threshold (for percent participation) of about 3 $\times$ 10^3 V/cm. Overall, these predictions are in reasonable agreement with what is known about the transmembrane voltage behavior.

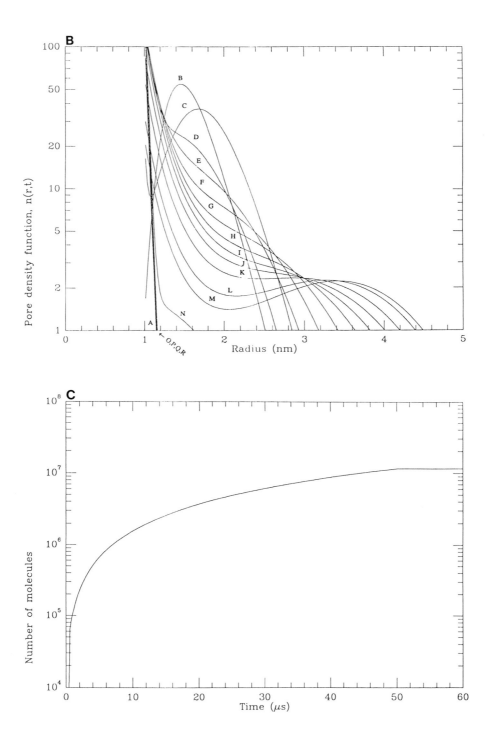

its collapse should also be governed by purely electrical interactions. By following the pore population over its development, the contribution of each transport mechanism can be estimated. For cell membranes a pressure difference will usually exist, so that pores of electrical origin may be further expanded. The striking very large pores visualized by rapid-freeze electron microscopy (Chang and Reese, 1990) may have been generated in this way. Their very large size (40–100 nm) and relatively long lifetime (10–40 ms) suggest that (1) they may contribute significantly to molecular transport, and (2) based on other isolated cell experiments (Benz and Zimmermann, 1989; Kinosita et al., 1988) and theoretical models (Powell et al., 1986; Weaver and Powell, 1989; Barnett and Weaver, 1991), these large pores cannot be the "primary pores" caused directly by elevated transmembrane voltages.

2. Metastable Pore Formation

The possible effects of metastable pores should also be considered, for example, by extending the model to include interactions that "trap" one or more of the transient

---

Figure 6   B. Corresponding behavior of $n(r,t)$. The predicted development of a pore population with a wide range of sizes is shown. Initially, before the transmembrane voltage has increased from zero (A), only a small number of almost minimum size pores are present (lower left corner; curve A). As $U(t)$ rises, the pore population rapidly increases, with both greater numbers and sizes of pores. Note the continued evolution of the pore population to include progressively larger pores while the pulse is on. The population peak rapidly reaches approximately 1.5 nm before the population partially collapses due to REB (curve B, 0.4 μs). Subsequently, during the initial collapse due to REB, most pores shrink to smaller radii, but some expand (shoulder in population on right). This subpopulation of larger pores is expected to be crucial to the transport of macromolecules. At the same time the transmembrane voltage has reached an approximate plateau while the pulse remains on. After the membrane has discharged, the pore population collapses completely, with shrinkage of pores occurring rapidly, followed by a more gradual disappearance of minimum size pores. Curves C through N are at times 0.63, 1.00, 1.58, 2.51, 3.98, 6.31, 10.0, 15.8, 25.1, and 39.8 μs (1/5 decade on a log scale), and curves K through R occur at 50–55 μs in 1-μs intervals. C. Predicted electrical drift contribution to the molecular transport of propidium iodide (PI) across one of the two membranes of a cubic cell. Note the break at $t = 50$ μs, where N_s ceases to increase because $U(t)$ falls rapidly at the pulse end, and this causes both the transport driving force and the pore population to rapidly decay. As in Fig. 5, PI was treated as a charged sphere. PI is highly charged, and therefore reasonably expected to have a significant electrical drift contribution. The extracellular bulk concentration is 80 μM, the value used an experiment with intact yeast and propidium iodide (Bartolette et al., 1989). For the purpose of estimating the Born energy of Eq. (10), the effective squared charge was taken to be $z_{s,eff}^2 = z_s^{1.5}$. On the order of 10^7 molecules are predicted to be taken up. Other simulations (not shown) predict that far fewer molecules are transported as the size and charge of the molecule is increased.

large pores. Although not yet understood mechanistically, there is growing experimental evidence that a significant number of cells retain an ability to take up large numbers of macromolecules long after $U(t)$ should have decayed to low values. Experiments that show a decaying number of cells, each with an essentially undiminished ability for large uptake (Tsoneva et al., 1990; Harrison et al., private communication) are particularly suggestive of metastable pores. For example, protrusion of cytoplasmic macromolecules into pores could provide a coulombic repulsion that prevents pore shrinkage to r_{min}. Similarly, the entry of a charged linear molecule (e.g., DNA) should provide a "foot-in-the-door" effect, in which the pore remains open until the molecule either is expelled or transported. Capture of membrane proteins at a pore's edge (Fig. 1) or interaction with the cytoskeleton might also lead to metastable pores.

3. Transport through Metastable Pores

After $U(t)$ has decayed by REB, a small residual transmembrane voltage may persist due to continuing ion pump activity or a diffusion potential. An order-of-magnitude estimate for the electrical drift contribution of a single, moderately large metastable pore is made by neglecting both the hindrance factor and Born energy repulsion in Eq. (15), and setting $U \, \Delta t_{\text{steady state}} = 1$ V–S (Table 6).

$$\overline{N}_{s,\text{metastable pore}} \approx \left[\frac{\eta_s z_s e c_s r}{\left(1 + \dfrac{h}{\pi r} \right)} \right] U \, \Delta t_{\text{steady state}}$$

$$\text{with} \quad r = h \quad \text{(illustrative pore size)} \tag{16}$$

Table 6

Combinations of $U_{\text{steady state}}$ and $\Delta t_{\text{steady state}}$ with the Same Electrically Driven Transport

$U_{\text{steady state}}$	$\Delta t_{\text{steady state}}$	$\approx$	Time for same molecular transport	
500 mV	100 μs	=	10^{-4} s	Electrically expanded pore
5 mV	10 ms	=	10^{-2} s	Metastable pore[a]
0.05 mV	1 s	=	1 s	Metastable pore[a]

[a]Pore-expanding transmembrane voltages are found experimentally to usually vanish within 10^{-6} to 10^{-3} s of a pulse end because of REB (Benz and Zimmermann, 1980; Kinosita et al., 1988; Powell et al., 1989), in agreement with theoretical predictions (Powell et al., 1986; Weaver and Powell, 1989; Barnett and Weaver, 1991).

Consider two extreme cases of molecular size. DNA is large polyelectrolyte with a high electrical mobility, $z_s e \eta_s \approx 10^{-8}$ V^2/m-s (Hartford and Flygare, 1975), because $z_s \approx 10^4$. If DNA is present extracellularly at an effective concentration of $c_s = 1$ μM, then Eq. (16) predicts $N_{DNA} \approx 6$ molecules. For small molecules with only a few charges but essentially the same electrical mobility, the same order of magnitude is predicted. Thus, if longer times are invoked for this size metastable pore, its drift contribution still cannot account for the observed transport (e.g., 10^8 molecules of propidium iodide by yeast). Diffusion estimates (using the Einstein relation to obtain the diffusion constant) are even smaller. Likewise, the convective contribution of electroosmosis (Dimitrov and Sowers, 1990) should become very small after the transmembrane voltage decays to low values. Nevertheless, small transmembrane voltages, large metastable pores, and long times may together contribute significantly to transport. In fact, several experiments indicate that a transport mechanism with a lifetime of order 10^2 s or longer can occur (Kinosita and Tsong, 1977; Tsoneva et al., 1990; Harrison et al., private communication). This incomplete evidence for metastable pores, taken in combination with the successful prediction of ion transport by a heterogeneous transient pore population, leads to two tentative conclusions: (1) a large number of transient pores occurring for short times ($t < 10^{-3}$ s) can quantitatively account for significant transport of ions and small molecules, and possibly large molecules, by electrical drift, but (2) a quantitative understanding of post-REB transport mechanisms ($t > 10^{-3}$ s) does not yet exist.

IV. Isolated Cell Electroporation Model

A. Cubic Cell Model for Electroporation

The topologies of planar and cell membranes are significantly different. Planar membranes can discharge only through the membrane (or the external circuit if the pulse generator remains connected). Cells can discharge either by conduction through the membrane or by an extracellular pathway. Thus cells have a short discharge time, τ_{DIS}, even for electric field pulses that do not cause electroporation. A cubic cell model for electroporation has been developed that assumes that (1) only a portion of the cell membrane significantly participates in electroporation, and (2) this portion can be approximated by two planar membranes (Weaver et al., submitted). This extension of the planar model provides a better description of electrical behavior, and therefore a more realistic basis for describing molecular transport (Fig. 6).

V. Related Experimental Issues

A. Importance of Direct Comparison to Experimental Results

Tests of theoretical models should involve direct comparisons of theoretical predictions and experimentally measurable quantities. For electrical behavior this should be most readily accomplished for artificial planar bilayer membranes, for which the geometry, membrane chemical and mechanical properties, and electrical exposure and measurement are fairly well defined. At the other extreme, living cells are of the greatest biological interest, have a much larger surface-to-volume ratio (facilitating molecular transport measurements), but are the most difficult to describe in terms of cellular membrane geometry, membrane properties, and electrical exposure and measurements. Quantitative determination of the number of fluorescent molecules transported into cells should be generally measurable (Bartoletti *et al.*, 1989). As shown in Fig. 7, the number of molecules taken up by individual cells of the yeast *Saccharomyces cerevisiae* has been determined for propidium iodide. In this case

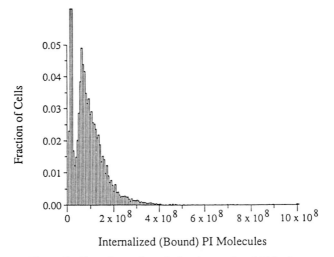

Internalized (Bound) PI Molecules

Figure 7   Experimental result for the uptake of PI by intact cells of the yeast *S. cerevisiae* (Bartoletti *et al.*, 1989). An order-of-magnitude comparison between this type of measurement and the theoretical model predictions can be made by estimating the order of magnitude of participating membrane area, and also the maximum change in transmembrane voltage caused by an electric field pulse.

Table 7

Possible Sources of Heterogeneity in Cell Electroporation

Source	Significance
Cell orientation	$\vec{E}_e$ is a directional quantity
Cell–cell separation	Perturbation of local field by nearby cells
Tissue heterogeneity	Perturbation of local field by tissue
Membrane composition	Composition variation within cell population[a]
Pore statistical behavior	Electroporation is fundamentally stochastic

[a]The observed universality of electroporation at about $U \approx 1$ V argues against significant membrane composition effects, but minor contributions to heterogeneity may be due to membrane properties.

$N_s = 1.0^{+0.71}_{-0.86} \times 10^8$ molecules for a 50-μs pulse of magnitude $E_e = 8 \pm 0.5$ kV/cm. For 70-kD FITC-labeled dextran, which has a small charge, and is therefore not expected to be predominantly transported by drift, $N_s = 1.4^{+1.0}_{-1.28} \times 10^5$ molecules under the same conditions.

B. Heterogeneity of Electroporation within a Cell Population

Cellular systems are well known to exhibit cell-to-cell wall variations in composition and function (Table 7). Further, the vectorial nature of the electric field suggests that cell orientation will be fundamentally important for electroporation. Thus, it is not surprising that electroporative phenomena are found to vary for the individual cells within an electrically pulsed population (Weaver *et al.*, 1988; Bliss *et al.*, 1988; Bartoletti *et al.*, 1989).

Theoretical models could explicitly include orientation effects in cell suspensions and in tissue, could investigate the fundamentally stochastic nature of pores, and might be able to consider the effect of variation in membrane properties, if these can be obtained from experimental studies. Experimental methods (e.g., flow cytometry and image analysis) that can quantitatively examine several thousands of individual cells are likely to be important to studies of electroporation.

C. Experimental Method Issues

To be useful for testing theoretical models, the issues noted in Table 8 are important.

Table 8
Experimental Methodology Issues Important to Testing Theories

Issue	Significance
Temperature rise, $\Delta\overline{T}$	Thermal, not electrical, effects may be dominant[a]
Electrical field at cell, E_e	$E_{nominal} \equiv V_0/L_{chamber} \geqslant E_e$ (Bliss et al., 1988)
Electric field uniformity	Interpretations difficult if E_e not the same for all cells
Number of cells studied	Confidence that cells are representative
Molecular uptake kinetics[b]	Fluorescence of molecules must change upon cell entry
Molecular uptake end point[b]	General, but remove extracellular molecules before measurement
Molecular release kinetics[b]	Limited to cells loaded with fluorescent molecules
Molecular release end point[b]	General, but requires cells to be loaded

[a]$\overline{T}$ is the spatially averaged temperature of the extracellular medium, ρ_e is its mass density, and c_e its specific heat. $d\overline{T}/dt = \sigma_e E_e^2/\rho_e c_e$ is often very large, particularly for small cells such as bacteria. Much smaller entities (e.g., liposomes) may be fundamentally inaccessible to electroporation in media with physiological conductivity, because ΔT_e becomes extremely large.

[b]With calibration quantitative fluorescence measurements of individual cells can be made using flow cytometry and image analysis (fluorescence videomicroscopy). Flow cytometry is well suited to making large numbers (10^4 to 10^5) of individual end-point measurements with low detection limits (e.g., 10^3 fluorescein molecules) (Shapiro, 1988; Bartoletti et al., 1989; Melamed et al., 1990). Image analysis can provide kinetic measurements for the limited case of loaded cells or vesicles. However, many imaging systems have temporal resolution of order 10^{-2} s (Inoue, 1986; Dimitrov and Sowers, 1990), significantly slower than the time scale (10^{-7} to 10^{-3} s) of the electrical behavior. More recently, pulsed laser image analysis with submicrosecond temporal resolution and spatial resolution for a single cell has been demonstrated, and has confirmed that short-timescale electrical behavior by using voltage-sensitive fluorescence dyes (Kinosita et al., 1988, 1991). A fundamental tradeoff governs the number of cells measured by image analysis: the number of pixels per cell versus the number of cells measured in a given time. Better spatial resolution results in fewer cells analyzed. For these reasons, flow cytometry and image analysis are complimentary.

Acknowledgments

This work supported partially by ONR contract N00014-87-K-0479, and by Army Research Office grant DAAL03-90-G-0218.

References

Abidor, I. G., Arakelyan, V. B., Chernomordik, L. V., Chizmadzhev, Y. A., Patushenko, V. F., and Tarasevich, M. R. (1979). Electric breakdown of bilayer membranes: I. The main experimental facts and their qualitative discussion. *Bioelectrochem. Bioenerg.* **6**, 37–52.

Barnett, A., and Weaver, J. C. (1991). Electroporation: A unified, quantitative theory of reversible electrical breakdown and rupture. *Bioelectrochem. Bioenerg.* **25**, 163–182.

Barnett, A., and Weaver, J. C. Molecular transport across a planar bilayer membrane due to electroporation caused by a short pulse: Prediction of the electrical drift contribution. (In preparation).

Bartoletti, D. C., Harrison, G. I., and Weaver, J. C. (1989). The number of molecules taken up by electroporated cells: Quantitative determination. *FEBS Lett.* **256**, 4–10.

Benz, R., and Conti, F. (1981). Reversible electrical breakdown of squid giant axon membrane. *Biochim. Biophys. Acta* **645**, 115–123.

Benz, R., and Zimmermann, U. (1980). Relaxation studies on cell membranes and lipid bilayers in the high electric field range. *Bioelectrochem. Bioenerg.* **7**, 723–739.

Benz, R., Beckers, F., and Zimmermann, U. (1979). Reversible electrical breakdown of lipid bilayer membranes: A charge-pulse relaxation study. *J. Membrane Biol.* **48**, 181–204.

Bliss, J. G., Harrison, G. I., Mourant, J. R., Powell, K. T., and Weaver, J. C. (1988). Electroporation: The distribution of macromolecular uptake and shape changes in red blood cells following a single 50 μsecond square wave pulse. *Bioelectrochem. Bioenerg.* **19**, 57–71.

Chang, D. C., and Reese, T. S. (1990). Changes in membrane structure induced by electroporation as revealed by rapid-freezing electron microscopy. *Biophys. J.* **58**, 1–12.

Chernomordik, L. V., and Chizmadzhev, Y. A. (1989). Electrical breakdown of lipid bilayer membranes: Phenomenology and mechanism. *In* "Electroporation and Electrofusion in Cell Biology" (E. Neumann, A. E. Sowers, and C. A. Jordan, eds), pp. 83–95. Plenum Press, New York.

Condon, E. U., and Odishaw, H. (eds.) (1958). "Handbook of Physics." McGraw-Hill, New York.

Crowley, J. M. (1973). Electrical breakdown of bimolecular lipid membranes as an electromechanical instability. *Biophys. J.* **13**, 711–724.

Deryagin, B. V., and Gutop, Y. V. (1962). Theory of the breakdown (rupture) of free films. *Kolloidn. Zh.* **24**, 370–374.

Dimitrov, D. S., and Jain, R. K. (1984). Membrane stability. *Biochim. Biophys. Acta* **779**, 437–468.

Dimitrov, D. S. (1984). Electric field-induced breakdown of lipid bilayers and cell membranes: A thin viscoelastic film model. *J. Membrane Biol.* **78**, 53–60.

Dimitrov, D. S., and Sowers, A. E. (1990). Membrane electroporation—Fast molecular exchange by electroosmosis. *Biochim. Biophys. Acta* **1022**, 381–392.

Farkas, D. L. (1989). External electrical field-induced transmembrane potentials in biological systems: Features, effects and optical monitoring. *In* "Electroporation and Electrofusion in Cell Biology" (E. Neumann, A. E. Sowers, and C. A. Jordan, eds.), pp. 409–431. Plenum Press, New York.

Fishman, H. M., and Macey, R. I. (1969). The N-shaped current-potential characteristic in frog skin: I. Time development during step voltage clamp. *Biophys. J.* **9**, 127–139.

Glaser, R. W., Leikin, S. L., Chernomordik, L. V., Pastushenko, V. F., and Sokirko, A. I. (1988). Reversible electrical breakdown of lipid bilayers: Formation and evolution of pores. *Biochim. Biophys. Acta* **940**, 275–287.

Hartford, S. L., and Flygare, W. H. (1975). Electrophoretic light scattering on calf thymus deoxyribonucleic acid and tobacco mosaic virus. *Macromolecules* **8**, 80–83.

Inoué, S. (1986). "Video Microscopy." Plenum Press, New York.

Jordan, P. C. (1983). Electrostatic modeling of ion pores: II. Effects attributable to the membrane dipole potential. *Biophys. J.* 41, 189–195.

Jordan, P. C. (1984). Effect of pore structure on energy barriers and applied voltage profiles: I. Symmetrical channels. *Biophys. J.* 45, 1091–1100.

Karplus, M., and Petsko, G. A. (1990). Molecular dynamics simulations in biology. *Nature* 347, 631–639.

Kinosita, K., Jr., and Tsong, T. Y. (1977). Formation and resealing of pores of controlled sizes in human erythrocyte membrane. *Nature* 268, 438–441.

Kinosita, K., Jr., Ashikawa, I., Saita, N., Yoshimura, H., Itoh, H., Nagayama, H., and Ikegami, A. (1988). Electroporation of cell membrane visualized under a pulsed-laser fluorescence microscope. *Biophys. J.* 53, 1015–1019.

Litster, J. D. (1975). Stability of lipid bilayers and red blood cell membranes. *Phys. Lett.* 53A, 193–194.

Lojewska, Z., Farkas, D. L., Ehrenberg, B., and Loew, L. L. (1989). Analysis of the effect of medium and membrane conductance on the amplitude and kinetics of membrane potentials induced by externally applied electrical fields. *Biophys. J.* 56, 121–128.

Melamed, E. R., Lindmo, T., and Mendelsohn, M. L. (1990). "Flow Cytometry and Sorting," 2nd ed. Wiley-Liss, New York.

Neumann, E., and Rosenheck, K. (1972). Permeability changes induced by electric impulses in vesicular membranes. *J. Membrane Biol.* 10, 279–290.

Neumann, E., Schaefer-Ridder, M., Wang, Y., and Hofschneider, P. H. (1982). Gene transfer into mouse lyoma cells by electroporation in high electric fields. *EMBO J.* 1, 841–845.

Neumann, E., Sowers, A., and Jordan, C. (eds.) (1989). "Electroporation and Electrofusion in Cell Biology." Plenum, New York.

Newman, J. (1966). Resistance for flow of current to a disk. *J. Electrochem. Soc.* 113, 501–502.

Parsegian, V. A. (1969). Energy of an ion crossing a low dielectric membrane: Solutions to four relevant electrostatic problems. *Nature* 221, 844–846.

Pastushenko, V. F., and Chizmadzhev, Y. A. (1982). Stabilization of conducting pores in BLM by electric current. *Gen. Physiol. Biophys.* 1, 43–52.

Pastushenko, V. F., Chizmadzhev, Y. A., and Arakelyan, V. B. (1979). Electric breakdown of bilayer membranes: II. Calculation of the membrane lifetime in the steady-state diffusion approximation. *Bioelectrochem. Bioenerg.* 6, 53–62.

Potter, H. (1988). Electroporation in biology: Methods, applications, and instrumentation. *Anal. Biochem.* 174, 361–373.

Powell, K. T., and Weaver, J. C. (1986). Transient aqueous pores in bilayer membranes: A statistical theory. *Bioelectrochem. Bioelectroenerg.* 15, 211–227.

Powell, K. T., Derrick, E. G., and Weaver, J. C. (1986). A quantitative theory of reversible electrical breakdown. *Bioelectrochem. Bioelectroenerg.* 15, 243–255.

Powell, K. T., Morgenthaler, A. W., and Weaver, J. D. (1989). Tissue electroporation: Observation of reversible electrical breakdown in viable frog skin. *Biophys. J.* 56, 1163–1171.

Renkin, E. M. (1954). Filtration, diffusion and molecular sieving through porous cellulose membranes. *J. Gen. Physiol.* 38, 225–243.

Shapiro, H. M. (1988). "Practical Flow Cytometry," 2nd ed. A. R. Liss, New York.

Sowers, A. E., and Lieber, M. R. (1986). Electropore diameters, lifetimes, numbers, and locations in individual erythrocyte ghosts. *FEBS Lett.* **205**, 179–184.

Stämpfli, R. (1958). Reversible electrical breakdown of the excitable membrane of a Ranvier node. *Ann. Acad. Brasil. Ciens.* **30**, 57–63.

Sugar, I. P. (1981). The effects of external fields on the structure of lipid bilayers. *J. Physiol. Paris* 77, 1035–1042.

Sugar, I. P., and Neumann, E. (1984). Stochastic model for electric field-induced membrane pores: Electroporation. *Biophys. Chem.* **19**, 211–225.

Taupin, C., Dvolaitzky, M., and Sauterey, C. (1975). Osmotic pressure induced pores in phospholipid vesicles. *Biochemistry* **14**, 4771–4775.

Tien, H. T. (1985). Planar bilayer lipid membranes. *In* "Progress in Surface Science" (S. G. Davison, ed.), vol. 19, 1–106.

Tsoneva, I., Tomov, T., Panova, I., and Strahilov, D. (1990). Effective production of electrofusion of hybridomas secreting monoclonal antibodies against Hc-antigen of *Salmonella. Bioelectrochem. Bioenerg.* **24**, 41–49.

Weaver, J. C., and Mintzer, R. A. (1981). Decreased bilayer stability due to transmembrane potentials. *Phys. Lett.* **86A**, 57–59.

Weaver, J. C., and Powell, K. T. (1989). Theory of electroporation. *In* "Electroporation and Electrofusion in Cell Biology" (E. Neumann, A. Sowers, and C. Jordan, eds.), pp. 111–126. Plenum Press, New York.

Weaver, J. C. (1990). Electroporation: A new phenomenon to consider in medical technology. *In* "Emerging Electromagnetic Medicine" (M. E. O'Connor, R. H. C. Bentall, and J. C. Monahan, eds.), pp. 81–102. Springer-Verlag, Heidelberg.

Weaver, J. C., Harrison, G I., Bliss, J. G., Mourant, J. R., and Powell, K. T. (1988). Electroporation: High frequency of occurrence of the transient high permeability state in red blood cells and intact yeast. *FEBS Lett.* **229**, 30–34.

Weaver, J. C., Barnett, A., and Bliss, J. G. An approximate theoretical model for isolated cell electroporation: Quantitative estimates of electrical behavior (submitted).

8

Mechanisms of Electroporation and Electrofusion

Arthur E. Sowers

Department of Biophysics, School of Medicine, University of Maryland at Baltimore, Baltimore, Maryland 21201

I. Introduction and Scope
II. Electroosmosis in Electropores
III. Electrofusion and Electroporation Protocols
IV. Criteria for Membrane Fusion
V. New Fusion Product Understanding and Its Implications for Electrofusion
VI. Factors That Influence Electrofusion
    A. Pulse Parameters
    B. Membrane Structure
    C. Ionic Strength
    D. Changes Induced by Metabolism
    E. Presence of Low Concentrations of Aqueous-Soluble Macromolecules
    F. Heterogeneity of Fusion Substrate Membranes
    G. Dielectrophoretic Force
    H. Other Effects
VII. Misconceptions
    References

I. Introduction and Scope

The purpose of this chapter is to summarize what has been learned about the mechanisms of electrofusion and electroporation using erythrocyte ghosts as model membranes. This chapter emphasizes general features of the processes, explains the methodological considerations necessary, lists the known factors and their roles, provides practical advice, and reviews other relevant background material. Our studies have been directed primarily at learning more about the electrofusion mechanism by identifying those variables and factors that play major or minor roles in determining whether or not a given pair of membranes in contact will fuse. Past speculations (Pilwat *et al.*, 1981) on the involvement of electropores in the fusion mechanism have required that effort also be applied to the mechanism of electroporation, since part of this phenomenon may be related to the electrofusion phenomenon. Our observations about the occurrence of electroosmosis (see below) in electropores suggest approaches that may be beneficial in electroporation applications.

Because viable cells are complex and generate a wide variety of physiological responses to various stimuli, we have chosen to work with erythrocyte ghost membranes (Dodge *et al.*, 1963). The erythrocyte ghost (i.e., free of most if not all hemoglobin and soluble enzymes) is a convenient model for cell membranes because (1) it is comparatively easy to obtain preparations that are free or nearly free of cytoplasmic elements that can interfere directly or indirectly with membrane properties, (2) much is known about the membrane and many protocols have been developed for use with it, (3) a single hemolytic hole is produced during lysis (Lieber and Steck, 1982a, 1982b), which can be used to load the cytoplasmic compartment with soluble molecules for labeling purposes, but that hole is unlikely to interfere with most electrofusion and electroporation experiments, (4) under a range of conditions, the membrane is spherical, which simplifies mathematical modeling, (5) ghost membrane preparations from erythrocytes from a variety of mammalian species have markedly different lipid compositions (Hanahan, 1969; Nelson, 1967), which provides a natural variable to study, (6) most have diameters in the 6–8 μm range, making them not only ideal for light microscopy but also not susceptible to the differences in pulse-induced transmembrane voltage that would be expected if their diameters were dissimilar (see below), (7) they have all representative components of more complex cells (integral proteins, lipids—including cholesterol—a glycocalyx, a spectrin-based cytoskeleton, and a trans-leaflet asymmetry), and (8) it does not need a viability test after pulse treatment.

While space limitations preclude extensive discussions of other aspects of electrofusion and electroporation, additional information can be obtained from other recent reviews (Bates *et al.*, 1987; Chassy *et al.*, 1988; Hofmann and Evans, 1986; Pohl *et al.*, 1984; Potter, 1988; Sowers 1989a, 1989d; Zimmermann, 1986), a recent monograph (Neumann *et al.*, 1989), and a handbook for hybridoma protocols (Borrebaeck and Hagen, 1989). Earlier reviews dealing with electrofusion (Zimmermann 1982) and electroporation (Tsong, 1983) also contain much material of significance that is often omitted in the later literature.

II. Electroosmosis in Electropores

Information about electropore effective diameter can be obtained by the use of probe molecules with different diameters (e.g., Kinosita and Tsong, 1977). Electropores (or locations of electropermeabilization) have other properties that should be measurable. These include, for example, densities (electropores per unit area), locations on a membrane, lifetimes, and heterogeneity. These properties are only approximately and indirectly known. Three reports (Mehrle *et al.*, 1985; Rossignol *et al.*, 1983; Sowers and Lieber, 1986; shown in order), using different methodologies, suggest that the each of the two hemispheres of a general cell experiences qualitatively different permeability increases (Fig. 1).

Our study (Sowers and Lieber, 1986) contradicted the direction but not the differences between the hemispheres. In that study, FITC-dextran loaded into erythrocyte ghosts was observed to appear as a plume or cloud just outside the negative-facing hemisphere of each ghost membrane in the first 100–200 ms after a pulse. The origin for this phenomenon became clearer when the possibility of electroosmosis (Bedzyk *et a.*, 1990; McLaughlin, 1989; McLaughlin and Mathias, 1985), a well-studied and understood phenomenon, was considered. It has been known for nearly a century that the electrical double layer—including a thin layer between the bulk aqueous liquid and the surface—exists in any solution at the planar solution/solid interface when the solid interface carries fixed charges and the solution contains mobile ions of both charges (Fig. 2). An electric field parallel to the interface will cause a net hydrodynamic flow in the appropriate direction as long as there is an imbalance in the numbers of the two charges in the layer of liquid adjacent to the charged surface. If electropores, which are expected to be induced closer to the "poles" of the cell that face the electrodes, are viewed as cylinders with an average net negative (from ionized headgroups of phospholipids and ionized amino acid side chains on integral proteins) charge on this surface and with their axis perpendicular to the plane of the membrane, then a hydrodynamic flow would be expected during

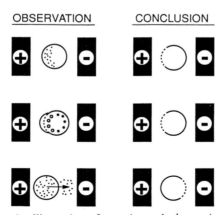

Figure 1 Illustration of experimental observation, spatial orientation, and conclusions regarding the permeabilization asymmetry effect (see text).

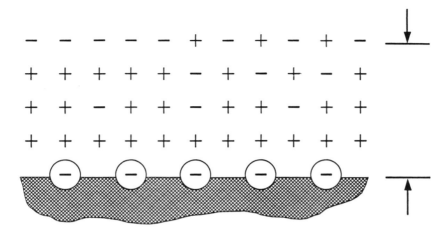

Figure 2 Charge density in an aqueous solution near a surface with fixed (negative) charges. Arrow shows thickness of electrical double layer where net polarity of counterions is positive and an electrically neutral bulk aqueous phase above the double layer.

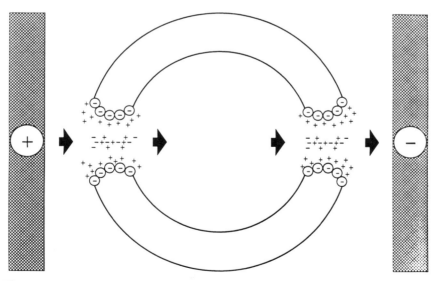

Figure 3 Illustration of basis for electroosmosis-induced hydrodynamic flow toward the negative electrode regardless of whether electropores are in positive-facing or negative-facing hemispheres.

the electric field pulse (Fig. 3). It was predicted and experimentally demonstrated that the overall permeabilization difference between both hemispheres would be less than originally thought if an electroporation experiment were conducted to take electroosmosis into account (Sowers, 1988). Thus the contradiction between our observation of increased permeability preferentially in the negative-facing hemisphere (Sowers and Lieber, 1986) and the observations of (Mehrle *et al.*, 1985; Rossignol *et al.*, 1983) is actually a manifestation of where the indicating marker is located before the pulse. Both hemispheres will be permeabilized in all cases, but if the marker is outside the cells then only permeabilization in the positive-facing hemisphere will be detectable, and vice versa if the marker is inside the cell. Although we recognize the possibility of asymmetric forces and some asymmetry in permeabilization, a properly conducted experiment demonstrated that the originally observed strong permeabilization asymmetry (Sowers and Lieber, 1986; and possibly Mehrle *et al.*, 1989) is actually an artifact (Sowers, 1988). We therefore think a major contribution to the understanding of the "classical" electropore and its properties is the reproducible and general observation of a aqueous liquid flow perpendicular to the plane of the membrane, which suggests that electroosmosis accompanies at least part of the electroporation process.

Further studies using intensified fluorescence microscopy to continuously monitor the intensity of the concentration of a soluble marker in single erythrocyte ghosts showed (Dimitrov and Sowers, 1990a) that there was a large-scale movement of marker during the pulse followed by a small-scale movement of marker at all times after the pulse (Table 1). Calculation showed that the rate of loss of marker during the rapid-short interval phase was $I/t = (1.0 - 0.67)/(0.16 \text{ s}) = 200\%/\text{s}$, while the rate of loss of marker during the slow-long interval was $I/t = 6\%/\text{s}$ (from Table 1). The phenomenon of video "lag" (Inoue, 1986) and a visually observable hydrodynamic inertia effect are two artifacts known to be present, however. An analysis of the effects of these artifacts on our data is likely to show a greater difference between the rate of loss during these two efflux phases. Nevertheless, the effect of the rate differences during the two phases is qualitatively consistent with what has been measured for transfection frequencies when DNA is added *before* compared to *after* the pulse treatment (Xie *et al.*, 1990). Hence, a larger transfection frequency may be due to the greater volume transport during the *electroosmosis-based* phase of electroporation compared to the postpulse *diffusion-based* phase of electroporation. This supports the practical recommendation that a greater amount of loading, or unloading, of a cell will take place *during* a pulse rather than *after* the pulse.

If electroosmosis occurs during electroporation, then it can be predicted that increases in buffer ionic strength and the presence of divalent cations, known to decrease the thickness of the double layer (McLaughlin, 1989; McLaughlin and Mathias, 1985), should thus decrease the volume flow during the pulse. This was also shown to be the case (Table 2) in semiquantitative measurements of the large-scale electroosmosis-based movement of a soluble fluorescent marker out of labeled

Table 1

Biphasic Electroporation-Induced Loss of Fluorescence from FITC Dextran-Loaded Erythrocyte Ghosts[a]

Time[b]	Condition	Instantaneous contents[c]
<0.00 s	Reference point	1.00
0.00 s	Single pulse[d] applied	N/A
0.16 s	Rapid loss stops	0.67
2.0 s	Slow loss continues	0.57

[a]From Dimitrov and Sowers (1990a).
[b]Time measured from continuously updated digital alphanumerics placed on each video frame and shifted to show 0.00 s as the reference point.
[c]Measured in terms of quantitative fluorescence in given video frame at center of erythrocyte ghost image. Intensity information train is composed of fluorescence optical path + Zeiss Venus camera + VCR + Colorado Video 321 video analyzer. Single-frame playback on the VCR was based on mechanical scanning of the same magnetic track on the video tape; and the analog meter measuring the output signal from the video analysis of the single video frame image had to be time-integrated to dampen signal noise. Overall system contrast transfer function was linear to within 10% measured range.
[d]Field strength 700 V/mm; decay half-time 1.0 ms.

Table 2

Relationship between Electroporation of 9–10 kD FITC-Dextran and Solution Buffer Strength and Presence of Divalent Cations[a] (Dimitrov and Sowers, 1990a).

Buffer	Strength	Loss due to pulse
$NaPO_4$	20 mM	20.6%
$NaPO_4$	60 mM	8.4%
$NaPO_4$	10 mM ⎫	3.8%
$MgCl_2$	10 mM ⎭	

[a]From Dimitrov and Sowers (1990a). Buffer (pH 8.5). Loss was determined with respect to prepulse intensity at center of spherical-shaped erythrocyte ghost *and* only during the electroosmosis-based phase of electroporation. ($NaPO_4$ = sodium phosphate from mixture of both monobasic and dibasic compounds.)

erythrocyte ghosts (Dimitrov and Sowers, 1990a). This is also consistent with at least two transfection studies in which the effect of an increase in buffer strength or divalent cation concentration resulted in smooth decreases in transfection (Neumann et al., 1982; Miller et al., 1988). It is acknowledged that there is other experimental evidence of much more complex phenomena during electric field-induced electroporation experiments (Cymbalyuk et al., 1988; Chernomordik et al., 1990). All data, including ours, suggest that electroporation protocols for transfection should include experiments to adjust buffer strength as it relates to ionic strength and to optimize survival/viability against transfection. Osmotic pressure can be provided by nonionic soluble molecules (e.g., sucrose).

While the pulse-induced permeability increase appears to be very rapid, the restoration of the prepulse permeability is slower, temperature dependent, and may not be completely reversible (Schwister and Deuticke, 1985; Serpersu et al., 1985).

III. Electrofusion and Electroporation Protocols

Electroporation simply requires that a means be provided to transiently expose cells to a strong electric field. This means that the minimum requirements include a chamber and a strong electric field pulse generator. The cells can be in suspension or attached to a substratum. However, before membranes can be electrofused, they must be brought into contact and maintained in contact until the pulse can be applied. Many methods are available. They include porous media (Sukharev et al., 1990), micromanipulation (Senda et al., 1979), chemical aggregation (Conrad et al., 1987; Lo et al., 1984; Lo and Tsong, 1989), dielectrophoresis (Burt et al., 1990; Iglesias et al., 1989; Pohl, 1978), and growth to confluence in a monolayer (Ishizaki et al., 1989; Teissie et al., 1982). The most common protocol involves the establishment of close membrane-membrane contact before the fusogenic pulse is applied. It is also possible to apply fusogenic electric field pulses *before* the membranes are brought into contact and still obtain fusion (Sowers, 1986, 1987). This phenomenon has been observed independently in two other laboratories (Montane et al., 1990; Teissie and Rols, 1986; Tsoneva et al., 1988) and is much less well understood or characterized. Additional discussion may be found in our previous reviews (Sowers, 1989a, 1989d).

IV. Criteria for Membrane Fusion

There is a prevalent misunderstanding with regard to word usage and understanding of the terms "cell fusion" and "membrane fusion" (Duzgunes and Bentz, 1988; Knutton and Pasternak, 1979; Morris et al., 1988). Two cells may combine to form one in a process that really has two parts (Fig. 4). The first is membrane fusion,

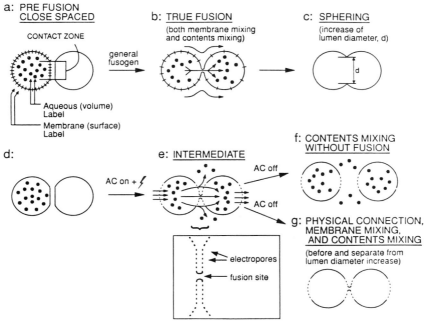

a: PRE FUSION
   CLOSE SPACED

CONTACT ZONE

general
fusogen

Aqueous (volume)
Label
Membrane (surface)
Label

b: TRUE FUSION
(both membrane mixing
and contents mixing)

c: SPHERING
(increase of
lumen diameter, d)

d

d:

AC on + ⚡

e: INTERMEDIATE

AC off

AC off

f: CONTENTS MIXING
   WITHOUT FUSION

electropores

fusion site

g: PHYSICAL CONNECTION,
   MEMBRANE MIXING,
   AND CONTENTS MIXING
(before and separate from
lumen diameter increase)

Figure 4   Macroscopically observable events as they rigorously relate to functional membrane fusion (a microscopic event). Two membranes, one labeled with both a lipid-soluble (spikes perpendicular to plane of membrane) and an aqueous-soluble label (solid circles), at "a" share close contact over a "contact zone" and fuse at "b" and allow membrane mixing (lateral two-dimensional diffusion) of lipid-soluble label and contents mixing (aqueous three-dimensional diffusion) of labels into originally unlabeled membrane on right. The single fusion site in "b" has a small lumen that expands, *after membrane fusion*, in diameter d, in "c" as the fused doublet changes, if unrestricted, in morphology toward the spherical (cell fusion) morphology. Electroosmosis-induced artifact that can propel aqueous molecules in left member of two membranes in "d" into originally unlabeled membrane in "e" through electropores as well as through fusion pores (see inset) can be distinguished in postpulse test for absence in "f" or presence in "g" of a physical connection between membranes by removing the dielectrophoretic force and waiting for Brownian motion to separate membranes not connected (Sowers, 1988).

which can occur but not be immediately obvious by, for example, phase contrast optics. The morphology of two fusing cells may eventually become spherical as they round up or otherwise form a syncytium (a mass of cytoplasm with multiple nuclei present surrounded by one plasma membrane). Membrane fusion can occur without cell fusion, but not vice versa (see Wojcieszyn *et al.*, 1983). The most rigorous evidence for membrane fusion would be provided if three criteria were satisfied: (1)

membrane mixing, (2) contents mixing, and (3) physical attachment. It has been shown that evidence for membrane mixing can appear but actually be an artifact (Szoka, 1987). We discovered (Sowers, 1988) that electroosmosis caused an artifactually high measure of fusion yield when contents mixing results were compared to membrane mixing results (cf. Fig. 4f and g). Adding an additional precautionary step to the assay (the criterion of physical attachment) eliminated the problem of this artifact, but then the assay was not a true contents mixing assay. Rather, the label used became, in effect, a membrane marker rather than a contents mixing marker. Further information on this problem will be found elsewhere (Chernomordik and Sowers, 1991). Indeed, careful reflection will reveal ways in which some of these criteria may be circumvented. Therefore, there is an absolute need to consider what is being measured in experiments involving a fusion assay. Assays must not only be accurate, but also be valid.

V. New Fusion Product Understanding and Its Implications for Electrofusion

Recent experiments (Chernomordik and Sowers, 1991) now clearly show that electrofusion of two spherical-shaped membranes in contact can proceed along two qualitatively different pathways and end in one of two classes of stable postfusion products: (1) single giant spheres, and (2) chains of polyspheres (Fig. 5). Regardless of the pathway and the criterion for fusion, both pathways will satisfy all three fusion criteria (see above). The factor responsible for these two classes of postfusion product correlates with physicochemical treatments (temperature, low ionic strength, and low pH) known to denature spectrin or cause it to be released from its attachment site on the cytoplasmic face of the plasma membrane. It is not known if the fusion products as induced in the first 10–200 ms after the pulse are identical or not. Thin-section electron microscopy of the stable polyspheres showed that the lumens had a stable perforated septum or diaphragm (Figs. 5 and 6) as a physical structure that was responsible for preventing the lumen diameter from expanding to the single giant sphere morphology. Thus, membrane fusion may not occur at a single point, but rather in a *fusion zone* where multiple fusion sites are induced. This is consistent with our finding of small vesicular fusion products inside fused mitochondrial inner membranes (Sowers, 1983). Internal vesiculation during membrane fusion was also reported in plant protoplasts (Vienken et al., 1983), but the time scale ($1\frac{1}{2}$–5 min) for this effect makes it more likely to be a physiological regulation response rather than a response due to a combination of membrane properties and a biophysical effect of the electric field on the membrane structure. This reveals that additional forces and experimental conditions may be involved in controlling postfusion morphologies.

The implications of the multiple fusion sites (Fig. 6) in the fusion zone and the

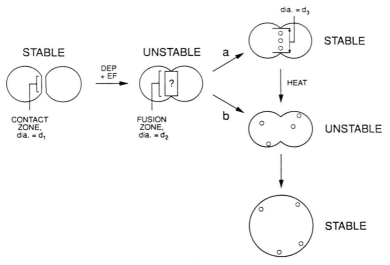

Figure 5 Pairs of erythrocyte ghosts sharing the same contact zone (left) will develop a fusion zone with an unstable but as yet undetermined ultrastructure (center), which can either expand in diameter (path a) and end in a doublet (upper right) in which the fusion zone (see Fig. 6) diameter stabilizes before forming a spherical endproduct morphology, or form an open lumen that expands in diameter (path b) and ends in a spherical morphology with small vesicles bound to the inner surface.

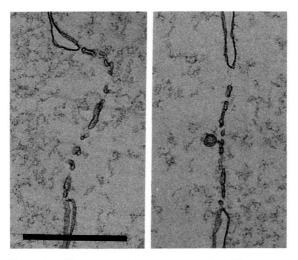

Figure 6 Thin-section electron micrographs of fusion zones illustrated in Fig. 5 show diaphragms of parallel double membranes perforated at approximately periodic intervals by fusion sites. Scale bar is 2.0 μm.

appearance of some periodicity in their location along the septum may be a clue to their formation. Fusion sites may be a consequence of a biophysical effect of the transmembrane electric field applied perpendicular to the plane of the membrane. Forces may be transmitted in the direction of the plane of the membrane, and modulated by the composition and properties of the membrane, and may be converted to a dissipation of this process at a site within some average unit area of the membrane. It is not known if area outside the contact zone (Figs. 4–6) comes into the area of the fusion zone during or after the pulse.

VI. Factors That Influence Electrofusion

Learning what factors have major or minor effects on the fusion process (as manifested in fusion yield, or some fusion-related property) provides clues to the mechanism. Our efforts have been directed at measuring effects on fusion yield as function of other physicochemical variables. Sometimes the effect of a particular variable is also dependent on other variables. The study of these variables is presently in progress. These variables are discussed individually and approximately in the order of the strengths of their effects as if the fusogen (the pulse) were held constant.

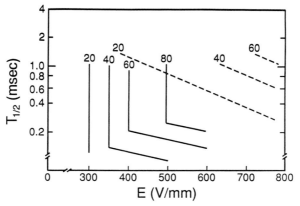

Figure 7   Single, exponentially decaying pulse-induced iso-fusion yield (numbers in percent) contours for rabbit (solid lines) and human (dotted lines) erythrocyte ghosts in sodium phosphate buffer (20 mM, pH 8.5) as a function of pulse field strength, E and pulse duration in terms of decay half-time (ms).

A. Pulse Parameters

The most critical parameters that control electrofusion yield are the specification of the electric field pulse in terms of its strength E (in the general range of 200–1000 V/mm) and duration. Duration must be accurately specified. Exponentially decaying pulses have a decay half-time, $T_{1/2}$ (usually in the range of 0.5–1.2 ms), or, for square-wave pulses, a pulse width (usually in the range 20–200 μs). Adjustment of these factors can have at least a two-order-of-magnitude effect on fusion yield (Fig. 7). Data have been obtained to show that a reciprocity effect is present in electroporation (Chu *et al.*, 1987) and electrofusion (Fig. 2 in Sowers, 1989d; Zhelev *et al.*, 1988). This means that, over some range, compensation for a reduction in pulse duration can be made by increasing pulse strength, and vice versa. Most experiments utilize either a square pulse waveform or an exponentially decaying pulse waveform. Some comparisons of data from exponential and square (Kubiniec *et al.*, 1990; Liang *et al.*, 1988; Saunders *et al.*, 1989) pulses in the same experimental paradigm have been published. Also, data from use of complex waveforms or pulse protocols exist (Chang, 1989). However, the final picture is unclear since other variables known to be important were not extensively studied.

B. Membrane Structure

Our studies show that substantially different fusion yields can be obtained when (1) the buffer is the same, and (2) the pulse is the same, but the membranes that are otherwise qualitatively *similar* in diameter are from *different* species (Sowers, 1989b), which are known to have different lipid compositions (Hanahan, 1969; Nelson, 1967). Erythrocyte ghost membranes from rabbits are not only more fusible by as much as one to two orders of magnitude than erythrocyte ghosts from humans, but are more fusible according to a different pulse regime (Fig. 7). While the fusion yield is independent of the pulse duration over one part of the plot, the other part of the plot shows that the fusion yield has a reciprocal strength/duration relationship. Thus it is essential that the observable variable (e.g., fusion yield) be rigorously measured as a function of *both* the pulse field strength *and* the pulse duration, because it will be the only way to discover the qualitative and quantitative dependencies of the fusion responses of the given membrane (Sowers, 1989a, 1989d). This author is not aware that similar phenomena have been reported to accompany electroporation studies, but the possibility suggests itself that transfection and survival should be measured against adjustments in *both* field strength and duration, despite the additional effort needed. Transfection, survival, and subsequent growth may be differently affected by pulse strength and duration. The factors that are responsible for the above effects are not known but could be linked with specific membrane components (e.g., lipid or integral protein) or some general membrane property (e.g., surface charge density, mechanical viscoelasticity), as would be expected to derive from the assemblage of all components together.

C. Ionic Strength

We have observed that fusion yield in erythrocyte ghosts has a peak with sodium phosphate buffer (pH 8.5) in the concentration range 20–30 mM (Sowers, 1989b). We suspect that the decrease in fusion yield at buffer concentrations >60 mM may be due to the well-known diminution in the dielectrophoretic force at higher solution conductivities and at concentrations <15 mM because repulsion (see Fig. 7 in Ohki, 1987) between charged surfaces will be greater at lower ionic strengths (McLaughlin, 1989; McLaughlin and Mathias, 1985). Since we observe fusion yield to diminish to zero or below practical detection limits around buffer (sodium phosphate, pH 8.5) strengths of 2–5 mM, we consider the ionic strength factor as another major factor in controlling fusion yield.

Inconsistencies in reports from other investigators using a variety of cultured (nucleated) cells indicate that still other factors play poorly understood roles. For example, Sukharev *et al.* (1990) showed that fibroblast electrofusion yield was proportional to the concentration of NaCl or KCl in the medium. On the other hand, studies using CHO (Chinese hamster ovary) cells (Blangero and Teissie, 1985; Rols and Teissie, 1989) showed a roughly *inversely* proportional relationship between fusion and ionic strength. It is necessary to realize, however, that phase optics-based visual assays were sometimes used in these studies, and events may be scored that are secondary to or partly separate from true fusion.

D. Changes Induced by Metabolism

We found that *in vitro* incubation of intact erythrocytes (which allowed glycolysis-based pathways and membrane potential generation to continue) at 0–4°C for various periods before ghost membranes were prepared showed that fusion yield was reproducibly modulated by a factor of 2–3, and in one case up to a factor of 10 (Sowers, 1989c). Since the effects of cytoplasmic and membrane metabolic pathways are arrested upon ghost preparation, the only cause of the change in fusion yield can be from some structural change made in the membrane itself as a consequence of the *in vitro* storage treatment.

E. Presence of Low Concentrations of Aqueous-Soluble Macromolecules

Low concentrations of residual hemoglobin and low concentrations of bovine serum albumin and dextran (70 kD) in the cytoplasmic compartment or in the compartment external to the ghost membranes were shown to enhance fusion yield by up to about 30% (Sowers, 1990). A similar modulating effect from the presence of soluble macromolecules (sucrose, polyethylene glycol) outside of cells has been reported for electroporation (Rols and Teissie, 1990). The mechanism by which this occurs is completely unknown.

F. Heterogeneity of Fusion Substrate Membranes

A pair of membranes in contact that are about to be fused can be identical (i.e., A + A = homofusion) or dissimilar (i.e., A + B = heterofusion). In heterofusion of rabbit + human erythrocyte ghosts, the fusion yields are midway between rabbit + rabbit and human + human homofusion (Sowers, 1989b). However, when adult chicken erythrocyte ghosts and chicken embryo erythrocyte ghosts were used in the three combinations, the result was more complex (Sowers and Kapoor, 1988).

G. Dielectrophoretic Force

Generally, the dielectrophoretic force, which holds membranes in pearl chains and presses them in such a way as to create flat parallel contact areas (or zones), is a complex but generally proportional function of the alternating electric field strength (Pohl, 1978; Sauer, 1983). We have shown that at higher dielectrophoretic field strengths, fusion yield is measurably higher (Fig. 8), although slightly (10%) so,

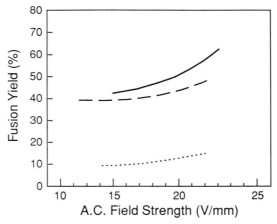

Figure 8 Fusion yield in rabbit erythrocyte ghosts as a function of buffer (sodium phosphate, pH 8.5) strength and alternating (60 Hz) electric field strength. Dotted line, 10 mM; dashed line, 20 mM; solid line, 60 mM. Alternating field strength is given in root-mean-square V/mm.

in a reproducible way and is unlikely to be a consequence of solution heating by the alternating current (Sowers, 1989b). However, this effect is also dependent in a complex way on buffer strength, and a full analysis remains to be done. A preliminary analysis has been published (Dimitrov et al., 1990).

H. Other Effects

Effects of pH on fusion yield are relatively small in erythrocyte ghosts (Fig. 5 in Sowers and Kapoor, 1988) and CHO cells (Blangero and Teissie, 1985), but larger for intact erythrocytes (Chang et al., 1989). Spectrin integrity appears to have a strong effect on fusion product morphology (see below) but much less of an effect on fusion yield (Chernomordik and Sowers, 1991). While early studies (Sowers, 1984; Zimmermann, 1982) sometimes used more than one pulse to achieve fusion when fusion yields were unacceptably low, we found that adjustment of field strength and decay half-time for optimum fusion yield (Sowers, 1989a, 1989d) would provide high yields with only one pulse.

VII. Misconceptions

A number of misconceptions have persisted in the literature. Early papers (e.g., Zimmermann, 1982) that included the use of dielectrophoresis to induce close membrane–membrane contact before the application of the fusogenic pulse have emphasized that (1) low-conductivity media, (2) high-frequency (10^6 Hz) alternating electric fields, and (3) a bulk nonhomogeneous field must be used or dielectrophoresis will not take place. We have shown these requirements to be conditional and not absolute. Further discussion, which is beyond the scope of this chapter, will be found in our earlier reviews (Sowers, 1989a, 1989d) and other papers (Burt et al., 1990; Dimitrov et al., 1990; Kaler and Jones, 1990). Often light optics microscopy has been used to monitor fusion. This is acceptable when there is a one-to-one relationship between membrane fusion and cell fusion. But if this condition is not fulfilled, then fusion yield may be underestimated. We have evidence that this is a critical factor (Sowers, 1988, 1989a; Chernomordik and Sowers, 1991) that requires rigorous consideration.

Acknowledgment

Supported by ONR grant N00014-J-89-1715.

References

Bates, G., Saunders, J., and Sowers, A. E. (1987). Electrofusion: Principles and applications. *In* "Cell Fusion," A. E. Sowers, ed., Plenum Press, New York, pp. 367–395.

134   Arthur E. Sowers

Bedzyk, M. J., Bommarito, G. M., Caffrey, M., and Penner, T. L. (1990). Diffuse-double layer at a membrane-aqueous interface measured with x-ray standing waves. *Science* **248**, 52–56.

Blangero, C., and Teissie, J. (1985). Ionic modulation of electrically-induced fusion of mammalian cells. *J. Membrane Biol.* **86**, 247–253.

Borrebaeck, C. A. K., and Hagen, I. (eds.) (1989). "Electromanipulation in Hybridoma Technology—A Laboratory Manual." Stockton Press, New York.

Burt, P. H. J., Pethig, R., Gascoyne, P. R. C., and Becker, F. F. (1990). Dielectrophoretic characterisation of friend murine erythroleukaemic cells as a measure of induced differentiation. *Biochim. Biophys. Acta* **1034**, 93–101.

Chang, D. C. (1989). Cell poration and cell fusion using an oscillating electric field. *Biophys. J.* **56**, 641–652.

Chang, D. C., Hunt, J. R., and Gao, P.-Q. (1989). Effects of pH on cell fusion induced by electric fields. *Cell Biophys.* **14**, 231–243.

Chassy, B. M., Mercenier, A., and Flickinger, J. (1988). Transformation of bacteria by electroporation. *TibTech* **6**, 303–309.

Chernomordik, L. V., and Sowers, A. E. (1991). Evidence that the spectrin network and a non-osmotic force control the fusion product morphology in electrofused erythrocyte ghosts. *Biophys. J.* in press.

Chernomordik, L. V., Sokolov, A. V., and Budker, V. G. (1990). Electrostimulated uptake of DNA by liposomes. *Biochim. Biophys. Acta* **1024**, 179–183.

Chu, G., Hayakawa, H., and Berg, P. (1987). Electroporation for the efficient transfection of mammalian cells with DNA. *Nucleic Acids Res.* **15**, 1311–1326.

Conrad, M. K., Lo, M. M. S., Tsong, T. Y., and Snyder, S. H. (1987). Bioselective cell-cell fusion for antibody production. *In* "Cell Fusion" (A. E. Sowers, ed.), Plenum Press, New York, pp. 427–439.

Cymbalyuk, E. S., Chernomordik, L. V., Broude, N. E., and Chizmadzhev, Yu. A. (1988). Electrostimulated transformation of *E. coli* cells pretreated by EDTA solution. *FEBS Lett.* **234**, 203–207.

Dimitrov, D. S., and Sowers, A. E. (1990a). Membrane electroporation—Fast molecular exchange by electroosmosis. *Biochim. Biophys. Acta* **1022**, 381–392.

Dimitrov, D. S., Apostolova, M. A., and Sowers, A. E. (1990). Attraction, deformation and contact of membranes by a 60 Hz sine wave electric field. *Biochim. Biophys. Acta* **1023**, 389–397.

Dodge, J. T., Mitchell, C., Hanahan, D. J. (1963). The preparation and chemical characterization of hemoglobin-free ghosts from human erythrocytes. *Arch. Biochem. Biophys.* **100**, 119–130.

Duzgunes, N., and J. Bentz. (1988). Fluorescence assays for membrane fusion. *In* "Spectroscopic Membrane Probes" (L. M. Loew, ed.), Vol. 1, CRC Press, Boca Ratan, FL, pp. 117–159.

Hanahan, D. J. (1969). Characterization of the erythrocyte membrane. *In* "Red Cell Membrane Structure and Function" (G. A. Jamieson and T. J. Greenwalt, eds.), Lippencott, Philadelphia, pp. 83–92.

Hofmann, G. A., and Evans, G. A. (1986). Electronic genetic-physical and biological aspects of cellular electromanipulation. *IEEE Eng. Med. Biol. Mag.* **6**, 6–25.

Iglesias, F. J., Santamaria, C., Lopez, M. C., and Domingues, A. (1989). Dielectrophoresis:

Behavior of microorganisms and effect of electric fields on orientation phenomena. *In* "Electroporation and Electrofusion in Cell Biology" (E. Neumann, A. E. Sowers, and C. A. Jordan, eds.), Plenum Press, New York, pp. 37–57.

Inoue, S. (ed.) (1986). "Video Microscopy." Plenum Press, New York.

Ishizaki, K., Chang, H. R., Eguchi, T., and Ikengae, M. (1989). High voltage electric pulses efficiently induce fusion of cells in monolayer culture. *Cell Struct. Func.* 14, 173–181.

Kaler, K. V. I. S., and Jones, T. B. (1990). Dielectrophoretic spectra of single cells determined by feedback-controlled levitation. *Biophys. J.* 57, 173–182.

Kinosita, Jr., K., and Tsong, T. Y. (1977). Formation and resealing of pores of controlled sizes in human erythrocyte membranes. *Nature* 268, 438–441.

Knutton, S., and Pasternak, C. A. (1979). The mechanism of cell-cell fusion. *Trends Biochem. Sci.* 4, 220–223.

Kubiniec, R. T., Liang, H., and Hui, S. W. (1990). Effect of pulse length and pulse strength on transfection by electroporation. *Biotechniques* 8, 16–20.

Liang, H., Prurucker, W. J., Stenger, D. A., Kubiniec, R. T., and Hui, S.-W. (1988). Uptake of fluorescence-labeled dextrans by 10T 1/2 fibroblasts following permeation by rectangular and exponential decay electric field pulses. *Biotechniques* 6, 550–558.

Lieber, M. R., and Steck, T. L. (1982a). A description of the holes in human erythrocyte membrane ghosts. *J. Biol. Chem.* 257, 11651–11659.

Lieber, M. R., and Steck, T. L. (1982b). Dynamics of the holes in human erythrocyte membrane ghosts. *J. Biol. Chem.* 257, 11660–11666.

Lo, M. M. S., and Tsong, T. Y. (1989). Producing monoclonal antibodies by electrofusion. *In* "Electroporation and Electrofusion in Cell Biology." (E. Neumann, A. E. Sowers, and C. A. Jordan, eds.), Plenum Press, New York, pp. 259–270.

Lo, M. M. S., Tsong, T. Y., Conrad, M. K., Strittmatter, S. M., Hester, L. D., and Snyder, S. H. (1984). Monoclonal antibody production by receptor-mediated electrically-induced cell fusion. *Nature (Lond.)* 310, 792–794.

McLaughlin, S., and Mathias, R. T. (1985). Electro-osmosis and the reabsorption of fluid in renal proximal tubules. *J. Gen. Physiol.* 85, 699–728.

McLaughlin, S. (1989). The electrostatic properties of membranes. *Annu. Rev. Biophys. Biophys. Chem.* 18, 113–136.

Mehrle, W., Zimmermann, U., and Hampp, R. (1985). Evidence for asymmetrical uptake of fluorescent dyes through electro-permeabilized membranes of *Avena* mesophyll protoplasts. *FEBS Lett.* 185, 89–94.

Mehrle, W., Hampp, R., and Zimmermann, U. (1989). Electric pulse induced membrane permeabilisation. Spatial orientation and kinetics of solute efflux in freely suspended and dielectrophoretically aligned plant mesophyll protoplasts. *Biochim. Biophys. Acta* 978, 267–275.

Miller, J. F., Dower, W. J., and Tompkins, L. S. (1988). High-voltage electroporation of bacteria: genetic transformation of *Campylobacter jejuni* with plasmid DNA. *Proc. Natl Acad. Sci. USA* 85, 856–860.

Montane, M.-H., Dupille, A., Alibert, G., and Teissie, J. (1990). Induction of a long-lived fusogenic state in viable plant protoplasts permeabilized by electric fields. *Biochim. Biophys. Acta* 1024, 203–207.

Morris, S. J., Bradley, D., Gibson, C. C., Smith, P. D., and Blumenthal, R. (1988). Use of membrane-associated fluorescence probes to monitor fusion of bilayer vesicles: Application to rapid kinetics using pyrene excimer/mpnpmer fluorescence. *In* "Spectroscopic Membrane Probes" (L. M. Loew, ed), CRC Press, Boca Ratan, FL, pp. 161–191.

Nelson, G. J. (1967). Lipid composition of erythrocytes in various mammalian species. *Biochim. Biophys. Acta* 144, 221–232.

Neumann, E., Sowers, A. E., and Jordan, C. (eds.) (1989). "Electroporation and Electrofusion in Cell Biology." Plenum Press, New York.

Neumann, E., Schaefer-Ridder, M., Wang, Y., and Hofschneider, P. H. (1982). Gene transfer into mouse lyoma cells by electroporation in high electric fields. *EMBO J.* 1, 841–845.

Ohki, S. (1987). Physico chemico factors underlying lipid membrane fusion. *In* "Cell Fusion" (A. E. Sowers, ed.), Plenum Press, New York, pp. 331–352.

Pilwat, G., Richter, H.-P., and Zimmermann, U. (1981). Giant culture cells by electric field-induced fusion. *FEBS Lett.* 133, 169–174.

Pohl, H. A. (1978). "Dielectrophoresis." Cambridge University Press, London.

Pohl, H. A., Pollock, K., and Rivera, H. (1984). The electrofusion of cells. *Int. J. Quant. Chem. Quant. Biol. Symp.* 11, 327–345.

Potter, H. (1988). Electroporation in biology: Methods, application, and instrumentation. *Anal. Biochem.* 174, 361–373.

Rols, M.-P., and Teissie, J. (1990). Modulation of electrically induced permeabilization and fusion of Chinese hamster ovary cells by osmotic pressure. *Biochemistry* 29, 4561–4567.

Rols, M.-P., and Teissie, J. (1989). Ionic-strength modulation of electrically induced permeabilization and associated fusion of mammalian cells. *Eur. J. Biochem.* 179, 109–115.

Rossignol, D. P., Decker, G. L., Lennarz, W. J., Tsong, T. Y., and Teissie, J. (1983). Induction of calcium-dependent, localized cortical granule breakdown in sea-urchin eggs by voltage pulsation. *Biochim. Biophys. Acta* 763, 346–355.

Sauer, F. A. (1983). Forces on suspended particles in the electromagnetic field. *In* "Coherent Excitations in Biological Systems" (H. Frohlich and F. Kremer, eds.), Springer Verlag, New York, pp. 134–144.

Saunders, J. A., Smith, C. R., and Kaper, J. M. (1989). Effects of electroporation pulse wave on the incorporation of viral RNA into tobacco protoplasts. *Biotechniques* 7, 1124–1131.

Schwister, K., and Deuticke, B. (1985). Formation and properties of aqueous leaks induced in human erythrocytes by electrical breakdown. *Biochim. Biophys. Acta* 816, 332–348.

Senda, M., Takeda, J., Abe, S., and Nakamura, T. (1979). Induction of cell fusion of plant protoplasts by electrical stimulation. *Plant Cell Physiol.* 20, 1441–1443.

Serpersu, E. H., Kinosita, K., Jr., and Tsong, T. Y. (1985). Reversible and irreversible modification of erythrocyte membrane permeability by electric field. *Biochim. Biophys. Acta* 812, 779–785.

Sowers, A. E. (1983). Fusion of mitochondrial inner membranes by electric fields produces inside-out vesicles: Visualization by freeze-fracture electron microscopy. *Biochim. Biophys. Acta* 735, 426–428.

Sowers, A. E. (1984). Characterization of electric field induced fusion in erythrocyte ghost membranes. *J. Cell Biol.* 99, 1989–1996.

Sowers, A. E. (1986). A long-lived fusogenic state is induced in erythrocyte ghosts by electric pulses. *J. Cell Biol.* 102, 1358–1362.

Sowers, A. E. (1987). The long-lived fusogenic state induced in erythrocyte ghosts by electric pulses is not laterally mobile. *Biophys. J.* **52**, 1015–1020.

Sowers, A. E. (1988). Fusion events and nonfusion contents mixing events induced in erythrocyte ghosts by an electric pulse. *Biophys. J.* **54**, 619–625.

Sowers, A. E. (1989a). The mechanism of electroporation and electrofusion in erythrocyte membranes. *In* "Electroporation and Electrofusion in Cell Biology" (E. Neumann, A. E. Sowers, and C. Jordan, eds.), Plenum Press, New York, pp. 229–256.

Sowers, A. E. (1989b). Electrofusion of dissimilar membrane fusion partners depends on additive contributions from each of the two different membranes. *Biochim. Biophys. Acta* **985**, 339–342.

Sowers, A. E. (1989c). Evidence that electrofusion yield is controlled by biologically relevant membrane factors. *Biochim. Biophys. Acta* **985**, 334–338.

Sowers, A. E. (1989d). The study of membrane fusion and electroporation mechanisms. *In* "Charge and Field Effects in Biosystems 2" (M. J. Allen, S. F. Cleary, and F. M. Hawkridge, eds.), Plenum Press, New York, pp. 315–337.

Sowers, A. E. (1990). Low concentrations of macromolecular solutes significantly affect electrofusion yield in erythrocyte ghosts. *Biochim. Biophys. Acta* **1025**, 247–251.

Sowers, A. E., and Kapoor, V. (1988). The mechanism of erythrocyte ghost fusion by electric field pulses. *In* "Proc. Int. Symp. Molecular Mechanism of Membrane Fusion" (S. Ohki, D. Doyle, T. D. Flanagan, S.-W. Hui, and E. Mayhew, eds.), Plenum Press, New York, pp. 237–254.

Sowers, A. E., and Lieber, M. R. (1986). Electropore diameters, lifetimes, numbers, and locations in individual erythrocyte ghosts. *FEBS Lett.* **205**, 179–184.

Sukharev, S. I., Bandrina, I. N., Barbul, A. I., Fedorova, L. I., Abidor, I. G., and Zelenin, A. V. (1990). Electrofusion of fibroblasts on the porous membrane. *Biochim. Biophys. Acta* **1034**, 125–131.

Szoka, F. C. (1987). Lipid vesicles: Model systems to study membrane-membrane destabilization and fusion. *In* "Cell Fusion" (A. E. Sowers, ed.), Plenum Press, New York, pp. 209–240.

Teissie, J., and Rols, M. P. (1986). Fusion of mammalian cells in culture is obtained by creating the contact between cells after their electropermeabilization. *Biochem. Biophys. Res. Commun.* **140**, 258–264.

Teissie, J., Knutson, V. P., Tsong, T. Y., and Lane, M. D. (1982). Electric pulse-induced fusion in 3T3 cells in monolayer culture. *Science* **216**, 537–538.

Tsoneva, I., Panova, I., Doinov, P., Dimitrov, D. S., and Stahilov, D. (1988). Hybriodima production by electrofusion: Monoclonal antibodies against the Hc antigen of *Salmonella*. *Stud. Biophys.* **125**, 31–35.

Tsong, T. Y. (1983). Voltage modulation of membrane permeability and energy utilization in cells. *Biosci. Rep.* **3**, 487–505.

Vienken, J., Zimmermann, U., Ganser, R., and Hampp, R. (1983). Vesicle formation during electrofusion of mesophyll protoplasts of *Kalanchoe daigremontiana*. *Planta* **157**, 331–335.

Wojcieszyn, J. W., Schlegel, R. A., Lumley-Sapanski, K., and Jacobson, K. A. (1983). Studies on the mechanism of polyethylene glycol-induced cell fusion using fluorescent membrane and cytoplasmic probes. *J. Cell Biol.* **96**, 151–159.

Xie, T.-D., Sun, L., and Tsong, T. Y. (1990). Study of mechanisms of electric field-induced

DNA transfection I: DNA entry by surface binding and diffusion through membrane pores. *Biophys. J.* **58,** 13–19.

Zhelev, D. V., Dimitrov, D. S., and Doinov, P. (1988). Correlation between physical parameters in electrofusion and electroporation of protoplasts. *Bioelectrochem. Bioenerg.* **20,** 155–167.

Zimmermann, U. (1982). Electric field-mediated fusion and related electrical phenomena. *Biochim. Biophys. Acta* **694,** 227–277.

Zimmermann, U. (1986). Electrical breakdown, electropermeabilization and electrofusion. *Rev. Physiol. Biochem. Pharmacol.* **105,** 75–256.

9

Interfacial Membrane Alteration Associated with Electropermeabilization and Electrofusion

J. Teissié and M.-P. Rols

Centre de Recherches de Biochimie et de Génétique Cellulaires du CNRS,
31062 Toulouse Cedex, France

I. Introduction

Introduction of foreign molecules inside the cell cytoplasm is naturally prevented by the plasma membrane. But this low permeability can be transiently suppressed by the application of short, high-intensity electric field pulses to the cell suspension

Guide to Electroporation and Electrofusion

(Neumann *et al.*, 1989). Such electropermeabilization is now routinely used in cell biology and biotechnology to introduce foreign activities into cells through the introduction of engineered plasmids coding for them (electrotransformation). First described for mammalian cells, this method is now used routinely for gene transfer in walled systems: bacteria and yeasts (Neumann *et al.*, 1989).

The protocols are in some cases very efficient but are obtained through an empirical approach. It is clear that the lack of results reported in many cases can just be due to a misuse of the method. One must indeed consider that very little is known about the molecular mechanisms implied in the process. All descriptions have used the lipid bilayer as a description of the biological membrane, taking advantage of the observation that pure lipid vesicles can be electropermeabilized (Teissié and Tsong, 1981). It was initially suggested that electrocompression of the layer leads to its breakdown (Crowley, 1973). This physical description was not able to explain the total reversibility of the process. Taking into account the lateral mobility of phospholipids, due to the free volume between them in bilayers, it was then proposed that the external field induced an expansion of these defects from a "hydrophobic" to a hydrophilic state (Chernomordik *et al.*, 1985). Such a description was able to explain a fast reversibility of the permeabilization, not the long-lived process that was experimentally observed. In fact, these models were always postulating that structured pores were needed to explain the permeabilization (Glaser *et al.*, 1988, Sugar and Neumann, 1984). A recent more detailed analysis of the electrical problems associated with such pores suggested the limits of such a description (Barnett, 1990). Membrane permeability can be explained by processes other than structured lipid pores (Zimmermann, 1986). Phase separation and mismatch in phospholipid protein interactions are other good candidates (Schwister and Deuticke, 1985; Cruzeiro-Hansson and Mouritsen, 1988).

In this chapter, we describe the different steps that are present in the creation of the transiently permeable state of the cell membrane. The observation of fusogenicity associated with the electropermeabilized character is a direct indication that the membrane solution interface is transiently altered in such a way that the organization of interfacial water layers is severely changed.

II. Electropermeabilization

Most of the experiments in our group are run by using the CNRS electropulsator marketed by Jouan (France). The most important feature of this system is the generation of square-wave pulses with duration between 5 μs and 24 ms with repetition frequencies between 0.1 and 10 Hz. Field strengths can be selected up to 9 kV/cm.

One major factor in our experimental procedures is to consider the viability of pulsed cells. This was systematically checked by observing their ability to grow 24 h after pulsing (in the case of mammalian cells).

In most experiments, Chinese hamster ovary cells were used as a model system (see for example Rols and Teissié, 1990a).

III. Permeabilization Quantification

The main consequence of electropermeabilization is to permit a free exchange of small molecules and of ions across the plasma membrane. This process can be assayed at the single cell or at the cell population level.

In the first case, the flow of exchanged molecules is directly measured on each cell observed under a microscope. This is easily done by following the accumulation of a fluorescent dye (Bartoletti *et al.*, 1989). It was recently reported that this flow is described by the mathematical expression (Rols and Teissié, 1990b)

$$\Phi(S) = KX(N,T)(1 - E_{p,r}/E) \, \Delta S$$

where S is the exchanged molecule (ions, ATP, fluorescent dye), $\Phi(S)$ is the flow, K is a parameter that is a function of the nature of the cell (physiological control) and of S (for a given cell, K decreases with an increase in the size of S), N is the number of pulses, T is the pulse duration, $X(N,T)$ describes that the level of permeation as a function of N and T, $E_{p,r}$ is the threshold field, which is a function of the cell strain and for a given strain, a function of the cell effective radius r, E is the field intensity, and ΔS is the concentration gradient of S between the cytoplasmic compartment and the external buffer. When the inflow of an exogenous molecule is observed, then the initial flow is proportional to the added external concentration of S.

It is experimentally time consuming to assay the inflow at the single cell level, and furthermore, sophisticated instrumentation is required. One must measure the change in concentration in the cytoplasm of one single cell by use of a microscope fitted with a photomultiplier (PM) tube. As cells differ from each other by their sizes and in some cases by their physiological identities, one must measure these flows on many cells in order to get a statistically significant information. If a physiologically homogeneous cell population (e.g., a permanent cell line in culture) is used, then, Φ_{pop}, the total inflow cumulated in each cell in the population or the leakage of ATP as observed by, for example, the luminescence of the luciferin/luciferase assay kit (Rols and Teissié, 1990b), is described by

$$\Phi_{pop} = \sum \Phi(r) = \sum KX(N,T)(1 - E_{p,r}/E) \, \Delta S$$

where only $E_{p,r}$ is a function of r. Then

$$\Phi_{pop} = KX(N,T) \, \Delta S \sum (1 - E_{p,r}/E)$$

But as

$$rE_{p,r} = E_{p,1}$$

where $E_{p,1}$ is a constant (this describes permeabilization as a consequence of modulation of the electric field membrane potential difference and occurs for a strain specific threshold),

$$\Phi_{pop} = KX(N,T)\,\Delta S \sum (1 - E_{p,1}/rE)$$

This expression shows that averaging over a population as obtained by measuring an optical density or the fluorescence emission of electroloaded cell suspensions will not detect that the inflow is not constant for different cells in population.

The most frequently used assay for a population is to detect the percentage of stained cells. Stained cells are empirically determined through their observation either with a microscope or with a cell analyzer. In fact, cells are considered as stained when the accumulated material is larger than an experimental detection threshold (Det), that is, when

$$\int_0^T \Phi dt > \mathrm{Det}$$

and T is the loading period following the pulse. It is clear that this criterion depends on K and r for given pulsing conditions (E, T, N). When using a physiologically homogeneous cell population, permeation is directly related to the size distribution. From these technical aspects of the observation of electropermeabilization, it is obvious that clearcut conclusions on the mechanisms of electropermeabilization can be obtained only by a direct observation of the cells (percentage of blue-stained cells when using the trypan blue assay) or by directly measuring the exchange of molecules between the cytoplasm and the external buffer [ATP as in Rols and Teissié (1990b), ions as in Schwister and Deuticke (1985)]. When such reliable assays are used, it is observed that permeabilization is dependent on the field strength. Cell permeabilization is detected only when the field is stronger than a given threshold and then increases. This dependence can be described by the slope of the change of the extend of permeabilization as a function of the field strength, that is, dP/dE ($P = 50\%$). As theoretically predicted above, these parameters are experimentally dependent on the assay, the cell line, and the electrical conditions (pulse duration and number). Any conclusion should be corrected by these experimental limitations of the permeabilization assay.

IV. Time Course of Events Associated with Electropermeabilization

A. Modulation of the Membrane Potential Difference

Under electropermeabilization conditions, the main effect of the electric field on the cell is a modulation of the membrane potential difference (MPD) due to the poor electrical conductivity of a cell membrane as compared to what is present either in the cytoplasm (in spite of the presence of organelles) or in the external medium (where buffers are used to keep the pH constant and ions are almost always present to keep the cells viable). The expression of the MPD, ΔV, has been derived from the Maxwell equations (Neumann *et al.*, 1989) and is a complex function of the specific conductivity of the membrane which is normally low in an intact cell, the specific conductivities of the pulsing buffer and of the cytoplasm, the membrane thickness, and the cell size. Thus,

$$\Delta V = f(\lambda_i)\, g(r)\, E\, \cos\,\theta$$

where λ_i designates the different conductivities and θ is the angle between the directions of the radius at the point considered on the cell surface and of the field (Neumann and Boldt, 1990; Lojewska *et al.*, 1989). this means that ΔV is not uniform on the cell surface (Kinosita *et al.*, 1988). It should be mentioned that in

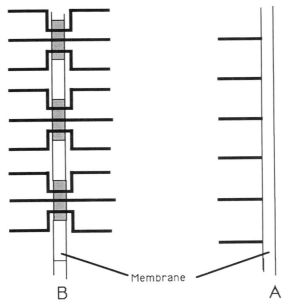

Figure 1   Electric field pattern close to the membrane. A. When the membrane is intact, the electric field lines (thick lines) are parallel when the external field is homogeneous (thin flat parallel electrodes). B. Following the induction and the expansion steps, patches of the membrane are brought to the permeabilized state (TPS are shown as dashed parts). The current is going to flow across the TPS. A complex pattern of electric field lines is present with some of them almost parallel to the membrane.

many cases, the cell surface is supposed to be invagination-free, an assumption that may differ from biological fact, since villi are present on the surface of most cells.

One must mention a problem that is present as soon as the cell membrane is permeabilized. As experimentally observed, ions can then cross the membrane, one must conclude that locally the membrane conductance is high but as a consequence ΔV must drop in this area due to the associated decrease in $f(\lambda_i)$. The induced membrane potential difference is then topologically much more difficult to describe than under nonpermeabilizing conditions. This means that the description of the molecular processes occurring during the cell membrane permeabilization are not triggered under constant MPD conditions even if the external electric field is kept constant. When the permeabilizing field is present, the membrane should be considered as a mosaic of areas with a low conductance mixed with defects (transient permeant structures, TPS) where the conductance is very high. The electric current density is going to be very high in the TPS. As shown in Fig. 1, such a description of the membrane during electropermeabilization implies that the electric field lines will be distorted in order to be focused inside the TPS.

This comment suggests that a lateral component of the electric field parallel to the cell surface should play a role in electropermeabilization. This role may involve an electrophoretic effect on membrane proteins.

B. A Very Short "Induction" Step Triggers the Process

Experimentally, electropermeabilization is detected by an increase of exchanged molecules across the membrane after pulsing the cells under suitable conditions. For a cell with a given size, this can be detected when the field is larger than a threshold value. This change is induced very quickly; the time limitation is in most cases due to the induction of the MPD changes by the field, which is dependent both on the cell size and on the conductivity of the pulsing buffer (Neumann and Boldt, 1990). This induction step is very fast, as shown by conductivity experiments on an RBC (red blood cell) suspension (Kinosita and Tsong, 1979). It occurs on the nanosecond time scale and may be due to the induction of kinks in the fatty acid chain regions of phospholipids. The involvement of phospholipids in the induction of electropermeabilization is indicated by the observation that pure phospholipid LUV (large unilamellar vesicles) can be electropermeabilized under conditions very similar to that which is observed with cells (Teissié and Tsong, 1981).

After correcting for the shifts in an apparent permeabilizing threshold due to the experimental assay (Rols and Teissié, 1990b), by using the extrapolation of the apparent permeabilization field threshold at infinite pulse duration, a characteristic threshold E_p is detected. It is the same regardless of the permeabilization assay that is used.

C. The Defects Expand during the Pulse

The extent of permeabilization (i.e., the magnitude of the induced in-flow of exchanged molecules) is under the control of the pulse duration and of the number of pulses (Rols and Teissié, 1990a, 1990b). This suggests that when the field is present, an expansion step is present. This was clearly shown directly by conductance experiments on RBC (Kinosita and Tsong, 1979) and on bilayer lipid membrane (BLM) (Chernomordik *et al.*, 1985) and indirectly by following the outflow of cytoplasmic ATP (Rols and Teissié, 1990b). This last study brings experimental evidence that it was only the part of the pulsed cell surface where the MPD was brought to the permeabilization threshold that was affected in this expansion step. The other conclusion was that this process occurred at the same rate everywhere in the affected area. As the magnitude of the calculated MPD does not play a role in the expansion step, it may be that the MPD is brought by the field to the permeabilizing value but cannot increase any further. This can be explained by our comments on the complex pattern of the electrical state of the membrane under electropermeabilizing conditions. This is in some ways a feedback effect of the permeabilization on its trigger, the MPD. The conclusion is that the expansion step occurs at constant MPD and the extent of its consequences (i.e., the increase in the flow across the permeabilized membrane after pulsing) is dependent only on the pulse duration and on the number of pulses. This is clearly shown by the $X(N,T)$ parameter in the description of the flow. Up to now, the molecular processes responsible for this expansion step remain unknown but will be tentatively discussed in the next part of this chapter. As mentioned above, the complexity of the electrical pattern in the immediate vicinity of the permeabilized membrane when the field is present makes a theoretical prediction difficult, even when dealing with pure lipid bilayers. The case of cell membranes is more complicated due to the presence of integral proteins. Studies of electropermeabilization-induced ATP leakage in mammalian cells showed that for small values of N, $X(T)$ is a linear function of T for short T values but that a saturation was present for longer pulse durations. For short pulse duration (less than 0.1 ms), $X(N)$ was a linear function of N, showing the cumulative character of the expansion step. A saturation effect is nevertheless detected when N is larger than 10. It appears that the field intensity E, when larger than the permeabilizing threshold E_p, does not play a role in the expansion step but nevertheless controls the magnitude of the induced exchange by increasing the fraction of the cell surface where the permeabilizing threshold in MPD is induced by the external field (Schwister and Deuticke, 1985). this is described by the $(1 - E_p/E)$ term in the expression of the flow.

This expansion step is nevertheless driven by the external field. The field intensity must remain larger than the permeabilizing threshold. If in a succession of pulses the field intensity is brought to a subthreshold value after being initially at a suprathreshold one, no further effect in the expansion is given by the latter pulses.

Studies on processes associated with electropermeabilization showed that in the range of interpulse intervals that we used, the length of the time interval between the pulses did not play a role even down to 100 μs. This observation presents a problem when describing electropermeabilization processes occurring under a time-dependent electric field (capacitance discharge system). One may extend the conclusions we obtained with square-wave pulses by suggesting that the field is effective in the expansion step only as long as it is larger than the permeabilizing value. If true this means that the expansion step is not going to be the same everywhere in the affected fraction of the cell surface. For large values of θ, the expansion step would last a shorter time than for smaller values (i.e., permeabilization would be more effective at the poles facing the electrodes). A description of the processes would be more difficult when using a capacitor discharge system.

D. A Very Fast Stabilization Process Follows the Pulse

As soon as the field is brought back to zero (in the case of square-wave pulses), a very fast annealing step is present. This was directly observed under the microscope by a stroboscopic approach (Kinosita *et al.*, 1988). From a study on the electric conductance of BLM (Chernomordik *et al.*, 1985), this was confirmed by the observation of a nonohmic behavior. Another indirect observation of this immediate post-field step is that high-molecular-weight molecules cross the membrane only as long as the field is present (Sowers, 1989; Sowers and Lieber, 1986; Dimitrov and Sowers, 1990).

E. The Long-Lived Permeabilized State Disappears Slowly

These different approaches all suggest that this stabilization step is very fast (in the millisecond time range) but is followed by a new organization of the membrane where the exchange of small molecules (molecular mass up to 2000 Da) remains very high. This is the classical electropermeabilized state of the pulsed cell membrane. This was observed to have a very long lifetime (minute time range) and to be under the control of the temperature. In the case of CHO (Chinese hamster ovary) cells, this permeant state was retained for up to at least 4 h by using a postpulse incubation at 4°C leaving the cell viability unaffected. This last fact was demonstrated by the ability of cells to grow when brought into the culture medium under optimal conditions (37°C). Permeation would disappear in a couple of minutes if pulsed cells were brought to 37°C just after pulsing. These observations suggest that electropermeabilization was a very complex process where many cell components other than membrane lipids were involved, even if, as mentioned above, the reaction is evoked at the level of phospholipids as shown by the electropermeabilization of

LUV (Teissié and Tsong, 1981). As the process is short-lived with LUV, the implication of proteins and of the cytoskeleton is predictable. This has been experimentally supported by the effect of the colchicin (Rols and Teissié, 1991).

The disappearance of the electropermeabilized state of the membrane has been shown to be a first-order process (Rols and Teissié, 1990b). The rate constant is not dependent on the intensity of the field that triggers the permeabilization. But it decreases (i.e., the time constant value becomes larger) with an increase in the pulse duration and in the number of repetitive pulses. These conclusions on the kinetics of resealing were obtained with square-wave pulses. They also apply when using exponentially decaying pulses. It should be mentioned that, for given pulse number and duration, electropermeabilization apparently lasts longer for higher field intensities. This is due to the fact that it is the flow that is measured, not the local perturbation. The flow is present across all the affected area, which is known to be larger for larger field intensities (Schwister and Deuticke, 1985). As a consequence, the higher the field, the larger the flow. In many cases, permeabilization is assayed by checking the percentage of cells that are stained (i.e., where the extent of staining is larger than a threshold). Due to this approach, as increasing the field will increase the flow, the permeabilized state will be detected for a longer time after pulsing if stronger fields are used. But when measuring at a single cell level the disappearance of permeabilization (i.e., the decay of the flow of exchanged molecules after pulsing), it was observed that it is not a function of the field intensity (Rols and Teissié, 1990b).

V. Electropermeabilized Cells Are Fusogenic

Since the early 1980s, it has been known that when cells in close contact were submitted to electric field pulses suited to induce a high level of permeabilization, they were observed to fuse in many cases. This process was apparently controlled by the electrostatic interactions between the cell surfaces, mainly due to proteins. It is not clear what molecular processes are involved in the fusion processes. It was shown that the cell fusion was a multistep process where membrane fusion was followed by a very slow step in which the cytoskeleton reorganization occured (Blangero et al., 1989).

In 1986, it was shown first for erythrocyte ghosts (Sowers, 1986) and then for mammalian cells (Teissié and Rols, 1986) that fusion can be obtained by bringing into contact cells that were electropermeabilized. This observation has been recently extended to plant protoplasts (Montané et al., 1990). This process is nevertheless controlled by the electrostatic interactions between cells (Rols and Teissié, 1989; Rols and Teissié, 1990c). This last point was missed in experiments where negative results were reported (Sukharev, 1989). This observation is of great importance in the understanding of electropermeabilization. When cells are brought into contact, several forces are present that, as a final result, prevent cells from spontaneously fusing (Marra and Israelachvili, 1985). Van der Waals forces attraction is balanced by the repulsive character of electrostatic interactions. But if cells are brought into closer contact (for example, after proteolytic treatment), a very strong repulsive force is present when the intercellular distance is of the order of nanometers: the hydration force, which prevents a more intimate contact between the two cell surfaces and as such hinders their fusion. Its origin is proposed to be due to the organization of the interfacial molecules. Being dipoles, water molecules in the vicinity of the cell membrane are submitted to a local electric field and are oriented with their dipoles facing the external phase. This layer is well ordered and generates a dipolar field that orders a second layer of water molecules. This ordering process is active for three to four successive layers. When cells are brought into contact, their respective water layers repulse each other. This is the origin of the hydration forces, which play a decisive role in preventing spontaneous cell fusion (Parsegian and Rau, 1984). In order to trigger cell fusion, the hydration force strength should be decreased. As a consequence, bringing the cell surface to a fusogenic state implies disordering interfacial water molecules. Such a long-lived disorder can only be obtained by destroying the origin of the order (i.e., the organization of the cell membrane–water interface). To explain the long-lived fusogenicity of electropermeabilized cells, we must conclude that an alteration of the interface has occured. The structural change of the interface associated with the high permeability induces the fusogenicity, suggesting that any physical effect playing a role in the process of cell fusion should also alter electropermeabilization. An implicit implication of this prediction is that a parallel effect should be present on the process that restores the normal unpermeable state. It is along that step that the molecular organization giving the repulsive hydration forces present when cells are in contrast is recreated.

VI. Molecular Alterations of the Cell Membrane during Electropermeabilization

Two classes of molecular alterations from electric field pulses have been detected. The first class is associated with events detected only as long as the pulse is present; the second is associated with the long-lived permeability to small molecules (and ions).

Electron microscopy shows the induction of cracks in erythrocyte ghosts (lifetime less than 1 s) (Stenger and Hui, 1986), a delayed creation of craterlike pores in red blood cells under hypoosmotic conditions, and the presence of long-lived blebs (Escande-Geraud et al., 1988; Gass and Chernomordik, 1990). An increase in the number of villi was detected (Escande-Geraud et al., 1988).

Conductance experiments are indicative of a high nonohmic state during the pulse (Chernomordik et al., 1985).

Contact-angle measurements on pulsed plant protoplasts show an increase in the hydrophobocity of the interface (Hahn-Hagerdal *et al.*, 1986).

NMR studies of phospholipid head groups and fatty acid chains are the only techniques giving direct access to the molecular consequences of electropermeabilization. In the case of pure phospholipid multilayered systems (MLV), a tilt of the head group is observed during the pulse but the hydrophobic region is not affected. In the case of CHO cells, the long-lived permeability is associated with a new and reversible organization of the head group with an alteration of the averaged position as referred to the plane of the membrane (Lopez *et al.*, 1988). Some similarities with the effect of low concentrations of polyethyleneglycol (PEG) are present.

These last two observations (hydrophobicity and NMR) are more direct evidence that electropermeabilization is associated with a modification of the membrane solution interface and that any proposed interpretation of its mechanism should take this into account.

VII. Chemical and Physical Alterations of the Membrane Affect the Expansion Step

The thermodynamical implications of these conclusions have been checked by altering the membrane order (Rols *et al.*, 1990) and the osmotic pressure of the pulsing buffer (Rols and Teissié, 1990a). CHO cells are used as a model system, taking advantage of their osmotic stability (their size and, as a consequence at a given field strength, the induced MPD are not affected by the osmotic pressure under the experimental conditions which they were used).

Ethanol prepulse incubation, lysolecithin insertion, change in the osmotic pressure of the pulsing buffer, change in its ionic content, and pretreatment of the cell surface by trypsin do not alter the permeation threshold E_p when the prepulse treatment does not directly affect the cell viability. Electropermeabilization is always induced for the same value of the cell MDP. Specific effects are detected along the expansion step as shown by the dependence of the induced permeabilization on the pulse duration (and number) at a given field strength or by the shape of the permeabilization versus field strength plot at given pulse duration and number (Fig. 2). The dP/dE is strongly affected. At a given field strength, longer pulses are needed in order to induce the same permeabilization (Fig. 3). Decreasing the membrane order lowers it, and increasing the order has the opposite effect. Hypoosmolarity, observed to decrease the membrane surface undulations and as such their repulsive effect on cell contacts, facilitates permeabilization (high dP/dE, same P at short pulses), and hyperosmolarity brings the opposite observations (Fig. 2 and 3).

These experimental results are the direct evidence that increasing repulsive forces between cells when in contact (intercellular effect) by a modification of the membrane (intracellular modification) hinders electropermeabilization. Hyperosmolarity and

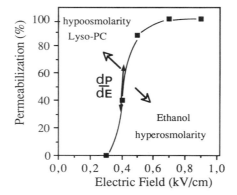

Figure 2 Permeabilization extent as a function of the field intensity. Permeabilization is always detected for the same field strength, whatever the cell treatment, but the slope *dP/dE* is affected. This plot was obtained with plated CHO cells. Permeabilization was assayed by tryptan blue staining.

decreasing the membrane order are both known to induce larger membrane undulations. This increase in the amplitude of fluctuations is the basis for the increase intensity of short-range cell–cell interactions.

As predicted, ethanol slows down the resealing step (i.e., increases the permeabilized state lifetime), and lysolecithin gives the opposite effect (Fig. 4). Resealing occurs more quickly under hypoosmolarity (Fig. 4).

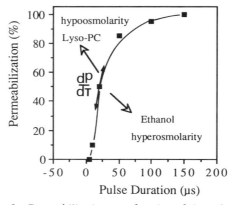

Figure 3 Permeabilization as a function of the pulse duration. Permeabilization increases with an increase in the pulse duration. This change (i.e., *dP/dT*) is affected by the chemical and physical treatment of pulsed cells. Permeabilization was assayed by trypan blue staining.

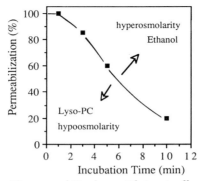

Figure 4   The permeabilized state lifetime is affected by the chemical treatment. Treatment facilitating electropermeabilization shortens the lifetime. Permeabilization was assayed by trypan blue staining. The lifetime was assayed by a delayed addition of the reporter dye after pulsing.

VIII. Conclusions

The present work shows that a key process in electropermeabilization is alteration of the membrane such that it loses its structured water layers. Other processes should play a role, such as electrostatic interactions or protein–protein contact. But their effect is to modulate the main process that gives the membrane its fusogenic character.

The present results are indicative that the reorganization of the interface occurs during the expansion step, in agreement with the proposed coalescence–percolation model of membrane electropermeabilization (Sugar *et al.,* 1987). Due to its analogy with the description of phase-transition phenomena, it is tempting to propose that more similarities should be present, as already suggested (Cruzeiro-Hansson and Mouritsen, 1988). As the core of the membrane is apparently not affected, as shown by the NMR studies on lipid systems or by the fluorescence studies on CHO cells (Lopez et al., 1988), this transition should affect specifically the interface region, maybe through an increase in fluctuations. This suggests that membrane permeability is under the control of its interface organization and that increased fluctuations in the glycerol region would increase it as thermodynamically predicted (Nagle and Scott, 1978).

Acknowledgments

Thanks are due to Mrs. Zalta for her help in cell culture and to Mrs. Maurel for her secretarial assistance. This work was supported by the CNRS and the French Association pour la Recherche sur le Cancer.

References

Barnett, A. (1990). The current voltage relation of an aqueous pore in a lipid bilayer membrane. *Biochim. Biophys. Acta* **1025**, 10–14.

Bartoletti, D. C., Harrison, G. I., and Weaver, J. C. (1989). The number of molecules taken up by electroporated cells: Quantitative determination. *FEBS Lett.* **256**, 4–10.

Blangero, C., Rols, M. P., and Teissié, J. (1989). Cytoskeletal reorganization during electric-field induced fusion of Chinese hamster ovary cells grown in monolayers. *Biochim. Biophys. Acta* **981**, 295–302.

Chernomordik, L. V., Kozlov, M. M., Melikyan, G. B., Abidor, I. G., Markin, V. S., and Chizmadzhev, Y. A. (1985). The shape of lipid molecules and monolayer fusion. *Biochim. Biophys. Acta* **812**, 643–655.

Crowley, J. M. (1973). Electrical breakdown of biomolecular lipid membranes as an electromechanical instability. *Biophys. J.* **13**, 711–724.

Cruzeiro-Hansson, L., and Mouritsen, O. G. (1988). Passive ion permeability of lipid membranes modelled via lipid-domain interfacial area. *Biochim. Biophys. Acta* **944**, 63–73.

Dimitrov, D. S., and Sowers, A. E. (1990). Membrane electroporation—Fast molecular exchange by electroosmosis. *Biochim. Biophys. Acta* **1022**, 381–392.

Escande-Geraud, M. L., Rols, M. P., Dupont, M. A., Gas, N., and Teissié, J. (1988). Reversible plasma membrane ultrastructural changes correlated with electropermeabilization in Chinese hamster ovary cells. *Biochim. Biophys. Acta* **939**, 247–259.

Gass, G. V., and Chernomordik, L. V. (1990). Reversible large scale deformations in the membranes of electrically treated cells: Electroinduced bleb formation. *Biochim. Biophys. Acta* **1023**, 1–11.

Glaser, R. W., Leikin, S. L., Chernomordik, L. V., Pastuchenko, V. F., and Sokirko, A. I. (1988). Reversible electrical breakdown of lipid bilayers: formation and evolution of pores. *Biochim. Biophys. Acta* **940**, 275–287.

Hahn-Hagerdal, B., Hosono, K., Zachrisson, A., and Bornman, C. H. (1986). Polyethylene glycol and electric field treatment of plant protoplasts: characterization of some membrane properties. *Physiol. Plants* **67**, 359–364.

Kinosita, K., and Tsong, T. Y. (1979). Voltage induced conductance in erythrocyte membranes. *Biophys. Biochim. Acta* **554**, 479–494.

Kinosita, K., Ashikawa, I., Saita, N., Yoshimura, H., Itoh, H., Nagayama, K., and Ikegami, A. (1988). Electroporation of cell membrane visualized under a pulsed-laser fluorescence microscope. *Biophys. J.* **53**, 1015–1019.

Lojewska, Z., Farkas, D. L., Ehrenberg, B., and Loew, L. M. (1989). Analysis of the effect of medium and membrane conductance on the amplitude and kinetics of membrane potentials induced by externally electric fields. *Biophys. J.* **56**, 121–128.

Lopez, A., Rols, M. P., and Teissié, J. (1988). 31P NMR analysis of membrane phospholipid organization in viable, reversibly electropermeabilized Chinese hamster ovary cells. *Biochemistry* **27**, 1222–1228.

Marra, J., and Israelachvilli, J. (1985). Direct measurements of forces between phosphatidylcholine and phosphatidylethanolamine bilayers in aqueous electrolyte solutions. *Biochemistry* **24**, 4608–4618.

Montane, M. H., Dupille, E., Alibert, G., and Teissié, J. (1990). Induction of a long-lived fusogenic state in viable plant protoplasts permeabilized by electric fields. *Biochim. Biophys. Acta* **1024**, 203–207.

Nagle, J. F., and Scott, H. L. (1978). Lateral compressibility of lipid mono- and bilayers. *Biochim. Biophys. Acta* **513**, 236–243.

Neumann, E., and Boldt, E. (1990). Membrane electroporation: The dye method to determine cell membrane conductivity. *In* "Membrane Biotechnology," C. Nicolau and D. Chapman, eds., A. R. Liss, New York, pp. 69–84.

Neumann, E., Sowers, A. E., and Jordan, C. (eds.) (1989). "Electroporation and Electrofusion in Cell Biology." Plenum, New York.

Parsegian, V. A., and Rau, D. C. (1984). Water near intracellular surfaces. *J. Cell. Biol.* **99**, 196s–202s.

Rols, M. P., and Teissié, J. (1989). Ionic-strength modulation of electrically induced permeabilization and associated fusion of mammalian cells. *Eur. J. Biochem.* **179**, 109–115.

Rols, M. P., and Teissié, J. (1990a). Modulation of electrically induced permeabilization and fusion of Chinese hamster ovary cells by osmotic pressure. *Biochemistry* **29**, 4561–4567.

Rols, M. P., and Teissié, J. (1990b). Electropermeabilization of mammalian cells: Quantitative analysis of the phenomenon. *Biophys J.* **58**, 1089–1098.

Rols, M. P., and Teissié, J. (1990c). Implications of membrane interface structural forces in electropermeabilization and electrofusion. *Bioelectrochem. Bioenerg.* **24**, 101–111.

Rols, M. P., and Teissié, J. (1991). Evidence for cytoskeleton implication in cell permeabilization and electrofusion. *In* "Proc. 47th Meeting of the Division de Chimie Physique: The Living Cell in Its 4 Dimensions," C. Troyanovsky, ed., pp. 251–260.

Rols, M. P., Dahhou, F., Mishra, K. P., and Teissié, J. (1990). Control of electric field induced cell membrane permeabilization by membrane order. *Biochemistry* **29**, 2960–2966.

Schwister, K., and Deuticke, B. (1985). Formation and properties of aqueous leaks induced in human erythrocytes by electrical breakdown. *Biochim. Biophys. Acta* **816**, 332–348.

Sowers, A. E. (1986). A long lived fusogenic state is induced in erythrocyte ghosts by electric pulses. *J. Cell Biol.* **102**, 1358–1362.

Sowers, A. E. (1989). The study of membrane electrofusion and electroporation mechanisms. *In* "Charge and Field effects in Biosystems 2," M. J. Allen, S. F. Cleary and F. M. Hawkridge, eds., Plenum, New York, pp. 315–337.

Sowers, A. E., and Lieber, M. R. (1986). Electropore diameters, lifetimes, numbers, and locations in individual erythrocytes ghosts. *FEBS Lett.* **205**, 179–184.

Stenger, D. A., and Hui, S. W. (1986). Kinetics of ultrastructural changes during electrically-induced fusion of human erythrocytes. *J. Membrane Biol.* **93**, 43–53.

Sugar, I. P., and Neumann, E. (1984). Stochastic model for electric field induced membrane pores—Electroporation. *Biophys. Chem.* **19**, 211–225.

Sugar, I. P., Forster, W., and Neumann, E. (1987). Model of cell electrofusion—Membrane electroporation, pore coalescence and coalescence. *Biophys. Chem.* **26**, 321–335.

Sukharev, S. I. (1989). On the role of intermembrane contact in cell electrofusion. *Bioelectrochem. Bioenerg.* **21**, 179–191.

Teissié, J., and Tsong, T. Y. (1981). Electric field induced transient pores in phospholipid lipid vesicles. *Biochemistry* **20**, 1548–1554.

Teissié, J., and Rols, M. P. (1986). Fusion of mammalian cells in culture is obtained by creating the contact between the cells after their electropermeabilization. *Biochem. Biophys. Res. Commun.* **140**, 258–266.

Zimmermann, U. (1986). Electrical breakdown, electropermeabilization and electrofusion. *Rev. Physiol. Biochem. Pharmacol.* **105**, 175–256.

10

Membrane Fusion Kinetics

Dimiter S. Dimitrov

Section on Membrane Structure and Function, National Cancer Institute, National
Institutes of Health, Bethesda, Maryland 20892

I. Introduction
II. Fusion Kinetics
    A. Fluorescence Videomicroscopy
    B. Spectrofluorimetry
III. Delays
    A. A Definition
    B. Determinants of Delays
    C. An Empirical Formula
IV. Rates of Fusion
    A. Individual Cells
    B. Populations of Cells
V. Fusion Yields
VI. Models of Fusion Kinetics
VII. Conclusion
    References

I. Introduction

Electrofusion is a nonspecific process that does not require membrane receptors. A
wide variety of membranes can be fused by electric fields (for general reviews see
Neumann et al., 1989; Sowers, 1987; Teissie and Rols, 1986; Lo et al., 1984; Berg
et al., 1983; Zimmermann et al., 1988). Viral fusion is a specific process that is
induced by specialized fusion proteins (Blumenthal, 1987). One example is the
human immunodeficiency virus (HIV). Its fusion with cell membranes is induced
by the highly specific interaction of the envelope viral glycoprotein with its membrane receptor (Dalgleish et al., 1984; Klatzmann et al., 1984).

Though it seems that electrofusion and viral fusion kinetics should be entirely different, this is not the case. Data for electrofusion (Dimitrov and Sowers, 1990a) and fusion induced by viral proteins (Morris *et al.*, 1989; Sarkar *et al.*, 1989) give similar kinetic dependencies consisting of three major parts: (1) lag times (delays), (2) fluorescence changes indicating fusion, and (3) plateaus that give the maximal fusion yields.

Delays are the time periods between the application of the fusion trigger and the first indication for fusion. They reflect lifetimes of intermediates in fusion. Thus, analysis of the delays in membrane fusion is a major source of information for fusion mechanisms. Fusion yields are related to individual variations in cell membranes with respect to their ability to fuse. Delays and fusion yields for electrofusion and viral fusion show both similarities and differences.

Similarity between electrofusion and viral fusion can be due to similarities in physicochemical and geometrical properties of membranes and intervening liquid layers. These are nonspecific factors. They can be estimated by formulas derived from laws of physical chemistry of surfaces and membranes (Dimitrov, 1983; Dimitrov and Jain, 1984). The differences can arise from the different forces that drive the fusion reactions. For electrofusion, the driving force depends mainly on the polarizability and conductivity of the membranes and the surrounding media, and the strength and duration of the electric pulse. In viral fusion, the driving force depends primarily on conformational changes of fusion proteins, which are determined in a complex way by interactions with specific ligands and receptors. Therefore, driving forces in viral fusion must be modeled for each system.

By comparing both types of fusion, insight into the role of specific and nonspecific factors in membrane fusion kinetics can be obtained. This is the basic goal of this chapter.

II. Fusion Kinetics

Kinetics of fusion has been monitored on both single cells and populations of cells. Fluorescence videomicroscopy was used for observation of single cells, while spectrofluorimetry was the preferred method for cell populations (Loyter et al., 1988). Quantitative correlations between the two types of experimental data are difficult to make (Chen and Blumenthal, 1989).

A. Fluorescence Videomicroscopy

Fluorescence videomicroscopy allows continuous monitoring of the fluorescent dye transfer from individual labeled membranes to unlabeled ones. Commonly, the fluorescence increase in the originally unlabeled membrane is recorded (Fig. 1). The increase in fluorescence is a function of time and spatial location on the membrane.

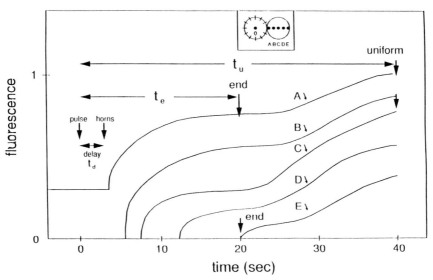

Figure 1   An example of fluorescence intensity changes for an originally unlabeled membrane after electrofusion with a labeled membrane (with permission from Dimitrov and Sowers, 1990a). The space locations of the points of measurement are shown in the inset. The distances from the contact with the labeled membrane are A, 0 μm (the contact); B, 1.7 μm; C, 3.3 μm (the center of the erythrocyte ghost); D, 5.0 μm; and E, 6.6 μm (the far end of the membrane). The fluorescence intensity is normalized to that in the center of the labeled membrane (point O in the inset). The fluorescence intensity at the membrane contact (A) is higher than zero before the pulse because of the close proximity of the labeled membrane. The increase in fluorescence begins after a delay at a point in time that coincides with a characteristic appearance of the fluorescence as "horns." The dye diffuses until reaching the far ends of the membranes (after time t_e) and uniform redistribution (after time t_u).

There are three major parts of the time dependencies: (1) a time lag (delay) between the application of the fusion trigger (the electric pulse) and the onset of fluorescence increase in the unlabeled membrane, (2) gradual increase of the fluorescence intensity, and (3) saturation. Similar dependencies were observed for individual cell fusion events induced by viral fusion proteins [see Fig. 3 in Sarkar *et al.* (1989) for fusion induced by the influenza hemagglutinin].

B. Spectrofluorimetry

Spectrofluorimetry is rarely used to measure electrofusion kinetics. Stenger and Hui (1988) monitored the mixing of the aqueous contents of erythrocyte ghosts during electrofusoin. The kinetics curves did not show any delays, within the time resolution of the method, which is about 0.1 s. There was a sharp increase in fluorescence, following application of single pulse, which reached a plateau.

Spectrofluorimetry is widely used to monitor kinetics of viral fusion (Loyter *et al.*, 1988). A typical kinetic curve is functionally very similar to that for single cells observed by fluorescence microscopy. It has a delay, an increase in fluorescence, and a plateau.

III. Delays

Delays in fusion vary widely for different systems. They can be less than 5 ms for the neurotransmitter release (Heuser *et al.*, 1979) or larger than 1 h for cell fusion induced by the HIV envelope proteins (D. Dimitrov, H. Golding, and R. Blumenthal, unpublished results). For planar–planar bilayer fusions the lifetimes of the events leading to fusion ("waiting time" for fusion) are in the range of seconds to minutes (Chernomordik *et al.*, 1987). For the same system with different experimental conditions, delays can vary by orders of magnitude. For example, delays in electrofusion of erythrocyte ghosts range from milliseconds to minutes (Dimitrov and Sowers, 1990a). Delays in viral fusion follow the same range (Morris *et al.*, 1989; Clague *et al.*, 1990; Hoekstra and Kok, 1989; Spruce *et al.*, 1989).

A. A Definition

It was suggested that cell fusion involves six possible stages (Rand and Parsegian, 1986): (1) stable membrane apposition, (2) triggering of fusion, (3) contact, (4) focused destabilization, (5) membrane coalescence, and (6) restabilization. The delay is the time period between the triggering of fusion and the membrane coalescence. Therefore, it includes the stages of making contact and focused destabilization in the scheme proposed by Rand and Parsegian (1986). What is the definition of the delay in membrane fusion from an experimental point of view? The answer to this question varies, depending on the experimental system. One definition is that the delay is the time between the application of the trigger of fusion and the first indication of fusion. While reasonable, this definition requires answering other questions. When exactly is the trigger applied? What is the response time of the measuring system? What is the minimal increase in the signal that should be assumed as an indication of fusion?

The definition used here is that the delay is the time interval between trigger application and first indication of fusion. The first indication of fusion occurs when the average signal increase exceeds the level of noise. Any comparison of delays obtained by different methods requires careful examination of the experimental conditions of measurements and definitions of delays.

In electrofusion the moment of trigger application can be precisely located in time. For many other systems the trigger is a chemical substance that requires time for diffusion. This can be an extended period of time and not easily measured. Another advantage of electrofusion is the use of precisely regulated AC fields to achieve close membrane apposition. This leads to reproducible and controllable initial starting positions of the membranes. In viral fusion this is achieved by prebinding the virus to target cells at low temperature. The initial separation of the membranes is, however, regulated by parameters, such as medium composition, that can also affect the fusion reaction.

B. Determinants of Delays

The major determinants of the delays in fusion are (1) the driving force of the fusion reaction and (2) the resistance of the system to changing its state. The driving force is proportional to the difference in the free energies of the system of two interacting membranes after the application of the trigger and after completion of fusion. Therefore, it depends on the respective enthalpies and entropies, and the temperature. Increasing the strength of the fusogen can increase the enthalpy and therefore the driving force, which can in turn lead to short delays. Above a certain limit, however, a strong fusogen can significantly increase the entropy. This can result in membrane fragmentation and lack of fusion.

The resistance of the system can be of two major parts: (1) existence of energy barriers and (2) "diffusion barriers." The energy barrier(s) can be due to repulsive intermembrane interactions and structural rearrangements leading to membrane continuity. The diffusion barrier depends on the temperature, the medium and the membrane viscosity, and membrane geometry, including initial intermembrane separation. Therefore, any systematic study of delays in fusion should address their dependence on the strength of the fusogen, temperature, viscosity, membrane geometry, and initial intermembrane separation.

A study of delays in electrofusion of erythrocyte ghosts (Dimitrov and Sowers, 1990a) showed that they decreased over the range 4 to 0.3 s with an increase in (1) the pulse strength from 0.25 to 1.0 kV/mm, (2) the pulse duration in the range, 0.073–1.8 ms, and (3) the dielectrophoretic force that brings the membranes into close apposition before triggering fusion. The delays decreased 2–3 times with an increase in temperature from 21 to 37°C. The Arrhenius plot yielded straight lines. The calculated activation energy, 17 kcal/mol, does not depend on the pulse strength. The delays were proportional to the medium viscosity (Dimitrov and Sowers, 1990a).

The delays in fusion between fibroblasts, expressing influenza hemagglutinin, and R18-labeled erythrocytes were on the order of seconds and decreased with decreasing pH in the range 5.4–4.9 (Sarkar et al., 1989). The activation energy is 18 kcal/mol for temperatures between 27 and 50°C. The delays in fusion of intact viruses with erythrocyte ghosts showed similar trends but were shorter than for cell fusion (Clague *et al.*, 1990). The delay in fusion of Sendai virus with erythrocyte ghosts increased proportionally with an increase in the medium viscosity (D. Dimitrov and R. Blumenthal, unpublished data).

C. An Empirical Formula

The data for delays in electrofusion of erythrocyte ghosts (Dimitrov and Sowers, 1990a) can be described by an empirical formula of Arrhenius type:

$$\text{Delay} = C\mu \, \exp(E/RT) \qquad (1)$$

where C is a constant, μ is medium viscosity, E is activation energy, R is the gas constant and T is absolute temperature. The constant C is inversely proportional to the driving force of the fusion reaction and proportional to the systems resistance to fusion. Therefore, it depends on all the factors that determine the driving force and the fusion resistance. The activation energy does not depend on the strength of the fusogen.

IV. Rates of Fusion

A. Individual Cells

Rates of fusion are commonly measured by taking the derivatives of the fluorescence change with respect to time. For individual cells they reflect the rate of diffusion of the dye through the intermembrane junction following fusion. There are two limiting cases: (1) the dye transfer through the intermembrane junction is fast compared to its lateral diffusion in the originally unlabeled membrane, and (2) the transfer through the intermembrane junction is slower than the diffusion.

In the first case the dye distribution is nonuniform. The rate of dye transfer is entirely determined by the diffusion coefficient of the dye in the originally unlabeled membrane and the membrane geometry. This is the case, shown in Fig. 1, for electrofusion of erythrocyte ghosts (Dimitrov, and Sowers, 1990a). The plot of the square of the distance between the diffusion front and the membrane contact zone versus time is linear. The slope gives $4D$, D being the diffusion coefficient. This is one method of measuring the lateral diffusion of fluorescent dyes. Therefore in this case the increase in fluorescence in the initial stage of the membrane transfer is not a measure of rate of fusion. It only shows that the rate of dye transfer through the intermembrane junction is faster than the rate of diffusion. After the diffusion front reaches the far end of the cell, the fluorescence intensity continues to increase until reaching the final equilibrium state. This stage also depends on the diffusion. The case of fast dye transfer through intermembrane junctions occurs when the total length of membrane contact by junctions is larger than the contact perimeter. If the intermembrane junctions are channels, the product of the number of channels and their average perimeter should be larger than the perimeter of the membrane contact.

When the rate of dye transfer through the fusion junction is lower than the diffusion rate, the increase of fluorescence is uniform. The fluorescence increase depends only on the properties of the fusion junction. The rate of increase reflects the number and size of the fusion junctions and their change with time. When the number and size of the fusion junctions are constant, the fluorescence increase should be linear with time in the beginning of the dye transfer. Any differences from this linear relationship should be attributed to changes of the size and number of the fusion junctions, including creation of new or closing of "old" junctions. Therefore, fluorescence changes after delays may contain important information about intermediates in fusion and their evolution with time.

In both cases, the dye should redistribute until an equilibrium is reached. If the membranes are the same, the fluorescence should be the same. In many cases the fluorescence intensity is not the same in the labeled and the fused membrane that was originally unlabeled. This may be due to at least two factors: (1) the dye may have a different solubility or quantum yield in the two membranes and (2) the fusion junctions may be disrupted after a certain period of time.

B. Populations of Cells

For a population of cells the rate of fusion can be defined in at least two ways. Either it is the number of cells fused per unit time or it is the total fluorescence change per unit time. In the first case fusion rates can be measured simply by counting the number of cells fused as a function of time. Evidently, in this case the rate of fusion is given by the individual differences in the lag times (delays). It will be equal to the number of fused cells that have delays within a time interval divided by that time interval. In the second case there will be an effect of the rate of dye transfer to the unlabeled cells. For example, the water-soluble dye can be transferred through pores by electroosmosis within milliseconds (Dimitrov, and Sowers, 1990b). Therefore, the measured rates of fluorescence increase can be higher than the actual rates of fusion. This could be the reason for the high rates and lack of delays in the kinetic curves obtained by using water-soluble dyes (Stenger and Hui, 1988, 1989).

V. Fusion Yields

Fusion yields are commonly defined as the maximal number of fused labeled cells normalized to the total number of labeled cells (Sowers, 1984). They can be obtained by integrating the rate of fusion with respect to time from zero to infinity. The fusion rates are, however, mainly determined by the individual differences of cells, which cause variations in delays. If all the cells were exactly the same and the conditions for fusion were exactly the same then the fusion yield would be either 0 or 100%. In the case of zero fusion, the delay will be infinite. In the other case (100% fusion) the delay will be a number greater than zero that will not vary within the cell population. Therefore, variations in fusion yields tell us how different cells are with respect to fusion properties. This may lead to the conclusion that delays and fusion yields are not necessarily related.

For a population of cells, however, a change in the parameters of the system that leads to a decrease of the delay commonly causes an increase in fusion yields. For example, in electrofusion of erythrocyte ghosts, an increase in pulse strength and duration decreases the delay and increases the fusion yield (Dimitrov and Sowers, 1990a). This means that with higher field strength pulses, more cells will fuse. Those that would have fused anyway will do so more quickly. This indicates a property of the cell membrane that differs within a cell population. A shorter delay at the stronger pulse is caused by the increase in the driving force of the fusion reaction. Therefore, these are two different effects, which can be separated. In support of this conclusion are several experimental data (Dimitrov and Sowers, 1990a): (1) Under strong pulses practically all cells fuse. However, the delay can change several times with increases in the pulse strength. (2) For weak pulses the delay does not change significantly. The fusion yield, however, changes several times with the change of the pulse. (3) The fusion yield changes significantly, while the delay does not, in buffers of different ionic strength.

VI. Models of Fusion Kinetics

In spite of their fundamental differences, electrofusion and viral fusion may have a similar time course of fusion. This similarity may arise from nonspecific factors, which determine the membranes' resistance to fuse. A simplistic view is that the driving force for both types of fusion is entirely different. The resistance to fuse, however, and the way to overcome the barriers, due to intermembrane repulsion and membrane stability, may be very similar.

One major difference between viral fusion and electrofusion is the kinetics of "commitment" to fusion (Morris et al., 1989). In terms of driving forces, it corresponds to the kinetics of the phenomena leading to an increase in the free energy of the system. In fusion of influenza virus this is the conformational change of the hemagglutinin after lowering the pH and consequent exposure of the hydrophobic domain. In electrofusion the "commitment" occurs during the electric pulse, which is very short. The conformational changes of the viral proteins are rather slow (seconds) compared to the duration of the electric pulse (microseconds). Therefore, the stage of "accumulation" of free energy can be long for the viral fusion and short for electrofusion.

The differences in the ways the free energy is accumulated and the final states of high free energy are the major causes for the differences between electrofusion and viral fusion kinetics. It is reasonable to speculate that during the subsequent stages of "usage" of that free energy to destabilize and fuse membranes, the decrease of the driving force with time will be different for both types of fusion. Therefore, any factor that affects the driving force during this stage will invoke a different response for electrofusion than for viral fusion.

These general considerations do not provide a detailed kinetic description of membrane fusion. There are several phenomenological models, which focus on different characteristics of fusion. One model describes the kinetics of conformational changes of the viral proteins as transitions to states of different potency to fusion (Blumenthal, 1988). Another model takes into account the binding stage in the viral fusion kinetics (Bentz *et al.*, 1988). Both models describe quantitatively kinetics of viral fusion. They use, however, phenomenological constants that should be determined by other models or physical theories. Some of these constants can be obtained by an approach based on the fluctuation wave mechanisms of local membrane approach, contact, destabilization, and fusion (Dimitrov, 1983; Dimitrov and Jain, 1984; Zhelev *et al.*, 1988). It is intended to describe the contribution of the nonspecific factors in membrane fusion. It allows an estimation of the delays if the driving force is known and the energy barriers are small. One of the formulas, for example, estimates the time t needed for the fluctuations in membrane shape to grow until reaching local molecular contacts. It reads (Dimitrov, 1983)

$$t = 24\mu T/h^3 P^3 \qquad \text{when} \quad B(dP/dh)/T^2 \ll 1 \qquad (2)$$

$$t = 9\mu(6B)^{1/2}/h^3(dP/dh)^{3/2} \qquad \text{when} \quad B(dP/dh)/T^2 \gg 1 \qquad (3)$$

where μ is the viscosity of the liquid layer between the membranes, T is the membrane tension, B the bending elasticity, P the driving pressure, which is the driving force per unit area, and h the intermembrane separation.

Equations (2) and (3) allow calculations of that part of the delay that is due to membrane approach if the approach occurs through fluctuation waves. Depending on the driving pressure and the parameters of the system, these equations give values ranging from milliseconds to hours. A major problem in their application is the calculation of the driving force as a function of the intermembrane separation. These equations predict the functional dependencies of the fusion kinetics based on parameters of the membrane system. For example, they predict that the delays should be proportional to the viscosity of the liquid layer between the membranes. This was confirmed for electrofusion of red blood cell ghosts (Dimitrov and Sowers, 1990a) and fusion of Sendai virus with erythrocytes (D. Dimitrov and R. Blumenthal, unpublished data).

VII. Conclusion

Membrane fusion is a multistage process. Dissecting steps of this process leads to understanding its mechanisms. Fusion kinetics is determined by the lifetimes of intermediates, which are characteristic for each of the fusion stages. The delays in fusion are a measure of the overall lifetime of the intermediates or the lifetime of the major rate-determining intermediate. Similarity between delays in electrofusion and viral fusion may reflect similarity in physicochemical and geometrical properties of membranes and intervening liquid films. Differences may arise from different forces that drive electrofusion and viral fusion.

Acknowledgments

I am grateful to Dr. A. Sowers for providing an opportunity to work in his laboratory, where some of the results used in this work were obtained, and for helpful discussions. I thank Dr. R. Blumenthal for the inspiration to measure and analyze delays in fusion and for the excitement of working with him in a beautiful environment. Thanks are due to Dr. J. Saunders for the helpful suggestions in preparing the final version of the manuscript. I appreciate the interesting comments and suggestions of Dr. M. Clague. Thanks are due to Dr. D. Batey for thoroughly reading the manuscript.

References

Bentz J., Nir, S., and Covell, D. G. (1988). Mass-action kinetics of virus-cell aggregation and fusion. *Biophys. J.* **54**, 449–462.
Berg, H., Bauer, E., Berg, D., Forster, W., Hamann, M., Jacob, H.-E., Kurischko, A., Muhlig, P., and Weber, H. (1983). Cell Fusion by electric fields. *Stud. Biophys.* **94**, 93–96.
Blumenthal, R. (1987). Membrane fusion. *Current Topics Membrane Transp.* **29**, 203–254.
Blumenthal, R. (1988). Cooperativity in viral fusion. *Cell Biophys.* **12**, 1–12.
Chen, Y., and Blumenthal, R. (1989). On the use of self-quenching fluorophores in the study of membrane fusion kinetics. The effect of slow probe redistribution. *Biophys. Chem.* **34**, 283–292.
Chernomordik, L. V., Sukharev, I. G., Popov, V. F., Pastushenko, A. V., Abidor, I. G., and Chizmadzhev, Y. A. (1987). The electrical breakdown of cell and lipid membranes: The similarity of phenomenologies. *Biochim. Biophys. Acta* **902**, 360–373.
Clague, M. J., Schoch, C., Zech, L., and Blumenthal, R. (1990). Gating kinetics of pH-activated membrane fusion of vesicular stomatitis virus with cells: Stopped flow measurements by dequenching of octadecylrhodamine fluorescence. *Biochemistry* **29**, 1303–1308.
Dalgleish, A. G., Beverley, P. C., Clapham, P. R., Crawford, D. H., Greaves, M. F., and Weiss, R. A. (1984). The CD4 (T4) antigen is an essential component of the receptor for the AIDS retrovirus. *Nature* **312**, 763–767.

Dimitrov, D. S., and Jain, R. K. (1984). Membrane stability. *Biochim. Biophys. Acta* 779, 437–468.

Dimitrov, D. S. (1983). Dynamic interactions between approaching surfaces of biological interest. *Progr. Surface Sci.* 14, 295–424.

Dimitrov, D. S., and Sowers, A. E. (1990a). A delay in membrane fusion: lag times observed by fluorescence microscopy of individual fusion events induced by an electric field pulse. *Biochemistry* 29, 8337–8344.

Dimitrov, D. S., and Sowers, A. E. (1990b). Membrane electroporation—Fast molecular exchange by electroosmosis. *Biochim. Biophys. Acta* 1022, 381–392.

Heuser, J. E., Reese, T. S., Dennis, M. J., Jan,Y., Jan, L., and Evans, L. (1979). Synaptic vesicle exocytosis captured by quick freezing and correlated with quantal transmitter release. *J. Cell Biol.* 81, 275–300.

Hoekstra, D., and Kok, J. W. (1989). Entry mechanisms of enveloped viruses. Implications for fusion of intracellular membranes. *Biosci. Rep.* 9, 273–305.

Klatzmann, D., Champagne, E., Chamaret, S., Gruest, J., Guetard, D., Hercend, T., Gluckman, J., and Montagnier, L. (1984). T-lymphocyte T4 molecule behaves as the receptor for human retrovirus LAV. *Nature* 312, 767–768.

Lo, M. M. S., Tsong, Y. T., Conrad, M. K., Strittmatter, S. M., Hester, L. H., and Snyder, S. H. (1984). Monoclonal antibody production by receptor-mediated electrically induced fusion. *Nature* 310, 792–794.

Loyter, A., Citovsky, V., and Blumenthal, R. (1988). The use of fluorescence dequenching methods to follow viral fusion events. *Methods Biochem. Anal.* 33, 128–164.

Morris, S. J., Sarkar, D. P., White, J. M., and Blumenthal, R. (1989). Kinetics of pH-dependent fusion between 3T3 fibroblasts expressing influenza hemagglutinin and red blood cells. *J. Biol. Chem.* 264, 3972–3978.

Neumann, E., Sowers, A. E., and Jordan, C. A. (eds.) (1989). "Electroporation and Electrofusion in Cell Biology." Plenum Press, New York.

Rand, R. P., and Parsegian, V. A. (1986). Mimicry and mechanism in phospholipid models of membrane fusion. *Annu. Rev. Physiol.* 48, 201–212.

Sarkar, D. P., Morris, S. J., Eidelman, O., Zimmerberg, J., and Blumenthal, R. (1989). Initial stages of influenza hemagglutinin-induced cell fusion monitored simultaneously by two fluorescent events: cytoplasmic continuity and lipid mixing. *J. Cell Biol.* 109, 113–122.

Sowers, A. E. (1984). Characterization of electric field-induced fusion in erythrocyte ghost membranes. *J. Cell Biol.* 99, 1989–1996.

Sowers, A. E. (1987). "Cell Fusion." Plenum Press, New York.

Spruce, A. E., Iwata, A., White, J. M., and Almers, W. (1989). Patch clamp studies of single cell-fusion events mediated by a viral fusion protein. *Nature* 342, 555–558.

Stenger, D. A., and Hui, S. W. (1988). Human erythrocyte electrofusion kinetics monitored by aqueous contents mixing. *Biophys. J.* 53, 833–838.

Stenger, D. A., and Hui, S. W. (1989). Electrofusion Kinetics: Studies using electron microscopy and fluorescence contents mixing. *In* "Electroporation and Electrofusion in Cell Biology" (E. Neumann, A. E. Sowers, and C. A. Jordan, eds.), Plenum Press, New York, pp. 167–180.

Teissie, I., and Rols, M. P. (1986). Fusion of mammalian cells in culture is obtained by creating the contact between cells after their electropermeabilization. *Biochem. Biophys. Res. Commun.* 140, 258–266.

Zhelev, D. V., Dimitrov, D. S., and Doinov, P. (1988). Correlation between physical parameters in electrofusion and electroporation of protoplasts. *Bioelectrochem. Bioenerg.* 20, 155–167.

Zimmermann, U., Arnold, W. M., and Mehrle, W. (1988). Biophysics of electroinjection and electrofusion. *J.Electrostatics* 21, 309–345.

11

Effects of Intercellular Forces on Electrofusion

Sek Wen Hui[1] and David A. Stenger[2]

[1]Membrane Biophysics Laboratory, Roswell Park Cancer Institute,
Buffalo, New York 14263

[2]Center for Bio/Molecular Science and Engineering, Naval Research Laboratory,
Washington, D. C. 20375

I. Introduction

After a decade of research and application of electrofusion, the forces involved in the fusion process are still not well understood. Despite the quantitative treatment of the force responsible for the membrane electric breakdown process, which leads to the permeation of a single cell membrane (Jeltsch and Zimmermann, 1979; Dimitrov, 1983; Needham and Hochmuth, 1989), there is still no satisfactory description of the force leading to the spontaneous breakdown and fusion of two adjacent membranes brought together prior to fusion. It is therefore desirable to have a testable, quantitative treatment of the fusion forces, from which more precise and efficient fusion protocols may be developed.

In the case of simultaneous breakdown of a pair of cell membranes of apposing cells, an additional force to that due to the supercritical membrane potential should be considered. This force arises from the additional dielectrophoretic (DEP) force between the two adjacent cells, induced by the applied electric pulse. The existence of this force is apparent in the observation of electrofusion of freely suspending cells. The application of a fusing pulse leads to a temporary compaction of pearl-chained cells along the applied field direction. In time-resolved freeze-fracture electron microscopy (Stenger and Hui, 1986), the gap between pearl-chained cells was seen to be further narrowed during and immediately after the pulse application. This force, in addition to that due to the pulse-induced supercritical membrane potential, which causes electromechanical breakdown of individual membranes, may contribute to the simultaneous breakdown and fusion of the pair of readily attached membranes. In addition, in the neighborhood of the cell–cell contact area, the local field strength is modified by the pulse-induced dipoles of these two cells. When this is taken into account, the local field strength is found to be slightly enhanced by the presence of the adjacent cells.

In this review, we will summarize our effort to understand the nature of the forces involved and to test the validity of our approach.

II. Theory

Consider a pair of tightly apposed membranes of two adjacent cells being forced together, either by using a centrifuge or by employing dielectrophoresis alignment (to form a pearl chain), and experiencing a higher field pulse. The higher electric field pulse induces further charge separation within the highly conductive cell, in the low-conductivity external medium, and induces charges on both sides of the practically insulated membranes, all at different rates. The time constants of each process are determined by the complex permittivities of the cell interior ($\varepsilon_i = \varepsilon_i - j\sigma_i/\omega$), the external medium ($\varepsilon_e = \varepsilon_e - j\sigma_e/\omega$), and the cell membranes ($\varepsilon_m = \varepsilon_m - j\sigma_m/\omega$) (Kaler and Jones, 1990). For simplicity, we consider the permittivity of the membrane to be much larger than that of either the cell interior or the exterior—that is, the membrane is a nearly perfect capacitor. Each cell would experience a maximal membrane potential V_m at opposite poles (Holzapfel *et al.*, 1982). Ignoring the endogenous membrane potential, which is much smaller than the induced one, we have

$$V_m = faE_{eff}/(1 + \omega^2\tau_m^2)^{1/2} \qquad (1)$$

where f is a geometrical factor, a is the semimajor axis of the cell along the field direction, ω is the angular frequency (components) of the applied pulse field, τ_m is the membrane charging time, given by

$$\tau_m = aC_m [1/\sigma_i + (f - 1)/\sigma_e] \qquad (2)$$

and C_m is the membrane capacitance. For spherical cells, $f = 1.5$ and a is the radius. The term E_{eff} is the local field strength as modified by the presence of the induced dipoles of the cells by the imposing pulse.

The pressure acting on each individual membrane by the induced membrane potential is

$$\pi_m = 0.5\varepsilon_0\varepsilon_e \, (faE_{eff}/h\sqrt{1 + \omega^2\tau_m^2})^2 \tag{3}$$

and

$$E_{eff} = E + P_{eff}/(16\pi\varepsilon_0\varepsilon_e a^3) \tag{4}$$

where P_{eff} is the effective induced polarization of the cells and h is the dielectric thickness of the membrane.

For spherical cells, P_{eff} may be expressed as

$$P_{eff} = (4/3)\pi a^3\varepsilon_0\varepsilon_e \, \text{Re}\{K_{eff}\}E = BE \tag{5}$$

and the complex effective permittivity of cells in a medium K_{eff} is

$$K_{eff} = 3(\varepsilon_i - \varepsilon_e)/(\varepsilon_i + 2\varepsilon_e) \tag{6}$$

The pressure acting on the pair of membranes forming a sandwich between cells, due to the induced cell dipole interaction, may be approximated by the capacitive force from the net induced charges on both sides of the membrane contact area A, that is, on the inner sides of the adjacent cells. The dielectric constant of the tight sandwich may be approximated by ε_m (discussed later).

$$\pi_d = P_{eff}^2/(8a^2\varepsilon_0\varepsilon_m A^2) \tag{7}$$

The total pressure exerted on the membrane at the contact area A during the pulse may be approximated as

$$\pi_t = \pi_m + \pi_d \tag{8}$$

Therefore, if the permittivities of the membranes, the external medium, and the cell interior are known, one may calculate the values of K_{eff} and P_{eff} by Eqs. (5) and (6), from which the pressures generated by the pulse may be derived by Eqs. (3), (4), (7), and (8). Thereby the mechanical breakdown threshold of the membrane pair may be predicted.

III. Calculation of the Pulse-Induced Pressure

A. Determination of the Effective Polarization of Cells in a Given Medium

It is usually impractical to determine the permittivities of cells and cell membranes. An alternative is to measure the frequency-dependent value of K_{eff} for a particular cell–medium combination. A convenient method to determine the K_{eff} value is to measure the dielectrophoretic force F_d on a typical (spherical) cell, using the relation

$$F_d = 2\pi a^3 \varepsilon_0 \varepsilon_e \, \text{Re}\{K_{eff}\} \, \nabla E^2 \qquad (9)$$

The force F_d may be measured by the Stokes method—at a terminal horizontal velocity v, the dielectrophoretic force on a spherical cell in a horizontal field gradient is counterbalanced by the viscosity drag ($6\pi a \eta v$) on the cell, η being the viscosity of the medium.

Alternatively, the force F_d may also be measured by the levitation method with a vertical field gradient—the dielectrophoretic force on a spherical cell in a vertical field gradient is counterbalanced by the gravitation pull ($4\pi a^3 \, \Delta \rho g/3$) on the cell, g being the gravitation constant. The difference in densities between the cell and medium, $\Delta \rho$, is determined by the sedimentation velocity at zero field gradient, using the Stokes equation (Kaler and Jones, 1990).

Since a, v and ε_e can be measured easily, and if the field gradient is known precisely, $\text{Re}\{K_{eff}\}$ can be determined with considerable accuracy. As mentioned previously, the value of $\text{Re}\{K_{eff}\}$ is dependent on frequency, and thus a spectrum must be obtained. For most electrofusion applications, $\sigma_i \gg \sigma_e$ and $\varepsilon_e > \varepsilon_i$, $\text{Re}\{K_{eff}\}$ is positive within a given frequency range bounded approximately by $1/\tau_m$ and $1/\tau_p$ $= \sigma_i/\varepsilon_i$ (Kaler and Jones, 1990). Within this frequency range, DEP force is positive, that is, the cells are attracted to regions of high field gradient. At frequencies below $1/\tau_m$, the value of $\text{Re}\{K_{eff}\}$ is negative, and hence so is the DEP force—that is, the cells are forced away from regions of high field gradient (Pohl, 1978; Sauer, 1985). Figure 1a and b show the pearl chaining of human erythrocytes (RBC) aligned in fields with frequencies below and above $1/\tau_m$. The spaces adjacent to the electrodes have higher field gradient.

In our setup to measure $\text{Re}\{K_{eff}\}$ by levitation, only positive DEP force is utilized. Therefore the values of $\text{Re}\{K_{eff}\}$ are measurable for the frequency range of positive value only (Stenger et al., 1990). The value of $\text{Re}\{K_{eff}\}$ in our levitation setup is inversely proportional to the square of the root mean square (rms) voltage required to levitate the sample cell (Kaler and Jones, 1990). Figure 2 shows measured levitation spectra of RBC and pronase-treated RBC (PRBC), from which the frequency-dependent $\text{Re}\{K_{eff}\}$ (hence P_{eff}) may be derived. Cells cannot be levitated beyond a certain frequency bandwidth defined by the conductivities of their suspension media. The bandwidth is bound by $1/\tau_m$ and $1/\tau_p$ as discussed above. The most significant difference between the spectra of intact and PRBC was the absence of a measurable low-frequency positive DEP response in the latter. Pronase treatment of RBC removes over 90% of the principal negative charge component of the glycocalyx, sialic acid. Since the external surface of the RBC lipid bilayer may be considered electroneutral, it is reasonable to conclude that it is glycocalyx depletion that lowers the low-frequency polarization capacity. A similar effect has been reported

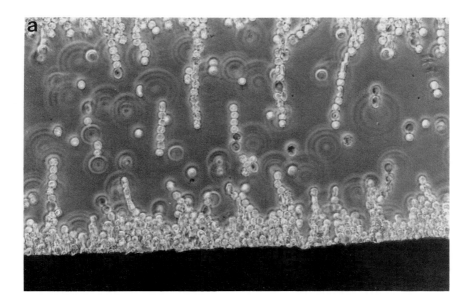

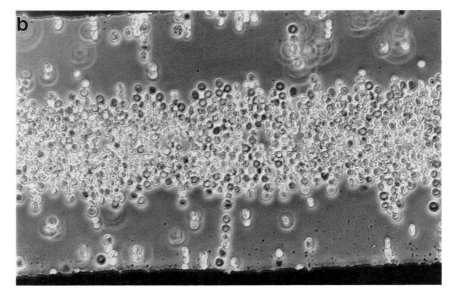

Figure 1   Dielectrophoretic (DEP) alignments of human erythrocytes (RBC) suspended in a 15-μS/cm medium. The alignment was made at (a) 1.5 MHz ($>1/\tau_m$) and (b) 65 kHz ($<1/\tau_m$) AC fields of amplitude 600 V/cm.

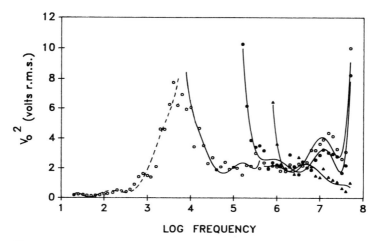

Figure 2 DEP levitation spectrum from 20 Hz to 50 MHz for intact erythrocytes in media of different conductivities. Pronase-treated erythrocytes have identical spectrum less the left (dotted line) portion. Open circles = 15 μS/cm, closed circles = 150 μS/cm, open triangles = 650 μS/cm.

when the cell wall is removed to produce canola plant protoplasts (Kaler and Jones, 1990).

B. Calculation of Pressures on the Membrane Pair

We are interested in calculating the pressure on the membranes in the contact area as a function of time during and immediately after the pulse application. To estimate the pressures as functions of time, using Eqs. (3)–(8), the parameter P_{eff} (hence $Re\{K_{eff}\}$) must be transformed from the frequency-domain to the time-domain values. For a Debye-type polarization, the value of P_{eff} is the sum of the instantaneous induced effective polarization and the decaying effective polarization induced previously. The decay time constant for induced polarization is τ_p (Holzapfel et al., 1982). Thus, we have

$$P_{eff}(t) = (1/\tau_p)B(t)E(t) * \exp(-t/\tau_p) \qquad (10)$$

where the asterisk indicates convolution. By applying the convolution theorem, $P_{eff}(t)$ may be calculated from the measured $Re\{K_{eff}(f)\}$ such that

$$P_{eff}(t) = (1/\tau_p) \, F[B(f)E(f)H(f)] \qquad (11)$$
$$= (4/3\tau_p)\pi a^3 \varepsilon_0 \varepsilon_e \, F[Re\{K_{eff}(f)\}E(f)H(f)]$$

where F indicates the Fourier transform operator, $E(f)$ is the power spectrum of the pulse, and $H(f)$ is the Fourier transform of the exponential function in Eq. (10). The calculation is described in Stenger *et al.* (1991). A typical $P_{eff}(t)$ of PRBC in a 150-µS/cm medium, in response to a rectangular pulse of 10 µs duration, is shown in Fig. 3. Since $Re\{K_{eff}(f)\}$ is nonzero only within a given bandwidth, the DC component of $P_{eff}(t)$ is insignificant. Only at the leading and the trailing edges of the pulse does $P_{eff}(t)$ make significant contribution.

The pressure generated in the initial 1–2 µs of the pulse is sufficient to reduce the gap between the two PRCBs to within 2–3 nm (Stenger *et al.*, 1991), such that the capacitive model is applicable. At this intercellular distance, the dielectric constant of the sandwich in Eq. (7) approaches ε_m. The space between untreated cells cannot be reduced to 2–3 nm (Stenger and Hui, 1986); therefore pronase treatment is necessary to fully utilize the π_d contribution (Stenger and Hui, 1986).

In the geometry of cells in a pearl chain next to a conducting electrode, the effective field is enhanced by the presence of neighboring cells and image charges that developed within the electrode; therefore the value of $P_{eff}(t)$ must be multiplied by such an enhancement factor (Jones *et al.*, 1986).

The pressure $\tau_m(t)$ and $\pi_d(t)$ may now be calculated by Eqs. (3) and (7). The total pressure is given by Eq. (8). The maximum pressure induced by the pulse is then compiled for different pulse conditions. The contributions of π_m and π_d, and

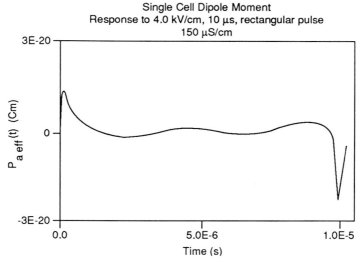

Figure 3   Calculated single cell effective dipole response to the single 4.0-kV/cm, 10-µs rectangular pulse for PRBC suspended in a 150-µS/cm medium.

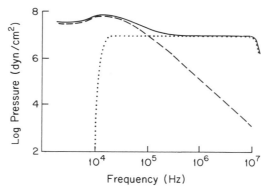

Figure 4   Calculated pressures π_m (dashed line), π_d (dotted line), and π_t (solid line) as functions of the applied pulse field frequency, for PRBCs aligned in three-cell chains next to a electrode, and suspended in a 15-μS/cm medium, in response to a 4-kV/cm sinusoidal pulse.

the total pressure π_t at different frequencies, for PRBC in a 15-μS/cm medium are plotted in Fig. 4. Because of the dependence of π_m on $1/\omega$, the high-frequency contribution due to membrane potential declines precipitously beyond $1/\tau_m$. This decline is compensated by the relatively constant π_d until $1/\tau_p$, at which point the effective polarization rapidly approaches zero. Ignoring the contribution of π_d results in underestimating the total pressure toward the high-frequency limit. For a rectangular pulse, the high-frequency components of the pressure at the leading and the trailing edges of the pulse, when the contribution of π_d is most significant, are thus underestimated.

IV. Experimental Verification

The mechanical breakpoint through the membrane contact area, which is comprised of a pair of tightly apposed membranes and a thin (2–3 nm) intermembranous spacing, is a set property of the membrane pair. The critical field strengths required to initiate fusion under various conditions then give a good gauge of the critical pressure involved.

Human erythrocytes diluted 1 : 10 from a pellet were treated with pronase E (3 mg/ml at 37°C for 30 min), and were suspended in 0.28 M inositol buffer adjusted to different conductivities by adding 1–3 mM of KCl. The cells were induced to form pearl chains of two to four cells in a 300-V/cm, 2.5-MHz field. A rectangular pulse, or a short train of an AC burst, of durations 10–50 μs, at different field strengths, was applied. The minimum field strengths required to initiate 50% fusion of three-cell chains were recorded as the critical fusion field strength E_c. Except for short-length pulses of duration <20 μs, which require higher critical field strength, the critical field strengths for most longer pulses are very similar, with slightly higher E_c for cells in lower-conductivity media. Higher-frequency AC bursts also required slightly higher E_c. The experimentally recorded E_c for 20-μs AC bursts of different frequencies at different medium conductivities are shown in Fig. 5.

Assuming a constant mechanical resistance of the membrane sandwich against the breakthrough pulse, the energy needed to thin the membrane sandwich during the pulse duration T is proportional to $\pi_t T$. We use an empirical value of 325 dyn-s/cm^2 as the critical work to break the membrane sandwich, in order to compare the calculated E_c for different pulse conditions. This value is based on similar calculated results corresponding to the experimental critical fusion threshold for a 4-kV/cm, 10-μs rectangular pulse, or an 8-kV/cm, 15-μs, 2.5-MHz AC burst in cells suspended in a 15-μS/cm medium. These events represent both extremes of

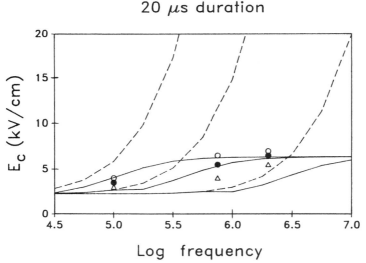

Figure 5   Theoretical threshold curves calculated by accounting for dipole interaction pressures (solid lines), without accounting for dipole interaction pressures (dashed lines), and experimentally obtained (data points) threshold field amplitudes for a 20-μs sinusoidal pulses as a function of frequency. The top, middle, and bottom solid line curves and the left, middle, and right dashed line curves correspond in each case to 15, 150, and 650-μS/cm media, respectively. Experimental data points: open circles, 15 μS/cm; closed circles, 150 μS/cm; and open triangles, 650 μS/cm.

the π_m and π_d contributions. This critical pressure is close to that derived from electromechanical measurements of Needham and Hochmuth (1989). With this breakdown pressure, we calculated E_c for various pulsing conditions. The E_c values for a 20-μs sinusoidal burst at three frequencies, in three media of different conductivities, are plotted in Fig. 5, with and without the contribution of π_d.

Given that the membrane mechanical resistance is derived from different conditions, the calculated and experimental values show remarkable agreement. Furthermore, the calculated curves prove that ignoring the contribution of π_d results in a gross underestimate of π_r, hence an overestimate of E_c.

To determine the importance of close membrane positioning for membrane destabilization, isolated, 6-carboxyfluorescein diacetate (CFDA) labeled cells were randomly suspended in the three media and exposed to single rectangular or sinusoidal pulses. Efflux of the probe was monitored by fluorescence microscopy. Video recordings showed that for rectangular pulses, the efflux occurred from the anode side of the cell, consistent with the contribution of a negative surface potential at the inner cell membrane. The threshold rectangular pulse amplitudes for observable CFDA leakage from >50% of the isolated PRBC are shown in Fig. 6. The CFDA efflux threshold amplitudes were much higher than those for fusion, due in part to the fact that complete lysis was necessary for detectable efflux. However, a very similar pulse duration dependence was observed at less than 30 μs, reflective of the fusion threshold curve. Consistent with our estimate for the dipole moment con-

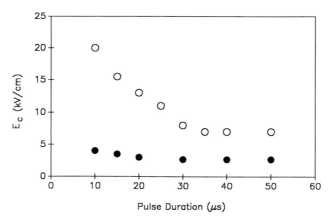

Figure 6   Threshold field amplitudes (open circles) for the efflux of CFDA induced by a single rectangular pulse of various durations. Cells are suspended in a 15-μS/cm medium. Threshold electric field amplitudes necessary for fusion, under the same condition, are shown as closed circles.

tribution of membrane destabilization were observations that (1) the threshold for the efflux of the contents was independent of external conductivity, and more significantly, (2) no sinusoidal pulse of frequencies 10^5 to 2.5×10^6 Hz caused CFDA efflux, even for 17.5 kV/cm (rms) fields of 150 μs duration. This observation reaffirms the theoretical assertion that π_m falls rapidly with increasing frequency so that isolated cells, having no dipole–dipole interactions, remain intact without the contribution of π_d.

V. Concluding Remarks

The fusion process in DEP-aligned cells apparently involves more than the electric breakdown of individual cell membranes. The contribution of the intercellular pressure, caused by the interaction between pulse-induced dipoles of adjacent cells, is an important factor. This is especially true for electrofusion under the conditions of low medium conductivity and high pulse frequency. The latter holds for short rectangular pulses when the high-frequency components of the pulse field at the leading and trailing edges account for a significant portion of the pulse. The π_d at the leading part of any pulse serves to bring the cells even closer than during the alignment, while the remaining part contributes to break the membrane sandwich, probably at discreet locations on the closely contacting area. Ultrastructural analysis supports this hypothesis (Stenger and Hui, 1986).

The clarification of parameters affecting electrofusion threshold leads us to appreciate more of the often-used but insufficiently understood mechanism. Although the above analysis is by no means an exhaustive account of all electric factors, let alone all factors controlling the fusion efficiency, it nevertheless brings us a step closer to designing rational protocols for improving electrofusion efficiency.

Acknowledgment

This work was supported by a grant GM30969 from the National Institutes of Health. The measurement of DEP levitation was done in collaboration with Dr. K. V. I. S. Kaler, University of Calgary.

References

Dimitrov, D. S. (1983). Dynamic interactions between approaching surfaces of biological interest. *Progr. Surface Sci.* 14, 295–424.

Holzapfel, C., Vienken, J., and Zimmermann, U. (1982). Rotation of cells in an alternating electric field: Theory and experimental proof. *J. Membrane Biol.* 67, 13–26.

Jeltsch, E., and Zimmermann, U. (1979). Particles in a homogeneous electrical field: A model for the electrical breakdown of living cells in a Coulter counter. *Bioelectrochem. Bioenerg.* 6, 349–384.

Jones, T. B., Miller, R. D., Robinson, K. S., and Fowlkes, W. Y. (1986). Multipolar interactions of dielectric spheres. *J. Electrostatics* **22**, 231–244.

Kaler, K. V. I. S., and Jones, T. B. (1990). Dielectrophoretic levitation of single cells determined by feedback-controlled levitation. *Biophys. J.,* **57**, 173–182.

Needham, D., and Hockmuth, R. M. (1989). Electro-mechanical permeabilization of lipid vesicles: Role of membrane tension and compressibility. *Biophys. J.* **55**, 1001–1009.

Pohl, H. A. (1978). "Dielectrophoresis." Cambridge University Press, London.

Sauer, F. A. (1985). *In* "Interactions between Electromagnetic Fields and Cells," A. Chiabrera, C. Dinolini, and H. P. Schwan, Plenum, New York, p. 181.

Stenger, D. A., and Hui, S. W. (1986). The kinetics of ultrastructural changes during electrically-induced fusion of human erythrocytes. *J. Membrane Biol.* **93**, 43–53.

Stenger, D. A., Kaler, K. V. I. S., and Hui, S. W. (1991). Dipole interactions in electro-fusion. *Biophys. J.* **59**, 1074–1084.

12

Dynamics of Cytoskeletal Reorganization in CV-1 Cells during Electrofusion

Qiang Zheng and Donald C. Chang[1]

Department of Molecular Physiology and Biophysics, Baylor College of Medicine, Houston, Texas 77030

I. Introduction

It is known from earlier studies that pulsed high-intensity electric fields can cause cells with contacted membranes to fuse (Senda *et al.*, 1979; Neumann *et al.*, 1980;

---

[1]Present address: Department of Biology, Hong Kong University of Science and Technology, Kowloon, Hong Kong.

Guide to Electroporation and Electrofusion

Zimmermann et al., 1981; Teissie et al., 1982). During recent years, this electro-fusion method has been used in many areas of biological research, including plant somatic hybridization (Bates et al., 1987), murine and human hybridoma and, monoclonal antibody production (Lo et al., 1984; Karsten et al., 1985; Pratt et al., 1987; Foung and Perkins, 1989). Electrofusion has also been shown to have many advantages over conventional virus- or polyethylene glycol- (PEG-) induced fusion methods (Zimmermann et al., 1985; Bates et al., 1987; Karsten et al., 1989).

Most previous studies of electrofusion have been focused on applications and development of improved techniques. The use of electrofusion to study the mechanisms of cell fusion is relatively new. For example, electrofusion of human erythrocytes (Stenger and Hui, 1986) and human erythrocyte ghosts (Sowers, 1984, 1989) has been used as a model system to study the mechanisms of membrane fusion. However, fusion of membranes is one of the major steps of cell fusion, especially for intact cells. Cell fusion mainly involves two steps: membrane fusion and cytoplasmic fusion. Unlike artificial membrane systems such as liposomes, intact cells have complicated and well-organized cytoplasms. To form a viable hybrid or syncytium, the fusing cells need to merge and reorganize their cytoplasms into one functional unit (cytoplasmic fusion). Sowers (1984) reported that even in human erythrocyte ghosts, membrane fusion might not lead to cytoplasmic fusion. In some case, cells with their membranes fused may separate at a later time (Gaertig and Iftode, 1989).

Through morphological and biochemical studies, it has been shown that cyto-plasmic filamentous proteins are a prominent part of intracellular organization (Lloyd, et al., 1986). The filamentous proteins, which we now call "cytoskeleton," mainly consist of three systems: microtubules, microfilaments, and intermediate filaments. These filament systems play crucial structural and organizational roles in cellular functions. For example, cell shape, cell motility, and even mRNA localization (Yisraeli et al., 1990) are thought to be controlled by the cytoskeleton. Since the cytoskeletal proteins play an important role in the organization of cytoplasmic structures and functions, the fusion process must involve a restructuring of the cytoskeletal system. The primary purpose of our study is to examine the role of cytoskeleton during cell fusion.

Involvement of cytoskeleton in cell fusion has been previously suggested in virus- and PEG-induced fusion (Holmes and Choppin, 1968; Wang et al., 1979; Kajstura, 1989). Recently, structural and dynamic changes of microtubule and actin networks during various stages of electrofusion have also been investigated (Zheng and Chang, 1989; Blangero et al., 1989; Zheng and Chang, 1990). In this chapter, a more detailed study of the dynamics of cytoskeletal structures during cell electrofusion is presented. The structures of microtubules, actin, and the intermediate filament (IF) vimentin in the fusing cells were examined by fluorescence microscopy. To increase the efficiency of electrofusion and to minimize undesirable side effects, we used an improved electrofusion method in which a radiofrequency (RF) electric field was used to induce membrane fusion, instead of a direct current (DC) electric field (Chang, 1989a, 1989b). Cells grown on substrata can be fused with high efficiency using this method. The method is also simple. After applying a single short electric pulse, cells began to fuse in minutes, and the whole fusion process was completed in 3–4 h. Changes of cytoskeletal structures were examined at various times after the initiation of fusion.

II. Materials and Methods

A. Cell Culture

In this study, CV-1 cells obtained from the American Type Tissue Collection (ATCC, Rockville, MD) were used throughout the experiments. CV-1 cells were maintained in culture in Dulbecco's modified Eagle's medium (DMEM) (Gibco, Gaithersberg, MD) supplemented with 10% fetal bovine serum (FBS) (Gibco) and 100 IU each of penicillin and streptomycin at 37°C. For electrofusion, the cells were plated on 12-mm-square glass coverslips (VWR Scientific, Inc., San Francisco) at 1 : 3 and allowed to grow for 12 h to reach about 50% confluence. This relatively low density of cells was used to reduce the occurrence of multiple-cell fusion.

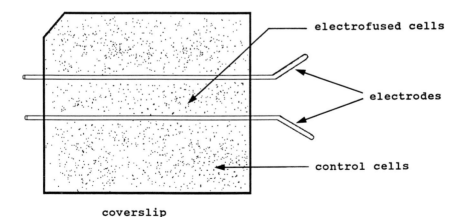

Figure 1   A schematic diagram showing the arrangement of electrodes by which a pulsed RF electric field can be applied to cells grown on a coverslip.

B. Electrofusion of Cells

The coverslip on which CV-1 cells became attached was first rinsed twice with the fusion medium (FM) to wash out the culture medium. The FM consisted of 2 mM HEPES, 280 mM mannitol, and 1 mM MgCl$_2$, with pH adjusted to 7.2. Then two parallel platinum electrodes, separated by a distance of 1.5 mm, were placed on top of the coverslip and a pulsed RF electric field (field strength 1 kV/cm, width 0.2 ms, frequency 100 kHz) was applied to the cells via the parallel electrodes (Fig. 1). Only the cells situated inside the region between the two electrodes were exposed to the electric field. The cells in the areas outside the electrodes were not subjected to electric field and thus served as controls. A few minutes after electrical treatment, the cells were returned to the normal culture medium and cultivated at 37°C with 5% CO$_2$. The cells were fixed for fluorescent staining at various times following the electric treatment (typically at 5, 15, 30, 60, 120, and 180 min).

C. Drug Treatment

Colchicine (Sigma Chemical Co., St. Louis, MO) was stored in a 100 mM stock solution in DMEM without FBS. Before electrofusion, CV-1 cells were first treated with colchicine (0.01, 0.1, 1 mM, respectively) for 1 h. Then the culture medium containing colchicine was washed out by FM. After electric treatment, the cells were returned to regular culture medium containing colchicine (0.01, 0.1, 1 mM respectively) for culture. At various time after the initiation of cell fusion (2, 3, or 4 h), the cells were fixed either by the method described below for fluorescence staining or with 0.25% glutaradehyde for observation by phase microscopy.

Taxol was obtained from the National Cancer Institute, and stored in a 2 mM stock solution in dimethyl sulfoxide (DMSO). CV-1 cells were treated with 10 µM taxol only after electrical treatment. After 2 or 3 h of incubation, the cells were fixed with 0.25% glutaradehyde.

D. Fluorescence Labeling

CV-1 cells in fusion progress were rapidly washed three times with PEMP medium (100 mM PIPES, 5 mM EGTA, 0.5 mM MgCl$_2$, 4% PEG 6000, pH adjusted to 6.9 with NaOH), and then fixed and permeabilized in a combined solution consisting of 3% paraformaldehyde, 1% DMSO, and 0.05% Triton X-100 in PEMP for 30 min at room temperature (Zheng and Chang, 1990). After fixation, the cells were further permeabilized with 0.5% Triton X-100 in phosphate-buffered saline (PBS) (20 mM KH$_2$PO$_4$, 46.7 mM Na$_2$HPO$_4$, 70 mM NaCl, pH 7.2) for 2 min and washed in PBS once.

1. Triple Staining of Microtubule, Actin, and Nuclei

The fixed and permeabilized cells were first incubated with the primary antibody (mouse anti-β-tubulin monoclonal antibody, Calbiochem Co., San Diego, CA, diluted 1 : 20 in PBS) at 37°C for 45 min in a humid chamber. Then the cells were washed in 0.05% Tween 20 in PBS once, and in double-distilled water twice, and incubated with a secondary antibody conjugated with fluorescein [rabbit anti-mouse immunoglobulin G (IgG)/H&L antiserum, Calbiochem Co., San Diego, CA, diluted 1 : 20 in PBS] at 37°C for 45 min. The nonbinding antibodies were washed out in 0.05% Tween 20 and double-distilled water.

After microtubule (MT) staining, microfilaments were labeled by exposing the cells to rhodamine-conjugated phalloidin (3.3 μM stock in methanol, Molecular Probes, Inc., Eugene, OR, diluted 1 : 100 in PBS) for 10 min at room temperature, and washed in PBS once.

Finally, to position the nuclei, the cells were incubated briefly with 1 μg/ml of the fluorochrome, bisbenzimide H33258 (Calbiochem Co.), in PBS for 2 min at room temperature, washed once in PBS, and washed twice in double-distilled water. The labeled cells were mounted in a medium containing Mowiol 4-88 (Calbiochem Co.) following the method of Osborn and Weber (1982).

2. Double Staining of MTs and Vimentin Intermediate Filaments

The fixed and permeabilized cells were first incubated with a mixture of primary antibodies [rabbit anti-tubulin polyclonal antibody, Sigma Chemical Co., St. Louis, MO, diluted 1 : 10 in PBS containing 1% bovine serum albumin (BSA); mouse anti-vimentin monoclonal antibody, Amersham International Co., UK, diluted 1 : 10 in PBS with 1% BSA] at 37°C for 45 min. Then the cells were washed in 0.05% Tween 20 in PBS once, and in double-distilled water twice, and incubated with a mixture of secondary antibodies (goat anti-rabbit antiserum conjugated with FITC, Sigma Chemical Co., St. Louis, MO, diluted 1 : 100 in PBS with 1% BSA; goat anti-mouse IgG/H&L conjugated with rhodamine, American Qualex International, Inc., La Mirada, CA, diluted 1 : 100 in PBS with 1% BSA) at 37°C for 45 min. After washing with 0.05% Tween 20 once, and with double-distilled water twice, the cells were mounted in Mowiol 4-88 medium.

E. Fluorescence Microscopy

Samples were examined under a Zeiss Axiophot microscope (Carl Zeiss, Inc., Germany) with epifluorescence attachment using either a 40× (N.A. = 1.3 oil) or a 63× (N.A. = 1.25 oil) objective. Images were recorded either on Kodak Ektachrome 400 color slide film or on Kodak T-max 400 black and white professional film.

III. Results

After attached CV-1 cells (grown on a glass coverslip) were exposed to a single pulse of radiofrequency (RF) electric field (1 kV/cm, 0.2 ms, 100 kHz), they began to fuse in a few minutes. The fusion process was completed in a few hours. An assay for the completion of fusion was that a large number of cells became multinucleate, and the nuclei in fusing cells had all aggregated (see Fig. 2). The fusion efficiency obtained by this method typically can reach up to 70% (see Table 1).

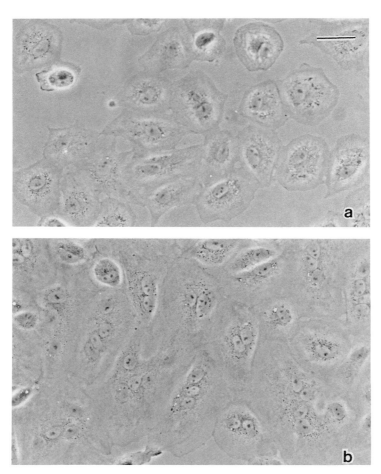

Figure 2   A schematic diagram showing the arrangement of electrodes by which a pulsed RF electric field can be applied to cells grown on a coverslip.

Table 1

Distribution of Nuclei Population in Cells with Various
Numbers of Nuclei[a]

	Relative population of nuclei (%) (N is number of nuclei per cell)		
	$N = 1$	$N = 2$–3	$N \geqslant 4$
Control	81.6 ± 1.3	13.9 ± 1.3	0
Electrofused	26.6 ± 3.8	48.7 ± 4.4	24.6 ± 0.5

[a]Statistical results (mean ± SE) were determined from three sets of experiments. Isolated cells with no fusion partner were not counted. Based on these data, the fusion yield is calculated to be 69%.

A. Changes of the Microtubule Organization during Fusion

Using an immunofluorescent staining technique, we have examined the changes of the microtubule (MT) network during successive stages of cell fusion (see Fig. 3). In control CV-1 cells (i.e., before electric treatment), a network of MTs can be seen in which the MTs radiated outward from the microtubule organizing center (MTOC) to the cell periphery (Fig. 3a). Such a pattern was typical for all interphase CV-1 cells in an attached state without electric treatment.

A few minutes after CV-1 cells were exposed to a pulse of RF electric field, a slightly diffused MT network was observed, and the overall immunofluorescent signal of MTs became slightly weaker than that of the control cells (see Fig. 3b). No clear MTs were observed at the newly formed cytoplasmic bridges at this time. However, about 15 min after the electric treatment, MTs were observed to gradually condense at the cytoplasmic bridges between the fusing cells. These MTs appeared as bundles running parallel to the axis of the bridges and formed a continuous network that connected the MTOCs in the neighboring fusing cells (Fig. 3c).

These parallel MT bundles became more extensive with the progression of fusion. A larger number of such MT bundles that condensed at the cytoplasmic bridges was observed during the next 15–30 min (Fig. 3d and e). A magnified view of such MT bundles is shown in Fig. 4. In some cases (particularly at narrow cytoplasmic bridges), the MT bundles were so dense that the cytoplasmic bridges appeared to be filled entirely with MTs (see arrowhead in Fig. 3d). Most interestingly, at about the same time, the nuclei of fusing cells began to aggregate toward each other.

The nuclei in the fusing cells continued to aggregate during the next few hours. These aggregating nuclei were always seen to be connected by a network of parallel MT bundles. The time required for nuclear aggregation was not uniform among all

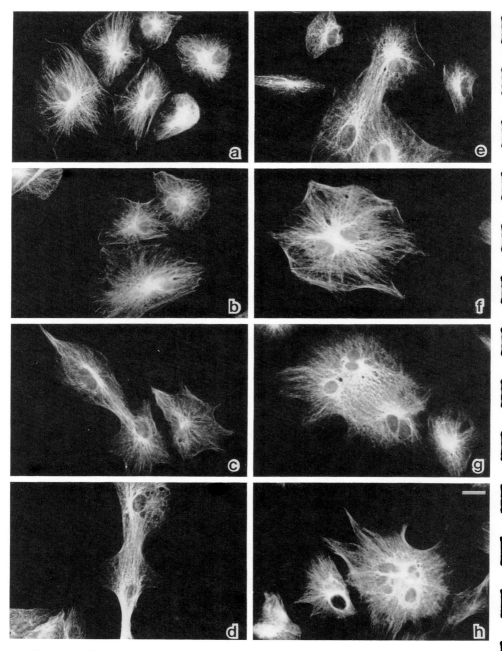

Figure 3 Distribution of microtubules in CV-1 cells at various stages of electrofusion. (a) Control CV-1 cells; (b) at $t = 5$ min, where t is the time following the electric treatment; (c) $t = 15$ min; (d, e) $t = 30$ min; (f) $t = 2$ h; (g, h) $t = 3$ h. Scale bar = 25 μm.

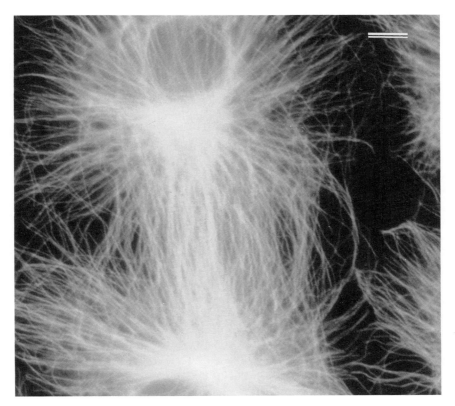

Figure 4   Magnified view of MT bundles in fusing CV-1 cells at 45 min after the initiation of fusion by electric pulsing. Scale bar = 5 μm.

fusing cells. If the fusion involved only two or three cells, the nuclear aggregation would be completed in about 2 h. At this point, MTs adopted a distribution pattern similar to that observed in the control cells, in which MT bundles were seen to radiate from a single MTOC shared by the aggregated nuclei (Fig. 3f). When the fusion involved multiple cells, the nuclear aggregation required more time to become complete. Some of the multicell fusion was completed in 3 h, and nuclei in such syncytia became completely aggregated. In such a case, the structure of the MTs would return to the normal radial distribution pattern with a single MTOC shared by the aggregated nuclei (Fig. 3h). However, some of the multicell fusion needed more time to become complete. Figure 3g shows a syncytium derived from five cells at 3 h after the electric initiation of fusion. Only two pairs of the nuclei had aggregated. The organization of MT apparently was shaped by the distribution of the nuclei; the MT system was mainly composed of networks of parallel MT bundles connecting the two pairs of aggregated nuclei.

B. Changes of Actin Structure during Fusion

The structures of F-actin in CV-1 cells during different stages of cell fusion were examined by labeling with rhodamine-phalloidin. In control CV-1 cells, most of the F-actin filaments were condensed into stress fibers (Fig. 5a), which appeared as thick bundles. Five minutes after the electric pulsation, when cytoplasmic bridges were first established by membrane fusion, part of the actin stress fibers had begun to disappear (Fig. 5b). During the following minutes, the stress fibers continued to fade away (Fig. 5c). At the same time, a more diffused form of F-actin began to appear. At about 30–60 min after the electrical pulse, most of the thick stress fibers normally seen in control cells had largely disappeared. The F-actin appeared either as a diffused form (Fig. 5e) or as bundles, which were thinner than the normal stress fibers (Fig. 5d). These new bundles usually tended to concentrate at the periphery of the fusing cells.

Within 1 h after the initiation of electrofusion, cytoplasms of the fusing cells generally would have largely merged. However, if the fusing partners were originally far apart, the merging of cytoplasms might require more time. In certain syncytia, we have observed the presence of very narrow cytoplasmic bridges at almost 1 h after the initiation of electrofusion. We found many arc-shaped cell edges in these fusing cells (Fig. 5e). These arc-shaped structures were often supported by very heavily condensed actin bundles (arrowhead, Fig. 5e).

As the fusion progressed further, the cellular content of the fusing cells eventually merged. The newly mixed cytoplasmic structures apparently underwent a reorganization. Some stress fibers began to reappear at about 90 min after the electric pulsation. When the syncytium was formed by fusing a small number of cells (e.g., two or three), the fusion process usually would be completed by 2 h. At that time the newly formed stress fibers would have the appearance of normal stress fibers seen in the control cells (Fig. 5f).

C. Changes of Intermediate Filament Network during Fusion

As a main class of filamentous proteins, intermediate filaments (IFs) were suggested to have important cellular functions (Geiger, 1987). We have examined the changes of intermediate filament network during electrofusion. In CV-1 cells, IFs consist mainly of vimentin proteins (Terasaki et al., 1986; Zheng and Chang, 1990). The correlation between the structures of vimentin IFs and MTs was examined using a double-labeling method.

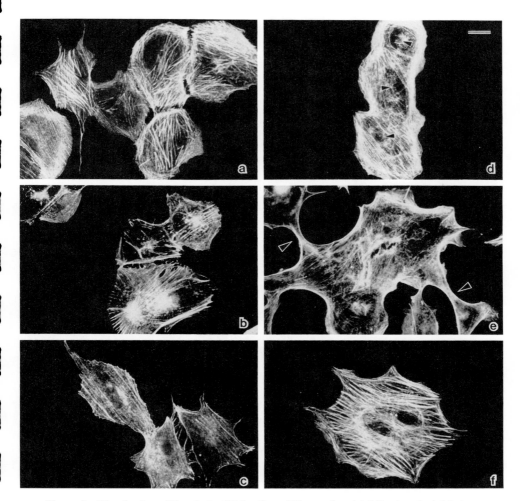

Figure 5   Distribution of F-actin in CV-1 cells at different time (*t*) following the initiation of electrofusion. (a) Control CV-1 cells; (b) *t* = 5 min; (c) *t* = 15 min; (d) *t* = 30 min; the positions of the three nuclei of the fusing cells are marked by arrowheads; (e) *t* = 60 min, arrowheads point to the narrow cytoplasmic bridges; (f) *t* = 2 h. Scale bar = 25 μm.

Vimentin IFs in control CV-1 cells distributed in a pattern similar to MTs; the proteins appeared as a filamentous network radiating from the perinuclear region outward to cell periphery (see Fig. 6a and b). Such a pattern was typical for all interphase CV-1 cells. But unlike MTs, vimentin IFs did not extend into cytoplasmic bridges at the early phase of electrofusion. It was almost 60 min after the initiation

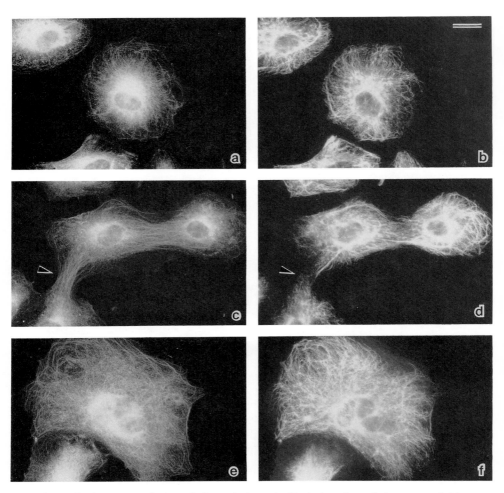

Figure 6 Structures of microtubules and vimentin IFs during electrofusion as revealed by a double labeling method. (a, c, e) MTs. (b, d, f) Vimentin IFs. (a, b) Control CV-1 cells. (c, d) Electrofused CV-1 cells at 60 min after initiation of fusion. Arrowheads point to a cytoplasmic bridge with MT bundles but without vimentin IFs. (e, f) A polykaryon derived from four CV-1 cells at 2 h after electric treatment. Scale bar = 20 μm.

of electrofusion before vimentin IFs were observed to appear in some of the cytoplasmic bridges and distributed along MT bundles (Fig. 6c and d). After this time, vimentin IFs were seen to gradually infiltrate into the cytoplasmic bridges along the MT bundles.

When the nuclei in fusing cells have completely or almost completely aggregated, dense vimentin IFs were observed to condense surrounding the aggregated nuclei. Vimentin IF patterns similar to control cells were seen after the nuclei aggregated, in which vimentin IFs radiated from the perinuclear region toward the cell periphery (Fig. 6e and f).

D. Effects of Drugs on Fusion

The morphological observation using fluorescence microscopy indicated that there was a clear relationship between the structures of MTs and nuclear aggregation during fusion. To further examine the role of MTs in nuclear aggregation, we conducted a series of pharmacological experiments. CV-1 cells were treated with colchicine before and after electrical treatment. A high concentration of colchicine, a microtubule-depolymerizing drug, was found to be capable of inhibiting the nuclear aggregation in fusing cells. Our results showed that although 0.01–1 mM of colchicine did not affect the membrane fusion, the nuclear aggregation was largely inhibited by colchicine (see Table 2). Compared with control cells, the nuclear aggregation was inhibited at a rate of about 50%. Colcemid, an analog of colchicine, has a similar effect on the nuclear aggregation. In these drug-treated electrofused cells, the cytoplasmic bridges still extended as in the control cells, but the bulk of cytoplasms were accumulated around the individual nuclei in fusing cells. The cytoplasmic bridges appeared to be very thin and without much cytoplasm (see Fig. 7b, arrowheads pointing toward the extended cytoplasmic bridges). Immunofluorescence microscopy showed that there were no MT bundles observed in the cytoplasmic bridges; mostly depolymerized and broken MTs were observed surrounding nuclei (see Fig. 8). On the other hand, the structures of F-actin were barely altered by the colchicine treatment, except that a slightly higher proportion of F-actins was concentrated at the cell periphery.

Colchicine treatment not only disrupted MTs, but also prevented vimentin IFs from extending into the cytoplasmic bridges even at the late stage of electrofusion.

Table 2

Effects of Drugs on Nuclear Aggregation in CV-1 Cells during Electrofusion

| | Percentage of aggregated nuclei in syncytia | | | | |
| | | Colchicine | | | |
Time	Control	0.01 mM	0.1 mM	1 mM	Taxol, 10 μM
2 h	70.1 ± 4.0	33.3 ± 4.0	34.6 ± 2.5	36.7 ± 5.2	39.3 ± 4.0
3 h	82.0 ± 3.9	31.8 ± 4.4	38.3 ± 8.7	38.1 ± 2.2	30.1 ± 8.5

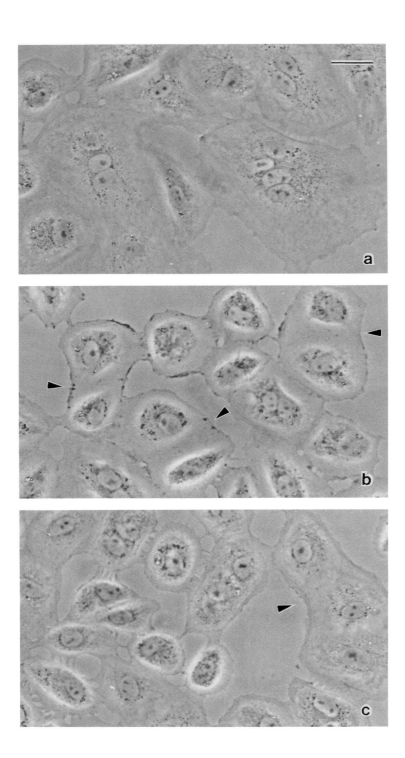

Thick bundles of vimentin IFs were observed to accumulate around the individual nuclei (Hynes and Destree, 1978).

Interesting results were also obtained from treatment with taxol, a microtubule-stabilizing drug. Applying 10 μM taxol in the culture medium after electric treatment could reduce the tendency of nuclear aggregation (see Table 2 and Fig. 7c), but did not affect the extension of cytoplasmic bridges. However, the morphology of taxol-treated cells was different from that of colchicine-treated cells. Unlike colchicine-treated cells, the bulk of each cytoplasmic mass was observed to merge in the cytoplasmic bridges between fusing cells.

IV. Discussion

Our study indicates that cell fusion induced by electric field is accompanied by dynamic changes in cytoskeletal structures. This finding suggests that the reorganization of the cytoskeletal system may play an important role in the mechanisms of cell fusion. In particular, aggregation of nuclei in the fusing cells seems to depend on the formation of parallel MT bundles between the nuclei. Results of drug treatment also confirmed that the nuclear aggregation in fusing cells was closely related to the structure of MTs. Colchicine treatment could prevent both formation of MT bundles and nuclear aggregation. More interesting, taxol, a microtubule-stabilizing drug, also could inhibit the nuclear aggregation. These results suggest that MTs play an active role in the nuclear movement during cell fusion. We suggested that the parallel MT bundles might appear to pull the nuclei together by sliding action. The involvement of MTs in the nuclear aggregation may be explained diagrammatically, as shown in Fig. 9: (a) In the control cells, the MT system is distributed mainly in the form of aster MTs with their minus ends connecting to the microtubule organizing center (MTOC) and their plus ends extending toward the cell periphery. (b) When the neighboring cells are induced to fuse their membranes, a cytoplasmic bridge is formed. (c) Because of the dynamic property of MTs, the aster MTs of the two fusing cells begin to extend into the cytoplasmic bridges by adding more MTs to their plus ends. (d) As MTs that originated from the two MTOCs continue to extend, they overlap to form parallel (or more correctly, anti-parallel) MT bundles. (e) A sliding actin between MTs (generated by some MT-associated motor proteins) can cause MTOCs to aggregate together. Thus, the neighboring nuclei can be pulled together by the MT bundles through their connection with the MTOCs. (f) After

---

Figure 7   Effects of drugs on the nuclear aggregation in CV-1 cells during electrofusion. Cells were fixed at 3 h after the initiation of electrofusion. (a) Electrofused CV-1 cells without drug treatment; (b) treated by 0.1 mM colchicine; (c) treated by 10 μM taxol. Arrowheads point toward the cytoplasmic bridges. Scale bar = 30 μm.

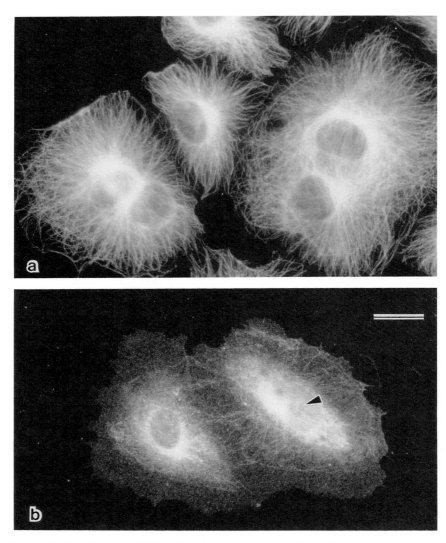

Figure 8 Distribution of MTs and nuclei in electrofused CV-1 cells with (b) and without (a) colcemid treatment. The cells were fixed for immunofluorescence at 1 h after electric pulsation. Arrowhead points to a nucleus surrounded by depolymerized MTs. Scale bar = 20 μm.

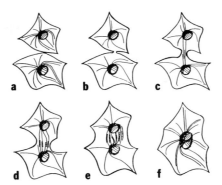

Figure 9   Diagram showing the process by which reorganization of microtubules may be involved in the aggregation of nuclei during electrofusion.

the nuclei have aggregated, the two MTOCs would merge with one another to become a single MTOC. This process is similar to the action of mitotic spindles that push the centrioles apart during mitosis. However, the MT motors here run in a reverse direction.

The result that vimentin IFs tended to continuously distribute along the parallel MT bundles in the later stages of fusion indicated vimentin IFs may also be involved in the nuclear aggregation; for example, vimentin IFs distributed along MT bundles might help MTs to move nuclei together. The time delay between the condensation of vimentin IFs and MTs in the cytoplasmic bridges may due to their difference in polymerization dynamics. Previous work showed that fluorescently labeled tubulin can incorporate into MT networks in minutes after it was microinjected into cells (Keith *et al.*, 1981; Saxton *et al.*, 1984; Kirschner and Mitchison, 1986), while microinjected biotinylated vimentin took hours to integrate into IF networks (Vikstron *et al.*, 1989).

Unlike MTs, F-actin does not seem to be directly involved in the movement of the nuclei. We observed no connection of F-actin bundles to any pair of aggregating nuclei throughout the entire fusion process. The reappearance of the stress fibers at the later stages of cell fusion also bears no relationship to the timing of nuclear aggregation. Furthermore, the pattern of the newly formed stress fibers has no correlation with the position or movement of the nuclei in fusing cells.

The role played by F-actin in cytoplasmic fusion is not as clear as that of microtubules. During the early stages of fusion, F-actin tended to condense at the edges of the cytoplasmic bridges. These newly condensed actin bundles sometimes evolved into dense, arc-shaped belts (see Fig. 5) that resemble the actin edge-bundles observed in 3T3 cells (Zand and Albrecht-Buehler, 1989). These actin edge-bundles are thought to be related to the anchoring structures of the cells; they can also provide mechanical support to the membrane of the cytoplasmic bridges. Furthermore, such actin edge-bundles can serve as a base from which the newly formed lamellipodia can extend outward. Such development would allow the cytoplasmic bridge to extend laterally.

V. Conclusion

We have demonstrated in this work that electric field-induced cell fusion is a useful model for studying the dynamic changes of cytoskeletal structures in the cell-fusion process. We found that all three kinds of cytoskeletal structures underwent dynamic reorganization during electrofusion. The reorganization of MTs may play an important role in cell fusion by providing the mechanical links needed for nuclear aggregation, while the question of whether vimentin intermediate filaments were also involved in such a nuclear aggregation is still unclear. The role of reorganization of F-actin is more complicated because of the diversity of its structures. Condensed actin bundles at the edges of cell and cytoplasmic bridges may help to extend the cytoplasmic bridges laterally to allow the merging of cytoplasms.

Acknowledgments

We thank John Hunt for his technical assistance and the staff of the Marine Biological Laboratory, Woods Hole, for their support. We also thank Dr. Julian P. Heath for his generous help. D.C.C. is a recipient of PHS grant NS25803.

References

Bates, G., Nea, L. J., and Hasenkampf, C. A. (1987). Electrofusion and plant somatic hybridization. In "Cell Fusion" (A. E. Sowers, ed.), Plenum Press, New York, pp. 479–496.

Blangero, C., Rols, M. P., and Teissie, J. (1989). Cytoskeletal reorganization during electric field-induced fusion of Chinese hamster ovary cells grown in monolayers. Biochim. Biophys. Acta 981:295–302.

Chang, D. C. (1989a). Cell poration and cell fusion using a pulsed radiofrequency electric field. In "Electroporation and Electrofusion in Cell Biology" (E. Neumann, ed.), Plenum Publishing, New York, pp. 215–227.

Chang, D. C. (1989b). Cell poration and cell fusion using an oscillating electric field. Biophys. J. 58, 641–652.

Foung, S. K., and Perkins, S. (1989). Electric field-induced cell fusion and human monoclonal antibodies. J. Immunol. Methods 116:117–122.

Gaertig, J., and Iftode, F. (1989). Rearrangement of the cytoskeleton and nuclear transfer in Tetrahymena thermophila cells fused by electric field. J. Cell Sci. 93, 691–703.

Geiger, B. (1987). Intermediate filaments: Looking for a function. *Nature* **329**, 392–393.

Holmes, K. V., and Choppin, P. W. (1968). On the role of microtubules in movement and alignment of nuclei in virus-induced syncytia, *J. Cell Biol.* **39**, 526–543.

Hynes, R. O., and Destree, A. T. (1978). 10 nm Filaments in normal and transformed cells. *Cell* 13:151–163.

Kajstura, J. (1989). F-actin is involved in the polyethylene glycol-induced fusion of chick embryo fibroblasts. *Folia Histochem. Cytobiol.* **27**, 19–24.

Karsten, U., Papsdorf, G., Roloff, G., Stolley, P., Abel, H., Walther, I., and Weiss, H. (1985). Monoclonal anti-cytokeratin antibody from a hybridoma clone generated by electrofusion. *Eur. J. Cancer Clin. Oncol.* **21**, 733–740.

Karsten, U., Stolley, P., Walther, I., Papsdorf, G., Weber, S., Conrad, K., and Pasternak-Kopp, J. (1989). Direct comparison of electric field-mediated and PEG-mediated cell fusion for the generation of antibody producing hybridomas. *Hybridoma* **7**, 627–633.

Keith, C. H., Feramisco, J. R., and Shelanski, M. (1981). Direct visualization of fluorescein-labeled microtubules in vitro and in microinjected fibroblasts. *J. Cell Biol.* **88**, 234–240.

Kirschner, M., and Mitchison, T. (1986): Beyond self-assembly: from microtubules to morphogenesis. *Cell* **45**, 329–342.

Lloyd, C. D., Hyams, J. S., and Warn, R. M. (eds.) (1986). "The Cytoskeleton: Cell Function and Organization." The Company of Biologists.

Lo, M. M., Tsong, T. Y., Conrad, M. K., Strittmatter, S. M., Hester, L. D., and Snyder, S. H. (1984). Monoclonal antibody production by receptor-mediated electrically induced cell fusion. *Nature* **310**, 792–794.

Neumann, E., Gerisch, G., and Opatz, K. (1980). Cell fusion induced by high electric impulses applied to *Dictyostelium*. *Naturwissenschaften* **67**, 414–415.

Osborn, M., and Weber, K. (1982). Immuno-fluorescence and immunocytochemical procedures with affinity purified antibodies: Tubulin-containing structures. *In* "Methods in Cell Biology," Academic Press, New York, vol. 24, Chapter 7.

Pratt, M., Mikhalev, A., and Glassy, M. C. (1987). The generation of Ig-secreting UC 729-6 derived human hybridomas by electrofusion. *Hybridoma* **6**, 469–477.

Saxton, W. M., Stemple, D. L., Leslie, R. J., Salmon, E. D., Zavortink, M., and McIntosh, J. R. (1984). Tubulin dynamics in cultured mammalian cells. *J. Cell Biol.* **99**, 2175–2186.

Senda, M., Takeda, J., Abe, S., and Nakamura, T. (1979). Induction of cell fusion of plant protoplasts by electrical stimulation. *Plant Cell Physiol.* **20**, 1441–1443.

Sowers, A. E. (1984). Characterization of electric field-induced fusion in erythrocyte ghost membranes. *J. Cell Biol.* **99**, 1989–1996.

Sowers, A. E. (1989). The mechanism of electroporation and electrofusion in erythrocyte membranes. *In* "Electroporation and Electrofusion in Cell Biology" (E. Neumann, A. E. Sowers, and C. A. Jordan, eds.), Plenum, New York, pp. 229–255.

Stenger, D. A., and Hui, S. W. (1986). Kinetics of ultrastructural changes during electrically induced fusion of human erythrocytes. *J. Membr. Biol.* **93**, 43–53.

Teissie, J., Knutson, V. P., Tsong, T. Y., and Lane, M. D. (1982): Electric pulse-induced fusion in 3T3 cells in monolayer culture. *Science* **216**, 537–538.

Terasaki, M., Chen, L. B., and Fujiwara, K. (1986). Microtubules and the endoplasmic reticulum are highly interdependent structures. *J. Cell Biol.* **103**, 1557–1568.

Vikstron, K. L., Borisy, G. G., and Goldman, R. D. (1989). Dynamic aspects of intermediate filament networks in BHK-21 cells. *Proc. Natl. Acad. Sci. USA* **86**, 549–553.

Wang, E., Cross, R. K., and Choppin, P. W. (1979). Involvement of microtubules and 10-nm filaments in the movement and positioning of nuclei in syncytia. *J. Cell Biol.* **38**, 320–337.

Yisraeli, J. K., Sokol, S., and Melton, D. A. (1990). A two-step model for the localization of maternal mRNA in *Xenopus* oocytes: Involvement of microtubules and microfilaments in the translocation and anchoring of Vg1 mRNA. *Development* **108**, 289–298.

Zand, M. S., and Albrecht-Buehler, G. (1989). What structures, besides adhesions, prevent spread cells from rounding up? *Cell Motility and the Cytoskeleton* **13**, 195–211.

Zheng, Q., and Chang, D. C. (1989). Changes in cytoskeletal structures during cell fusion induced by a radio-frequency electric field. *J. Cell Biol.* **109**, 4(2):271a.

Zheng, Q., and Chang, D. C. (1990). Dynamic changes of microtubule and actin structures in CV-1 cells during electrofusion. *Cell Motility and the Cytoskeleton* **17**, 345–355.

Zimmermann, U., Scheurich, P., Pilwat, G., and Benz, R. (1981). Cells with manipulated functions: New perspectives for cell biology, medicine, and technology. *Angew. Chem. Int. Ed. Engl.* **20**, 325–344.

Zimmermann, U., Vienken, J., Halfmann, J., and Emeis, C. C. (1985). Electrofusion: A novel hybridization technique. *In* "Advances in Biotechnological Processes," Vol. 4 (A. Mizrahi and A. L. van Wezel, eds.), Liss, New York, pp. 79–150.

Part II

Applications of Electroporation and Electrofusion in Current Research

13

Gene Transfer into Adherent Cells Growing on Microbeads

Huntington Potter and Stefan W. F. Cooke

Department of Neurobiology, Harvard Medical School, Boston, Massachusetts 02115

I. Introduction

Since its inception less than 10 years ago, electroporation as a means of introducing macromolecules into various cell types has become the method of choice in many situations. The traditional method of gene transfer, by uptake of calcium phosphate/DNA coprecipitates (Graham and Van der Eb, 1973), works well with fibroblasts, but has proved difficult to apply to undifferentiated mammalian cells, such as lymphocytes and neurons, and is completely unsuited to plant cells, parasites, or bacteria. However, because it is a physical rather than a biochemical technique, electroporation has a much wider applicability, has been successfully used in essentially all cell types—animal, plant, and microbial—and causes less perturbation of the target cells and electroporated molecules than alternative approaches. In addition to yielding a high frequency of permanent or transient transfectants, electroporation is also substantially easier to carry out than alternative techniques and is highly reproducible.

Although electroporation is effective in a wide variety of cell types, each situation requires slightly different conditions that depend on the special characteristics of the target cell type. Therefore a comprehensive description of all the methods and applications of electroporation is now far beyond the scope of one chapter. The conditions given below and in the methods section of this volume (Chapter 27) should be taken as a useful starting point for extending electroporation to other applications.

II. Advantages of Electroporation for Gene Transfection

In essence, electroporation makes use of the fact that the cell membrane acts as an electrical capacitor, which is unable (except through ion channels) to pass current. Subjecting membranes to a high-voltage electric field results in their temporary breakdown and the formation of pores that are large enough to allow macromolecules, as well as smaller molecules such as ATP, to enter or leave the cell. The reclosing of the membrane is a natural decay process, which can be delayed by keeping the cells at 0°C. Following closure, the exogenous DNA is then free to enter the nucleus and be transcribed in a transient fashion and, at a lower frequency, become integrated into the host genome to generate a permanently transfected cell line.

After electroporation, the exogenously added DNA initially appears to be free in the cell cytoplasm and nucleoplasm (Bertling et al., 1987), rather than being incorporated into phagocytic vesicles, as is the case of DNA taken up as $CaPO_4$ or DEAE-dextran coprecipitates (Graham and Van der Eb, 1973; Sussman et al., 1984). This may explain why electroporation can, in some cells, result in a lower level of mutation of transfected DNA when compared with most traditional gene transfer methods (compare, for example, Drinkwater and Klinedinst, 1986; Bertling et al., 1987; with Calos et al., 1983; Razzaque et al., 1983). Only microinjection results in a similarly low spontaneous mutation frequency of exogenously added DNA (Thomas and Capecchi, 1987). Mechanistically, electroporation can be considered effectively as a mass microinjection procedure. The amount of DNA that can be introduced into the nuclei of electroporated mammalian cells, for instance, is in the range of 0.5 pg, corresponding to 10^4 DNA molecules or 8% of total endogenous host DNA (Bertling et al., 1987). The maximum size of the DNA molecules that can be introduced by electroporation is at least 150 kb (Knutson and Yee, 1987).

Another feature distinguishing electroporation from calcium phosphate coprecipitation for transfection of eukaryotic cells is the state of the *integrated* DNA in the permanent cell lines that arise after selection in appropriate antibiotic media. In the case of calcium phosphate, the amount of DNA taken up and integrated into the genome of each transfected cell is in the range of 3 million bp. As a result, the transfected DNA often integrates as large tandem arrays containing many copies of the transfected DNA. The advantage of this procedure is evident primarily when one wishes to transfect whole genomic DNA into recipient cells and select for some phenotypic change such as malignant transformation. Here, a large amount of DNA integrated per recipient cell is essential. In contrast, electroporation can be adjusted to yield 1 to about 20 copies of an inserted gene (Potter *et al.*, 1984; Toneguzzo *et al.*, 1986, 1988) or a large amount of genomic DNA (Jastreboff, *et al.*, 1987). For gene expression studies low copy number is advantageous, since it allows one to know the particular copy responsible for the gene expression being observed. Even when few copies of an electroporated marker gene are introduced, cotransfection of a nonselectable gene of interest is very efficient (Toneguzzo *et al.*, 1988).

III. Applications of Electroporation in Molecular and Cellular Biology

In addition to generating permanently transfected cell lines carrying genetically engineered genes of interest, electroporation can also be used to introduce DNA into cells for transient expression assays. For this application, the bacterial gene coding for chloramphenicol acetyltransferase (CAT) can be linked to eukaryotic transcription promoter and enhancer sequences. While linearized DNA was found to increase the yield of permanently transfected cells, for transient expression it is preferable to leave the DNA covalently closed and supercoiled, possibly to both reduce degradation and promote transcription initiation. Gene expression is assayed by measuring CAT enzyme activity by virtue of its ability to acetylate [^{14}C]chloramphenicol and change its mobility during thin-layer chromatography (Gorman *et al.*, 1982).

Although the most widespread application of electroporation has been for gene transfer, the technique also can be used to introduce proteins, metabolites, and other small molecules into recipient cells. For instance, actin filaments can be labeled in living carrot cells (Traas *et al.*, 1987) and primary chick corneal fibroblasts (K. Daniels and E. D. Hay, personal communication) after electroporation in the presence of rhodaminyl lysine phallotoxin. This has allowed the visualization of much finer actin filaments during all phases of the cell cycle than was previously possible with fixed cells. K. Daniels (personal communication) has also succeeded in electroporating neuronal cells in developing chick neural tube embedded in agar, a model for studies of primary tissue. Nucleoside triphosphates and other nucleoside analogs (Sokoloski *et al.*, 1986; Knight and Scrutton, 1986), as well as inositol lipids (Van Haastert *et al.*, 1989), can also be introduced into living cells, allowing a number of experiments on intracellular metabolism to be carried out more directly. Finally, various proteins, including antibodies, can be introduced into cells by electroporation (see, for example, Chakrabarti *et al.*, 1989; Berglund and Starkey, 1989), allowing specific intracellular proteins to be labeled and/or inactivated.

Thus a wide variety of molecules that normally cannot be transported across the plasma membrane can be introduced into enough cells to carry out biochemical analyses of the resulting changes in cell metabolism. In such investigations, electroporation is being used as essentially a mass scale version of microinjection and should have fewer adverse effects on cell metabolism than chemical permeabilization.

IV. Potential Applications to Human Gene Therapy

Perhaps the most exciting application of electroporation in mammalian cells is for purposes of gene therapy. Because of their very high efficiency, retroviral vectors have been primarily used to introduce DNA into primary cells prior to their reintroduction into the organism (see, for example, Williams *et al.,* 1984). However, improvements in electroporation make it almost as efficient as retroviral infection for gene transfer, without the potential disadvantages that might arise from using a retrovirus. When a retrovirus enters a target cell, it integrates essentially randomly in the genome and thus has potential for introducing mutational damage by the mere fact of its insertion. In addition, the transcription promoter within the long-terminal repeats (LTRs) of an integrating retrovirus can result in the expression of nearby genes. If the virus integrates adjacent to an oncogene, malignant transformation of the target cell can result. Finally, because the retrovirus still retains the capability of excising from the genome and reintegrating or potentially even reinfecting other cells, provided it is supplied *in trans* with the appropriate proteins (as might occur if the cells are infected by a second retrovirus of a viable type), there is no guarantee that the retrovirus-transfected cells represent a safe and stable means of introducing an engineered gene into a living organism.

In contrast, these drawbacks are lessened by using electroporation for gene transfer. First, there are no long-terminal repeats associated with the introduced genes. Second, there is good evidence that electroporated genes can recombine with their homologous host gene (Smithies *et al.,* 1985; Thomas and Capecchi, 1987). The resultant cell actually acquires the wild-type normal version of its mutant gene at the exact location in the chromosome where it should normally reside, thus reducing the potential mutagenetic effects of random insertion. The successful use of electroporation to introduce genes into the germline of mice is extensively discussed in Chapter 14 by Reid and Smithies in this volume.

V. Electroporation of Adherent Cells on Microbeads

As reviewed above (see also Potter, 1988), and amply illustrated by the other reports in this volume, electroporation has been successfully used to introduce macromolecules into a wide variety of cell types. Although most easily applied to cells that grow normally in suspension, such as lymphocytes, electroporation can also be used to introduce macromolecules into adherent cells that have first been removed from their substrate by trypsin or ethylenediamine tetraacetic acid (EDTA). For instance, fibroblasts (reviewed in Potter, 1988) and neurons (PC12 cells) (Potter and Montminy, 1986) can be removed from their substrates and electroporated in suspension for either transient or stable expression.

However, many situations exist where it would be advantageous to be able to introduce macromolecules into adherent cells still attached to their substrate. For instance, many adherent cells, such as neurons or endothelial cells and even fibroblasts, have a unique morphology when attached to their substrates. In addition, they may show importantly different morphologies and behavior when attached to different kinds of substrates. For instance, neurons "growing" (sending out processes) on collagen or tissue culture plastic behave quite differently from neurons growing on laminin.

A number of transfection approaches are available for introducing macromolecules into cells still attached to a substrate. The classical calcium phosphate–DNA precipitation technique (Graham and Van der Eb, 1973) was developed for introducing DNA into adherent cells. However, this procedure, as discussed above, can be quite toxic and is unsuccessful in many delicate cell types. Lipofectin is also applicable to adherent cells but is expensive and can be toxic to some cells. On the other hand, there is no *a priori* reason why electroporation should not be equally applicable to substrate-bound cells as to suspended cells, provided an appropriate means can be devised for exposing the cells to the electric fields. One such approach is to lay a pair (or a series) of parallel wires onto the substrate surface just above the cells to serve as electroporation electrodes. Dr. Chang elsewhere in this volume presents the results of such a procedure (see Chapter 19).

We have taken a different approach to electroporating adherent cells attached to their substrates, which makes use of available methodology and instrumentation in a new way that will be easily adapted in laboratories already carrying out electroporation in the traditional manner. Specifically, we have been able to electroporate DNA into cells attached to the surface of microbeads in suspension. The results are still preliminary, but the first indication is that the electroporation efficiency is as at least as high as for the same cells in suspension. Because plastic microbeads are very easy to manipulate, come in various types applicable to almost all adherent cells, and can be kept in suspension for short periods of time, it is straightforward to wash a sample of microbeads carrying cells in appropriate electroporation buffer, introduce the beads at any desired concentration into the electroporation chamber in suspension, and carry out the electroporation in the same manner as for suspended cells alone. As long as the concentration of microbeads is not huge, so that their contribution to the total volume remains small, drastic changes in electroporation parameters seem, by our preliminary experiments, not to be necessary. The procedure should in principle be applicable to any adherent cell type for both transient and stable expression. As with all electroporation experiments, however, optimization for each cell type and, in this case, probably for high concentrations of microbeads would be advisable for obtaining high transfection efficiencies.

To test and demonstrate the applicability of electroporation of cells on microbeads, we introduced a plasmid carrying an RSV promoter-driven β-galactosidase gene into COS-1 cells growing on Cytodex 1 (Pharmacia) microbeads. The microbeads were first prepared according to the manufacturer's instructions. COS-1 cells growing in tissue culture dishes were harvested by standard procedures, added to the resuspended microbeads, and placed in a bacterial or tissue culture dish. Microscopic examination after several hours indicated good adherence and appropriate morphology of the cells on the microbeads.

The following day, the microbeads together with their medium were removed from the petri dish and electroporated with the Bio-Rad Gene Pulser according to the protocol described elsewhere in this volume (see Chapter 29).

Microscopic examination of the COS cells on the microbeads immediately following electroporation indicated the usual slightly grainy appearance often seen in cells following traditional electroporation, but the cells were still attached in approximately the same frequency as before electroporation. Within several hours the cells had completely recovered (except for those that had died and fallen off the microbeads), and by the next day the cells had the appearance of normal, untransfected COS cells. Forty-eight to 72 h following electroporation, the microbeads together with their cells were harvested by settling and were stained for the presence of β-galactosidase according to standard methods (see Chapter 28 by MacGregor, this volume).

The successful use of standard equipment to electroporate cells growing on microbead substrates should be generally applicable to all adherent cells and offers many advantages over electroporation of trypsinized cells in suspension, as discussed above. In addition, for studies of stably transfected lines, the microbeads offer an additional advantage in that they serve essentially as micro microculture wells. That is, once a cell is growing on a microbead, its descendents tend to stay on that microbead, and thus a clone develops. If the electroporation is carried out on cells stuck to microbeads at a ratio of approximately one per microbead, then each microbead will, after growth for a few days or a week (depending on the division time of the cell), contain a clone of a single transfected cell. These clones can be recovered if a selectable marker is introduced with the DNA. Alternatively, the cells can be stained with antibodies and the positive microbeads recovered either visually (for instance by virtue of their fluorescent antibody-labeled cells) or by the application of antibody panning.

In summary, we have succeeded in adapting electroporation to the transfection of DNA into adherent cells growing on a substrate. This procedure should be particularly useful for studying the effect of introducing macromolecules into cells such as neurons, whose specific morphology on their substrate is important for their function.

Acknowledgments

I am grateful to the many colleagues who, over the years, have communicated their electroporation results to me prior to publication. The work in the laboratory is supported by grants from the National Institutes of Health and the Alzheimer's Disease and Related Disorders Association.

References

Berglund, D. L., and Starkey, J. R. (1989). Isolation of viable tumor cells following introduction of labelled antibody to an intracellular oncogene product using electroporation. *J. Immunol. Methods* **120**, 79–87.

Bertling, W., Hunger-Bertling, K., and Cline, M. J. (1987). Intranuclear uptake and persistence of biologically active DNA after electroporation of mammalian cells. *J. Biochem. Biophys. Methods* **14**, 223–232.

Calos, M. P., Lebkowski, H. S., and Botchan, M. R. (1983). High mutation frequency in DNA transfected into mammalian cells. *Proc. Natl. Acad. Sci. USA* **80**, 3015–3019.

Chakrabarti, R., Wylie, D. E., and Schuster, S. M. (1989). Transfer of monoclonal antibodies into mammalian cells by electroporation. *J. Biol. Chem.* **264**, 15494–15500.

Drinkwater, N. R., and Klinedinst, D. K. (1986). Chemically induced mutagenesis in a shuttle vector with a low-background mutant frequency. *Proc. Natl. Acad. Sci. USA* **83**, 3402–3406.

Gorman, C. M., Moffat, L. F., and Howard, B. H. (1982). Recombinant genomes which express chloramphenicol acetyltransferase in mammalian cells. *Mol. Cell. Biol.* **2**, 1044–1051.

Graham, L. F., and Van der Eb, A. (1973). A new technique for the assay of infectivity of human adenovirus 5 DNA. *Virology* **52**, 456–567.

Jastreboff, M. M., Ito, E., Bertino, J. R., and Narayanan, R. (1987). Use of electroporation for high-molecular-weight DNA-mediated gene transfer. *Exp. Cell. Res.* **171**, 513–517.

Knight, D., and Scrutton, M. (1986) Gaining access to the cytosol: The technique and some applications of electropermeabilization. *Biochem. J.* **234**, 497–506.

Knutson, J. C., and Yee, D. (1987). Electroporation: Parameters affecting transfer of DNA into mammalian cells. *Anal. Biochem.* **164**, 44–52.

Potter, H. (1988). Electroporation in biology: Methods, applications, and instrumentation. *Anal. Biochem.* **174**, 361–373.

Potter, H., and Montminy, M. (1986). Introduction of cloned genes into PC 12 pheochromocytoma cells by electroporation. *Disc. Neurosci.* **3**, 138–143.

Potter, H., Weir, L., and Leder, P. (1984) Enhancer-dependent expression of human κ immunoglobulin genes introduced into mouse pre-B lymphocytes by electroporation. *Proc. Natl. Acad. Sci. USA* **81**, 7161–7165.

Razzaque, A., Mizusawa, H., and Seidman, M. M. (1983). Rearrangement and mutagenesis in a shuttle vector plasmid after passage in mammalian cells. *Proc. Natl. Acad. Sci. USA* **80,** 3010–3014.

Smithies, O., Gregg, R. G., Boggs, S. S.,Koralewski, M. A., and Kucherlapati, R. S. (1985). Insertion of DNA sequences into the human chromosome β-globin locus by homologous recombination. *Nature* **317,** 230–234.

Sokoloski, J. A., Jastreboff, M. M., Bertino, J. R., Sartorelli, A. C., and Narayanan, R. (1986). Introduction of deoxyribonuclease triphosphates into intact cells by electroporation. *Anal. Biochem.* **158,** 272–277.

Sussman, D. J., and Milman, G. (1984). Short-term, high-efficiency expression of transfected DNA. *Mol. Cell. Biol.* **4,** 1641.

Thomas K. R., and Capecchi, M. R. (1987). Site-directed mutagenesis by gene targeting in mouse embryo-derived stem cells. *Cell* **51,** 503–512.

Toneguzzo, F., Hayday, A. C., and Keating A. (1986). Electric field mediated DNA transfer: transient and stable gene expression in human and mouse lymphoid cells. *Mol. Cell. Biol.* **6,** 703–706.

Toneguzzo, F., Keating, A., Lilly, S., and McDonald, K. (1988). Electric field-mediated gene transfer: Characterization of DNA transfer and patterns of integration in lymphoid cells. *Nucleic Acids Res.* **16,** 5515–5532.

Traas, J. A., Doonan, J. H., Rawlins, D. J., Shaw, P. J., Watts, J., and Lloyd, C. W. (1987). An actin network is present in the cytoplasm throughout the cell cycle of carrot cells and associates with the dividing nucleus. *J. Cell. Biol.* **105,** 387–395.

Van Haastert, P. J., De Vries, M. J., Penning, L. C., Roovers, E., Van der Kaay, J., Erneux, C., and Van Lookeren Campagne, M. M. (1989). Chemoattractant and guanosine 5'-[gamma-thio]triphosphate induce the accumulation of inositol 1,4,5-trisphosphate in *Dictyostelium* cells that are labelled with [^{3}H]inositol by electroporation. *Biochem. J.* **258,** 577–86.

Williams, D. A., Lemischka, I. R., Nathan, D. G., and Mulligan, R. C. (1984). Introduction of new genetic material into pluripotent haematopoietic stem cells of the mouse. *Nature* **310,** 476–480.

14

Gene Targeting and Electroporation

Laura H. Reid and Oliver Smithies

Department of Pathology, University of North Carolina, Chapel Hill,
North Carolina 27599

I. Introduction

DNA introduced into mammalian cells usually integrates into random sites in the cellular genome by nonhomologous recombination. However, introduced DNA can occasionally be directed to specific sites of integration by homologous recombination between sequences present in that DNA and in a chromosomal locus. This protocol, now referred to as gene targeting, was first demonstrated by Smithies *et al.* in 1985. Since then gene targeting has developed into a powerful tool in molecular biology because the directed integration of introduced DNA allows one to create planned chromosomal alterations in mammalian cells.

Homologous recombination between a native chromosomal gene and exogenous DNA is, however, a rare event in mammalian cells, occurring approximately 1000-fold less frequently than nonhomologous recombination (Smithies *et al.*, 1985; Thomas and Capecchi, 1987). Consequently, gene targeting in mammalian cells is dependent on effective methods of transformation, since only a small fraction of cells taking up the DNA will integrate it at the specific target locus. Electroporation provides a fast and inexpensive method of introducing exogenous DNA into mammalian cells. It can easily produce large numbers of transformed cells, often with cell types resistant to other transformation protocols (Potter *et al.*, 1984). For these reasons, electroporation is the preferred method of introducing plasmids in gene targeting experiments.

In this report, we describe several gene targeting experiments using electroporation, with particular reference to the methods used to isolate targeted mammalian cells. We also discuss how gene targeting in combination with embryonic stem (ES) cell technology can lead to the generation of mice with predetermined mutations.

II. Types of Gene Targeting Constructs

Gene targeting experiments in mammalian cells use two types of targeting constructs, both originally described in yeast (Hinnen *et al.*, 1978) and illustrated in Fig. 1. These constructs are referred to either as O- and Ω-type, based on the shape of the linear target DNA when pictorially aligned with its homologous chromosomal locus (Gregg and Smithies, 1986), or as insertion- and replacement-type, based on the consequence of their integration into the chromosomal locus (Thomas and Capecchi, 1987). Each construct usually contains both homologous regions of DNA, which have the same nucleotide sequence as the target gene, and nonhomologous regions of DNA, which are not represented in the cellular genome. The geometrical arrangement of these homologous and nonhomologous blocks determines the outcome of the targeting event. In O-type (insertion) recombination, a single crossover event inserts the targeting construct into the target locus so that the recombinant locus has a duplication of the homologous DNA. In Ω-type (replacement) recombination, a double crossover event leads to replacement of the chromosomal DNA with the targeting construct. Gene targeting has been observed with both types of constructs at approximately equal frequencies in mammalian cells (Thomas and Capecchi, 1987).

III. Methods of Isolating Targeted Cells

A typical targeting experiment may electroporate 10^7 cells to generate 10^3 colonies that have integrated the exogenous DNA at random sites in the cellular genome, and only one colony that has integrated the DNA at the specific target locus. The isolation of the rare targeted cells from among the mass of nontargeted and nontransformed cells can be difficult. Currently, two protocols are used for the isolation procedure. Both methods rely upon the activity of selectable sequences located either in the target gene or on the introduced DNA.

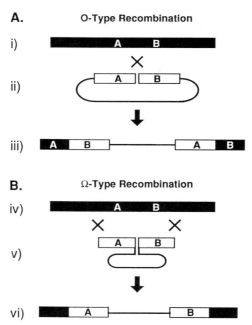

Figure 1   Construct designs for gene targeting. Homologous recombination with an O-type construct (A) and an Ω-type construct (B) are illustrated. (A) A single crossover event inserts the O-type construct (ii) into the chromosomal locus (i) to create a duplication in the recombinant locus (iii). (B) A double crossover event exchanges homologous sequence in the chromosomal target (iv) and in the Ω-type construct (v). No duplication of the homologous DNA is generated in the recombinant locus (vi). Homologous regions are aligned and represented as lettered boxes; nonhomologous regions are represented as thin lines. Crosses indicate the sites of homologous recombination.

A. Direct Selection

Targeted cells can sometimes be isolated by a selectable alteration in the activity of the target gene itself. For example, the recombination event may alter the expression of the target gene to confer resistance to a selective medium. In these direct selection experiments, the targeting DNA is introduced by electroporation, and treated cells are plated in selective medium that prevents the growth of cells that are not transformed or that have integrated the DNA at a random locus. All surviving colonies should therefore contain the targeted alteration.

The hypoxanthine phosphoribosyltransferase (*hprt*) locus has been a frequent target in direct selection experiments since cells that exhibit the presence or absence of *hprt* activity can be selected *in vitro* by their growth in hypoxanthine/aminopterin/thymidine (HAT) medium or 6-thioguanine (6-TG) medium, respectively. This selectable phenotype was utilized in targeting experiments by Doetschman *et al.* (1987), later repeated by Thompson *et al.* (1989), that corrected an *hprt* mutation in the mouse ES cell line, E14TG2a. These male-derived cells are HPRT⁻ as the result of a spontaneous deletion that removed the promoter, first two exons, and an undetermined length of upstream DNA from the single, X-linked *hprt* locus (Hooper *et al.*, 1987). As shown in Fig. 2, the E14TG2a cells were transformed with an O-type targeting construct that contained about 5 kb of homology to the mutant *hprt* locus arranged so that homologous recombination between the construct and chromosomal sequences inserted the missing *hprt* promoter and exon DNA. Suc-

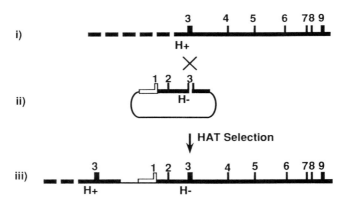

Figure 2   Correction of a mutant *hprt* locus by gene targeting. Homologous recombination at the mutant *hprt* locus in the HPRT⁻ E14TG2a mouse ES cells is illustrated. A single crossover event inserts the O-type targeting DNA (ii) into the mutant *hprt* locus (i) to create an intact *hprt* gene in the recombinant locus (iii). The HPRT⁺ targeted cells are isolated by their ability to survive HAT medium. Boxes represent exons; thick lines represent *hprt* genomic DNA; thin lines represent nonhomologous sequences; interrupted lines represent genomic DNA normally several kilobasepairs upstream of the *hprt* locus. The targeting construct contains *hprt* DNA derived from the human gene (open boxes) and *hprt* DNA derived from the mouse gene (filled boxes). A *Hind*III site is present in the target locus (indicated by H +), but is removed from the targeting DNA (indicated by H −) due to a 4-bp insertion. See Doetschman *et al.* (1987) for more details.

cessfully targeted cells contained an intact *hprt* gene and were isolated by their ability to survive in HAT medium.

Similar selective methods have been used in gene targeting experiments that corrected characterized mutations at the adenine phosphoribosyltransferase locus in Chinese hamster ovary cells (Adair *et al.*, 1989) and mutations at immunoglobulin loci in hybridoma cells (Baker *et al.*, 1988; Baker and Shulman, 1988; Smith and Kalogerakis, 1990). In these latter experiments, the activity of the immunoglobulin loci in transformed cells was determined by complement-mediated lysis assays, rather than by growth in selective medium. Direct selection can also be used to isolate cells with newly created mutations at the target locus. Thomas and Capecchi (1987) and Doetschman *et al.* (1988) isolated ES cells inactivated at the *hprt* locus by gene targeting based, in part, on their survival in 6-TG medium.

B. Positive and Negative Enrichment

Few genes have phenotypes that are directly selectable *in vitro*. Thus, indirect methods of isolating targeted cells are required for most gene targeting experiments. One method uses positive selection for a marker gene included in the targeting DNA to help isolate targeted cells. This technique enriches for the targeted population since cells that are not transformed are eliminated. In this way, only the colonies that contain the targeting DNA are screened for its incorporation at the target locus.

Once the enrichment phase has been accomplished by selection there still remains the problem of identifying the targeted cells within the population of transformed cells. Usually, novel genomic bands generated by homologous recombination are used to distinguish the rare targeted cells from the majority of cells that have integrated the DNA at random locations. In the original gene targeting experiment by Smithies *et al.* (1985), a *supF* gene, which suppresses amber mutations, was included in the targeting DNA. Targeted cells were detected by an assay that depended on rescuing a diagnostic fragment in amber-containing bacteriophages. An improvement on this genome analysis was developed by Kim and Smithies (1988) using the polymerase chain reaction (PCR). This procedure relies upon the prudent selection of two PCR primers. One primer is complementary to sequences present in the introduced DNA. The other primer is complementary to sequences present in the chromosomal locus. Effective PCR amplification is only possible when these primers are juxtaposed after homologous recombination at the target locus. Targeted cells are therefore distinguished from nontargeted cells by their ability to amplify the diagnostic PCR band.

Figure 3 shows an example of the positive selection and PCR protocols as they were used by Koller and Smithies (1989) to isolate mouse ES cells that contained a targeted mutation at the β2-microglobulin (*β2m*) gene. Normal mouse ES cells

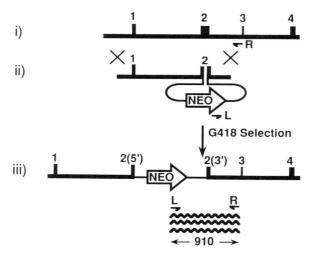

Figure 3   Inactivation of the β2m locus by gene targeting. A double crossover event replaces the normal β2m DNA (i) with the neo-disrupted DNA on the Ω-type targeting construct (ii). Transformed cells are isolated in G418 medium and individually screened for homologous recombination. Only targeted cells with the recombinant β2m locus (iii) can amplify the diagnostic PCR fragment (wavy lines) between the two primers (L and R). Boxes represent exons; thick lines represent β2m genomic DNA; thin lines represent nonhomologous sequences; arrow represents the neo gene. See Koller and Smithies (1989) for more details.

were transformed with an Ω-type targeting construct that contained a neomycin phosphotransferase (neo) gene, which confers resistance to the antibiotic G418, inserted into the second exon of a 5-kb β2m gene fragment. Cells transformed with the targeting DNA were selected in G418 medium and individually screened by PCR for the recombinant fragment. With this protocol, β2m targeted cells were identified at a frequency of about 1 per 100 G418-resistant colonies.

The enrichment provided by positive selection can be increased when the selectable gene insert lacks promoter or polyadenylation signals (Jasin and Berg, 1988). In the absence of these regulatory sequences, the marker gene will be expressed only when it is inserted into genome sites that provide or do not require the missing regulatory signals. Such sites are rare in the mammalian genome and are usually associated with active genes. In this way, fewer nonhomologous transformants are included in the screening procedure.

Another method of eliminating the nonhomologous transformants is the positive

A.   **Targeted Integration**

i)

ii)

iii)

B.   **Random Integration**

iv)

v)

vi)

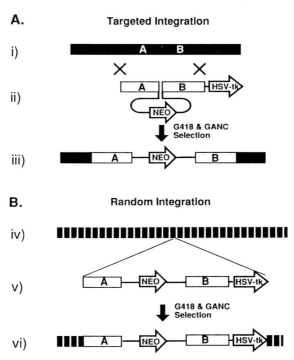

Figure 4 Positive and negative selection for gene targeting. Targeted integration (A) and random integration (B) of PNS targeting DNA are illustrated at two chromosomal loci. One locus (heavy black lines) has homology to the targeting DNA; the other locus (interrupted line) has no homology to the targeting DNA. The *neo* and HSV-*tk* genes used to isolate the targeted cells are represented by arrows. (A) Homologous recombination between the target chromosomal locus (i) and the PNS targeting DNA (ii) introduces the *neo* gene, but not the HSV-*tk* gene into the recombinant locus (iii). Recombinant cells are G418-resistant and GANC-resistant. (B) Nonhomologous recombination between an unrelated chromosomal locus (iv) and the ends of the PNS targeting DNA (v) inserts both the *neo* and the HSV-*tk* genes into the recombinant locus (vi). Recombinant cells are G418-resistant, but GANC-sensitive. This protocol was originally described by Mansour *et al.* (1988).

and negative selection (PNS) protocol developed by Mansour *et al.* (1988) and illustrated in Fig. 4. With this method, targeted cells are isolated by a combination of positive selection for cells that contain the targeting DNA and negative selection against cells that have randomly incorporated it into their genome. The PNS method requires two selectable genes. The *neo* gene is often used as the positive selection marker inserted into the middle of the targeting DNA. The Herpes simplex virus thymidine kinase (HSV-*tk*) gene, which produces toxic metabolites of ganciclovir (GANC) when expressed in cells, has been most frequently used as the negative selection marker. This gene is added to one or both ends of an Ω-type targeting construct.

In PNS experiments, the targeting DNA is introduced by electroporation and treated cells are grown in medium containing G418 and GANC. Nontransformed cells are eliminated by the G418 selection. Since the HSV-*tk* gene is located outside the region of DNA homology, it will not be integrated at the target locus during homologous recombination (see Fig. 4A). Targeted cells will therefore lose the HSV-*tk* gene and will survive in GANC selection. In contrast, transformed but nontargeted cells will die in GANC selection, since the HSV-*tk* gene will be integrated with the targeting DNA during nonhomologous recombination (see Fig. 4B). Thus, only targeted cells and those random transformants that contain mutations in the HSV-*tk* gene will survive both G418 and GANC selection. With this double enrichment procedure, targeted cells have been isolated at frequencies ranging from 1 per 10 to 1 per 1000 of the G418- and GANC-resistant colonies (Mansour *et al.*, 1988; Johnson *et al.*, 1989; DeChiara *et al.*, 1990; McMahon and Bradley, 1990; Thomas and Capecchi, 1990).

C. Nonselectable Alterations

As shown in Table 1, the direct selection protocol as well as the positive and negative enrichment techniques have allowed the isolation of many cell lines modified by gene targeting. Yet enrichment techniques are restricted to alterations that are compatible with the insertion of a selectable gene into the target locus. Future gene targeting experiments will no doubt attempt to create more subtle modifications in the target locus. Thus, new enrichment methods are being developed to assist in isolating targeted cells with nonselectable alterations.

1. IN-OUT Targeting Design

As mentioned in Section II, O-type gene targeting generates duplicate regions of the homologous DNA at the target locus (see Fig. 1B). Like other direct repeats, this duplication may be unstable in targeted cells, so spontaneous recombination between the repeats can lead to the excision of one copy of the duplicated DNA. Such recombinant cells have only one copy of the duplicated alleles at the target

Table 1
Genes Modified by Gene Targeting in
Mammalian Cells

Target gene	Reference
c-abl	Schwartzberg et al. (1990)
aprt	Adair et al. (1989)
aP2	Johnson et al. (1989)
adipsin	Johnson et al. (1989)
CD4	Jasin et al. (1990)
dhfr	Zheng and Wilson (1990)
en-2	Joyner et al. (1989)
c-fos	Johnson et al. (1989)
β-globin	Smithies et al. (1985)
box 1.1	Zimmer and Gruss (1989)
hprt	Thomas and Capecchi (1987)
	Doetschman et al. (1987)
	Doetschman et al. (1988)
	Thompson et al. (1989)
Ig κ	Baker and Shulman (1988)
Ig μ	Baker et al. (1988)
	Smith and Kalogerakis (1990)
IGF-II	Dechiara et al. (1990)
int-1	Thomas and Capecchi (1990)
	McMahon and Bradley (1990)
int-2	Mansour et al. (1988)
β2m	Koller and Smithies (1989)
	Zijlstra et al. (1989)
MHC E_α	Brinster et al. (1989)
N-myc	Charron et al. (1990)
RPII215	Steeg et al. (1990)

locus. The choice of which copy remains after the excision should be random. Thus, by designing the initial targeting DNA correctly, it should be possible to create characterized mutations that are retained after a spontaneous excision reaction.

This protocol, referred to as IN-OUT gene targeting, was originally described in yeast (Scherer and Davis, 1979). Its ability to introduce small, nonselectable alterations in mammalian cells has recently been demonstrated by Valancius and Smithies (1991). These investigators first corrected HPRT⁻ ES cells using the O-type targeting plasmid constructed by Doetschman et al. (1987) (illustrated in Fig. 2). The genomic sequences present in this particular targeting plasmid differ from the homologous chromosomal DNA by a restriction enzyme alteration: a HindIII site in the endogenous hprt locus was removed from the introduced plasmid by a 4-bp insertion (see Fig. 2). Thus, the HPRT⁺ targeted cells generated during this first, "IN," step, contained an inexact duplication of the homologous DNA at the *hprt* locus. During the second, "OUT," step, spontaneous HPRT⁻ revertants were isolated from cultures of the HPRT⁺ targeted cells at a frequency of 1 per 10^6 plated on 6-TG medium. Further analysis revealed that 23 of the 26 6-TG-resistant colonies examined had been generated by an accurate excision reaction. In addition, 19 of 20 6-TG-resistant colonies retained the restriction enzyme alteration introduced on the original targeting DNA. These 6-TG revertants are identical to the original HPRT⁻ ES cells except for the 4-bp restriction enzyme alteration.

The IN-OUT experiments by Valancius and Smithies (1991) relied upon the selectable phenotype of the target *hprt* locus to isolate the targeted cells during the IN step and those spontaneous recombinants arising in the OUT step that contain a subsequent excision event. At nonselectable loci, these cells could be isolated with the aid of a doubly selectable gene or a combination of genes located in the non-homologous region of the O-type targeting DNA. The *hprt* minigene that we have constructed (Reid *et al.*, 1990) is a good candidate for the doubly selectable gene in IN-OUT experiments of this type, since both its introduction and removal from cells can be detected in selective medium, and because ES cells deficient in the X-linked *hprt* gene are available (Hooper *et al.*, 1987; Kuehn *et al.*, 1987).

Figure 5 illustrates the design of an IN-OUT targeting procedure that, in conjunction with the *hprt* minigene, could be used at a nonselectable locus. During the IN step, the *hprt* gene and the desired alteration are introduced into the target locus of HPRT⁻ cells using an O-type targeting vector to produce an inexact duplication of the homologous DNA. HPRT⁺ targeted cells are identified by their growth in HAT medium and by PCR analysis to detect a recombinant fragment. During the OUT step, spontaneous recombination between the duplicated regions removes the *hprt* gene and one copy of the homologous DNA. These HPRT⁻ recombinant cells are isolated by their growth in 6-TG medium. Depending on the site of the crossover event, the homologous DNA remaining in the final recombinant cells may contain either the sequences present in the original target locus or the desired alteration introduced on the target DNA. Genome analysis of the 6-TG-resistant colonies would be necessary to identify those cells that had acquired the mutant DNA.

2. Cotransformation Protocols

Several investigators have reported that cells simultaneously transformed with two unlinked fragments often integrate both fragments into their genome (Wigler *et al.*, 1979; Boggs *et al.*, 1986; Toneguzzo *et al.*, 1986, 1988). This cotransformation procedure has been used to isolate cells that incorporate nonselectable exogenous DNA by screening populations of cells transformed with a co-introduced, selectable sequence (Newman *et al.*, 1983; Kuhn *et al.*, 1984). The same procedure might be useful in gene targeting to isolate cells with nonselectable gene alterations. If non-selectable targeting DNA was co-introduced with an unrelated selectable gene, then

A.

i)

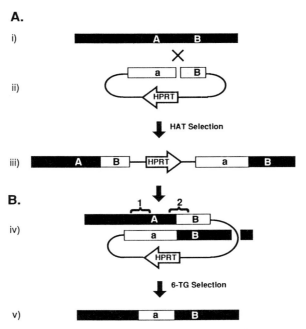

ii)

HAT Selection

iii)

B.

iv)

6-TG Selection

v)

Figure 5   IN-OUT targeting design. A scheme for the IN-OUT protocol that combines gene targeting (A) and an excision reaction (B) at a nonselectable locus is illustrated. (A) The IN step. A single crossover inserts the O-type targeting DNA (ii) into the chromosomal locus (i). The targeting DNA contains the *hprt* minigene (arrow) and two regions of homology (open boxes), one of which has a small, nonselectable mutation (a). Transformed cells are isolated in HAT medium and screened for homologous recombination. Targeted cells are HPRT$^+$ and contain an inexact duplication of the homologous regions (AB and aB) at the recombinant locus (iii). (B) The OUT step. The duplicated regions of the targeted cell can align (iv) so that homologous recombination between them removes the *hprt* gene and one of the duplicated regions. Recombinant cells are isolated in 6-TG medium. If the excision crossover occurs in region 2, then the final recombinant locus is identical to the original target locus (outcome not shown). If the excision crossover occurs in region 1, then the final recombinant locus has the desired mutation. Note that the only difference between the original target locus (i) and the desired recombinant locus (v) is the subtle change from "A" to "a" in one region of homology.

a high proportion of the cells that integrated the selectable gene might also contain the targeting DNA, possibly integrated at the target locus. Thus, targeted cells could be enriched in the cotransformed population without requiring the creation of a selectable alteration at the target locus.

We have tested this possibility, and have demonstrated that targeted cells can be isolated using the cotransformation protocol (Reid *et al.*, 1991). After the coelectroporation of *hprt* targeting DNA and a *neo* gene into mouse ES cells, *neo* transformed colonies were isolated in G418 medium and screened for a gene targeting event at the *hprt* locus. Hprt targeted cells were identified within the population of *neo* transformants at a frequency of 1 per 70 G418-resistant colonies. In parallel experiments using the same targeting construct, targeted *hprt* cells were found at a frequency of approximately 1 per 5500 nonselected colonies. Thus, we observed an 80-fold enrichment for the targeted cells within the colonies transformed with co-introduced DNA as compared to the nonselected colonies (1/70 versus 1/5500). This enrichment was obtained despite the surprising finding that targeted cells were rarely co-transformed (see Section V). Cotransformation protocols may therefore be useful in isolating targeted cells with nonselectable alterations.

IV. Transferring Targeted Alterations to the Mouse Germ Line

Many successful gene targeting experiments have now been performed in mouse ES cells. These totipotent cells are derived from the inner cell mass of mouse blastocysts (Evans and Kaufman, 1981). They can be grown for extended periods of time *in vitro* on feeder layers [or in the presence of leukemia inhibitory factor (Smith *et al.*, 1988; Williams *et al.*, 1988)] without differentiation, yet they retain their ability to resume normal development *in vivo* when injected into mouse blastocysts and implanted into pseudopregnant mice. The chimeric pups born from the injected blastocysts contain a mixture of tissues that are derived both from the introduced ES cells and from the host blastocyst. If the chimeric animals contain germ cells derived from the introduced ES cells, then these animals can transmit the ES cell genotype to their offspring. Thus, genetically altered ES cells can be used to generate genetically altered mice (Koller *et al.*, 1989; Thompson *et al.*, 1989).

Several genes inactivated by targeting experiments have been established in the mouse germ line (Schwartzberg *et al.*, 1989; Zijlstra *et al.*, 1989; DeChiara *et al.*, 1990; Koller *et al.*, 1990; McMahon and Bradley, 1990; Stanton *et al.*, 1990; Thomas and Capecchi, 1990). Mice homozygous for these mutations can be highly informative with respect to the function of the target gene. In some cases, the loss of gene activity is associated with grossly abnormal phenotypes. For example, McMahon and Bradley (1990) and Thomas and Capecchi (1990) have generated mice homozygous for an *int-1* gene inactivated by gene targeting. Both groups of investigators reported that most of the *int-1*-deficient animals died *in utero* or soon after birth.

The single surviving homozygote, observed by Thomas and Capecchi, exhibited loss of balance and uncoordinated movement. Examination of tissues in this animal and several homozygous embryos revealed an underdeveloped brain morphology, especially in the cerebellum and mesencephalon. These data suggest a prominent role for the *int-1* protein in the development of brain tissue.

In some cases, the loss of gene activity is not associated with an obviously abnormal phenotype but may nevertheless be very informative in mice. For example, animals homozygous for the *β2m* null mutation described earlier (see Fig. 3) have superficially normal phenotypes. Yet they do not produce MHC class I antigens and are deficient in one class of T cells (Koller *et al.*, 1990; Zijlstra *et al.*, 1990). This deficiency apparently does not inhibit normal embryo development, but may compromise immune responses in adult mice. Not surprisingly, the response of these animals to various immunological challenges is proving to be complex and of considerable interest. Likewise, mice lacking *hprt* gene activity appear healthy (Hooper *et al.*, 1987; Kuehn *et al.*, 1987), although this deficiency in humans leads to Lesch-Nyhan syndrome, a severe neurological disorder. The ability of these *hprt⁻* mice to avoid the neurological defects observed in humans may help unravel the etiology of the human disorder or suggest alternate metabolic pathways that compensate for the *hprt* deficiency.

V. DNA Integration in Targeted Cells

Early experiments by Boggs *et al.* (1986) and Toneguzzo *et al.* (1988) demonstrated that cells transformed by electroporation usually contained single copies of the exogenous DNA. Such single-copy integrations are an advantage in gene targeting experiments since undesirable side effects are likely to be reduced when only the targeted gene is changed. This integration pattern, however, contrasts with the frequent concatemers and multiple integration sites observed after calcium phosphate precipitation and microinjection protocols (Perucho *et al.*, 1980; Folger *et al.*, 1982).

Most targeted cell lines contain only a single copy of the targeting DNA (Smithies *et al.*, 1985; Doetschman *et al.*, 1987; Thomas and Capecchi, 1987; Baker and Shulman, 1988; Joyner *et al.*, 1989; Koller and Smithies, 1989), although occasionally randomly integrated copies of the DNA are observed (Adair *et al.*, 1989). Until recently, we and others had assumed that this pattern of DNA integration was a consequence of the electroporation procedure used to introduce the target DNA. Indeed, experiments by Toneguzzo *et al.* (1988) supported this view by demonstrating that the integration pattern of DNA can be altered by the electroporation conditions. However, we have recently obtained data suggesting that single-copy integrations during gene targeting are not simply related to the conditions used for electroporation (Reid *et al.*, 1991). We have found that, under constant conditions of electroporation and DNA concentration, 75% of the nontargeted cells were cotransformed, but only 4% of the targeted cells had integrated a second DNA molecule. Thus, cells in which DNA has been integrated by homologous recombination are approximately 20-fold less likely to have taken up a second DNA molecule into their genomes than are cells in which DNA has been integrated nonhomologously. In addition, we found that the DNA that had integrated by nonhomologous recombination was often present as concatemers at multiple sites within the genome, whereas the DNA that had integrated by homologous recombination was observed only as a single copy at the target locus. These results suggest that the single-site integration observed in targeted cells may be related more to the mechanism of recombination than to the electroporation conditions.

VI. Conclusion

Targeted modifications of a number of genes in mammalian cells have been achieved through the introduction of gene targeting constructs by electroporation. When created in mouse ES cells and transferred to the mouse germ line, these genetic alterations are extremely useful for the analysis of individual gene function *in vivo*. Thus, the combination of electroporation, gene targeting, and ES cell technology offers a new approach for studying the mammalian genome.

Acknowledgments

Our research is supported by grants HL37001 and GM20069 from the National Institutes of Health.

References

Adair, G. M., Nairn, R. S., Wilson, J. H., Seidman, M. M., Brotherman, K. A., MacKinnon, C., and Scheerer, J. B. (1989). Targeted homologous recombination at the endogenous adenine phosphoribosyltransferase locus in Chinese hamster cells. *Proc. Natl. Acad. Sci. USA* **86**, 4574–4578.
Baker, M. D., and Shulman, M. J. (1988). Homologous recombination between transferred and chromosomal immunoglobulin κ genes. *Mol. Cell. Biol.* **8**, 4041–4047.
Baker, M. D., Pennell, N., Bosoyan, L., and Shulman, M. J. (1988). Homologous recombination can restore normal immunoglobulin production in a mutant hybridoma cell line. *Proc. Natl. Acad. Sci. USA* **85**, 6432–6436.
Boggs, S. S., Gregg, R. G., Borenstein, N., and Smithies, O. (1986). Efficient transformation and frequent single site, single copy insertion of DNA can be obtained in mouse erythroleukemia cells transformed by electroporation. *Exp. Hematol.* **14**, 988–994.

Brinster, R. L., Braun, R. E., Lo, D., Avarbock, M. R., Oram, R., and Palmiter, R. D. (1989). Targeted correction of a major histocompatibility class II MHC E_α gene by DNA microinjected into mouse eggs. *Proc. Natl. Acad. Sci. USA* **86**, 7087–7091.

Charron, J., Malynn, B. A., Robertson, E. J., Goff, S. P., and Alt, F. W. (1990). High-frequency disruption of the *N-myc* gene in embryonic stem and pre-B cell lines by homologous recombination. *Mol. Cell. Biol.* **10**, 1799–1804.

DeChiara, T. M., Efstratiadis, A., and Robertson, E. (1990). A growth-deficiency phenotype in heterozygous mice carrying an insulin-like growth factor II gene disrupted by targeting. *Nature* **345**, 78–80.

Doetschman, T., Gregg, R. G., Maeda, N., Hooper, M. L., Melton, D. W., Thompson, S., and Smithies, O. (1987). Targeted correction of a mutant *hprt* gene in mouse embryonic stem cells. *Nature* **330**, 576–578.

Doetschman, T., Maeda, N., and Smithies, O. (1988). Targeted mutation of the *hprt* gene in mouse embryonic stem cells. *Proc. Natl. Acad. Sci. USA* **85**, 8583–8587.

Evans, M. J., and Kaufman, M. H. (1981). Establishment in culture of pluripotential cells from mouse embryos. *Nature* **292**, 154–156.

Folger, K. R., Wong, E. A., Wahl, G., and Capecchi, M. R. (1982). Patterns of integration of DNA microinjected into cultured mammalian cells: Evidence for homologous recombination between injected plasmid DNA molecules. *Mol. Cell. Biol.* **2**, 1372–1387.

Gregg, R. G., and Smithies, O. (1986). Targeted modification of human chromosomal genes. *Cold Spring Harbor Symp. Quant. Biol.* **51**, 1093–1099.

Hinnen, A., Hicks, J. B., and Fink, G. R. (1978). Transformation of yeast. *Proc. Natl. Acad. Sci. USA* **75**, 1929–1933.

Hooper, M., Hardy, K., Handyside, A., Hunter, S., and Monk, M. (1987). *hprt*-deficient (Lesch-Nyhan) mouse embryos derived from germline colonization by cultured cells. *Nature* **326**, 292–295.

Jasin, M., and Berg, P. (1988). Homologous integration in mammalian cells without target gene selection. *Genes. Dev.* **2**, 1353–1363.

Jasin, M., Elledge, S. J., Davis, R. W., and Berg, P. (1990). Gene targeting at the human *CD4* locus by epitope addition. *Genes. Dev.* **4**, 157–166.

Johnson, R. S., Sheng, M., Greenberg, M. E., Kolodner, R. D., Papaioannou, V. E., and Spiegelman, B. M. (1989). Targeting of nonexpressed genes in embryonic stem cells via homologous recombination. *Science* **245**, 1234–1236.

Joyner, A. L., Skarnes, W. C., and Rossant, J. (1989). Production of a mutation in mouse *en-2* gene by homologous recombination in embryonic stem cells. *Nature* **338**, 153–156.

Kim, H.-S., and Smithies, O. (1988). Recombinant fragment assay for gene target based on the polymerase chain reaction. *Nucleic Acids Res.* **16**, 8887–8903.

Koller, B. H., and Smithies, O. (1989). Inactivating the β2-microglobulin locus in mouse embryonic stem cells by homologous recombination. *Proc. Natl. Acad. Sci. USA* **86**, 8932–8935.

Koller, B. H., Hagemann, L. J., Doetschman, T., Hagaman, J. R., Huang, Shiu, Williams, P. J., First, N. L., Maeda, N., and Smithies, O. (1989). Germ-line transmission of a planned alteration made in a hypoxanthine phosphoribosyltransferase gene by homologous recombination in embryonic stem cells. *Proc. Natl. Acad. Sci. USA* **86**, 8927–8931.

Koller, B. H., Marrack, P., Kappler, J. W., and Smithies, O. (1990). Normal development of mice lacking β2m and deficient in MHC Class I proteins and CD8⁺ T cells. *Science* **248**, 1227–1230.

Kuehn, M. R., Bradley, A., Robertson, E. J., and Evans, M. J. (1987). A potential animal model for Lesch-Nyhan syndrome through introduction of HPRT mutations into mice. *Nature* **326**, 295–298.

Kuhn, L. C. McClelland, A., and Ruddle, F. H. (1984). Gene transfer, expression, and molecular cloning of the human transferrin receptor gene. *Cell* **37**, 95–103.

Mansour, S. L., Thomas, K. R., and Capecchi, M. R. (1988). Disruption of the proto-oncogene *int-2* in mouse embryo-derived stem cells: A general strategy for targeting mutations to non-selectable genes. *Nature* **336**, 348–352.

McMahon, A. P., and Bradley, A. (1990). The *Wnt-1 (int-1)* proto-oncogene is required for development of a large region of the mouse brain. *Cell* **62**, 1073–1085.

Newman, R., Domingo, D., Trotter, J., and Trowbridge, I. (1983). Selection and properties of a mouse L-cell transformant expressing human transferrin receptor. *Nature* **304**, 643–645.

Perucho, M., Hanahan, D., and Wigler, M. (1980). Genetic and physical linkage of exogenous sequences in transformed cells. *Cell* **22**, 309–317.

Potter, H., Weir, L., and Leder, P. (1984). Enhancer dependent expression of human immunoglobulin genes introduced into mouse pre-B lymphocytes by electroporation. *Proc. Natl. Acad. Sci. USA* **81**, 7161–7165.

Reid, L. H., Gregg, R. G., Smithies, O., and Koller, B. H. (1990). Regulatory elements in the introns of the human *hprt* gene are necessary for its expression in embryonic stem cells. *Proc. Natl. Acad. Sci. USA* **87**, 4299–4303.

Reid, L. H., Shesely, E. G., Kim, H. S., and Smithies, O. (1991). Co-transformation and gene targeting in mouse embryonic stem cells. *Mol. Cell. Biol.* **11**, 2769–2777.

Scherer, S., and Davis, R. W. (1979). Replacement of chromosome segments with altered DNA sequences constructed *in vitro*. *Proc. Natl. Acad. Sci. USA* **76**, 4951–4955.

Schwartzberg, P. L., Goff, S. P., and Robertson, E. J. (1989). Germ-line transmission of a *c-abl* mutation produced by targeted gene disruption in ES cells. *Science* **246**, 799–803.

Schwartzberg, P. L., Robertson, E. J., and Goff, S. P. (1990). Targeted gene disruption of the endogenous *c-abl* locus by homologous recombination with DNA encoding a selectable fusion protein. *Proc. Natl. Acad. Sci. USA* **87**, 3210–3214.

Smith, A. G., Heath, J. K., Donaldson, D. D., Wong, G. G., Moreau, J., Stah, M., and Rogers, D. (1988). Inhibition of pluripotential embryonic stem cell differentiation by purified polypeptides. *Nature* **336**, 688–690.

Smith, A. J. H., and Kalogerakis, B. (1990). Replacement recombination events targeted at immunoglobulin heavy chain DNA sequences in mouse myeloma cells. *J. Mol. Biol.* **213**, 415–435.

Smithies, O., Gregg, R. G., Boggs, S. S., Koralewski, M. A., and Kucherlapati, R. S. (1985). Insertion of DNA sequences into the human chromosomal β-globin locus by homologous recombination. *Nature* **317**, 230–234.

Stanton, B. R., Reid, S. W., and Parada, L. F. (1990). Germ line transmission of an inactive N-*myc* allele generated by homologous recombination in mouse embryonic stem cells. *Mol. Cell. Biol.* **10**, 6755–6758.

Steeg, C. M., Ellis, J., and Bernstein, A. (1990). Introduction of specific point mutations into RNA polymerase II by gene targeting in mouse embryonic stem cells: Evidence for a DNA mismatch repair mechanism. *Proc. Natl. Acad. Sci. USA* **87**, 4680–4684.

Thomas, K. R., ad Capecchi, M. R. (1987). Site-directed mutagenesis by gene targeting in mouse embryo-derived stem cells. *Cell* **51**, 503–512.

Thomas, K. R., and Capecchi, M. R. (1990). Targeted disruption of the murine *int-1* proto-oncogene resulting in severe abnormalities in midbrain and cerebellar development. *Nature* **346**, 847–848.

Thompson, S., Clarke, A. R., Pow, A. M., Hooper, M. L., and Melton, D. W. (1989). Germ line transmission and expression of a corrected *hprt* gene produced by gene targeting in embryonic stem cells. *Cell* **56**, 313–321.

Toneguzzo, F., Hayday, A. C., and Keating, A. (1986). Electric field-mediated DNA transfer: Transient and stable gene expression in human and mouse lymphoid cells. *Mol. Cell. Biol.* **6**, 703–706.

Toneguzzo, F., Keating, A., Glynn, S., and McDonald, K. (1988). Electric field-mediated gene transfer: Characterization of DNA transfer and patterns of integration in lymphoid cells. *Nucleic Acids Res.* **16**, 5515–5532.

Valancius, V., and Smithies, O. (1991). Testing an "IN-OUT" targeting procedure for making subtle genomic modifications in mouse embryonic stem cells. *Mol. Cell. Biol.* **11**, 1402–1407.

Wigler, M., Sweet, R., Sim, G. K., Wold, B., Pellicer, A., Lacy, E., Maniatis, T., Silverstein, S., and Axel, R. (1979). Transformation of mammalian cells with genes from procaryotes and eucaryotes. *Cell* **16**, 777–785.

Williams, R. L., Hilton, D. J., Pease, S., Willson, T. A., Stewart, C. L., Gearing, D. P., Wagner, E. F., Metcalf, D., Nicola, N. A., and Gough, N. M. (1988). Myeloid leukaemia inhibitory factor maintains the developmental potential of embryonic stem cells. *Nature* **336**, 684–687.

Zheng, H., and Wilson, J. (1990). Gene targeting in normal and amplified cell lines. *Nature* **344**, 170–173.

Zijlstra, M., Li, E., Sajjadi, F., Subramani, S., and Jaenisch, R. (1989). Germ-line transmission of a disrupted β2-microglobulin gene produced by homologous recombination in embryonic stem cells. *Nature* **342**, 435–438.

Zijlstra, M., Bix, M., Simister, N. E., Loring, J. M., Raulet, D. H., and Jaenisch, R. (1990). β2-Microglobulin deficient mice lack CD4$^-$8$^+$ cytolytic T cells. *Nature* **344**, 709–711.

Zimmer, A., and Gruss, P. (1989). Production of chimaeric mice containing embryonic stem (ES) cells carrying a homeobox *Hox1.1* allele mutated by homologous recombination. *Nature* **338**, 150–153.

15

Pollen Electrotransformation for Gene Transfer in Plants

James A. Saunders, Benjamin F. Matthews, and Sally L. Van Wert

Plant Sciences Institute, Beltsville Agricultural Research Center, United States
Department of Agriculture, Beltsville, Maryland 20705

I. Introduction

Genetic engineering of crop plants promises to increase plant productivity and the quality of the plant product. Numerous genes encoding highly abundant plant proteins ($>1\%$ of the total protein) have been cloned. This is due to the relative ease with which these proteins can be purified to homogeneity, allowing antibodies to be produced for screening of cDNA expression libraries (Burr *et al.*, 1982; Chandler *et al.*, 1983; Spena *et al.*, 1983). Several genes encoding much less abundant proteins ($<0.1\%$ of the total protein) have also been cloned (Peleman *et al.*, 1989; Hesse *et al.*, 1989). Genetic transformation in higher plants, however, has been limited by the lack of suitable techniques to bridge the gap between the cloning of a desirable gene and expression of that foreign gene in an intact, differentiated plant. Most methods that are currently in use for gene transfer in plants have important limitations. The major procedures include protoplast fusion, electroporation, microprojectile bombardment, *Agrobacterium*-mediated gene transfer, microinjection of DNA, direct uptake of DNA into microspores or embryos, and pollen electrotransformation.

DNA can be transferred directly into plant protoplasts through either somatic cell fusion or the uptake of isolated DNA. The use of isolated plant protoplasts for gene transfer avoids the formidable barrier of the plant cell wall. Electroporation, in particular, has been successful in introducing DNA into plant protoplasts (Fromm *et al.*, 1985, 1986). The use of electroporation for plant gene transfer has become a rapid, reliable, and reproducible technique for introducing specific genes into plant germplasm lines. These methods work quite well, but are of restricted value because many economically important crop plants are not readily regenerated from protoplasts into whole plants. Furthermore, tissue culture protocols for regeneration usually require several months of sterile culture. Major crop plants, such as corn, wheat, rice, and soybean, are still difficult to regenerate from protoplasts, although limited success has been achieved with some of these crops (Fromm *et al.*, 1986; Rhodes *et al.*, 1988; Shimamoto *et al.*, 1989). For example, transient gene expression has been reported in electroporated rice, corn, wheat, and barley tissue sections (Dekeyser *et al.*, 1990), and both rice (Wang *et al.*, 1988) and corn calli (Gordon-Kamm *et al.*, 1990) have been transformed by microprojectile bombardment.

Agrobacterium tumefaciens, harboring the tumor-inducing (Ti) plasmid, is a popular gene transfer vehicle in dicots (Horsch *et al.*, 1985), but is a poor system for gene transfer into monocots. Modification of the Ti plasmid DNA has been successfully used to introduce foreign genes into some members of the Solanaceae (Shahin and Simpson, 1986; Fillatti *et al.*, 1987) and a few other plant families (Damm *et al.*, 1989). *Agrobacterium tumefaciens,* however, is unable to infect many economically important crops, like those of the Graminaceae (Fraley *et al.*, 1983; Barton *et al.*, 1983; Broglie *et al.*, 1984).

Microinjection has also been used by some laboratories to insert DNA into plant cells (Crossway, *et al.*, 1986). This method works, but is tedious, requiring a high level of skill and specialized equipment, and, as a consequence, it not applicable to many situations.

Two methods that show great potential utilize pollen as the transformation vehicle. In one system, pollen is bombarded with tungsten particles coated with the DNA of interest (Klein *et al.*, 1988) and the transformed pollen grains are placed on stigmas of emasculated flowers. Plants grown from the resultant seed are screened to identify transformants. To date, transient expression of DNA has been observed in transformed pollen grains (Twell *et al.*, 1989), but no transformed plants have been generated. In the other system, pollen is allowed to germinate, until the pollen tube begins to extrude from the pollen grain. The cell wall of the rapidly expanding pollen tube is either absent or markedly thinner than that of the pollen grain (Cass and Peteya, 1979; Picton and Steer, 1982) and can potentially allow the entry of DNA into the tube. Foreign DNA is forced through the pollen tube membrane by electroporation, and the electroporated pollen is used to pollinate emasculated flowers. Again, plants from the resultant seed are screened for transformants.

The potential advantages of pollen transformation are numerous. No protoplast or cell culture is involved because transformed seed is obtained directly from the plant. Neither pollen transformation method involves protoplast regeneration, lengthy tissue culture techniques, or the use of *Agrobacterium*. Since plants are obtained directly from seed, it is less likely that vigor and fertility will be compromised as compared to transformed plants obtained through regeneration from protoplasts. The risk of tissue culture-induced variation is also minimized. Gametic transformation is obtained instead of chimeric transformation. The latter occurs with microprojectile transformation of intact tissues. Pollen transformation offers the promise of a convenient, economical procedure for rapidly producing genetically engineered plants that can be successfully applied to pollinating crop plants without the prerequisite of detailed tissue culture studies. This chapter will discuss pollen electrotransformation, the introduction of DNA into germinating pollen by electroporation, and its application to gene transfer in plants.

II. Current Research on Electroporation

The incorporation of DNA into both plant and animal cells through the process of electroporation has become an accepted technique for convenient gene transfer (Potter *et al.*, 1984; Bates *et al.*, 1987; Matthews and Saunders, 1989; Saunders *et al.*, 1989b). Electroporation has been successfully used to introduce DNA into plant protoplasts (Fromm *et al.*, 1985, 1986), intact plant cells (Morikawa *et al.*, 1986), plant tissue (Dekeyser *et al.*, 1990), animal cells (Wong and Neumann, 1982; Potter *et al.*, 1984), fibroblasts (Liang *et al.*, 1988), mammalian red blood cells (Tsong and Kinosita, 1985; Chu *et al.*, 1987), and yeast (Weber *et al.*, 1981), among others. There have also been several recent investigations that have examined the introduction of viral RNA into plant protoplasts (Hibi *et al.*, 1986; Okada *et al.*, 1986). Electroporation has been successfully accomplished in several plant species using both commercial and "homemade" electroporation devices (Mischke *et al.*, 1986; Boston *et al.*, 1987; Saunders *et al.*, 1989a).

Evidence for the success of transformation after electroporation has been obtained by radioactive labeling of DNA (Tsong and Kinosita, 1985), transient gene expression (Potter *et al.*, 1984; Smithies *et al.*, 1985), and the formation of stable transformants (Riggs and Bates, 1986; Stopper *et al.*, 1985).

The success of the electroporation method is dependent, in part, on optimizing parameters relative to the membrane, the DNA, and the electric field. Two different DC high-voltage pulse wave forms have been utilized for these experiments. Using the square-wave pulse, both the amplitude and the duration of the pulse can be accurately controlled. The amplitude of the wave form can be controlled for the exponential wave; however, the discharge rate of the pulse can only be modified in a general manner. For optimum pollen electrotransformation, a number of the electroporation parameters have been evaluated, including the field strength and pulse duration. The field strength of the pulse is controlled by two components: the applied current, and the electrode gap. To produce an effective electroporation pulse, minimal threshold levels for both the pulse duration and the pulse field strength must be exceeded. Our data suggest that the field strength of the pulse is inversely related to the pulse duration in such a way that, over a limited range, one variable may be increased as the other is decreased and a reversible pore may still be induced (Saunders *et al.*, 1989b; Abdul-Baki *et al.*, 1990). The applied electrical field induces the formation of pores or transient holes in the cell membrane through which the DNA can enter the cell. The time the pores remain open is a variable characteristic that may be controlled in part by the applied field strength of the DC pulse. Okada *et al.* (1986) suggested that the transient pores may remain open for as long as 10 min following the treatment and that the size of the pore may exceed 30 nm in diameter. Liang *et al.* (1988) suggested that pore induction, size, and frequency are controlled by pulse amplitude, whereas the length of time the pores remain open is controlled by pulse duration. Benz and Zimmerman (1981) indicated that different cell types require different field strengths to induce pores because of differences in membrane composition and osmotic properties. Pulses that do not meet the minimum field strength threshold may not induce any pore formation, and excessive field strengths lead to irreversible breakdown of the cell membranes.

III. Previous Reports on Pollen Transformation

The concept of the use of pollen to effect genetic modification of subsequent progeny has been cited in the literature for some time, and the term pollen transformation was coined in the 1970s (Hess, 1987). Several investigators have suggested that DNA, when added to pollen as a solution or a paste, is capable of being taken up and expressed in progeny. De Wet *et al.* (1985) and Ohta (1986), both using corn, reported that foreign DNA added to compatible pollen could produce seed that phenotypically expressed characteristics of the foreign DNA. Pandey (1978, 1980) described experiments in which pollen of *Nicotiana* was enucleated by irradiation, and was used to successfully pollinate flowers. This report provided evidence that pollen transformation may be possible using the enucleated pollen as a transformation system. In addition, Hess (1987) describes a series of reports using petunia and corn that indicate some DNA uptake may occur in pollen exposed to exogenously added DNA. Unfortunately, none of these reports described any mechanism for introducing the DNA into the pollen, nor did they obtain conclusive molecular evidence that the gene transfer had actually occurred. These deficiencies were pointed out by Sanford *et al.* (1985) and others who were unable to repeat the results of these pioneer studies, thus leaving the process of pollen transformation in a state of skepticism. Additional complications were added by Matousek and Tupy (1985) and Roeckel *et al.* (1988), who described the release of very active DNA nucleases from germinating pollen grains. These studies implied that foreign DNA, present in a mixture of germinating pollen, will be degraded within a few minutes. Dekeyser *et al.* (1990) avoided the nuclease problem by preincubating excised tissues in electroporation buffer for several hours to leach active nucleases prior to transfer to fresh medium containing plasmid DNA. Roeckel *et al.* (1988), however, were unable to eliminate nuclease activity by washing pollen grains several times prior to germination. As a result of these negative reports, there was little research on pollen transformation until a mechanism was found, namely electroporation, to rapidly incorporate exogenously supplied DNA into the pollen grain prior to its degradation by nucleases.

IV. Pollen Biology

The pollen grain is the carrier of the male gametes, or their progenitor, the generative nucleus. It is virtually a dehydrated organism with a moisture content ranging from 6 to 60%, a haploid chromosome number, and an inactive metabolism (Gaude and Dumas, 1987). Pollen grains have complex walls with two distinct layers, an external ornamented layer, the exine, and an internal layer, the intine. They have more than one nucleus and are ready to germinate after a period of hydration. The grain itself contains the vegetative, or tube nucleus, and a generative nucleus. The generative nucleus gives rise mitotically to the two male gametes sometime between the pollen grain's formation in the anther and just before the pollen tube reaches the embryo sac of the ovule.

The pollen tube arises from the pollen grain upon germination. The pollen grain cell wall is absent from the pollen tube (Cass and Peteya, 1979; Picton and Steer, 1982). The apical zone of the pollen tube is single-layered. An inner layer appears in the subapical zone and may increase in thickness beyond this zone. The inner tube wall contains callose, microfibrils of cellulose, and "noncellulosic" microfibrils with "pectinlike" properties (Kroh and Knuiman, 1982). This pollen tube wall is a much less substantial structure than that of a mature plant cell or the pollen grain.

The stigma is the receptive surface on the style in angiosperm flowers. Upon germinating, the generative nucleus of the pollen grain is conveyed down the style via the pollen tube and enters the ovary through the micropyle. Fertilization occurs when a male gamete fuses with an egg cell in an ovule (Stanley and Linsken, 1974).

Pollen is readily available in sufficient quantities to be amenable to genetic manipulation; however, its cell wall represents a formidable barrier for investigations of pollen transformation. Either the wall of the pollen grain or the germinating pollen tube is apparently necessary for successful fertilization of the ovule, due to the presence of specific stigmatic recognition factors involved in pollen tube formation. Production of these recognition factors is controlled by expression of the "S" supergene complex, which involves pollen wall/stigmatal surface interactions (Gaude and Dumas, 1987). Upon germination in simple culture media, the pollen wall forms specific site recognition factors necessary for subsequent pollen tube elongation and fertilization of the ovule. The absence of the pollen cell wall on the germinated pollen tube imparts an incompatibility reaction and fertilization of the ovule does not take place; thus, removal of the pollen wall is not a viable technique for pollination. In the early stages of pollen germination the incompatibility reaction does not occur (in compatible species) even though the developing pollen tube has virtually no cell wall. It is the compatibility and lack of a substantial cell wall in young pollen tubes that can be utilized for pollen transformation.

V. Electrotransformation of Pollen

A. Overview

Pollen of tobacco (*Nicotiana gossei* L. Domin) is being used to examine the feasibility of using germinating pollen as a model system to achieve pollen electrotransformation. The genetically modified pollen would then be used to pollinate emasculated female flowers, producing genetically modified seed directly from the plant. This gene transfer technique can be applied to a wide variety of plant species without regard to specific host compatibility or the ability of these species to grow in tissue culture.

The use of germinating pollen coupled with electroporation as a method of gene transfer was preceded by work of Mishra *et al.* (1987), who, upon observing the release of fluorescent dyes from electroporated pollen, concluded that the induced membrane permeability may allow the uptake of macromolecules. The pollen tube has a less substantial cell wall than the pollen grain or a plant cell, presenting a situation analogous to a plant protoplast in which the cell wall has been enzymatically removed, allowing for direct introduction of foreign DNA into the gametic cell by electroporation. Pollen transformation, however, minimizes tissue culture procedures encountered in protoplast and *Agrobacterium* transformation.

B. Pollen Storage and Germination

Tobacco plants (*N. gossei* L. Domin) were grown in a greenhouse supplemented with fluorescent lamps to extend the light period, to provide a source of pollen. Pollen

Table 1

Effect of Storage Conditions on Pollen Germination

Storage time (wks)	Pollen germination (%) at storage temperature			
	26°C	5°C	−20°C	−70°C
1	78.2	91.3	94.8	87.6
2	76.1	88.9	92.3	85.5
3	20.5	74.0	89.2	78.4
4	27.3	68.3	76.2	82.3

was collected by hand shaking the flower in the morning before the corollas close, and was pooled. Tobacco pollen collected in this manner has a moisture content of approximately 13% (w/w), exhibits a very high level of purity, and has germination rates up to 90%. Pollen can be stored frozen at −20 or −70°C for several weeks without losing significant viability (Table 1) (Abdul-Baki *et al.*, 1990).

The standard conditions for germinating the pollen prior to electroporation are as follows: Four milligrams of pollen is placed in a 1-ml plastic cuvette in 300 μl of germination medium (0.29 M sucrose, 1.27 mM Ca(NO$_3$)$_2$, 0.16 mM H$_3$BO$_3$, and 1 mM KNO$_3$, pH 5.2 adjusted with H$_3$BO$_3$; Dickinson, 1968). The pollen is suspended in this medium and placed on a shaker at 60 rpm and 30°C for 30–60 min until the pollen tube extrudes about one diameter's length from the pollen grain (Fig. 1). This germination medium also works well with lily and alfalfa pollen; however, it does not work well with corn or barley. Adjustments in osmotic and medium components may be necessary for individual species. Due to the fragility of the germinating pollen and its tendency to stick to surfaces of glass and plastic

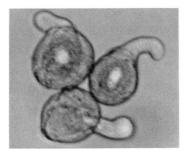

Figure 1   Tobacco pollen germinated for 1 h.

ware, mechanical manipulations during germination are kept to a minimum. *Nicotiana gossei* pollen remains viable for several days in germination medium, and the pollen tube grows at least 1 cm in length from a 60-μm pollen grain. However, preincubated pollen with long germination tubes is not able to achieve successful pollination and fertilization (Section IV). Thus, the tobacco pollen should be incubated in the germination medium for less than 60 min.

C. Electroporation of Pollen

Saunders *et al.* (1989b) showed that the shape of the electroporation wave can have a significant effect on the transformation efficiency of protoplast populations being treated. In general, square-wave generators can provide transformation efficiencies over a broader range of electroporation conditions than can exponential pulse generators. Based on this research we chose to use a square-wave generator for pollen electrotransformation.

The uptake of DNA into germinating pollen, while maintaining pollen viability and fertilization potential, has been examined in tobacco. Parameters affecting both the uptake of DNA into pollen by electroporation and pollen viability after electroporation include field strength and duration of the pulse, resealing time of the induced pore, number of pulses delivered and time between pulses, pollen germination and storage temperatures, length of time for pollen germination, pollen and DNA concentration, and buffer composition and additives, among others. These types of optimization studies are essential for the effective utilization of pollen as a viable recipient of DNA by electroporation.

To evaluate the feasibility of using electroporated pollen as an effective transformation vector of foreign DNA we chose to initially concentrate on the most obvious question: can DNA be taken up by pollen grains through electroporation procedures without killing the pollen grain? The simplest way to answer this question, without complicating the experiment with additional variables concerning lack of expression, was to use radioisotopically labeled DNA as an effective indicator of successful DNA uptake.

To circumvent the physical barrier of the cell wall and maintain the recognition factors necessary for fertilization, the pollen was first allowed to germinate, as described in Section V,B, and the DNA was introduced directly into the germinating pollen tube by electroporation. The cuvette used for germination is also the electroporation chamber. After electroporation the pollen grains are allowed to remain in the same vessel to minimize further mechanical manipulations while the pores reseal.

For our DNA uptake studies the 4 mg of tobacco pollen was collected, germinated, and electroporated in a 1-ml sterile cuvette containing 300 μl germination medium as described previously (Abdul-Baki *et al.*, 1990). The pollen was treated with DNase I (2 units/ml) 40 min after the electroporation pulse to remove unincorporated DNA. The pollen was collected on filter paper and the filter was thoroughly washed with germination medium (15 ml). Radioactivity in washed pollen and in the filtrate was measured in a liquid scintillation counter using ScintiVerse II (Fisher Scientific) as a cocktail.

Field strength, expressed as kV/cm, is influenced by several factors including the voltage or amplitude of the applied pulse, the gap or distance between the two electrodes of the electroporation chamber, and the resistance of the electroporation chamber. The resistance was constant between experiments as long as the same electroporation chamber was used but would be altered by changes in the conductivity of the electroporation medium and the concentration of pollen grains in the medium.

The amount of DNA taken up by *N. gossei* pollen is strongly dependent upon the electric pulse field strength used in electroporation (Abdul-Baki *et al.*, 1990). Field strengths below 4 kV/cm using an 80-μs pulse are not effective in inducing pores as measured by the uptake of labeled DNA. There is a major increase in DNA uptake when the field strength is increased from 6 to 8 kV/cm. The optimum field strength of 8–9 kV/cm has no detectable adverse effects on the growth of the pollen grain. Although there was no deleterious effect of the field strength on pollen survival when field strengths of up to 10 kV/cm were applied, pollen survival does decrease from over 90% at 10 kV/cm to 70% at 15 kV/cm. Thus, there is a reasonably broad range over which the field strength does not diminish pollen survival, yet allows electroporation of DNA into the pollen tube. The field strengths necessary for uptake of DNA into pollen are much higher than optimal conditions for protoplast electroporation and are attributed to the presence of the pollen cell wall as well as the higher conductivity of the electroporation medium used for the germinating pollen (Saunders *et al.*, 1989a). The induction of pores through electroporation is directly related to the diameter of the cell used (Bates *et al.*, 1987). Thus larger cells reach the critical voltage for pore formation at a lower field strength than smaller cells. It should be pointed out that tobacco pollen grains are larger than most animal cells and comparable to plant protoplasts, yet they require a higher field strength than mammalian cells and plant protoplasts for optimum electroporation.

To ascertain the effect of the pulse duration and of multiple pulses on DNA uptake by tobacco pollen, the pollen was subjected to pulses of 9 kV/cm for durations up to 80 μs. The uptake of DNA increased with increasing pulse length up to 80 μs, which at 9 kV/cm represents the optimum electroporation profile for the uptake of DNA utilizing a single pulse (Abdul-Baki *et al.*, 1990). The application of multiple pulses within 2 s of the first caused a reduction in the DNA uptake by the germinating pollen, regardless of the duration of the pulse.

Since pore formation by electroporation is a reversible process, the time required to reseal the induced pores becomes an important variable for the successful incorporation of foreign DNA. Higher rates of incorporation of DNA were obtained when the electroporated germinating tobacco pollen grains were allowed to remain undisturbed in the electroporation medium containing the labeled DNA following the pulse. Maximal DNA uptake was attained 45 min after electroporation. Apparently, this incubation period allows for the resealing of pores induced in the cell membrane during electroporation. This slow resealing time is in sharp contrast to the rapid estimates of the time it takes for pore resealing in other studies using different cell types (Coster and Zimmermann, 1975). Several reports have suggested that more than one type of pore may be formed during the electroporation process and the lifetime of different pores may extend over several minutes (Sowers, 1984).

To test the effect of membrane-modifying agents on the uptake of DNA during the resealing process, several different chemicals have been examined (Abdul-Baki *et al.*, 1990). The addition of dimethyl sulfoxide (DMSO), a membrane-modifying agent, to the tobacco pollen electroporation medium at final concentrations of 5 and 8% (v/v) results in reduced germination, lysis of the pollen, and very limited pollen tube growth. The addition of bovine serum albumin (BSA), at a final concentration of 16% (w/v), to the electroporation medium immediately following the pulse reduced DNA uptake from 6.6 to 1.3%. This decrease in DNA uptake may be caused by the ability of the BSA to clog or seal the pores induced by the electroporation, thus preventing DNA uptake. When the pollen is electroporated and incubated without BSA at various temperatures regimes, maximal DNA uptake occurs at 35°C (Abdul-Baki *et al.*, 1990).

DNA uptake by electroporated tobacco pollen is also dependent on the pollen concentration in the electroporation medium. Uptake of DNA increased as the pollen concentration increased up to 4 mg/300 μl of medium, and then declined at higher pollen concentrations (Abdul-Baki *et al.*, 1990). Observations on plant protoplasts, animal cells, and microorganisms provide evidence that cell concentration, size, osmotic conditions, and membrane properties all may influence the electroporation process (Bates *et al.*, 1987; Chu *et al.*, 1987). Our results, indicating that DNA uptake is dependent on the pollen concentration, are in agreement with those reported for other cells (Chu *et al.*, 1987). The decline in DNA uptake at high pollen concentrations demonstrates that DNA is in fact incorporated into the pollen rather than adsorbed on the pollen surface. Indeed, extensive washing of the pollen with DNase for 2.5 h did not remove the DNA from the germinating pollen.

The effect of nucleases, released by germinating pollen, on the integrity of added foreign DNA has also been investigated. The plasmid pBI221 (20 μg) (Fig. 2B) was added to germinated tobacco pollen and the suspension electroporated using the optimal conditions described (Abdul-Baki *et al.*, 1990). The pollen was pelleted by centrifugation 1, 4, 10, and 60 min after the addition of the plasmid. After the pollen was washed twice with germination medium, DNA was extracted and run

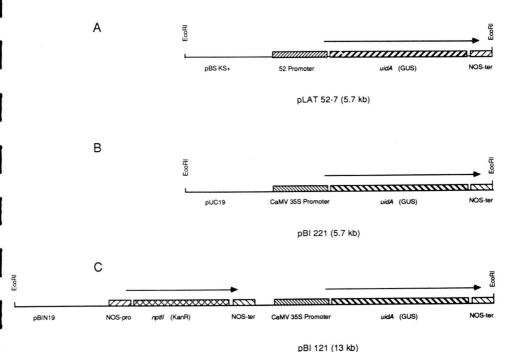

Figure 2 Plasmids pLAT52-7, pBI221, and pBI121 linearized with *Eco*R1. (A) pLAT52-7 containing the gene fusion 52-*uidA*-NOS3'. (B) pBI221 containing the gene fusion CaMV35S-*uidA*-NOS3'. (C) Plasmid pBI121 containing the gene fusions NOS5'-*nptII*-NOS3' and CaMV35S-*uidA*-NOS3'.

on an agarose gel (Fig. 3). There was no pBI221 DNA present in the pollen DNA extracts of the control; however, the plasmid DNA was present and intact in the pollen DNA extracts and in the supernatants of treatments harvested 1, 4, and 10 min after addition of pBI221. Although some degradation of the plasmid was present, the pollen DNA was intact. The first evidence of nuclease activity does not appear until 60 min after addition of pBI221, when the plasmid in the supernatant appears as a degraded smear of DNA (data not shown). Adding 2 μl DNase I (2 units/μl; Sigma) along with the plasmid DNA to the pollen resulted in the total degradation of the added pBI221 within 1 min. Thus, it seems clear that nucleases released from the germinating pollen do not significantly degrade added DNA within 10 min of its addition. This is contrary to the results suggested by Matousek and Tupy (1985) and Roeckel *et al.* (1988).

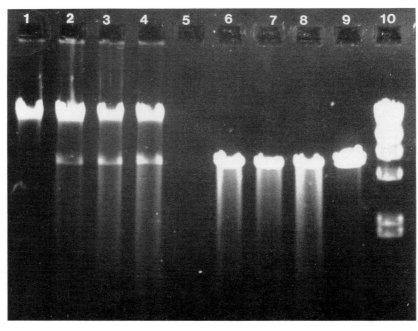

Figure 3 Effect of nucleases released by germinating pollen on added pBI221. Ethidium bromide-stained agarose gel of DNA isolated from germinating pollen (lanes 1–4) and its germination medium (lanes 5–8): untreated pollen (1 and 5) 1 min after addition of pBI221 (2 and 6), 4 min after addition of pBI221 (3 and 7), 10 min after addition of pBI221 (4 and 8), pBI221 (9), lambda-DNA digested with *Hin*dIII (10).

D. Detection and Transient Expression of Electroporated DNA in Pollen

Mulcahy (1986) has indicated that germinating pollen may be ideal for examining transient gene expression due to the high percentage of structural genes that are expressed during germination. Germinating pollen has been used to examine transient expression in a system using two different DNA constructs. Both plasmids contain promoters fused to the coding region of the *uidA* gene, encoding the marker enzyme β-glucuronidase (GUS) (Jefferson, 1985). Plasmid pLAT52-7 (Twell *et al.*, 1989) (Fig. 2A) contains a pollen-specific promoter from tomato, while pBI221 (Fig. 2B) contains the cauliflower mosaic virus CaMV 35S promoter. The plasmids were electroporated into germinating *N. gossei* pollen using the optimal electroporation conditions demonstrated with [³H]DNA. Both plasmids had been linearized by digestion with *Eco*RI. Successful transfer of these plasmids can be monitored by assaying GUS activity in pollen histochemically and fluorometrically (Jefferson *et* *al.,* 1987) and by Southern blot analysis (Southern, 1975) of DNA extracted from pollen.

For the histochemical assays, plasmid DNA (30 μg) was added to germinated pollen prior to electroporation. Blue color, indicative of a positive transformation, was visible in pollen electroporated with pLAT52-7 3 h after electroporation. Color development was greater in the grains than in the pollen tubes, regardless of the promoter used. In controls, no color development was seen until 24 h after the treatment, and even then the faint blue color was equal in both the pollen grains and tubes. This color development is probably due to nonspecific oxidation of the substrate 5-bromo-4-chloro-3-indolyl-beta-D-glucuronic acid. At 24 hr, the pollen electroporated with pLAT52-7 was very dark blue, while pollen electroporated with pBI221 was a darker blue than the controls, but not as dark as those treated with pLAT52-7. More of the viable pLAT52-7-treated pollen (close to 100%) was blue compared to pBI221-treated pollen (75%). Viability was significantly affected by these treatments (~50%), probably due to the presence of dimethyl formamide in the reaction assay medium. Thus, for the histochemical GUS assay, transient expression in germinating pollen is seen when electroporated in the presence of pLAT52-7 or pBI221; however, expression, based on color development, is greater in pollen electroporated in the presence of pLAT52-7. The difference in GUS expression, as seen by the blue color development, seems to be due to the effectiveness of the promoter on the construct electroporated into the germinating pollen.

Transient GUS activity was also demonstrated using a fluorometric assay (Jefferson

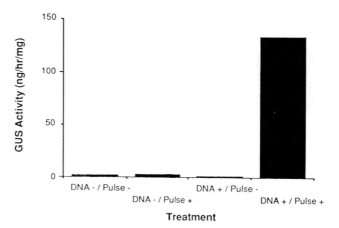

Figure 4   Expression of GUS in pollen 24 h after electroporation with pBI221 DNA. GUS activity was assayed fluorometrically and the specific activity (nM/h/mg) was determined for all samples.

et al., 1987) of pollen incubated in the germination medium for 24 h after electroporation. Increases in GUS activity as high as 60-fold greater than controls were detected fluorometrically in electroporated pollen with pBI221 (Fig. 4) (Matthews *et al.*, 1990).

Further evidence for the uptake of plasmid DNA by pollen was provided by Southern hybridization analysis of DNA extracted from pollen 1 and 24 h after electroporation in the presence of pBI221. The pollen was treated with DNase and extensively washed before analysis. *EcoRI* digested DNA from pollen treatments and controls were compared by agarose gel electrophoresis (Fig. 5A). Control pollen contained no pBI221. A band of DNA of the same size as pBI221 was present in DNA extracted from pollen treated with pBI221 and incubated for 1 h. The plasmid DNA appeared as a distinct band in pollen extracts 1 h after electroporation but not 24 h after the electroporation. A Southern blot of this gel hybridized with ^{32}P-labeled pBI221 shows the control pollen DNA did not hybridize with the plasmid while the treated pollen DNA did hybridize (Fig. 5B) (Matthews *et al.*, 1990). Some of the hybridizing DNA was of high molecular weight, suggesting that pBI221 had integrated into the genome of *N. gossei* by 24 hr after electroporation. These data conclusively show that pBI221 was taken up by pollen via electroporation,

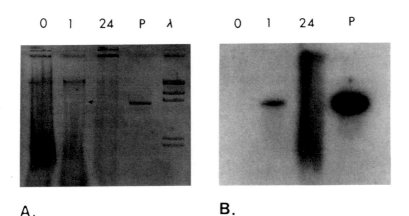

A.   **B.**

Figure 5   Agarose gel of *EcoRI* digested DNA extracts from pollen electroporated in the presence of linearized pBI221 (A) and Southern blot of gel hybridized with ^{32}P-labeled pBI221 (B). Untreated control (0), DNA from pollen treated with pBI221 linearized DNA an incubated for 1 h (1) or 24 h (24) after electroporation with linearized pBI221, linearized pBI221 DNA (P), and lambda-DNA digested with *Hin*dIII (λ). (Published by permission of Springer Verlag, from Matthews *et al.*, 1990).

remained intact, and provided transient expression of GUS activity, suggesting the incorporation of PBI221 into the pollen genome.

VI. Analysis of Transformed Plants

Although incorporation of DNA into pollen and expression of DNA in pollen are extremely important criteria to establish that pollen electrotransformation may be an effective gene transfer mechanism, the ultimate goal of the research is to produce viable seed with a genetically modified genome. This final question was answered with the demonstration that electroporated and genetically modified tobacco pollen could fertilize receptive emasculated flowers to produce viable seed, and that resultant plants from the seed were genetically modified.

Viable seed production was demonstrated for tobacco, alfalfa, and corn. We subjected the germinating pollen to electroporation procedures and then pollinated emasculated flowers with the treated pollen by pipetting aliquots of the treated pollen onto receptive stigmas. It is clear that the number of seed produced from each flower should be increased to bring seed production to normal levels (Table 2); however, the seed that did develop had more than a 90% viability rating when germinated on moistened filter paper. The greatest production of viable tobacco seed was produced by applying electroporated pollen to flowers that had been emasculated 4 days prior to pollination (Abdul-Baki *et al.*, 1990). We have analyzed *N. gossei* plants from seed derived from flowers fertilized with pollen electroporated with pBI221. Leaves from some of these plants expressed higher amounts of β-glucuronidase activity than the control (data not shown). Southern blot analysis confirmed the presence of *uidA* in leaf DNA from some plants showing high GUS activity (Fig. 6). DNA from other plants with high GUS activity did not hybridize to the *uidA* probe from pBI221.

Table 2

Production of Viable Seed from Flowers
Pollinated with Electroporated Pollen Grains

Treatment	Plant	Viable seed produced per flower
Dry pollen	Corn	159
Electroporated	Corn	72
Dry pollen	Tobacco	332
Electroporated	Tobacco	214
Dry pollen	Alfalfa	4
Electroporated	Alfalfa	2

Pollen transformation using the vector pBI221 (Clone-Tech) (Fig. 2C) has also been accomplished. In addition to the *uidA* gene, pBI121 carries the neomycin phosphotransferase II gene (*nptII*), which confers resistance to kanamycin. This plasmid provides a double screen for transformants. Putatively transformed seed is selected after surface sterilization by germination on Murashige and Skoog medium (Murashige and Skoog, 1962) containing 50 μg/ml kanamycin. Kanamycin-resistant plants can be assayed for GUS activity, and Southern blot analysis can confirm the presence of transforming sequence in their DNAs. This transformation screen is much less labor-intensive than screening large numbers of grown plants for GUS activity by the fluorometric assay alone.

Gene transfer via electroporation of pollen and normal fertilization techniques to produce transformed seeds needs further testing to assess its frequency, reproducibility, stability, and applicability to other economically valuable crop plants. We are currently investigating how pollen concentration, germination time, the con-

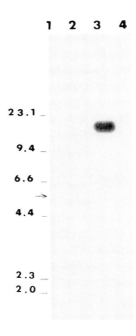

Figure 6   Southern blot of an agarose gel of *Eco*RI digested DNA from leaves of plants with high GUS activity. The blot was probed with the *uidA* gene from PBI221. Lambda *Hind*III molecular weight markers, arrow indicates size of linearized pBI221 (lane 1), DNA from control (lane 2), DNA from plants with high GUS activity (lanes 3 and 4).

centration of added DNA, and the physical form of the DNA affect seed production and transformation efficiency in tobacco. Further investigation into the stability of expression of the marker genes in the R1 and subsequent generations should confirm stable inheritance through pollen electrotransformation.

VII. Summary

Early investigations on pollen transformation were circumspect because there was no mechanism for moving the DNA into the pollen before degradation by nucleases released from the germinating pollen and there were no suitable marker genes to determine whether successful gene transfer had occurred. With the development of electroporation techniques and suitable marker DNA, experiments to transform pollen can be conducted and monitored with more assurance. Three critical components exist in pollen electrotransformation: (1) the ability of the pollen to take up DNA by electroporation without affecting pollen viability; (2) the ability of the pollen to carry the DNA in an expressible form; and (3) the ability of the pollen to fertilize the ovules to form viable seed and plants containing expressible DNA. We have definitively shown that added DNA is taken up by electroporated pollen, is both transiently expressed and incorporated into the pollen genome, and is present in plants derived from the pollination process.

Electroporation procedures may be enhanced through the use of various chemical membrane modifiers that may facilitate the incorporation of the DNA into the gametic cell; however, our preliminary experiments were not promising. Higher levels of expression of the incorporated DNA in transformed plants may be accomplished by using gene constructs containing stronger promoters. Similarly, we should be able to demonstrate tissue specific expression of the transforming DNA by using promoters from genes showing tissue-specific expression.

Acknowledgments

We would like to thank Viseth Ngauy for her assistance with this project.

References

Abdul-Baki, A. A., Saunders, J. A., Matthews, B. F., and Pittarelli, G. W. (1990). DNA uptake by electroporation of germinating pollen grains. *Plant Sci.* 70, 181–190.
Barton, K. A., Binns, A. N., Matzke, A. J. M., and Chilton, M. D. (1983). Regeneration of intact tobacco plants containing full length copies of genetically engineered T-DNA, and transmission of T-DNA to R1 progeny. *Cell* 32, 1033–1043.

Bates, G. W., Saunders, J. A., and Sowers, A. E. (1987). Electrofusion: Principles and applications. *In* "Cell Fusion" (A. E. Sowers, ed.), pp. 367–395. Plenum Press, New York.

Benz, R., Zimmermann, U., Wecker, E. (1981). High electric fields effects on the cell membranes of *Halicystis parvula*: A charge-pulse study. *Planta* 152, 314–318.

Boston, R. S., Becwar, M. R., Ryan, R. D., Goldsbrough, P. B., Larkins, B. A., and Hodges, T. K. (1987). Expression from heterologous promoters in electroporated carrot protoplasts. *Plant Phys.* 78, 742–746.

Broglie, R., Coruzzi, G., Fraley, R. T., Rogers, S. G., Horsch, R. B., Niedermeyer, J. G., Fink, C. L., Flick, J. S., and Chua, N. H. (1984). Light-regulated expression of a pea ribulose-1,5-bisphosphate carboxylase small subunit gene in transformed plant cells. *Science* 224, 838–843.

Burr, B., Burr, F. A., St. John, T. P., Thomas, M., and Davis, R. W. (1982). Zein storage protein gene family of maize. *J. Mol. Biol.* 154, 33–49.

Cass, D. D., and Peteya, D. J. (1979). Growth of barley pollen tubes *in vivo*. I. Ultrastructural aspects of early tube growth in the stigmatic hair. *Can. J. Bot.* 57, 386–396.

Chandler, P. M., Higgins, T. J. V., Randall, P. J., and Spencer, D. (1983). Influence of sulfur nutrition on developmental patterns of some major pea seed proteins and their mRNAs. *Plant Physiol.* 75, 651–657.

Chu, G., Hayakawa, H., and Berg, P. (1987). Electroporation for the efficient transfection of mammalian cells with DNA. *Nuclei Acids Res.* 15, 1311–1326.

Coster, H. G. L., and Zimmermann, U. (1975). Dielectric breakdown in the membrane of *Valonia utricularis*. *Biochim. Biophys. Acta* 382, 410–418.

Crossway, A., Hauptli, H., Houck, C., Irvine, J., Oaks, J., and Perani, L., (1986). Micromanipulation techniques in plant biotechnology. *Biotechniques* 4, 320–334.

Damm, B., Schmidt, R., and Willmitzer, L. (1989). Efficient transformation of *Arabidopsis thaliana* using direct gene transfer to protoplasts. *Mol. Gen. Genet.* 217, 6–12.

Dekeyser, R. A., Claes, B., De Rycke, R. M. U., Habets, M. E., Van Montagu, M. C., and Caplan, A. B. (1990). Transient gene expression in intact and organized rice tissues. *Plant Cell* 2, 591–602.

De Wet, J. M. J., Bergquist, R. R., Harlan, J. F., Brink, D. E., Cohen, C. E., Newell, C. A., and De Wet, A. E. (1985). Exogenous gene transfer in maize (*Zea mays*) using DNA-treated pollen. *In* "Experimental manipulation of ovule tissues" (G. P. Chapman, S. H. Mantell, and R. W. Daniels, eds.), pp. 197–209. Longman, London.

Dickinson, D. B. (1968). Rapid starch synthesis associated with increased respiration in germinating lily pollen. *Plant Phys.* 43, 1–8.

Fillatti, J., Kiser, J., Rose, R., and Comai, L. (1987). Efficient transfer of a glyphosate tolerance gene into tomato using a binary *Agrobacterium tumefaciens* vector. *Biotechnology* 5, 726–730.

Fraley, R. T., Rogers, S. G., Horsch, R. B., Sanders, P. R., Flick, J. S., Adams, S. P., Bittner, M. L., Brand, L. A., Fink, C. L., Fry, J. S., Galluppi, G. R., Goldberg, S. B., Hoffman, N. L., and Woo, S. C. (1983). Expression of bacterial genes in plant cells. *Proc. Natl. Acad. Sci. USA* 80, 4803–4807.

Fromm, M., Taylor, L. P., and Walbot, V. (1985). Expression of genes transferred into monocot and dicot plant cells by electroporation. *Proc. Natl. Acad. Sci. USA* 82, 5824–5828.

Fromm, M. E., Taylor, L. P., and Walbot, V. (1986). Stable transformation of maize after gene transfer by electroporation. *Nature* **319**, 791–793.

Gaude, T., and Dumas, C. (1987). Molecular and cellular events of self-incompatibility. *Int. Rev. Cyt.* **107**, 333–336.

Gordon-Kamm, W. J., Spencer, T. M., Mangano, M. L., Adams, T. R., Daines, R. J., Start, W. G., O'Brien, J. V., Chambers, S. A., Adams, W. R. Jr., Wiletts, N. G., Rice, T. B., Mackey, C. J., Kreuger, R. W., Kausch, A. P., and Lemaux, P. G. (1990). Transformation of maize cells and regeneration of fertile transgenic plants. *Plant Cell* **2**, 603–618.

Hess, D. (1987). Pollen-based techniques in genetic manipulation. *Int. Rev. Cytol.* **107**, 367–395.

Hesse, T., Feldwisch, J., Balshusemann, D., Baun, G., Puype, M., Vanderkerckhove, J., Lobler, M., Klambt, D., Schell, J., and Palme, K. (1989). Molecular cloning and structural analysis of a gene from *Zea mays* coding for a putative receptor for the plant hormone auxin. *EMBO J.* **2**, 1801–1805.

Hibi, T., Kano, H., Sugiura, M., Kazami, T., and Kimura, S. (1986). High efficiency electro-transfection of tobacco mesophyll protoplasts with tobacco mosaic virus RNA. *J. Gen. Virol.* **67**, 2037–2042.

Horsch, R. B., Fry, J. E., Hoffman, N. L., Eichholtz, D., Rogers, S. G., and Fraley, R. T. (1985). A simple and general method for transferring genes in plants. *Science* **227**, 1229–1231.

Jefferson, R. A. (1985). DNA transformation of *Caenorhabditis Elegans*: Development and application of a new gene fusion system. Ph.D. Dissertation, University of Colorado, Boulder, CO.

Jefferson, R. A., Kavanagh, T. A., and Bevan, M. W. (1987). GUS fusions: β-Glucuronidase as a sensitive and versatile gene fusion marker in higher plants. *EMBO J.* **6**, 3901–3907.

Klein, T. M., Wolf, E. D., Wu, R., and Sanford, J. C. (1988). High-velocity microprojectiles for delivering nucleic acid into living cells. *Nature* **327**, 70–73.

Kroh, M., and Knuiman, B. (1982). Ultrastructure of cell wall and plugs of tobacco pollen tubes after chemical extraction of polysaccharides. *Planta* **154**, 241–250.

Liang, H., Purucker, W. J., Stenger, D. A., Kubiniec, R. T., and Hui, S. W. (1988). Uptake of fluorescence-labeled dextrans by 10T 1/2 fibroblasts following permeation by rectangular and exponential electric field pulses. *BioTechniques* **6**, 550–558.

Matousek, J., and Tupy, J. (1985). The release and some properties of nuclease from various pollen species. *J. Plant Physiol.* **119**, 169–178.

Matthews, B. F., Abdul-Baki, A. A., and Saunders, J. A. (1990). Expression of a foreign gene in electroporated pollen grains of tobacco. *Sex. Plant Reprod.* **3**, 147–151.

Matthews, B. F. and Saunders, J. A. (1989). Gene transfer in plants. *In* "Biotic Diversity and Germplasm Preservation, Global Imperatives" (L. Knutson and A. K. Stoner, eds.), pp. 275–291. Kluwer Press, Dordrecht, Netherlands.

Mischke, B. S., Saunders, J. A., and Owens, L. D. (1986). A versatile low-cost apparatus for cell electrofusions and electrophysiological treatments. *J. Biochem. Biophys. Methods* **13**, 65–75.

Mishra, K. P., Joshua, D. C., and Bhatia, C. R. (1987). *In vitro* electroporation of tobacco pollen. *Plant Sci.* **52**, 135–139.

Morikawa, H., Iida, A., Matsuri, C., Ikegami, M., and Yamada, Y. (1986). Gene transfer into intact plant cells by electroinjection through cell walls and membranes. *Gene* 41, 121–124.

Mulcahy, D. L. (1986). Gametophytic gene expression. *In* "Genetic Approach to Plant Biochemistry" (A. D. Blowstein and P. J. King, eds.), pp. 247–258. Springer-Verlag, New York.

Murashige, T., and Skoog, F. (1962). A revised medium for rapid growth and bioassays with tobacco tissue cultures. *Physiol. Plant* 15, 473–497.

Ohta, Y. (1986). High-efficiency genetic transformation of maize by a mixture of pollen and exogenous DNA. *Proc. Natl. Acad. Sci. USA* 83, 715–719.

Okada, K., Nagata, T., and Takebe, I. (1986). Introduction of functional RNA into plant protoplasts by electroporation. *Plant Cell Physiol.* 27, 619–626.

Pandey, K. K. (1978). Gametic gene transfer in *Nicotiana* by means of irradiated pollen. *Genetica* 49, 53–69.

Pandey, K. K. (1980). Further evidence for egg transformation in *Nicotiana*. *Heredity* 45, 15–29.

Peleman, J., Boerjan, W., Engler, G., Seurinck, J., Botterman, J., Alliotte, T., Van Montagu, M., and Inze, D. (1989). Strong cellular preference in the expression of a housekeeping gene of *Arabidopsis thoeliana* encoding *S*-adenosylmethionine synthetase. *Plant Cell* 1, 81–93.

Picton, J. M., and Steer, M. W. (1982). A model for the mechanism of tip extension in pollen tubes. *J. Theor. Biol.* 98, 15.

Potter, H., Weir, L., and Leder, P. (1984). Enhancer-dependent expression of human k immunoglobulin genes introduced into mouse pre-B lymphocytes by electroporation. *Proc. Natl. Acad. Sci. USA* 81, 7161–7165.

Rhodes, C. A., Pierce, D. A., Mettler, I. J., Mascarenhas, D., and Detmer, J. (1988). Genetically transformed maize plants from protoplasts. *Science* 240, 204–207.

Riggs, C. D., and Bates, G. W. (1986). Stable transformation of tobacco by electroporation: Evidence for plasmid concatenation. *Proc. Natl. Acad. Sci. USA* 84, 5602–5606.

Roeckel, P., Heizmann, P., Dubois, M., and Dumas, C. (1988). Attempts to transform *Zea mays* via pollen grains, Effect of pollen and stigma nuclease activities. *Sex. Plant Reprod.* 1, 156–163.

Sanford, J. C., Skubik, K. A., and Reisch, B. I. (1985). Attempted pollen-mediated plant transformation employing genomic donor DNA. *Theor. Appl. Genet.* 69, 571–574.

Saunders, J. A., Matthews, B. F., and Miller, P. D. (1989a). Plant gene transfer using electrofusion and electroporation. *In* "Electroporation and Electrofusion in cell Biology" (E. Neumann, A. Sowers, and C. Jordan, eds.), pp. 343–354. Plenum Press, New York.

Saunders, J. A., Smith, C. R., and Kaper, J. M. (1989b). Effects of electroporation pulse wave on the incorporation of viral RNA into tobacco protoplasts. *BioTechniques* 7, 1124–1131.

Shahin, E. A., and Simpson, R. B. (1986). Gene transfer system for potato. *Hort. Sci.* 21(5), 1199–1201.

Shimamoto, K., Terada, R., Izawa, T., and Fujimoto, H. (1989). Fertile transgenic rice plants regenerated from transformed protoplasts. *Nature* 338, 274–276.

Smithies, O., Gregg, R. G., Boggs, S. S., Koralewaski, M. A., and Kucherlapati, R. S. (1985). Insertion of DNA sequences into the human chromosomal B-globin locus by homologous recombination. *Nature* 317, 230–234.

Southern, E. (1975). Detection of specific sequences among DNA fragments separated by gel electrophoresis. **J. Mol. Biol. 98,** 503–517.

Sowers, A. E. (1984). Characterization of electric field-induced fusion in erythrocyte ghost membranes. *J. Cell Biol.* **99,** 1989–1996.

Spena, A., Viotti, A., and Pirrotta, V. (1983). Two adjacent genomic zein sequences: Structure, organization and tissue-specific restriction patterns. *J. Mol. Biol.* **169,** 799–811.

Stanley, R. G., and Linskens, H. F. (1974). "Pollen Biology Biochemistry Management." Springer-Verlag, New York.

Stopper, H., Zimmermann, U., and Wecker, E. (1985). High yields of DNA transfer into mouse L-cells by electropermeabilization. *Z. Naturforsch.* **40c,** 929–932.

Tsong, T. Y., and Kinosita, K. (1985). Use of voltage pulses for the pore opening and drug loading, and subsequent resealing of red blood cells. *Biblio. Haematol.* **51,** 108–114.

Twell, D., Klein, T. M., Fromm, M. E., and McCormick, S. (1989). Transient expression of chimeric genes delivered into pollen by microprojectile bombardment. *Plant Physiol.* **91,** 1270–1274.

Wang, Y.-C., Klein, T. M., Fromm, M., Cao, J., Sanford, J. C., and Wu, R. (1988). Transient expression of foreign genes in rice, wheat and soybean cells following particle bombardment. *Plant Mol. Biol.* **11,** 433–439.

Weber, H., Forester, W., and Jacob, H. E. (1981). Parasexual hybridization of yeasts by electric field stimulated fusion of protoplasts. *Curr. Genet.* **4,** 165–166.

Wong, T. K., and Neumann, E. (1982). Electric field mediated gene transfer. *Biochem. Biophys. Res. Commun.* **107,** 584–587.

16

Electrofusion of Plant Protoplasts and the Production of Somatic Hybrids

George W. Bates

Department of Biological Science, Florida State University, Tallahassee, Florida 32306

I. Introduction

Among higher organisms, plants exhibit a remarkable degree of developmental flexibility. Tissues and cells isolated from virtually any plant part can be grown in culture and induced to proliferate new organs; ultimately, entire plants can be recovered from these cultures. Because of this ability of plant cells to regenerate in tissue culture, the fusion of somatic plant cells makes possible the formation of hybrids between sexually incompatible plant species. Even with early developments on somatic cell fusion of plants in the 1970s (Carlson *et al.*, 1972) it was hoped that somatic hybridization might allow the transfer of agriculturally important genes from wild species into crop plants. This is still the goal of researchers working on plant cell fusion, however, success has come more slowly than originally hoped. The exciting news is that the two decades of research on the fusion of plant protoplasts are finally beginning to bear fruit. In the last two years field tests of the first somatic hybrid plants with commercial potential have begun. These initial successes have been made possible by developments in plant molecular genetics, plant tissue culture, and the methodology of plant cell fusion—including electrofusion.

Fusion of plant cells requires that the cells first be stripped of their cellulosic cell walls by enzymatic digestion. The resulting naked protoplasts are quite fragile and susceptible to physical and chemical damage. For this reason the early reports of the electrofusion of plant protoplasts (Senda et al., 1979; Zimmermann and Scheurich, 1981) were regarded as a novelty by plant tissue culturists, who doubted the protoplasts would remain viable after exposure to the high-voltage fields used in electrofusion. This issue was ultimately settled in 1985 when two independent studies demonstrated that electrically fused plant protoplasts not only remained viable but could be grown into somatic hybrid plants (Bates and Hasenkampf, 1985; Kohn et al., 1985). These successes stimulated much more work on plant electrofusion. A brief overview of the primary literature reveals that more than 40 articles were published on the electrofusion of plant cells in the last five years. This figure represents about a threefold increase over the previous five years. Indeed, in the last few years electrofusion has become established as one of the two principal methods used for the fusion of somatic plant cells. Presently, about one-third of all the laboratories working on the somatic cell fusion of plants utilize electrofusion, and most of the rest work with some variation of the polyethylene glycol-induced fusion technique or the Ca^{2+}/high-pH technique. This chapter reviews recent developments in plant electrofusion and its application to plant somatic hybridization.

II. Factors Affecting Protoplast Electrofusion

The yield of fusion products is affected by a variety of physical, chemical, and biological factors, including pulse parameters, composition of the fusion medium, the method used for protoplast isolation, cell type, and cell size. These factors must be carefully considered in order to obtain maximum yields of viable fusion products.

Dielectrophoretic alignment of plant protoplasts is effective over a broad range of AC fields, but most workers utilize fields of 0.5–1.5 kHz, 100–200 V/cm, regardless of protoplast type or source. However, the DC pulse settings that give optimal fusion can vary depending on cell source. For square-wave pulses, the effective range extends from single or multiple pulses of 10 μs, 1000 V/cm, to pulses of 50 μs, 2000 V/cm. Protoplasts isolated from leaves are generally found to fuse at lower field strengths than protoplasts isolated from roots or suspension cultures (Tempelaar and Jones, 1985; Tempelaar et al., 1987), and, under optimal conditions, fusion efficiency is generally higher for leaf protoplasts (Tempelaar and Jones, 1985; Tempelaar et al., 1987). Suspension culture protoplasts, however, are more resistant to pulse-induced lysis and may have better long-term viability following fusion (G. W. Bates, unpublished observation).

As expected from theory, larger-diameter protoplasts exhibit lower DC voltage thresholds for fusion (Mehrle et al., 1990). This fact may pose difficulties when fusion between protoplasts of markedly different size is attempted. Also, protoplast preparations typically contain a population of cells with at least a twofold variation in diameter. Thus, pulses that give optimum fusion for the population may result in lysis of the largest cells, while the smallest cells remain unfused.

The method used for protoplast isolation is a generally overlooked factor in electrofusion and may be quite important. The commercially available enzymes used to digest plant cell walls are crude preparations that contain different levels of cellulases and other polysaccharidases, and are contaminated with a variety of enzymes including proteases. Nea and Bates (1987) found that electrofusion yields varied greatly for carrot protoplasts isolated with different commercial enzyme preparations even though all the enzymes tested appeared to completely remove the cell wall. Proteases and dispase have also been found to enhance protoplast electrofusion (Ruzin and McCarthy, 1986; Nea and Bates, 1987; Neal et al., 1987). Again, this indicates that cell surface preparation can be an important factor in electrofusion.

The addition of certain small molecules to the fusion medium can enhance electrofusion. Calcium ions (0.5–1.0 mM) have been frequently reported to promote fusion and reduce pulse-induced cell lysis (Watts and King, 1984; Tempelaar et al., 1987; Nea and Bates, 1987; Hibi et al., 1988). Spermine and spermidine have also been reported to promote fusion, although the effect of these polyamines on fusion is less reproducible than that of calcium (Ruzin and McCarthy, 1986; Tempelaar et al., 1987). Fusogens such as dimethyl sulfoxide (DMSO), glycerol, and lysolecithin promote protoplast electrofusion; however, at concentrations where these compounds are effective they markedly reduce protoplast viability and increase pulse-induced protoplast lysis (Ruzin and McCarthy, 1986; Nea and Bates, 1987).

When electrofusion is optimized, 40–70% of the protoplasts can be observed to undergo fusion (Watts and King, 1984; Bates, 1985; Tempelaar and Jones, 1985; Tempelaar et al., 1987). However, conditions leading to maximum fusion can substantially reduce cell viability and favor fusion events involving three or more cells (Bates, 1985). If the goal of protoplast fusion is the production of somatic hybrids, the percentage of cells fused is not as important as the recovery of viable binucleate heterokaryons. Fusion products formed from multiple cells tend to have reduced viability in culture and, when they do grow, result in the regeneration of highly polyploid and aneuploid plants. In most plant electrofusion systems the maximum recovery of binucleate heterokaryons is 10% or less of the cells recovered after fusion (Bates, 1985; Hamill et al., 1987; Hibi et al., 1988).

There are two difficulties in producing high yields of binucleate heterokaryons. First, high cell densities result in high fusion yields through the formation of long pearl chains. This circumstance inevitably leads to a preponderance of multinucleate fusion products. Second, cell contacts generated by dielectrophoresis occur more frequently between protoplasts of the same type than they do between different protoplasts. This is especially true when protoplasts of different size are being fused. As a result, homofusions are favored over heterofusions. Few attempts have so far been made to direct fusion toward the formation of heterofusion products. Tempelaar *et al.* (1987) examined the effect of altering the surface charge of one fusion partner by coating these protoplasts with spermine prior to fusion. However, no increase in heterofusions was observed. In principal, antibodies to cell-specific, cell-surface antigens could be used to direct fusion (Lo *et al.*, 1984). However, plant cell-surface antigens are still poorly characterized and appropriate cell-specific antigens are not yet available. In an intriguing recent report, Mehrle *et al.* (1989) found that when electrofusion was carried out under microgravity conditions, during parabolic space flight, yields of heterofusion products were stimulated 10-fold. In this experiment, fusion was sought between cells of very different densities. Presumably the increased yield of heterofusions was due to the more homogeneous mixing of cells in the fusion chamber under microgravity conditions. However, overall fusion yields were also stimulated.

In addition to the percentage of binucleate heterokaryons formed, the total number of fusion products recovered is also important. For most plant species, protoplast culture requires a cell density of at least 5×10^4 protoplasts/ml. Less than 10% (in many cases less than 1%) of these protoplasts grow into colonies from which plants can be regenerated. To satisfy this cell density requirement, most laboratories working with electrofusion now use preparative fusion chambers that hold 0.25 to 1.0-ml samples of protoplasts. Many different chamber configurations are in use, including chambers with multiple small wires or meander-shaped chambers (Bates, 1985; Mehrle *et al.*, 1989), flow-through chambers (Hibi *et al.*, 1988), and chambers with flat-plate electrodes (Watts and King, 1984; Hamill *et al.*, 1987).

III. Somatic Hybridization by Electrofusion

Since 1985, a large number of studies have documented the recovery of somatic hybrid plants by electrofusion (Table 1). The species listed in Table 1 are a representative sampling of those plants that are amenable to protoplast culture, a fact that by itself suggests that electrofusion is a nontoxic procedure. Members of the nightshade family, particularly species of *Nicotiana* and *Solanum,* have been worked with most frequently because protoplasts from these species are particularly easy to culture. Refinements in culture techniques are rapidly expanding the list of plant species that can be grown from protoplasts, and, in parallel with these developments, new somatic hybrid combinations are being formed by electrofusion. Recent success in producing somatic hybrids of *Oryza sativa* (Toriyama and Hinata, 1988), *Rudbeckia* (Al-Atabee *et al.*, 1990), and *Medicago* (Damiani *et al.*, 1988; Gilmour *et al.*, 1989) are prime indications that electrofusion is not restricted to plants that are easy to grow in tissue culture.

Success in the production of somatic hybrid plants by electrofusion can be attributed to two things: first, technical improvements in the fusion process and in

Table 1

Species Combinations in Which Somatic Hybrid Calli or
Plants Have Been Regenerated Following Electrofusion

Species combination	Level of regeneration achieved	Reference
Intraspecific fusions		
Nicotiana plumbaginifolia (+) *N. plumbaginifolia*	Plants	de Vries *et al.*, 1987
N. tabacum (+) *N. tabacum*	Plants	Koop and Schweiger, 1985
Oryza sativa (+) *O. sativa*	Plants	Toriyama and Hinata, 1988
Intrageneric fusions		
N. tabacum (+) *N. plumbaginifolia*	Plants	Bates and Hasenkampf, 1985
N. tabacum (+) *N. repanda*	Plants	Bates, 1990a
N. tabacum (+) *N. glutinosa*	Plants	Bates, 1990b
N. glauca (+) *N. langsdorffii*	Plants	Morikawa *et al.*, 1987
Solanum tuberosum (+) *S. brevidens*	Plants	Fish *et al.*, 1988a
Solanum tuberosum (+) *S. chacoense*	Plants	Matthews and Saunders, 1989
S. tuberosum (+) *S. phreja*	Calli	Puite *et al.*, 1986
S. melanogena (+) *S. torvum*	Plants	Sihachakr *et al.*, 1989
S. melanogena (+) *S. khasianum*	Calli	Sihachakr *et al.*, 1988
Lycopersicon esculentum (+) *L. peruvianum*	Plants	Wijbrandi *et al.*, 1990
Medicago sativa (+) *M. borealis*	Calli	Gilmour *et al.*, 1989
M. sativa (+) *M. arborea*	Calli	Damiani *et al.*, 1988
Rudbeckia hirta (+) *R.laciniata*	Plants	Al-Atabee *et al.*, 1990
Intergeneric fusions		
Nicotiana plumbaginifolia (+) *Hyoscyamus muticus*	Calli	Kishinami and Widholm, 1987
Oryza sativa (+) *Echinochola oryzicola*	Plants[a]	Terada *et al.*, 1987
O. sativa (+) *Porteresia coarctata*	Calli	Finch *et al.*, 1990
Rudbeckia purpurea (+) *Tithonia rotundifolia*	Calli	Al-Atabee *et al.*, 1990
Rudbeckia purpurea (+) *Dimorphoteca aurantiaca*	Calli	Al-Atabee *et al.*, 1990

[a]Plants were necrotic and did not grow to maturity.

chamber design have lead to the recovery of greater numbers of viable heterokaryons, and second, breakthroughs in the development of selectable-marker genes have improved our ability to isolate somatic hybrid cell lines from mass tissue cultures. In early work on plant somatic hybrids, hybrids were identified on the basis of morphological characters or screenable markers such as complementation between two chlorophyll-deficient mutants leading to the restoration of chlorophyll production (Melchers and Labib, 1974). Fusion of cells from species with different hormone requirements in tissue culture (Morikawa *et al.*, 1987) and fusion of species combinations leading to hybrid vigor in tissue culture (Austin *et al.*, 1985) have also been effective approaches for the recovery of somatic hybrids. Selection can also be based on the complementation of auxotrophic mutants, for example, nitrate reductase-deficient mutants (Kohn et al., 1985; Negrutiu et al., 1986; de Vries et al., 1987). The difficulty with all these approaches is they restrict fusion experiments to cell lines where specific mutants or genetic backgrounds are available, and such lines are not available for most plant species. Elegant studies by Koop and Schweiger (1985) and by Koop and Spangenberg (1989) show that electrofusion can be used to fuse preselected pairs of Nicotiana protoplasts that have been positioned in microdroplets of fusion medium. The fused protoplasts were then cultured individually, again in microdrops, and plants have been regenerated. This approach eliminates the need for a postfusion selection scheme; however, it is only practical for the few plant species whose protoplasts can be grown in microdrop cultures.

A great deal of effort has been put into the development of more general selection methods. One popular approach in the mid 1980s was the fusion of protoplasts that had been stained with two different fluorescent dyes. Fusion products could be identified by the presence of both dyes in a single cell, and the fusion products could be isolated manually by use of micropipets (Patnaik et al., 1982) or by flow sorting (Galbraith et al., 1984). Recent developments in plant transformation have now provided much easier selection methods for recovering somatic hybrids. Plants can now be readily transformed with plasmids carrying antibiotic-resistance genes such as genes for kanamycin resistance or hygromycin resistance (Klee and Rogers, 1989). Fusion of protoplasts isolated from two species that carry different antibiotic-resistance genes (Komari et al., 1989; Thomas et al., 1990), or one species with an antibiotic-resistance gene and one with some other selectable or screenable marker (Bates et al., 1987; Agoudgil et al., 1990; Wijbrandi et al., 1990), provide a general approach for the recovery of somatic hybrids from tissue cultures. An additional advantage of these antibiotic-resistance genes is that they provide convenient molecular markers for genetic analysis of somatic hybrid plants.

A. Hybrid Fertility and the Transfer of Nuclear Genes

The overall goal of plant somatic hybridization is the transfer of useful genes between plant species that cannot be interbred through sexual crosses. Although Table 1 indicates that somatic hybrids can be readily recovered, the capacity of these hybrids to undergo normal development is frequently poor. In many cases hybrid cell lines (calli) have been recovered but it has not been possible to induce the differentiation of plants from these calli. In other cases hybrid plants have been regenerated but the plants are sterile. There are exceptions to this generality. For example, the fusion of Solanum tuberosum and S. brevidens protoplasts has resulted in the formation of fertile somatic hybrid plants (Austin et al., 1985; Fish et al., 1988a). These two species cannot be crossed sexually and the crop species, S. tuberosum (potato), is susceptible to several viral and fungal diseases for which resistance genes are found in the wild species, *S. brevidens*. The somatic hybrids between these species have been found to be resistant to potato leaf roll virus and potato virus Y. Field tests of these plants are underway (Austin *et al.*, 1986; Fish *et al.*, 1988b). This example notwithstanding, most somatic hybrids that combine the nuclear genes of two cross incompatible species are sterile. Unless fertile hybrid plants can be recovered, the value of somatic hybridization is going to be limited to those crops that can be vegetatively propagated.

Difficulties with regenerating plants from somatic hybrid calli and the sterility of the somatic hybrid plants themselves could be the result of polyploidy and aneuploidy, which are frequently found in plant tissue cultures and somatic hybrids. Fusion of protoplasts isolated from haploid plants has been explored as a way of controlling the ploidy level of somatic hybrids (Pehu *et al.*, 1989; Toriyama and Hinata, 1988). However, polyploid and aneuploid plants were still recovered in addition to euploids. An alternate approach to improving the fertility of somatic hybrid plants is to deliberately induce the partial elimination of one chromosome set following protoplast fusion. The rationale for this approach is that genetically asymmetric hybrid plants, which have the complete genome of a recipient species plus a few chromosomes or chromosome fragments from a donor species, might be expected to have improved fertility when compared with hybrids having the complete nuclear genomes of two unrelated species.

Chromosome elimination can be induced by irradiating the donor protoplasts prior to fusion. However, if irradiation and fusion are followed by selection for a nuclear encoded gene of the donor, then hybrids can be obtained that contain a complete set of chromosomes from the recipient species plus a partial set of donor chromosomes (Fig. 1). Several recent studies confirm the merit of this approach. Asymmetric hybrids that are partially fertile have been produced by both electro-fusion and PEG fusion from several different interspecific combinations in *Nicotiana* (Bates *et al.*, 1987; Bates, 1990a, 1990b; Famelaer *et al.*, 1989) and two intergeneric fusions [*Nicotiana* (+) *Daucus*, Dudits *et al.* (1987) and *Nicotiana* (+) *Atropa*, Gleba *et al.* (1988)]. However, other studies have produced asymmetric hybrids that are sterile (Gupta *et al.*, 1984). The degree of radiation-induced chromosome elimination varies greatly. In some cases hybrids are obtained that have only one or a few donor chromosomes (Gupta *et al.*, 1984; Dudits *et al.*, 1987; Famelaer *et al.*, 1989; Piastuch and Bates, 1990), whereas in other experiments hybrids are recovered that have more than half the original set of donor chromosomes (Gleba *et al.*, 1989; Yamashita *et al.*, 1989; Bates, 1990b; Wijbrandi *et al.*, 1990). Both extreme and partial chromosome elimination may be observed in the same experiment (Famelaer, 1989; Bates, 1990a; Sjodin and Glimelius, 1989). Hybrids with more thorough elimi-nation of donor chromosomes may be more fertile than those that retain large numbers of donor chromosomes. Thus, ongoing work seeks to define and control the factors that control chromosome elimination.

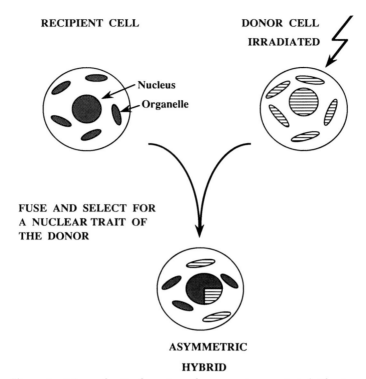

RECIPIENT CELL

**DONOR CELL
IRRADIATED**

Nucleus

Organelle

**FUSE AND SELECT FOR
A NUCLEAR TRAIT OF
THE DONOR**

**ASYMMETRIC
HYBRID**

Figure 1   Scheme for the formation of asymmetric somatic hybrids. Irradiation inactivates the nucleus of the donor cell. If fusion is followed by selection for a nuclear encoded trait of the donor, such as antibiotic resistance, hybrids are recovered that contain a complete recipient genome plus some nuclear DNA from the donor. If no irradiation is applied, the hybrids recovered often have most of the nuclear genes of both species. Irradiation does not affect the transmission of organelles. Thus, the hybrids initially have mixed populations of mitochondria and chloroplasts from both donor and recipient cells. As the hybrids grow in culture, the chloroplasts sort out to one pure type (donor or recipient) and the mitochondria undergo genetic recombination.

B. Transfer of Organellar Genes

Agriculturally important traits such as disease resistance, drought tolerance, and cold hardiness are important candidates for interspecific transfer by protoplast fusion. The genes for these traits generally seem to be nuclear encoded. However, the genes for several other important traits are cytoplasmically encoded, and there has been considerable success in transferring these genes between plant species by protoplast fusion.

Plants have two organelles, in addition to the nucleus, with their own genomes: mitochondria and chloroplasts. Chloroplast-encoded genes are known that effect photosynthetic rates and sensitivity to herbicides, as for example atrazine (Menczel *et al.*, 1986). Genes in the mitochondrion control male sterility (Belliard *et al.*, 1979). This cytoplasmic male sterility (CMS) results from a development defect in anther and pollen formation, and is important agriculturally in the production of high-yielding F1 hybrids. The transfer of cytoplasmically encoded traits between plant species is a very active field of research. Recently, a DNA delivery system capable of transforming chloroplasts was demonstrated (Svab *et al.*, 1990); however, this technique is far from routine. No free-DNA-mediated transformation system is available for plant mitochondria. The usual approach for the transfer of cyto-plasmically encoded traits, especially between sexually incompatible species, is by protoplast fusion.

Following protoplast fusion, the chloroplasts of the two species sort out (probably by random segregation) during subsequent cell divisions. Thus plants are regenerated that possess a hybrid nucleus but a pure population of chloroplasts from one or the other parental species. For example, the fusion of *Solanum tuberosum* and *S. brevidens* protoplasts was found to result in somatic hybrid plants 55% of which had chloro-plasts from *S. brevidens* and 45% of which had *S. tuberosum* chloroplasts (Pehu et al., 1989). The recovery of plants with mixed populations of chloroplasts or genetically recombined chloroplasts is very rare (Medgyesy *et al.*, 1985).

By contrast with chloroplasts, mitochondria often undergo genetic recombination following protoplast fusion. Thus somatic hybrid plants are recovered that contain a novel mitochondrial genome that differs from both parental species (Belliard *et al.*, 1979). Some somatic hybrids may also be recovered that have the mitochondria of one or the other parental species. However, mitochondrial recombination is more frequent than segregation.

The transmission of organelle and nuclear genes can be manipulated by several different approaches. Selectable marker genes provide one approach for controlling the transmission of organelles in somatic hybridization experiments. Plants resistant to streptomycin, spectinomycin, chloramphenicol, and lincomycin are readily iso-lated (Fluhr *et al.*, 1985). Resistance proves to be due to mutations in chloroplast DNA; thus, protoplast fusion followed by selection for antibiotic resistance can be used to produce somatic hybrids with a specific chloroplast type (Menczel *et al.*, 1981). Organelle-encoded mutations that could be used for directing the transmis-sion of mitochondria have been more difficult to obtain. The only antibiotic-resistant mutant isolated so far that involves a mutation in mitochondrial DNA is an oli-gomycin-resistant mutant of *N. tabacum* (Durand and Harada, 1989). Although CMS is a mitochondrially encoded trait, it is not expressed until late in floral development and thus cannot be used as a selectable marker for the recovery of somatic hybrids.

A more general approach for controlling the transmission of chloroplast encoded traits is to inactivate one of the protoplasts with iodoacetate prior to fusion (Nehls,

1978). As seen in Fig. 2, this results in the recovery of somatic hybrids that contain
the chloroplasts of the cell line not treated with iodoacetate. However, iodoacetate
does not seem to be as effective in controlling the transmission of mitochondrial
genes. Prefusion treatment with the specific mitochondrial inactivator rhodamine-
6-G is effective in controlling the mitochondrial composition of somatic hybrids in
some cases, but not in others (Bottcher *et al.*, 1989).

As explained earlier, the nuclear composition of plant somatic hybrids can be
controlled by X- or gamma-irradiation of one protoplast type prior to fusion. As
shown in Fig. 2, when irradiation and fusion are followed by selection for an

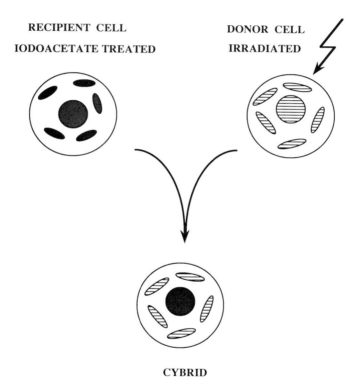

CYBRID

Figure 2   General scheme for the formation of cybrids. The nucleus
of the donor cell is inactivated by irradiation, whereas the organelles
of the recipient cell are inactivated with iodoacetate. After fusion and
selection for an organelle-encoded trait of the donor, cybrid plants
are recovered that contain the nucleus of the recipient and the chloroplasts of the donor. In some cases they cybrids may contain only
donor mitochondria; in others they will contain genetically recombined mitochondria.

organelle-encoded trait of the irradiated cell, the hybrids recovered generally contain the nucleus of the unirradiated recipient cells and the organelles of the donor (Zelcer *et al.*, 1978). Such plants are known as cybrids. Cybridization is becoming increasingly important method for transferring cytoplasmically encoded traits. When irradiation of donor protoplasts is combined with iodoacetate or rhodamine treatment of the recipient protoplasts, fusion often results in recovery of plants containing the nucleus of the recipient plus the cytoplasm of the donor.

Cybridization is being used primarily for the transfer of cytoplasmic male sterility because of the agricultural importance of this mitochondrial trait. For example, Yang *et al.* (1989) have transferred CMS between two cytotypes of rice by electrofusion. Similar experiments have been done with PEG fusion in tobacco (Zelcer *et al.*, 1978) and other species.

Another approach to cybridization is the fusion of experimentally modified cell types such as cytoplasts and karyoplasts. Protoplasts can be enucleated by centrifuging them through a dense Percoll cushion (Griesbach and Sink, 1983). Karyoplasts (nuclei surrounded by a plasma membrane) are recovered from under the cushion and cytoplasts (enucleated protoplasts) remain above the cushion. Fusion between cytoplasts and karyoplasts of different species or genotypes can be used for the transfer of cytoplasmically encoded traits. One of the unique opportunities afforded by electrofusion for cybridization is the individual fusion of preselected pairs of cytoplasts and karyoplasts, as demonstrated by Koop and Spangenberg (1989). They have developed a computerized system for placing pairs of protoplasts (or one karyoplast and one cytoplast) in a 500-nl droplet of mannitol. These protoplasts are then fused using a pair of positionable microelectrodes. In this way up to 200 pairs of preselected protoplasts can be fused in 1 h. The fused protoplasts are then transferred to tiny droplets of culture medium and can be regenerated into plants. Electrofusion of karyoplasts and vacuolated protoplasts was also demonstrated by Mehrle *et al.* (1989) in their study of fusion under microgravity conditions. However, the regeneration of plants from these fusions was not reported.

IV. Comparison of Electrofusion with PEG-Induced Fusion

Many early papers pointed to the high yields of fusion products that can be obtained by electrofusion as evidence that electrofusion is more efficient than fusion by PEG. It was also frequently suggested that electrofusion was less toxic because it avoids the use of agents such as PEG, high pH, and high concentrations of calcium. However, few of these studies made a direct comparison of the two techniques. When comparisons were made, considerably more effort was often put into optimizing one technique than the other. Recent work on electrofusion and PEG-induced fusion, including several direct comparisons between these techniques, have clarified this situation. The claim that electrofusion is less toxic than PEG is difficult to substantiate. Both procedures kill protoplasts, and protocols that enhance fusion also increase cell killing. Mesophyll cells of certain species can be quite susceptible to membrane damage by PEG. For example, mesophyll protoplasts of *Medicago* and *Solanum* species can be easily damaged by PEG (L. B. Johnson, personal communication; Fish *et al.*, 1988a) and have been successfully fused by electric fields (Damiani *et al.*, 1988). However, modifications of the PEG fusion procedure have overcome the toxicity problem, and *Medicago* somatic hybrids have now been formed by PEG-induced fusion as well (Thomas *et al.*, 1990). Critical examination of all the literature on the fusion of plant protoplasts suggests that protoplasts from all species and cell types can be successfully fused by either technique.

The efficiency of fusion, whether it is induced by PEG or by electric fields, varies from 1% to more than 50% depending on the study. In terms of somatic hybridization, however, the raw percentage of cells fused is not as important as the number of hybrid colonies recovered after each fusion experiment. Negrutiu *et al.* (1986) have compared electrofusion and all the commonly used PEG fusion protocols. They examined the total number of somatic hybrid colonies recovered after the fusion of 10^5 protoplasts of two complementing auxotrophic mutants of *N. plumbaginifolia*, and concluded that electrofusion gave as many colonies as the best PEG fusion protocols. The efficiencies of the optimized techniques were comparable but electrofusion was found to be "technically easy, and quicker than PEG-fusion" (Negrutiu *et al.*, 1986). Similar conclusions have been reached by other groups (Han San *et al.*, 1990; Fish *et al.*, 1988a). It should be pointed out that most preparative electrofusion chambers now in use will handle samples of up to 10^6 protoplasts. Samples of this size are larger than those generally used for PEG fusion. Scaling up PEG fusion to larger volumes can result in reduced fusion efficiency; thus electrofusion is generally found to give a higher throughput of fused protoplasts than PEG fusion (Gilmour *et al.*, 1989).

Given that the toxicity and fusion efficiency of electrofusion and PEG fusion are comparable, the choice of which technique to use often becomes simply a matter of convenience and availability of the necessary equipment. However, there remains one area in which these fusion techniques differ substantially. Treatment of protoplasts with PEG causes the protoplasts to clump in multicellular masses and to adhere to the bottom of the fusion vessel. This clumping makes it very difficult to isolate heterokaryons after PEG fusion with micropipetes or to fuse preselected pairs of cells, as Koop and Spangenberg (1989) demonstrated could be done following electrofusion. When hybrids are sought between species for which no selectable markers are available, the visual identification and manual isolation of heterokaryons can still be a practical approach for recovering somatic hybrids.

References

Agoudgil, S., Hinnisdaels, S., Mouras, A., Negrutiu, I., and Jacobs, M. (1990). Metabolic complementation for a single gene function associated with partial and total loss of donor DNA in interspecific somatic hybrids. *Theor. Appl. Genet.* **80**, 337–342.

Al-Atabee, J. S., Mulligan, B. J., and Power, J. B. (1990). Interspecific somatic hybrids of *Rudbeckia hirta* and *R. laciniata* (Compositae). *Plant Cell Rep.* **8**, 517-520.

Austin, S., Baer, M. A., and Helgeson, J. P. (1985). Transfer of resistance to potato leaf roll virus from *Solanum brevidens* into *Solanum tuberosum* by somatic fusion. *Plant Sci.* **39**, 75–82.

Austin, S., Ehlenfeldt, M. K., Baer, M. A., and Helgeson, J. P. (1986). Somatic hybrids produced by protoplast fusion between *S. tuberosum* and *S. brevidens:* Phenotypic variation under field conditions. *Theor. Appl. Genet.* **71**, 682–690.

Bates, G. W. (1985). Electrical fusion for the optimal formation of protoplast heterokaryons in *Nicotiana.Planta* **165**, 217–241.

Bates, G. W. (1990a). Asymmetric hybridization between *Nicotiana tabacum* and *N. repanda* receipient protoplast fusion: Transfer of TMV resistance. *Theor. Appl. Genet.* **80**, 481–487.

Bates, G. W. (1990b). Transfer of tobacco mosaic virus resistance by asymmetric protoplast fusion. *In* "Progress in Plant Cellular and Molecular Biology" (H. J. J. Nijkamp, L. H. W. Van Der Plas, and J. Van Aartrijk, eds.), Kluwer Academic Press, Dordrecht, pp. 275–291.

Bates, G. W., and Hasemkampf, C. A. (1985). Culture of plant somatic hybrids following electrical fusion. *Theor. Appl. Genet.* **70**, 227–233.

Bates, G. W., Hasenkampf, C. A., Contolini, C. L., and Piastuch, W. C. (1987). Asymmetric hybridization in *Nicotiana* by fusion of irradiated protoplasts. *Theor. Appl. Genet.* **74**, 718–726.

Belliard, G., Vedel, F., and Pelletier, G. (1979). Mitochondrial recombination in cytoplasmic hybrids of *Nicotiana tabacum* by protoplast fusion. *Nature* **281**, 401–403.

Bottcher, U. F., Aviv, D., and Galun, E. (1989). Complementation between protoplasts treated with either of two metabolic inhibitors results in somatic-hybrid plants. *Plant Sci.* **63**, 67–77.

Carlson, P. S., Smith, H. H., and Dearing, R. D. (1972). Parasexual interspecific plant hybridization. *Proc. Nat. Acad. Sci. USA* **69**, 2292–2294.

Damiani, F., Pezzotti, M., and Arcioni, S. (1988). Electric field mediated fusion of protoplasts of *Medicago sativa* L. and *Medicago arborea* L. *J. Plant Physiol.* **132**, 474–479.

de Vries, S. E., Jacobsen, E., Jones, M. G. K., Loonen, A. E. H. M., Tempelaar, M. J., Wijbrandi, J., and Feenstra, W. J. (1987). Electrofusion of biochemically well characterized nitrate reductase deficient *Nicotiana plumbaginifolia* mutants. Studies on optimization and complementation. *Plant. Sci.* **51**, 105–112.

Dudits, D., Maroy, E., Praznovsky, T., Olah, Z., Gyorgyey, J., and Cella, R. (1987). Transfer of resistant traits from carrot into tobacco by asymmetric somatic hybridization: Regeneration of fertile plants. *Proc. Natl. Acad. Sci. USA* **84**, 8434–8438.

Durand, J., and Harada, H. (1989). Interspecific protoplast fusion in *Nicotiana* provides evidence for a mitochondrial determinism of oligomycin-resistance. *Plant Sci.* **62**, 263–272.

Famelaer, I., Gleba, Y. Y., Sidorov, V. A., Kaleda, V. A., Parakonny, A. S., Boryshuk, N. V., Cherep, N. N., Negrutiu, I., and Jacobs, M. (1989). Intrageneric asymmetric hybrids between *Nicotiana plumbaginifolia* and *Nicotiana sylvestris* obtained by gamma fusion. *Plant Sci.* **61**, 105–117.

Finch, R. P., Slamet, I. H., and Cocking, E. C. (1990). Production of heterokaryons by the fusion of mesophyll protoplast of *Porteresia coarctata* and cell suspension-derived protoplasts of *Oryza sativa:* A new approach to somatic hybridization in rice. *J. Plant. Physiol.* **136**, 592–598.

Fish, N., Karp, A., and Jones, M. G. K. (1988a). Production of somatic hybrids by electrofusion in *Solanum. Theor. Appl. Genet.* **76**, 260–266.

Fish, N., Steele, S. H., and Jones, M. G. K. (1988b). Field assessment of dihaploid *Solanum tuberosum* and *S. brevidens* somatic hybrids. *Theor. Appl. Genet.* **76**, 880–886.

Fluhr, R., Aviv, D., Galun, E., and Edelman, M. (1985). Efficient induction and selection of chloroplast-encoded antibiotic-resistant mutants in *Nicotiana. Proc. Natl. Acad. Sci. USA* **82**, 1485–1489.

Galbraith, D. W., Afonso, C. L., and Harkins, K. R. (1984). Flow sorting and culture of protoplasts: Conditions for high-frequency recovery, growth and morphogenesis from sorted protoplasts of suspension cultures of *Nicotiana. Plant Cell Rep.* **3**, 151–155.

Gilmour, D. M., Davey, M. R., and Cocking, E. C. (1989). Production of somatic hybrid tissues following chemical and electrical fusion of protoplasts from albino cell suspensions of *Medicago sativa* and *M. borealis. Plant Cell Rep.* **8**, 29–32.

Gleba, Y. Y., Hinnisdaels, S., Sidorov, V. A., Kaleda, V. A. Parokonny, A. S., Boryshuk, N. V., Cherep, N. N., Negrutiu, I., and Jacobs, M. (1988). Intergeneric asymmetric hybrids between *Nicotiana plumbaginifolia* and *Atropa belladonna* obtained by "gammafusion." *Theor. Appl. Genet.* **76**, 760–766.

Griesbach, R. J., and Sink, K. C. (1983). Evacuolation of mesophyll protoplasts. *Plant Sci. Lett.* **30**, 297–301.

Gupta, P. P., Schieder, O., and Gupta, M. (1984). Intergeneric nuclear gene transfer between somatically and sexually incompatible plants through asymmetric protoplast fusion. *Mol. Gen. Genet.* **197**, 30–35.

Hamill, J. D., Watts, J. W., and King, J. M. (1987). Somatic hybridization between *Nicotiana tabacum* and *Nicotiana plumbaginifolia* by electrofusion of mesophyll protoplasts. *J. Plant. Physiol.* **129**, 111–118.

Han San, L., Vedel, F., Sihachakr, D., and Remy, R. (1990). Morphological and molecular characterization of fertile tetraploid somatic hybrids produced by protoplast electrofusion and PEG-induced fusion between *Lycopersicon esculentum* Mill. and *Lycopersicon peruvianum* Mill. *Mol. Gen. Genet.* **221**, 17–26.

Hibi, T., Kano, H., Sugiura, M., Kazami, T., and Kimura, S. (1988). High-speed electrofusion and electro-transfection of plant protoplasts by a continuous electro-manipulator. *Plant Cell Rep.* **7**, 153–157.

Kishinami, I., and Widholm, J. M. (1987). Auxotrophic complementation in intergeneric hybrid cells obtained by electrical and dextran-induced protoplast fusion. *Plant Cell Physiol.* **28**, 211–218.

Klee, H. J., and Rogers, S. G. (1989). Plant gene vectors and genetic transformation: Plant transformation systems based on the use of *Agrobacterium tumefaciens. In* "Cell Culture and Somatic Cell Genetics of Plants" (J. Schell and I. K. Vasil, eds.), Vol. 6, pp. 1–24, Academic Press, Orlando, FL.

Kohn, H., Schieder, R., and Schieder, O. (1985). Somatic hybrids in tobacco mediated by electrofusion. *Plant Sci.* **38**, 121–128.

Komari, T., Saito, Y., Nkaido, F., and Kumashiro, T. (1989). Efficient selection of somatic hybrids in *Nicotiana tabacum* L. using a combination of drug-resistance markers introduced by transformation. *Theor. Appl. Genet.* **77**, 547–552.

Koop, H. U., and Schweiger, H. G. (1985). Regeneration of plants after electrofusion of selected pairs of protoplasts. *Eur. J. Cell Biol.* **39**, 46–49.

Koop, H. U., and Spangenberg, G. (1989). Electric field-induced fusion and cell reconstitution with preselected single protoplasts and subprotoplasts of higher plants. *In* "Electroporation and Electrofusion in Cell Biology" (E. Neumann, A. E. Sowers, and C. A. Jordan, eds.), pp. 355–366, Plenum Press, New York.

Lo, M. M. S., Tsong, T. Y., Conrad, M. D., Strittmatter, S. M., Hester, L. D., and Snyder, S. H. (1984). Monoclonal antibody production by receptor-mediated electrically induced cell fusion. *Nature* **310**, 792–794.

Matthews, B. F. and Saunders, J. A. (1989). Gene transfer in plants. *In* "Biotic Diversity and Germplasm Preservation, Global Imperatives," Vol 13, pp. 275–291. Kluwer Academic, Dordrecht.

Medgyesy, P., Fejes, E., and Maliga, P. (1985). Interspecific chloroplast recombination in a *Nicotiana* somatic hybrid. *Proc. Natl. Acad. Sci. USA* **82**, 6960–6964.

Mehrle, W., Hampp, R., Naton, B., and Grother, D. (1989). Effects of microgravitation on electrofusion of plant cell protoplasts. *Plant Physiol.* **89**, 1172–1177.

Mehrle, W., Naton, B., and Hampp, R. (1990). Determination of physical membrane properties of plant cell protoplasts via the electrofusion technique: Prediction of optimal fusion yields and protoplast viability. *Plant Cell Rep.* **8**, 687–691.

Melchers, G., and Labib G. (1974). Somatic hybridisation of plants by fusion of protoplasts. *Mol. Gen. Genet.* **135**, 277–294.

Menczel, L., Nagy, F., Kiss, Z. R., and Maliga, P. (1981). Streptomycin resistant and sensitive somatic hybrids of *Nicotiana tabacum* + *Nicotiana knightiana:* Correlation of resistance to *N. tabacum* plastids. *Theor. Appl. Genet.* **59**, 191–195.

Menczel, L., Polsby, L. S., Steinback, K. E., and Maliga, P. (1986). Fusion-mediated transfer of triazine-resistant chloroplasts: Characterization of *Nicotiana tabacum* cybrid plants. *Mol. Gen. Genet.* **205**, 201–205.

Morikawa, H., Kumashiro, T., Kusakari, K., Iida, A., Hirai, A., and Yamada, Y. (1987). Interspecific hybrid plant formation by electrofusion in *Nicotiana. Theor. Appl. Genet.* **75**, 1–4.

Nea, L. J., and Bates, G. W. (1987). Factors affecting protoplast electrofusion efficiency. *Plant Cell Rep.* **6**, 337–340.

Nea, L. J., Bates, G. W., and Gilmer, P. J. (1987). Facilitation of electrofusion of plant protoplasts by membrane-active agents. *Biochim. Biophys. Acta* **897**, 293–301.

Negrutiu, I., De Brouwer, D., Watts, J. W., Sidorov, V. I., Dirks, R., and Jacobs, M. (1986). Fusion of plant protoplasts: A study using auxotrophic mutants of *Nicotiana plumbaginifolia,* Viviani. *Theor. Appl. Genet.* **72**, 279–286.

Nehls, R. (1978). The use of metabolic inhibitors for the selection of fusion products of higher plant protoplasts. *Mol. Gen. Genet.* **166**, 117–118.

Patnaik, G., Cocking, E. C., Hamill, J., and Pental, D. (1982). A simple procedure for the manual isolation and identification of plant heterokaryons. *Plant Sci. Lett.* **24**, 105–110.

Pehu, E., Karp, A., Moore, K., Steele, S., Dunckley, R., and Jones, M. G. K. (1989). Molecular,cytogenetic and morphological characterization of somatic hybrids of dihaploid *Solanum tuberosum* and diploid *S. brevidens. Theor. Appl. Genet.* **78**, 696–704.

Piastuch, W. C., and Bates, G. W. (1990). Chromosomal analysis of *Nicotiana* asymmetric somatic hybrids by dot blotting and in situ hybridization. *Mol. Gen. Genet.* **222**, 97–103.

Puite, K. J., Roest, S., and Pijnacker, L. P. (1986). Somatic hybrid potato plants after electrofusion of diploid *Solanum tuberosum* and *Solanum phureja. Plant Cell Rep.* **5**, 262-265.

Ruzin, S. E., and McCarthy, S. C. (1986). The effect of chemical facilitators on the frequency of electrofusion of tobacco mesophyll protoplasts. *Plant Cell Rep.* **5**, 342–345.

Senda, M., Takeda, J., Abe, S., and Nakamura, T. (1979). Induction of cell fusion of plant protoplasts by electrical stimulation. *Plant Cell Physiol.* **20**, 1441–1443.

Sihachakr, D., Haircour, R., Serraf, I., Barrientos, E., Herbreteau, C., Ducreux, G., Rossignol, L., and Souvannavong, V. (1988). Electrofusion for the production of somatic hybrid plants of *Solanum melongena* L. and *Solanum khasianum* C. B. Clark. *Plant Sci.* **57**, 215–223.

Sihachakr, D., Haircour, R., Chaput, M. H., Barrientos, E., Ducreux, G., and Rossignol, L. (1989). Somatic hybrid plants produced by electrofusion between *Solanum melongena* L. and *Solanum torvum* Sw. *Theor. Appl. Genet.* **77**, 1–6.

Sjodin, C. and Glimelius K. (1989). Transfer of resistance against *Phoma lingam* to *Brassica napus* by asymmetric somatic hybridization combined with toxin selection. *Theor. Appl. Genet.* **78**, 513–520.

Svab, Z., Hajdukiewicz, P., and Maliga, P. (1990). Stable transformation of plastids in higher plants. *Proc. Natl. Acad. Sci. USA* **87**, 8526–8530.

Tempelaar, M. J. and Jones, M. G. K. (1985). Fusion characteristics of plant protoplasts in electric fields. *Planta* **165**, 205–216.

Tempelaar, M. J., Duyst, A., De Vlas, S. Y., Krol, G., Symmonds, C., and Jones, M. G. K. (1987). Modulation and direction of the electrofusion response in plant protoplasts. *Plant Sci.* **48**, 99–105.

Terada, R., Kyozuka, J., Nishibayashi, S., and Shimamoto, K. (1987). Plantlet regeneration from somatic hybrids of rice (*Oryza sativa* L.) and barnyard grass (*Echinochloa oryzicola* Vasing). *Mol. Gen. Genet.* **210**, 39–43.

Thomas, M. R., Johnson, L. B., and White, F. A. (1990). Selection of interspecific somatic hybrids of *Medicago* by using *Agrobacterium*- transformed tissues. *Plant Sci.* **69**, 189–198.

Toriyama, K., and Hinata, K. (1988). Diploid somatic-hybrid plants regenerated from rice cultivars. *Theor. Appl. Genet.* **76**, 665–668.

Watts, J. W., and King, J. M. (1984). A simple method for large-scale electrofusion and culture of plant protoplasts. *Biosci. Rep.* **4**, 335–342.

Wijbrandi, J., Van Capelle, W., Hanhart, C. J., Van Loenen Martinet-Schuringa, E. P., and Koornneef, M. (1990). Selection and characterization of somatic hybrids between *Lycopersicon esculentum* and *Lycopersicon peruvianum. Plant Sci.* **70**, 197–208.

Yamashita, Y., Terada, R., Nishibayashi, S., and Shimamoto, K. (1989). Asymmetric somatic hybrids of *Brassica:* Partial transfer of *B. campestris* genome in *B. oleraceae* by cell fusion. *Theor. Appl. Genet.* **77**, 189–194.

Yang, Z.-Q., Shikanai, T., Mori, K., and Yamada, Y. (1989). Plant regeneration from cytoplasmic hybrids of rice (*Oryza sativa* L.). *Theor. Appl. Genet.* **77**, 305–310.

Zelcer, A., Aviv, D., and Galun, E. (1978). Interspecific transfer of cytoplasmic male sterility by fusion between protoplasts of normal *Nicotiana sylvestris* and X-ray irradiated protoplasts of male-sterile *N. tabacum. Z. Pflanzenphysiol.* **90**, 397–407.

Zimmermann, U., and Scheurich, P. (1981). High frequency fusion of plant protoplasts by electric fields. *Planta* **151**, 26–32.

17

Electrotransformation of Bacteria by Plasmid DNA

Jack T. Trevors,[1] Bruce M. Chassy,[2] William J. Dower,[3] and Hans P. Blaschek[2]

[1]Department of Environmental Biology, University of Guelph, Guelph, Ontario, Canada N1G 2W1

[2]Department of Food Science, University of Illinois, Urbana, Illinois 61801

[3]Affymax Research Institute, Palo Alto, California 94304

I. Development of Bacterial Electrotransformation

Progress in the genetic analysis of bacteria is often dependent on the availability and development of transformation methods. Development of methods for transformation is especially important in research conducted with bacterial species that are not naturally competent (competence is the ability of a bacterium to take up plasmid, chromosomal, or phage DNA). Approximately 15 naturally occuring transformation systems have been described (Mercenier and Chassy, 1988). The systems of natural competence of *Haemophilus influenzae, Bacillus subtilis, Streptococcus sanguis,* and *Streptococcus pneumoniae* have been carefully studied and are useful in genetic analysis of these bacteria (Mercenier and Chassy, 1988). However, the majority of bacterial species are not naturally competent. With these, alternative methods such as conjugation, plasmid labeling/mobilization, $CaCl_2$-induced pseudo-competence, or transformation of protoplasts followed by regeneration must be employed. Often these methods are highly strain specific and useful only with a limited number of well-characterized bacterial strains.

This chapter describes a relatively new method for bacterial transformation that is rapidly gaining acceptance. Electroporation is the term often applied to the transformation of eukaryotic and prokaryotic cells by electrical pulses of high amplitude and short duration in the presence of exogenous DNA. Pulses of electricity permeabilize bacteria to the influx of DNA, but, in the absence of direct evidence for the existence of actual pores in the bacterial membranes, it is probably more correct to use the terms electrotransformation or electropermeabilization. Setting semantics aside, it is clear that electrotransformation is a rapid and simple method that is applicable to a wide variety of bacteria. The method has opened the way to genetic analysis and manipulation of the majority of bacteria thus far studied.

The physical principles underlying electrofusion and electroporation of eukaryotic cells have been described in Chapters 1–12. Chapter 13 describes the applications of electroporation to eukaryotic cells. Although largely unnoticed, Shivarova *et al.* (1983) reported the transformation of protoplasts of *Bacillus cereus*; by electroporation. Electroporation gave a 10-fold advantage in frequency over the protoplast method previously used with *B. cereus*; the need for the tedious preparation and regeneration of protoplasts remained. What was needed was a method for the electroporation of whole untreated bacterial cells.

Reports that whole washed cells of the gram-negative bacterium *Escherichia coli* (Dower, 1987), and the gram-positive bacteria *Lactococcus lactis* (Harlander, 1987), *Lactobacillus casei* (Chassy and Flickinger, 1987), and *Streptococcus thermophilus* (Somkuti and Steinberg, 1987) could be transformed by electroporation in the presence of plasmid DNA sparked renewed interest in electrical methods of transformation. To date, more than 100 species of bacteria have been successfully transformed; a partial list appears in Table 1. Chapter 30 contains protocols for the electroporation of gram-positive and gram-negative bacteria. A number of reviews on bacterial electroporation have appeared in the last 4 years (Chassy *et al.*, 1988; Dower, 1990; Luchansky *et al.*, 1988; Miller, 1988; Roberts, 1990a, 1990b; Shigekawa and Dower, 1988; Solioz and Bienz, 1990) Electroporation methods for bacteria have also found their way into widely used laboratory manuals (Sambrook *et al.*, 1989; Sheen, 1989). A variety of commercial pulse generators suitable for bacterial electroporation is now available (see Chapter 35). The advantages of electrotransformation over other methods of bacterial transformation are listed in Table 2.

Transfection of eukaryotic cells by electroporation became an established procedure before methods for bacteria were developed (Potter, Chapter 13). Bacteria proved more difficult to electrotransform because they are fundamentally different

Table 1
Reported Electrotransformation of Bacterial Species

Actinobacillus pleuropneumoniae	Lalonde *et al.* (1989)
Agrobacterium rhizogenes	Mattanovich *et al.* (1989)
	Mersereau *et al.* (1990)
	Nagel *et al.* (1990)
	Wen-Jun and Forde (1989)
Agrobacterium tumefaciens	Mattanovich *et al.* (1989)
	Nagel *et al.* (1990)
	Wen-Jun and Forde (1989)
	Wirth *et al.* (1988)
Bradyrhizobium japonicum	Guerinot *et al.* (1990)
Campylobacter coli	Miller *et al.* (1988)
Campylobacter jejuni	Miller *et al.* (1988)
Citrobacter freundii	Wirth *et al.* (1989)
Enterobacter aerogenes	Wirth *et al.* (1989)
Erwini carotovora	Wirth *et al.* (1989), Ito *et al.* (1988)
Escherichia coli	
spheroplasts	Cymbalyuk *et al.* (1988), Dahlman and Harlander (1988),
(strain not mentioned)	Hofmann (1988), Larimer and Mural (1990)
wild-type	Wirth *et al.* (1989)
strain 3132 (wild-type)	Taketo (1988)
stain BB	Taketo (1988)
strain C	Taketo (1988, 1989)
strain C-600	Marcus *et al.* (1990)
strain DH1	Taketo (1988), Heery *et al.* (1989)
strain DH10B	Li (1990), Smith *et al.* (1990)
strain DH5α	Dower *et al.* (1988), Smith *et al.* (1990)
strain DH5αF'IQ	Smith *et al.* (1990)
strain DS941	Summers and Withers (1990)
strain HB101	Dower (1987), Fiedler and Wirth (1988), Kim and Blaschek
	(1989), Wirth *et al.* (1989), Smith *et al.* (1990), Speyer
	(1990), Haynes and Britz (1990)
strain JC8679	Summers and Withers (1990)
strain JM105	Fiedler and Wirth (1988), Wirth *et al.* (1988), Jacobs *et al.*
	(1990)
strain JM107	Dower (1987)
strain JM109	Smith *et al.* (1990), Taketo (1988)
strain K12A	Taketo (1989)
strain K803	Jacobs *et al.* (1990)
strain LE392	Dower (1987), Dower *et al.* (1988), Taketo (1988)
strain M5361	Takahashi and Kobayashi (1990)
strain MC1061	Calvin and Hanawalt (1988), Dower (1990), Smith *et al.* (1990),
	Willson *et al.* (1988)
strain MC1061/P3	Sheen (1989)
strain MV1190	Dower (1987)
strain p678-5 4	Marcus *et al.* (1990)

(*continued*)

Table 1

Continued

Escherichia coli	
strain TG1	Heery and Dunican (1989)
strain WA321	Fiedler and Wirth (1988), Wirth *et al.* (1989)
strain WM1100	Dower (1990)
strain XL1 BLUE	Taketo (1988)
Haemophilus pleuropneumoniae	see *Actinobacillus pleuropneumoniae*
Klebsiella aerogenes	Trevors (1990)
Klebsiella oxytoca	Wirth *et al.* (1989)
Klebsiella pneumoniae	Wirth *et al.* (1989)
Proteus mirabilis	Wirth *et al.* (1989)
Proteus vulgaris	Wirth *et al.* (1989)
Pseudomonas aeruginosa	Smith and Iglewski (1989), Smith *et al.* (1990)
Pseudomonas chlororaphis	Wirth *et al.* (1989)
Pseudomonas oxalaticus	Wirth *et al.* (1989)
Pseudomonas putida	Fiedler and Wirth (1988), Vehmaanperä (1989), Trevors (1990), Trevors and Starodub (1990)
Rhodospirillum molischianum	Wirth *et al.* (1989)
Salmonella typhimurium	Wirth *et al.* (1989), Taketo (1988), Dahlman and Harlander (1988)
Serratia plymuthica	Wirth *et al.* (1989)
Serratia marcescens	Wirth *et al.* (1989)
Vibrio cholerae	Marcus *et al.* (1990)
Xanthomonas campestris	Wirth *et al.* (1989)
Yesinia enterocolitica	Conchas and Carniel (1990)
Yersinia pseudotuberculosis	Conchas and Carniel (1990)
Yersinia pestis	Conchas and Carniel (1990)
	Gram-positive bacteria
Bacillus amyloliquefaciens	Wirth *et al.* (1989)
Bacillus anthracis	Bartkus and Leppla (1989)
Bacillus cereus	Shivarova *et al.* (1983), Berg *et al.* (1984), Somkuti and Steinberg (1988), Luchansky *et al.* (1988a, 1988b), Schurter *et al.* (1989), Belliveau and Trevors (1989, 1990)
Bacillus circulans	Lian-ying and Wang (1984)
Bacillus sphaericus	Taylor and Burke (1990)
Bacillus subtilis	Wirth *et al.* (1989), Bone and Ellar (1989), Brigidi *et al.* (1989), Schurter *et al.* (1989), Smith *et al.* (1990)
Bacillus thuringiensis	Mahillon *et al.* (1989), Bone and Ellar (1989), Luchansky *et al.* (1988a, 1988b), Schurter *et al.* (1989)
Bordetella pertussis	Zealey *et al.* (1988)
Bordetella parapertussis	Zealey *et al.* (1988)
Brevibacterium ammoniagenes	Dunican and Shivnan (1989)
Brevibacterium lactofermentum	Dunican and Shivnan (1989), Haynes and Britz (1990)
Clostridium acetobutylicum	Oultram *et al.* (1988)
Clostridium perfringens	Allen and Blaschek (1988, 1990), Kim and Blaschek (1989), Phillips-Jones (1990), Scott and Rood (1989)
Corynebacterium callunae	Dunican and Shivnan (1989)
Corynebacterium glutamicum	Dunican and Shivnan (1989), Haynes and Britz (1990)

(continued)

Table 1
Continued

Cytophaga johnsonae	J. L. Pate (personal communication)
Deinococcus radiodurans	Smith *et al.* (1990)
Enterococcus faecalis	Somkuti and Steinberg (1987), Fiedler and Wirth (1988), Luchansky *et al.* (1988a, 1988b)
Enterococcus hirae	Solioz and Bienz (1990), Solioz and Waser (1990)
Lactobacillus acidophilus	Luchansky *et al.* (1988a, 1988b, 1989)
Lactobacillus casei	Chassy and Flickinger (1987), Luchansky *et al.* (1988a, 1988b), Natori *et al.* (1990), Smith *et al.* (1990)
Lactobacillus fermentum	Luchansky *et al.* (1988a, 1988b)
Lactobacillus helveticus	Hashiba *et al.* (1990)
Lactobacillus lactis	MacIntyre and Harlander (1989)
Lactobacillus plantarum	Aukrust and Nes (1988), Badii *et al.* (1989), Luchansky *et al.* (1988a, 1988b)
Lactobacillus reuteri	Luchansky *et al.* (1988a, 1988b)
Lactococcus lactis subsp. *cremoris*	Dahlman and Harlander (1988), Powell *et al.* (1988), van der Lelie *et al.* (1988), de Vos *et al.* (1989)
Lactococcus lactis subsp. *lactis*	Harlander (1987), Dahlman and Harlander (1988), Langella and Chopin (1989), Luchansky *et al.* (1988a, 1988b), McIntyre and Harlander (1989a, 1989b), Powell *et al.* (1988), van der Lelie *et al.* (1988)
Leuconostoc dextranicum	Luchansky *et al.* (1988a, 1988b)
Leuconostoc lactis	Luchansky *et al.* (1988a, 1988b)
Leuconostoc paramesenteroides	David *et al.* (1989)
Listeria innocua	Alexander *et al.* (1990), Luchansky *et al.* (1988a, 1988b)
Listeria ivanovii	Alexander *et al.* (1990)
Listeria monocytogenes	Alexander *et al.* (1990), Luchansky *et al.* (1988a, 1988b), Park and Stewart (1990)
Listeria seeligeri	Alexander *et al.* (1990)
Pediococcus acidilactici	Luchansky *et al.* (1988a, 1988b)
Propionibacterium jensenii	Luchansky *et al.* (1988a, 1988b)
Staphylococcus aureus	Fiedler and Wirth (1988), Augustin and Götz (1990), Luchansky *et al.* (1988a, 1988b)
Staphylococcus auricularis	Fiedler and Wirth (1988)
Staphylococcus carnosus	Augustin and Götz (1990)
Staphylococcus epidermidis	Fiedler and Wirth (1988), Augustin and Götz (1990)
Staphylococcus sciuri	Fiedler and Wirth (1988)
Staphylococcus staphylolyticus	Augustin and Götz (1990)
Staphylococcus warneri	Fiedler and Wirth (1988)
Streptococcus strain NZ1240	van der Lelie *et al.* (1988)
Streptococcus cremoris	see *Lactococcus lactis* subsp. *cremoris*
Streptococcus lactis	see *Lactococcus lactis* subsp. *lactis*
Streptococcus mutans	Lee *et al.* (1989)
Streptococcus pyogenes	Dahlman and Harlander (1988), Suvorov *et al.* (1988)
Streptococcus sanguis	Ito *et al.* (1988), Somkuti and Steinberg (1989)
Streptococcus thermophilus	Somkuti and Steinberg (1987, 1988), Johnson *et al.* (1990)
Streptomyces lividans	MacNeil (1987)

(continued)

Table 1

Continued

	Cyanobacteria
Anabaena	Thiel and Poo (1989)
Fremyella diplosiphon	Bruns *et al.* (1989)
Nostoc	Thiel and Poo (1989)
Synechococcus	Thiel and Poo (1989)
	Other bacteria
Acholeplasma laidlawii	Lorenz *et al.* (1988)
Borrelia burgdorferi	Sambri *et al.* (1990)
Mycobacterium (BCG)	Snapper *et al.* (1988)
Mycobacterium smegmatis	Snapper *et al.* (1988)

from eukaryotic cells in two ways that significantly influence successful electroporation: (1) prokaryotes are physically smaller than eukaryotes and (2) unlike eukaryotes (other than plant cells), bacteria have thick cell walls. The effects of bacterial cell wall and cell size on electroporation will be discussed throughout this chapter. The bacteria are broken down into two major groups, gram-negative and gram-positive, on the basis of presence of an outer membrane and the structure of the cell wall. This major structural difference directly influences electrotransformation; the gram-negatives have generally proven easier to transform at high frequencies by electrical methods than have the gram-positives. This chapter will discuss factors that contribute to successful transformation of bacteria, barriers that occur with certain strains, and a number of aspects of the method that have not been thoroughly researched.

Table 2

Advantages of Electrotransformation of Bacteria

1. Rapid, inexpensive, reproducible.
2. Small volumes can be used.
3. Recipient cells usually require no special treatment(s).
4. Recipient cells can be prepared and stored at −70°C.
5. Use of highly purified plasmid DNA not necessary.
6. Wide range of plasmid DNA concentrations can be used.
7. Transformation frequencies are generally high.
8. Transformation efficiency independent of plasmid size for many species.
9. Applicable to the majority of gram-negative and gram-positive species.
10. A variety of commercial pulse generators and auxiliary equipment available.
11. An extensive applications literature already exists.

II. The Application of Electroporation to Bacteria

A. Gram-Negative Bacteria

In general, the gram-negative bacteria can be efficiently transformed by electrical methods (Table 1). The general success of electrotransformation experienced with the gram-negative bacteria seems to justify the conclusion that the relatively thin cell wall of these bacteria does not serve as a barrier to transforming DNA. The small size of most gram-negative bacteria (diameter ~ 1 μm) requires the application of higher field strength, usually on the order of 8–17 kV/cm, than is necessary for the transfection of eukaryotic cells (diameters 10 μm or more) by electroporation. The electrotransformation of a large number of strains of *Escherichia coli* that are frequently used in research has been reported (Table 1). Although frequencies and efficiencies vary from strain to strain, electrotransformation has become the best technique available for the introduction of DNA into *E. coli*. Under optimal conditions efficiencies of greater than 10^{10} transformants/μg DNA and frequencies of greater than 80% transformed cells can be obtained (Dower *et al.*, 1988). Large libraries have been prepared by electroporation of *E. coli* cells (Dower, Chapter 18; Böttger, 1988; Dynes and Firtel, 1989). Ligation mixtures may be used to directly transform *E. coli* and other bacteria (Böttger, 1988; Heery and Dunican, 1989; Jacobs *et al.*, 1990; Willson and Gough, 1988).

Table 1 contains references to 28 additional species of gram-negative bacteria that can be electrotransformed. Prior to the introduction of electroporation, the majority of these strains could not be transformed, or could be transformed only at low frequencies or efficiencies. Three examples of the value of electrotransformation methods can be taken from results obtained in one of our laboratories (J.T. Trevors). We were interested in the introduction of plasmids into *Pseudomonas* spp., *Klebsiella aerogenes*, and *Azotobacter vinelandii* to allow further studies on the stability of plasmids in strains after their introduction into soil and water samples. We were also interested in the study of novel, large plasmids postulated to have a role in metal-resistance in *Pseudomonas* and *Bacillus* species. *Klebsiella aerogenes* NCTC 418 had been previously transformed with pBR322 at a frequency of 5 × 10^{-6} (Sterkenburg *et al.*, 1984). *Azobacter vinelandii* had been also transformed at low frequency (Glick *et al.*, 1985). *Klebsiella aerogenes* NCTC418 and *P. putida* CYM318 were transformed via high-voltage electrotransformation with plasmids pBR322 (4.363 kb) and pRK2501 (11.1 kb), respectively. Both plasmids are useful cloning vectors in gram-negative bacteria (Bolivar and Backman 1979; Kahn *et al.*, 1979) and can be introduced by transformation into competent *E. coli* strains. The broad-host-range plasmid pRK2501 was also introduced into *A. vinelandii* and *P. fluorescens* R2f using electrotransformation (Trevors, 1990). Both *K. aerogenes* NCTC 418 (pBR322) and *P. fluorescens* R2f (pRK2501) transformants were then used in ecological investigations on organism survival and plasmid stability when these strains were introduced into non-sterile soil and water samples (Trevors *et al.*, 1989; van Elsas *et al.*, 1989). The 74-kb silver-resistance pKK1 plasmid originally described by Haefeli *et al.* (1984) could also be introduced in *P. putida* CYM318 using electrotransformation (Trevors and Starodub, 1990a, 1990b). Prior to the introduction of electrotransformation, studies like those described above would have been impossible.

B. Gram-Positive Bacteria

Many species of gram-positive bacteria for which methods of transformation were not previously available were found to be easily transformable by electroporation (Table 1). Among the gram-positives, natural competence is found in *Bacillus subtilis*, *S. pneumoniae*, *S. mutans*, and *S. sanguis* (Mercenier and Chassy, 1988). Table 1 lists 40 species of gram-positive bacteria that can be transformed by electroporation; the majority had not been transformed, or had been transformable only by tedious protoplast techniques, prior to the introduction of electrotransformation. Electroporation has made possible the genetic manipulation of a great variety of gram-positive bacteria. However, the frequencies and efficiencies of transformation observed with gram-positives are generally lower than those observed for gram-negatives. At times, only a few transformant colonies can be obtained. Seldom is more than 1% of the population transformed. An efficiency of 10^6 transformants/μg of DNA is a very satisfactory result with a gram-positive organism. Currently achievable frequencies and efficiencies are adequate for most research purposes, except in those cases involving the preparation of large libraries or the search for infrequent genetic events.

What cannot be seen in Table 1 is that electrotransformation of some species of gram-positives has not yet proven possible. As is true for gram-negatives, the pulse amplitudes presently available in commercial generators do not appear to impose a limitation on transformability of gram-positives. Less certain is the role of the more dense gram-positive cell wall as a barrier to transforming DNA. As discussed below, growth conditions and agents that negatively influence cell wall synthesis and structure frequently enhance the transformation of gram-positives. Considerable research activity is focused on ascertaining conditions required for the transformation of specific strains of interest. Most of this effort is directed at an empirical search for satisfactory, or optimal, conditions for a selected strain. The findings may do little to shed insight into the mechanism of bacterial electroporation or the underlying cases for inability to electrotransform. Systematic mechanistic studies are needed in order to explain the factor(s) that negatively affect the success of electroporation of gram-positives.

New frontiers in the molecular biology of industrially significant gram-positive bacteria have been opened by the introduction of electrotransformation methods.

The bacilli were the first gram-positive bacteria to be transformed by electroporation. Shivarova *et al.* (1983) used total plasmid DNA from *B. thuringiensis* subsp. *galleriae* harboring pUB110 (a kanamycin-neomycin-resistance plasmid) to transform protoplasts of *B. cereus* 9592 by electroporation. The transformation frequency was 10-fold higher than obtained with the *B. cereus* protoplast transformation procedure. An early study by Lian-ying and Wong (1984) described an electrofusion experiment using *B. circulans* and *E. coli* C600. Electrotransformation methods have now been applied to at least seven species of *Bacillus*. Numerous lactic acid bacteria have been reported to be transformable by electroporation. The transformation of commercially significant clostridia, brevibacteria, and coryneforms has also been reported (Table 1).

C. Special Applications

It has long been recognized (see earlier chapters) that electropermeabilization of eukaryotic cells could be used for the introduction of many kinds of molecules (i.e., proteins, dyes, chromophores, DNA) into cells; the efflux of metabolites and macromolecules from permeabilized cells is also a well-documented phenomenon. Bacterial permeabilization has not been extensively studied (Teissié, Chapter 9). Efflux and isolation of plasmid DNA from bacterial cells after electroshocking have been reported (Calvin and Hanawalt, 1988; Heery *et al.*, 1989; Kilbane and Bielaga, 1990; Li *et al.*, 1990; Kim *et al.*, 1990). In at least one case, it has been observed that the efflux of plasmid DNA resulted in the appearance of plasmid-cured isolates after the application of electroporation pulses (Heery *et al.*, 1989). The simultaneous efflux and influx of DNA that occurs during the permeabilized state has been used to mediate direct plasmid transfer between strains (Summers and Withers, 1990; Kilbane and Bielaga, 1990). The simultaneous entry, or cotransformation, of two nonidentical plasmids into a single cell has been observed after electrotransformation (Bone and Ellar, 1989). Taketo (1988) has reported that RNA can be transfected into *E. coli*.

The recovery of plasmid DNA together with cellular DNA and RNA implies that electroporation generates pores large enough to allow for leakage of cellular material. The greater recovery of plasmid DNA compared with genomic DNA could be taken to mean that plasmid DNA is weakly bound to the cell membrane, and that weakly bound plasmid DNA may be preferentially released during pore formation. Alternatively, the folded chromosome-membrane complex maybe too large to escape the cell subsequent to permeabilization. During electroporation-mediated plasmid DNA recovery from *Clostridium acetobutylicum* NCIB 6444, single-stranded (ss) pDM6 plasmid DNA but no double-stranded plasmid or genomic DNA were recovered (Kim *et al.*, 1990). These results suggest that single-stranded pDM6 is not tightly bound to the cell membrane and may easily pass through the transient pore. ssDNA may have a configuration that allows for its easy passage during electroporation. More recent evidence indicates that pDM6 is complexed with protein and is a filamentous viruslike particle (Kim and Blaschek, 1991) that can be recovered from the supernatant of *C. acetobutylicum* NCIB 6444. It seems unlikely that chromosomal DNA can exit cells through pores induced by electroporation. Chromosomal DNA appears only after extensive cell lysis induced be repeated high-amplitude pulses (Calvin and Hanawalt, 1988).

III. Factors Affecting Electrotransformation

A number of factors affect bacterial electroporation. In this section four major variables are considered. Since bacterial electroporation is a relatively new technique, much of the evidence presented is in the form of anecdotal observations. Table 3 lists some of the limitations that may adversely affect the ability to obtain high frequencies and efficiencies with all bacteria.

A. Growth and Preparation of Cells

The growth medium used for preparing cells for electroporation can directly influence the results obtained, but no systematic studies of this phenomenon have been carried out. It has been reported that improved frequencies of transformation are achieved when *Lactococcus lactis* is cultured in defined medium (McIntyre and Harlander, 1989). Choice of growth medium may effect a number of physiological parameters. Although not yet demonstrated experimentally, one possible direct effect of medium is on the density and quantity of the bacterial cell wall. It seems reasonable to assume that cells with dense accumulations of wall material will be more difficult

Table 3

Factors That May Affect Electrotransformation

---

1. Strain to strain variability within a species is commonly observed.
2. Growth medium used to culture recipient cells may influence efficiency of electrotransformation.
3. Maximum size of plasmid DNA that can be introduced may have a limit.
4. Electrical parameters (amplitude, time constant) are not predictable *a priori*.
5. Effects of bacterial morphology, dimensions, and state of aggregation unexplored.
6. Cuvette geometry and size have not been systematically studied for influence on electrotransformation.
7. No strategies for circumventing restriction barriers have been devised.

to electrotransform. Wall density may be a more significant problem in the attempted transformation of gram-positive bacteria than it is with the generally easily transformed gram-negatives.

Addition of agents to the growth media that affect wall structure has been used to enhance the electrotransformation of gram-positive bacteria. L-Threonine is known to inhibit cell wall cross-link synthesis in gram-positive bacteria. Threonine can be used to prepare cells that are more sensitive to the action of muralytic enzymes (Chassy and Giuffrida, 1980; Chassy et al., 1988). Incorporation of threonine into bacterial growth media has been observed to stimulate transformation frequencies as much as 10-fold (for example, see Park and Stewart 1990). High concentrations of glycine (1–4%) also can be used to gently inhibit cell wall synthesis. It has been reported that the addition of glycine and sucrose (as an osmotic stabilizer) results in the accumulation of an highly transformable population of L. lactis cells (Holo and Nes, 1989). Even more striking, is the observation that the efficiency of transformation of Listeria monocytogenes can be raised from 300 to 8×10^5 transformants/μg DNA by incorporation of low concentrations of the cell wall synthesis inhibitor penicillin in the growth medium (Park and Stewart, 1990). Gentle treatment of cells with muralytic enzymes prior to electrotransformation has been used to "soften" cells of gram-positives to increase transformability (Powell et al., 1988).

A survey of published procedures reveals no general agreement on the appropriate point in the cell cycle to harvest cells for electrotransformation. For maximum efficiency with E coli, exponentially growing cells perform best (Dower, 1990). For some species, late exponential or stationary-phase cells perform best (McIntyre and Harlander, 1989). An empirical evaluation must be made for each strain to determine the optimal point at which to harvest the cells. Similarly, studies should include an evaluation of several cultivation media and temperatures.

After harvest, cells should be washed free of medium. In most cases, it is desirable to lower the ionic strength of the cell suspension by repeated washes in water or very dilute buffers to reduce conductivity. Attempts to transform a conductive suspension with high-amplitude pulses lead to arcing in the sample chamber. This effect can be minimized by chilling the cuvettes on ice. It has been demonstrated that E. coli transforms best at reduced temperatures (Dower, 1990). The electroporation of chilled samples also helps avoid potential deleterious effects of temperature jumps that can occur during electrotransformation. Nonetheless, there may exist strains that will benefit from electroporation at higher temperatures.

Wall-less eukaryotic cells must be protected in suspension by the inclusion of an osmotic protectant; 0.15 M NaCl or 0.27 M sucrose is often employed for this purpose. The original applications of electroporation to bacteria made use of the same sucrose-containing electroporation buffers used for eukaryotic cells. It is now known that electroporation of many strains of bacteria does not require an agent such as sucrose to be present in the electroporation medium. Ultrapure water is a superior electroporation medium for many bacteria. The transformation of other strains is stimulated by inclusion of sucrose, glycerol, sorbitol, polyethylene glycol (PEG), and a number of similar agents (Dower *et al.*, Chapter 32). The exact role of these agents is unclear, but it is unlikely they function solely as classical osmo-protectants during the electroporation process since lysis does not usually occur in their absence. The agents may serve to balance the high internal turgor pressure found in bacteria with that of the external milieu by increasing the external osmotic pressure. It is also possible that these hydrophilic molecules exert a direct effect on the physical state of the cell wall peptidoglycan.

B. DNA

The majority of reports on electroporation of bacteria have employed plasmid DNA for transformation. However, phage DNA can also be used to transfect via electroporation (Chassy and Flickinger, 1987; Lorenz *et al.*, 1988; Snapper *et al.*, 1988; Heery and Dunican, 1989). The influence of DNA conformation on transformation has been investigated by a number of researchers (Powell *et al.*, 1988; Augustin and Gotz, 1990; Conchas and Carniel, 1990; Dower, 1990; Park and Stewart, 1990). In general, it appears that negatively supercoiled plasmid DNA molecules give rise to the highest frequencies, although relaxed circular DNA can be used to transform at high efficiency. Transformation of *E. coli* with relaxed, covalently closed, circular (ccc) DNA lowers the frequency of transformation to about 60% of that observed with native cccDNA (Leonard and Sedivy, 1990). An identical experiment performed with *Listeria monocytogenes* resulted in a reduction in transformation frequency to 42% of that observed with cccDNA (Park and Stewart 1990). Lower frequencies may be observed for linear molecules, ligation mixtures, and phage DNA. However, electroporation experiments can be carried out with DNA that is not highly purified if care is taken to remove interfering substances; DNA prepared by the alkaline lysis technique of Birnboim and Doly (1979) is adequate for most purposes. It should be noted that the effects of high-amplitude electric pulses on plasmid DNA have not been reported. It is uncertain if supercoiled (CCC) plasmid DNA is converted to open circular (OC) or linear DNA during electroporation, or if any other perturbation of DNA structure occurs. In principle, covalent bonds should not be broken during electroporation (Sowers and Lieber, 1986; Sugar and Newmann, 1984).

The relationship of plasmid size to transformation efficiency and frequency has been studied for a number of bacterial strains (Chassy and Flickinger, 1987; Belliveau and Trevors, 1990; Brigidi *et al.*, 1990; Dower, 1990; Marcus *et al.*, 1990; McIntyre and Harlander, 1989; Powell *et al.*, 1988; Smith *et al.*, 1990; Leonard and Sedivy, 1990). The size of the plasmid used to transform *E. coli* has little influence on transformation efficiency for plasmids as large as 100 kb (Leonard and Sedivy, 1990). Gram-positive bacteria may be slightly more size selective than gram-negative bacteria. For example, plasmid size does not appear to have a significant effect on the transformability of *Clostridium perfringens* within the size range of 7.6–26.5 kb (Allen and Blaschek, 1990; Table 4). The results obtained with gram-positive bacteria may be summarized as follows: for plasmid sizes smaller than 10–15 kb no size discrimination is apparent, in the range of 15–25 kb a strain- and method-dependent selectivity for smaller plasmids may begin to emerge, and as transforming plasmid size increases above 25 kb frequencies and efficiencies usually decline. Care must be taken in interpreting such results, since the plasmids being compared may have differing replicons, markers, and encode additional genes as size increases. Furthermore, most studies do not take into account the decreasing numbers of molecules per microgram of DNA present in solution as plasmid size increases (see Leonard and Sedivy, 1990).

Even though decreases in overall efficiency are noted, it is possible to introduce large DNA fragments into gram-positive bacteria by electroporation. To our knowledge the most notable example occurred with the transformation of pGB130 into *Bacillus cereus* 5 (Belliveau and Trevors, 1990; Izaki, 1981). In the course of studies on mercury resistance in *B. cereus* 5, plasmid pGB130 (195 Mdal) was identified in an environmental strain of *B. cereus* 5 that was originally isolated by Izaki (1981). Although reported to be resistant to mercuric chloride via mercuric reductase enzyme activity ($Hg^{2+} \rightarrow Hg^0$), the location (plasmid or chromosomal) of the genetic determinant(s) responsible for this activity was never located. Transformation of the

Table 4

Transformation of *Clostridium perfringens* 3624A-Rif[r]/Str[r] with a Variety of Plasmids Using the Optimized Conditions[a]

Designation	Plasmid DNA size (kb)	Selection[b]	Transformation efficiency[c]
pHR106	7.9	Cm	2.0×10^3
pAK201	8.0	Cm	9.2×10^4
pIP401	52.0	Tc	7.1
pAMβ1	26.5	Em	4.8×10^3
pVA1	11.0	Em	1.7×10^4
pVA677	7.6	Em	6.7×10^3

[a]Late-logarithmic cells, 3×10^8 CFU/ml cell input, 1 μg/ml plasmid DNA, 800 μl cuvette sample volume, E-buffer (0.27 M sucrose, 1 mM MgCl$_2$, 5 mM Na$_2$HPO$_4$; pH 7.4) and a 3-h expression period were utilized.

[b]Selection was carried out on TGY-agar plates containing either 20 μg of chloramphenicol (Cm)/ml, 5 μg of tetracycline (Tc)/ml, or 25 μg of erythromycin (Em)/ml.

[c]Calculated as transformants/μg DNA. Values represent the averages of duplicate determinations. (Results from Allen and Blaschek, 1990.)

130 Mdal pGB130 plasmid into *B. cereus* 569 is one of the first reports of a large plasmid introduced by electrotransformation. Presumptive Hg^R transformants were selected by plating on LB agar supplemented with 25 μM $HgCl_2$. This concentration of $HgCl_2$ completely inhibited colony growth of the recipient without the pGB130 plasmid (Belliveau and Trevors, 1990). Mercuric reductase activity was assayed spectrophotometrically in cell-free extracts of *B. cereus* 569 (pGB130) transformants. Activity in transformants was as high as in the *B. cereus* 5 wildtype (Belliveau and Trevors, 1990). Hg^R transformants were also resistant to phenyl mercuric acetate (PMA). This was not expected as the *B. cereus* 5 wildtype was originally reported sensitive to PMA. Upon further testing, it was also discovered that *B. cereus* 5 was resistant to PMA. This illustrates the value of electrotransformation in studying the biology of novel large plasmids.

A number of other factors related to the transforming DNA influence or control success in electrotransformation. It has been reported that a phenomenon not unlike incompatibility can prevent the transformation of plasmids into strains of *Streptococcus cremoris* (*Lactococcus lactis*) harboring related plasmids (van der Lelie *et al.*, 1988). A number of criteria must be met for transformation to succeed: (1) for plasmid transformation, plasmid replication-essential genes must express and function in the recipient, (2) marker genes must be efficiently expressed,and (3) the expression of entrant DNA-encoded traits must not interfere with host functions or physiology. Other obstacles to transformation, such as the presence of a restriction-modification system, are not directly circumvented by the technique of electroporation. For strains that can be effectively permeabilized, it may be possible to introduce sufficient DNA by electrotransformation to saturate restriction barriers that cannot be overcome with other methods. Electrotransformation can be used to detect the presence of a restriction-modification system once transformants are obtained through retransformation with modified DNA.

C. Parameters of Electroporation

As noted previously, theoretically, the small diameter of bacterial cells requires that higher field strengths be applied to achieve the 0.5–1.0 V depolarization across the cell membrane that is necessary for transient permeabilization. In practice, bacterial electroporation requires about 10- to 20-fold higher field strengths than are used for electrotransfection of tissue culture cells. Typically, pulses of 5–12 kV/cm are employed. The optimal pulse duration (time constant) is usually in the range of 2.5–7.5 msec. It has been suggested that the electropulse may serve to provide a driving force for the electrophoresis of DNA through the barrier presented by the cell wall (Neuman *et al.*, Chapter 6). It has been suggested also that electroporation is a multistep process (Leonardo and Sedivy, 1990); however, direct evidence of this phenomenon is lacking. Although the discussion here focuses on results obtained with capacitive discharge generators, square-wave generators may also be used to advantage with bacteria (Teissié, Chapter 9).

It is impossible to predict *a priori* what voltage setting and time constant will give the best results. An estimation of the range in which to begin experimentation can be derived from an understanding of the sensitivity of the strain to electric discharges. An experiment can be performed to determine a "voltage-kill curve." If DNA is not in short supply, a transformation can be attempted in each tube simultaneously. Electroporated cell suspensions are plated on selective and non-selective counting media. The numbers of transformants, if any, are scored from a count of the colonies appearing on the selective medium. Surviving cells are

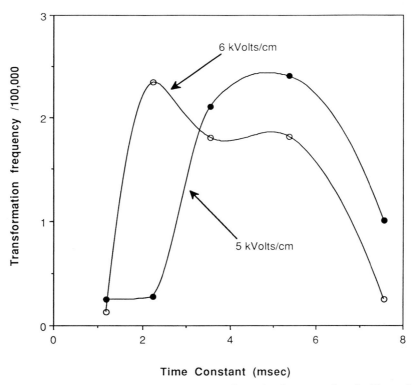

Figure 1   The effect of time constant on transformation frequency. *Lactobacillus casei* cells (10^8) were exposed to electric pulses of 5 kV/cm (●) or 6 kV/cm (○) for varying intervals using a BTX T-100 pulse generator, an external resistive timing bridge supplied by BTX, and 0.5-mm stainless steel flatpacks. The transforming DNA was 0.1 μg of pLP825 (for pLP825 see Chassy and Flickinger, 1987; results from B. M. Chassy, unpublished data).

enumerated from a count of the colonies on the nonselective plates. There is a unique
kill curve for each chosen pulse length. Longer pulses lead to decreased viability.
Electroporation usually occurs when some cells are killed. As voltage is increased
transformation frequency may continue to increase until the point at which the
numbers of cells killed exceeds the increases in number of transformants. Figure 1
shows the relationship between time constant, voltage, and yield of transformants
for *Lactobacillus casei*. Note that there exists an optimal time constant for each of
the two voltages studied. The effect of increasing pulse amplitude (expressed as
voltage) on the frequency of transformation of *L. casei* at a fixed time constant (2.5
ms) is shown in Fig. 2. While the data presented in the two preceding figures are
representative of those observed for other strains of bacteria, each organism displays
a virtually unique pattern of sensitivity to electricity, and optimal parameters for

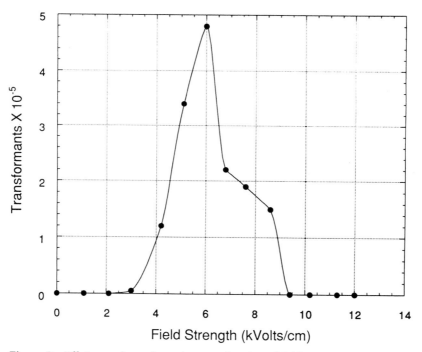

Figure 2 Efficiency of transformation as a function of field strength. *Lactobacillus
casei* cells (10^8) were exposed to electric pulses of the strengths indicated with a 2.5-
ms time constant using a BTX T-100 pulse generator, an external resistive timing
bridge supplied by BTX, and 0.5-mm stainless steel flatpacks. The transforming DNA
was 0.1 μg; of pLP825 (for pLP825 see Chassy and Flickinger, 1987; results from B.
M. Chassy, unpublished data).

transformation. Comparisons and conclusions drawn from the literature are complicated by the use in different laboratories of numerous strains, differing growth and electroporation media, a variety of transforming plasmids, several designs of commercial and homemade pulse generators, and a wide variety of time constants and/or voltages.

D. Effects of Electroporation on Cells

We do not have a good mechanistic or theoretical description of the processes underlying bacterial electroporation. Many questions remain unanswered. It is generally assumed that pores are formed in the bacterial membrane as a result of the electroporation pulse. Perhaps owing to their small size, and the fact that bacteria are covered by a dense wall and outer membrane (gram-negatives), visual evidence of pores has not yet been reported. That pores themselves play a role in electrotransformation is also unproven. Information describing the average number of pores per cell, variability in pore sizes, duration the pores remain open, and the distribution of pores over the surface of the cell is lacking as well. The permeabilization of bacteria is of short duration; addition of DNA within a few seconds after the electric discharge will not produce transformants (B. M. Chassy and J. Flickinger, unpublished). Similarly, addition of DNAse immediately after transformation will not interfere with the appearance of transformants (Teissié, Chapter 9; Taketo, 1988).

As previously noted, both DNA, small metabolites like ATP (Teissié, Chapter 9) and predominant intracellular ions like potassium (K^+) can leak from microbial cells after electropermeabilization. The impact of substantial loss of cytosolic components could be profound. At the same time, buffer components are free to enter the cells along with DNA. This latter possibility may be one reason why ultrapure water is often useful as an electroporation medium.

It is not known if membrane proteins are affected by high-voltage discharges. Considering the magnitude and duration of the electrical discharge, it would not be unreasonable to hypothesize an adverse effect on membrane protein structure and/or function. Charged protein molecules might be expected to be removed electrophoretically from the cell membrane during electroporation. It is even possible that protein movement *per se* forms the target site for initiation of permeabilization (electropore formation). One would expect the charged proteins of the cell membrane to respond more dramatically to an electric field than the lipid bilayer.

It has been suggested some time ago for eukaryotic cells that, during the electroporation pulse, the phospholipids and proteins on the cell membrane are rearranged in an electrical field and subsequently generate transient pores on the cell membrane (Benz and Zimmerman, 1981; Sugar and Newmann, 1984). Cell death following electroporation is believed to occur because the size of the pores generated exceeds the limit of the resealing capability of the porous cell membrane (Benz and Zimmerman, 1981). For this reason, only a subpopulation of cells may be transformed. This subpopulation may have pores that are just large enough for DNA uptake, but do not exceed the critical range for cell viability. If such a subset exists for bacteria, it can be concluded that the proportion of cells in this subset is substantially greater for gram-negative bacteria than it is for gram-positives. Transformation efficiencies for gram-positive bacteria reach a maximum with increasing DNA concentration that is well below the maximum observed for gram-negatives; above this maximal DNA concentration, no further increase in number of transformants is observed. The transformation frequency at which DNA saturation kinetics occurs with gram-positive bacteria almost always reflects the transformation of a small fraction of the cells. In some cases a subpopulation of electrotransformable cells cannot be demonstrated. Essentially 100% of the cells can be transformed in the case of the gram-negative bacterium *E. coli* (Dower, 1990; Chapter 18). It is not known if this reflects differences in the cell membranes or cell walls, or both. It is tempting to postulate that the fundamental difference in cell wall structure between these two classes of bacteria accounts for the differing subpopulations of transformable cells. The effectiveness of glycine and penicillin treatments in raising the transformation efficiency and frequency observed with a number of gram-positive bacteria implies that the cell wall of these bacteria presents a barrier to DNA entry; however, direct evidence to support this hypothesis is lacking.

Electroporation may have other unexplained effects on recipient cells that warrant further investigation. The only known effect of electricity on *E. coli,* other than the killing effect, is the observation by Ellis (1985) that a spherical strain of *E. coli* C grown on a solid minimal salts medium through which an electrical current was passed became rod shaped or occasionally produced larger filaments. It was observed that after an electroporation experiment with *E.coli* C600 as a recipient for a plasmid, a nontransformed strain had lost the ability to metabolize rhamnose (J. T. Trevors, unpublished observations). A parallel situation was encountered when it was observed that plasmid pVA677-containing transformants of *C. acetobutylicum* obtained by electroshocking did not produce butanol (A. Y. Kim and H. P. Blaschek, unpublished data). The ability to produce butanol by these transformants was regained following heat-shocking during subculturing. The loss of butanol-producing capability by *C. acetobutylicum* transformants may be a consequence of a temporary cellular physiological change in response to electroshock-induced stress rather than a genetic alteration. Since the energy supplied by electroporation is not considered sufficient to break covalent bonds (Sowers and Lieber, 1986; Sugar and Newmann, 1984), the inability to produce butanol may be a stress-induced physiological response. The loss of butanol production capability in *C. acetobutylicum* may be a useful marker for measurement of electroporation-induced stress. An understanding of the cellular response to electroporation-induced stress and the relationship between the electroshocking and the solvent-producing capability of *C. acetobutylicum* requires further study.

Table 5 summarizes a few of the practical aspects of our general lack of under-

Table 5

Factors That Have Not Been Fully Investigated or Resolved

Potential problems	Comments
1. Mutation(s) produced in chromosomal and plasmid DNA in some recipient cell types	High-voltage pulses may produce indirect mutations through shock effects on recipient cells
2. Temporary alteration in membrane structure and function	Possible membrane damage that allows leakage of cytosolic components and entry of medium or buffer components into cells; Loss of metabolic or transport functions
3. Successful transformation but cells are critically damaged	Difficult to recover transformants
4. Optimal electrotransformation conditions may be strain and plasmid dependent	Closely related species may have different optimum electroporation conditions; method empirical
5. No information available about existence, number, distribution, location, lifetime and diameter of pores	Unknown if mechanistic information would be useful in transformation of difficult strains

standing of the mechanism of bacterial electroporation. The introduction of electroporation as a tool for bacterial genetics has been remarkably rapid. Perhaps now that the power and the impact of the method are becoming fully appreciated, further research will bring insight into the unanswered questions noted here.

Acknowledgments

Research by J. T. Trevors is supported by the Natural Sciences and Engineering Research Council (NSERC) of Canada. Sincere appreciation is expressed to S. Sprowl and J. Hogeterp for typing portions the manuscript.

The authors would like to recognize and thank Connie Rickey of Biorad Laboratories, who compiled Table 1. The assistance of Cynthia Murphy in verifying the literature cited is appreciated.

Research by H. P. Blaschek is supported by the Illinois Corn Marketing Board, State of Illinois Competitive Value-Added Program, University of Illinois Agricultural Experiment Station, UIUC Research Board, and the University of Illinois Foundation.

References

Alexander, J. E., Andrews, P. W., Jones, D., and Roberts, I. S. (1990). Development of an optimized system for electroporation of *Listeria* species. *Lett. Appl. Microbiol.* **10**, 179–181.

284   Jack T. Trevors *et al.*

Allen, S. P., and Blaschek, H. P. (1988). Electroporation-induced transformation of intact cells of *Clostridium perfringens*. *Appl. Environ. Microbiol.* **54**, 2322–2324.

Allen, S. P., and Blaschek, H. P. (1990). Factors involved in the electroporation-induced transformation of *Clostridium perfringens*. *FEMS Microbiol. Lett.* **70**, 217–220.

Augustin, J., and Gotz, F. (1990). Transformation of *Staphylococcus epidermidis* and other staphylococcal species with plasmid DNA by electroporation. *FEMS Microbiol. Lett.* **66**, 203–208.

Aukrust, T., and Nes, I. F. (1988). Transformation of *Lactobacillus plantarum* with the plasmid pTV1 by electroporation. *FEMS Microbiol. Lett.* **52**, 127–131.

Badii, R., Jones, S., and Warner, P. J. (1989). Sphaeroplast and electroporation-mediated transformation of *Lactobacillus plantarum*. *Lett. Appl. Microbiol.* **9**, 41–44.

Bartkus, J. M., and Leppla, S. H. (1989). Transcription regulation of the protective antigen gene of *Bacillus anthracis*. *Infect. Immun.* **57**, 2295–2300.

Belliveau, B. H., and Trevors, J. T. (1989). Transformation of *Bacillus cereus* vegetative cells by electroporation. *Appl. Environ. Microbiol.* **55**, 1649–1652.

Belliveau, B. H., and Trevors, J. T. (1990). Mercury resistance determined by a self-transmissible plasmid in *Bacillus cereus* 5. *Biol. Metals* **3**, 188–196.

Benz, R., and Zimmerman, U. (1981). The re-sealing process of lipid bilayers after reversible electrical breakdown. *Biochim. Biophys. Acta* **640**, 169–178.

Berg, H., Augsten, K., Bauer, E., Förster, W., Jacob, H.-E., Mühlig, P., and Weber, H. (1984). Possibilities of cell fusion and transformation by electrostimulation. *J. Electroanal. Chem. Bioelectrochem. Bioenerg.* **12**, 119–133.

Bialkowska-Hobrzanska, H., Jaskot, D., and Berzel, R. (1990). Plasmid-mediated transformation of staphylococci by electroporation. *Abstr. Annu. Meet. Am. Soc. Microbiol.* **157**.

Birnboim, H. C., and Doly, J. (1979). A rapid alkaline extraction procedure for screening recombinant plasmid DNA. *Nucleic Acids Res.* **7**, 1513–1523.

Bolivar, F., and Backman, B. (1979). Plasmids of *Escherichia coli* as cloning vectors. *Methods in Enzymology*, vol. 68, pp. 245–267. Academic Press, New York.

Bone, E. J., and Ellar, D. J. (1989). Transformation of *Bacillus thuringiensis* by electroporation. *FEMS Microbiol. Lett.* **58**, 171–178.

Böttger, E. C. (1988). High-efficiency generation of plasmid cDNA libraries using electro-transformation. *BioTechniques* **6**, 878–880.

Brigidi, P., De Rossi, E., Bertarini, M. L., Riccardi, G., and Matteuzzi, D. (1990). Genetic transformation of intact cells of *Bacillus subtilis* by electroporation. *FEMS Microbiol. Lett.* **67**, 135–138.

Bryant, D. L., and Blaschek, H. P. (1988). Buffering as a means for increasing growth and butanol production by *Clostridium acetobutylicum*. *J. Ind. Microbiol.* **3**, 49–55.

Bruns, B. U., Briggs, W. R., and Grossman, A. R. (1989). Molecular characterization of phycobilisome regulatory mutants of *Fremyella diplosiphon*, *J. Bacteriol.* **171**, 901–908.

Calvin, N. M., and Hanawalt, P. C. (1988). High efficiency transformation of bacterial cells by electroporation. *J. Bacteriol.* **170**, 2796–2801.

Chassy, B. M., and Flickinger, J. L. (1987). Transformation of *Lactobacillus casei* by electroporation. *FEMS Microbiol. Lett.* **44**, 173–177.

Chassy, B. M., and Giuffrida, A. (1980). A method for the lysis of gram-positive asporogenous bacteria with lysozyme. *Appl. Environ. Microbiol.* **39**, 153–158.

Chassy, B. M., Mercenier, A., and Flickinger, J. (1988). Transformation of bacteria by electroporation. *Trends Biotechnol.* 6, 303–309.

Conchas, R. F., and Carniel, E. (1990). A highly efficient electroporation system for transformation of *Yersinia*. *Gene* 87, 133–137.

Cymbalyuk, E. S., Chernomordik, L. V., Broude, N. E., and Chizmadzhev, Y. E. (1988). Electro-stimulated transformation of *E. coli* cells pre-treated by EDTA solution. *FEBS Lett.* 234, 203–207.

Dahlman, D., and Harlander, S. (1988). Electroporation of intact microbial cells. *Abstr. Annu. Meet. Am. Soc. Microbiol.* 88, 155.

David, S., Sirnons, G., and de Vos, W. M. (1989). Plasmid transformation by electroporation of *Leuconostoc paramesenteroides* and its use in molecular cloning. *Appl. Environ. Microbiol.* 55, 1483–1489.

de Vos, W. M., Vos, P., de Haard, H., and Boerrigter, I. (1989). Cloning and expression of the *Lactococcus lactis* subsp. *cremoris* SK11 gene encoding an extracellular serine protease. *Gene* 85, 169–176.

Dower, B. (1987). Electro-transformation of intact bacterial cells. *Mol. Biol. Rep.* 1, 5.

Dower, W. J. (1990). Electroporation of bacteria: A general approach to genetic transformation. *In* "Genetic Engineering—Principles and Methods," vol. 12, pp. 275–296. Plenum Publishing, New York.

Dower, W. J., Miller, J. F., and Ragsdale, C. W. (1988). High efficiency transformation of *E. coli* by high voltage electroporation. *Nucleic Acids Res.* 16, 6127–6145.

Dynes, J. L., and Firtel, R. A. (1989). Molecular complementation of a genetic marker in *Dictyostelium* using a genomic DNA library. *Proc. Natl. Acad. Sci. USA* 86, 7966–7970.

Dunican, L. K., and Shivnan, E. (1989). High frequency transformation of whole cells of amino acid producing coryneform bacteria using high voltage electroporation. *Bio/Technology* 7, 1067–1070.

Ellis, H. W. (1985). Morphological changes in spherical *E. coli* induced by a DC electric field. *Microbios* 42, 119–124.

Fiedler, S., and Wirth, R. (1988). Transformation of bacteria with plasmid DNA by electroporation. *Anal. Biochem.* 170, 38–44.

Glick, B. R., Brooks, H. E., and Pasternak, J. J. (1985). Transformation of *Azotobacter vinelandii* with plasmid DNA. *J. Bacteriol.* 162, 276–279.

Guerinot, M. L., Morisseau, A., and Klapatch, T. (1990). Electroporation of *Bradyrhizobium japonicum*. *Mol. Gen. Genet.* 221, 287–290.

Haefeli, C., Franklin, C., and Hardy, K. (1984). Plasmid-determined silver resistance in *Pseudomonas stutzeri* isolated from a silver mine. *J. Bacteriol.* 158, 389–392.

Harlander, S. K. (1987). Transformation of *Streptococcus lactis* by electroporation. In "Streptococcal Genetics," (J. J. Ferretti and R. Curtiss III, eds., pp. 229–233. American Society of Microbiologists, Washington, DC.

Hashiba, H., Takiguchi, R., Ishii, S., and Aoyama, K. (1990). Transformation of *Lactobacillus helveticus* subsp. *jugurti* with plasmid pLHR by electroporation. *Agric. Biol. Chem.* 54, 1537–1541.

Haynes, J. A., and Britz, M. L. (1990). The effect of growth conditions of *Corynebacterium glutamicum* on the transformation frequency obtained by electroporation. *J. Gen. Microbiol.* 136, 255–263.

Heery, D. M., and Dunican, L. K. (1989). Improved efficiency M13 cloning using electroporation. *Nucleic Acids Res.* 17, 8006.

Heery, D. M., Powell, R., Gannon, F., and Dunican, L. K. (1989). Curing a plasmid from *E. coli* using high-voltage electroporation. *Nucleic Acids Res.* 17, 10131–10133.

Hofmann, G. A. (1988). The OPTIMIZOR®: A graphic pulse analyzer to monitor electro cell fusion and electroporation experiments. *BioTechniques* 6, 996–1002.

Hofmann, G. A., and Evans, G. A. (1986). Electronic, genetic, physical and biological aspects of cellular electromanipulation. *IEE Eng. Med. Biol. Mag.* 5, 6–25.

Holo, H., and Nes, I. (1989). High frequency transformation, by electroporation, of *Lactococcus lactis* subsp. *cremoris* grown with glycine in osmotically stabilized media. *Appl. Environ. Microbiol.* 55, 3119–3123.

Hülsheger, H., and Niemann, E.-G. (1980). Lethal effects of high-voltage pulses on *E. coli* K12. *Radiat. Environ. Biophys.* 18, 281–288.

Hülsheger, H., Potel, J., and Niemann, E.-G. (1981). Killing of bacteria with electric pulses of high field strength. *Radiat. Environ. Biophys.* 20, 53–65.

Ito, K., Nishida, T., and Izaki, K. (1988). Application of electroporation for transformation in *Erwinia carotovora*. *Agric. Biol. Chem.* 52, 293–294.

Izaki, K. (1981). Enzymatic reduction of mercurous and mercuric ions in *Bacillus cereus*. *Can. J. Microbiol.* 27, 192–197.

Jacobs, M., Wnendt, S., and Stahl, U. (1990). High-efficiency electro-transformation of *Escherichia coli* with DNA from ligation mixtures. *Nucleic Acids Res.* 18, 1653 (1990).

Johnson, T. L., Solaiman, D. K. Y., Somkuti, G. A., and Steinberg, D. H. (1990). Introduction of a cholesterol oxidase gene into *Streptococcus thermophilus* by electrotransformation. *Abstr. Annu. Meet. Am. Soc. Microbiol.* 156.

Kahn, M., Kolter, R., Thomas, C., Figurski, D., Meyer, R., Remaut, E., and Helinski, D. R. (1979). Plasmid cloning vectors derived from plasmids ColE1,F,. R6K and RK2. *Methods Enzymol.* 68, 368–380.

Kilbane, J. J., and Bielaga, B. A. (1990). Electroporation-mediated DNA purification and conjugation. *Abstr. Annu. Meet. Am. Soc. Microbiol.* 156.

Kim, A. Y., and Blaschek, H. P. (1989). Construction of an *Escherichia coli–Clostridium perfringens* shuttle vector and plasmid transformation of *Clostridium perfringens*. *Appl. Environ. Microbiol.* 55, 360–365.

Kim, A. Y., Vertes, A. A., and Blaschek, H. P. (1990). Isolation of a single-stranded plasmid from *Clostridium acetobutylicum* NCIB 6444. *Appl. Environ. Microbiol.* 56, 1725–1728.

Kim, A. Y., and Blaschek, H. P. (1991). Isolation and characterization of a filamentous viruslike particle from *Clostridium acetobutylicum* NCIB 6444. *J. Bacteriol.* 173, 530–535.

Knight, D. E. (1981). Rendering cells permeable by exposure to electric fields. *Tech. Cell. Physiol.* P113, 1.

Lalonde, G., Miller, J. F., Tompkins, L. S., and O'Hanley, P. (1989). Transformation of *Actinobacillus pleuropneumoniae* and analysis of R factors by electroporation. *Am. J. Vet. Res.* 50, 1957–1960.

Lalonde, G., Miller, J., Tompkins, L., Falkow, S., and O'Hanley, P. (1988). Electroporation of plasmid DNA into *Haemophilus pleuropneumoniae*. *Abstr. Annu. Meet. Am. Soc. Microbiol.* 88, 104.

Langella, P., and Chopin, A. (1989). Effect of restriction-modification systems on transfer of foreign DNA into *Lactococcus lactis* subsp. *lactis*. *FEMS Microbiol. Lett.* **59**, 301–306.

Larimer, F. W., and Mural, R. J. (1990). Megabase cloning in *E. coli*. Abstracts of papers presented at the 1990 meeting on Genome Mapping and Sequencing. Cold Spring Harbor Laboratory, Cold Spring Harbor, NY, p. 99.

Lee, S. F., Progulske-Fox, A., Erdos, G. W., Piacentini, D. A., Ayakawa, G. Y., Crowley, P. J., and Bleiweis, A. S. (1989). Construction and characterization of isogenic mutants of *Streptococcus mutans* deficient in major surface protein antigen P1 (I/II). *Infect. Immun.* **57**, 3306–3313.

Leonardo, E. D., and Sedivy, J. M. (1990). A new vector for cloning large eukaryotic DNA segments in *Escherichia coli*. *Bio/Technology* **8**, 841–844.

Li, S. J., Landers, T. A., and Smith, M. D. (1990). Electroporation of plasmids into plasmid-containing *Escherichia coli*. *Focus* **12**, 72–74.

Lian-ying, L. Y., and Wong, T. (1984). Study in transfer plasmid PBCI by electric pulse fusion in butirocin producing strain *Bacillus circulans* 3342 and *Escherichia coli* C600. *Chin. J. Antibiot.* **9**, 450–454.

Lorenz, A., Just, W., da Silva Cardoso, M., and Klotz, G. (1988). Electroporation-mediated transfection of *Acholeplasma laidlawii* with mycoplasma virus L1 and L3 DNA. *J. Virol.* **62**, 3050–3052.

Luchansky, J. B., Kleeman, E. G., Raya, R. R., and Klaenhammer, T. R. (1989). Genetic transfer systems for delivery of plasmid deoxyribonucleic acid to *Lactobacillus acidophilus* ADH: Conjugation, electroporation, and transduction. *J. Dairy Sci.* **72**, 1408–1417.

Luchansky, J. B., Muriana, P. M., and Klaenhammer, T. R. (1988a). Application of electroporation for transfer of plasmid DNA to *Lactobacillus, Lactococcus, Leuconostoc, Listeria, Pediococcus, Bacillus, Staphylococcus, Enterococcus,* and *Propionibacterium*. *Mol. Microbiol.* **2**(5), 637–646.

Luchansky, J. B., Muriana, P. M., and Klaenhammer, T. R. (1988b). Electrotransformation of Gram-positive bacteria. *Bio-Rad Bull.* **1350**, 1–3.

MacNeil, D. J. (1987). Introduction of plasmid DNA into *Streptomyces lividans* by electroporation. *FEMS Microbiol. Lett.* **42**, 239–244.

Mahillon, J., Chungjatupornchai, W., Docodk, J., Dierickx, S., Michiels, F., Perferoen, M., and Joos, J. (1989). Transformation of *Bacillus thuringiensis* by electroporation. *FEMS Microbiol. Lett.* **60**, 205–210.

Marcus, H., Ketley, J. M., Kaper, J. B., and Holmes, R. K. (1990). Effects of DNase production, plasmid size, and restriction barriers on transformation of *Vibrio cholerae* by electroporation and osmotic shock. *FEMS Microbiol. Lett.* **68**, 149–154.

Mattanovich, D., Rüker, F., da Câmara Machado, A., Laimer, M., Regner, F., Steinkellner, H., Himmler, G., and Katinger, H. (1989). Efficient transformation of *Agrobacterium* spp. by electroporation. *Nucleic Acids Res.* **17**, 6747.

MacIntyre, D. A., and Harlander, S. K. (1989a). Genetic transformation of intact *Lactococcus lactis* subsp. *lactis* by high-voltage electroporation. *Appl. Environ. Microbiol.* **55**, 604–610.

McIntyre, D. A., and Harlander, S. K. (1989b). Improved electroporation efficiency of intact *Lactococcus lactis* subsp. *lactis* cells grown in defined media. *Appl. Environ. Microbiol.* **55**, 2621–2626.

Mercenier, A., and Chassy, B. M. (1988). Strategies for the development of bacterial transformation systems. *Biochimie* 70, 503–517.

Mersereau, M., Pazour, G. J., and Das, A. (1990). Efficient transformation of *Agrobacterium tumefaciens* by electroporation. *Gene* 90, 149–151.

Miller, J. F. (1988). Bacterial electroporation. *Mol. Biol. Rep. Bio-Rad.* 5, 1–40.

Miller, J. F., Dower, D. J., and Tompkin, L. S. (1988). High-voltage electroporation of bacteria: genetic transformation of *Campylobacter jejuni* with plasmid DNA. *Proc. Natl. Acad. Sci. USA* 89, 856–860.

Nagel, R., Elliott, A., Masel, A., Birch, R. G., and Manner, J. M. (1990). Electroporation of binary Ti plasmid vector into *Agrobacterium tumefaciens* and *Agrobacterium rhizogenes*. *FEMS Microbiol. Lett.* 67, 325–328.

Natori, Y., Kano, Y., and Imamoto, F. (1990). Genetic transformation of *Lactobacillus casei* by electroporation. *Biochimie* 72, 265–269.

Neumann, E., and Wierth, P. (1986). Gene transfer by electroporation. *Am. Biotechnol. Lab.* 4, 10–15.

Oultram, J. D., Loughlin, M., Swinfield, T.-J., Brehm, J. K., Thompson, D. E., and Minton, N. P. (1988). Introduction of plasmids into whole cells of *Clostridium acetobutylicum* by electroporation. *FEMS Microbiol. Lett.* 56, 83–88.

Park, S. F., and Stewart, G. S. A. B. (1990). High efficiency transformation of *Listeria monocytogenes* by electroporation of penicillin-treated cells. *Gene* 94, 129–132.

Phillips-Jones, M. K. (1990). Plasmid transformation of *Clostridium perfringens* by electroporation methods. *FEMS Microbiol. Lett.* 66, 221.

Potter, H. (1988). Electroporation in biology: methods, applications, and instrumentation. *Anal. Biochem.* 174, 361–373.

Powell, I. B., Achen, M. G., Hillier, A. J., and Davidson, B. E. (1988). A simple and rapid method for genetic transformation of lactic streptococci by electroporation. *Appl. Environ. Microbiol.* 54, 655–660.

Roberts, S. S. (1990a). Electroporation: galvanizing cells into action. *J. NIH Res.* 2, 93–94.

Roberts, S. S. (1990b). Short-circuiting the defenses of cells. *J. NIH Res.* 2, 95.

Sale, A. J. H., and Hamilton, W. A. (1968). Effects of high electric fields on microorganisms, III. Lysis of erythrocytes and protoplasts. *Biochim. Biophys. Acta* 163, 37.

Sambri, V., and Lovett, M. A. (1990). Survival of *Borrelia burgdofferi* in different electroporation buffers. *Microbiologica* 13, 79–83.

Sambrook, J., Fritsch, E. F., and Maniatis, T. (1989). Transformation of *E. coli* by high-voltage electroporation (electrotransformation). *In* "Molecular Cloning: A Laboratory Manual," 2nd ed. book 1, p. 1.75, Cold Spring Harbor Laboratory Press, Cold Spring Harbor, NY.

Schurter, W., Geiser, M., and Mathé, D. (1989). Efficient transformation of *Bacillus thuringiensis* and *Bacillus cereus* via electroporation: Transformation of acrystalliferous strains with a cloned delta-endotoxin gene. *Mol. Gen. Genet.* 218, 177–181.

Scott, P. T., and Rood, J. I. (1989). Electroporation-mediated transformation of lysostaphin-treated *Clostridium perfringens*. *Gene* 82, 327–333.

Sheen, J. (1989). High-efficiency transformation by electroporation. *In* "Current Protocols in Molecular Biology," suppl. 5 (F. M. Ausubel, R. Brent, R. E. Kingston, D. D. Moore, J. G. Seidman, J. A. Smith, and K. Struhl, eds), pp. 184–187, John Wiley & Sons, New York.

Shigekawa, K., and Dower, W. J. (1988). Electroporation of eukaryotes and prokaryotes: A general approach to the introduction of macromolecules into cells. *Biotechniques* 6, 742–751.

Shivarova, N., Foster, W., Jacob, H. E., and Grigorova, R. (1983). Microbiological implications of electric field effects. VII. Stimulation of plasmid transformation of *Bacillus cereus* protoplasts by electric field pulses. *Z. Alleg. Microbiol.* 23, 595–599.

Smith, A. W., and Iglewski, B. H. (1989). Transformation of *Pseudomonas aeruginosa* by electroporation. *Nucleic Acids Res.* 17, 10509.

Smith, M., Jessee, J., Landers, T., and Jordan, J. (1990). High efficiency bacterial electroporation: 1 × 10^{10} *E. coli* transformants/mg. *Focus* 12, 38–40.

Snapper, S. B., Lugosi, L., Jekkel, A., Melton, R. E., Kieser, T., Bloom, B. R., and Jacobs, W. R., Jr. (1988). Lysogeny and transformation in mycobacteria: Stable expression of foreign genes. *Proc. Natl. Acad. Sci. USA* 85, 6987–6991.

Solioz, M., and Bienz, D. (1990). Bacterial genetics by electric shock. *Trends Biochem. Sci.* 15, 175–177.

Solioz, M., and Waser, M. (1990). Efficient electrotransformation of *Enterococcus hirae* with a new *Enterococcus–Escherichia coli* shuttle vector. *Bochimie* 72, 279–283.

Somkuti, G. A., and Steinberg, D. H. (1987). Genetic transformation of *Streptococcus thermophilus* by electroporation. *In* "Proc. 4th European Congress on Biotechnology 1987," vol. 1 (O. M. Neijssel, R. R. van der Meer, and K. Ch. A. M. Luyben, eds), p. 412, Elsevier, Amsterdam.

Somkuti, G. A., and Steinberg, D. H. (1988). Genetic transformation of *Streptococcus thermophilus* by electroporation. *Biochimie* 70, 579–585.

Somkuti, G. A., and Steinberg, D. H. (1989). Electrotransformation of *Streptococcus sanguis* Challis. *Curr. Microbiol.* 19, 91–95.

Sowers, A. E. and Lieber, M. R. (1986). Electropore diameters, lifetimes, numbers and locations in individual erythrocyte ghosts. *FEBS Lett.* 205, 170–184.

Speyer, J. F. (1990). A simple and effective electroporation apparatus. *Biotechniques* 8, 28–30.

Sterkenburg, A., Prozée, G. A. P., Leegwater, P. A. J., and Woulters, J. T. M. (1984). Expression and loss of the pBR322 plasmid in *Klebsiella aerogenes* NCTC 418, grown in chemostat cultures. *Antonie Leeuwenhoek J. Microbiol.* 50, 397–404.

Sugar, I. P., and Newmann, E. (1984). Stochastic model for electric field-induced membrane pore electroporation. *Biophys. Chem.* 19, 211–225.

Summers, D. K., and Withers, H. L. (1990). Electrotransfer: Direct transfer of bacterial plasmid DNA by electroporation. *Nucleic Acids Res.* 18, 2192.

Suvorov, A., Kok, J., and Venema, G. (1988). Transformation of group A streptococci by electroporation. *FEMS Microbiol. Lett.* 56, 95–100.

Takahashi, N., and Kobayashi, I. (1990). Evidence for the double-strand break repair model of bacteriophage 1 recombination. *Proc. Natl. Acad. Sci. USA* 87, 2790–2794.

Taketo, A. (1988). DNA transfection of *Escherichia coli* by electroporation. *Biochim. Biophys. Acta* 949, 318–324.

Taketo, A. (1989). RNA transfection of *Eschericha coli* by electroporation. *Biochim. Biophys. Acta* 1007, 127–129.

Taylor, L. D., and Burke, W. F., Jr. (1990). Transformation of an entomopathic strain of *Bacillus sphaericus* by high voltage electroporation. *FEMS Microbiol. Lett.* 66, 125–128.

Terracciano, J. J., Rapaport, E., and Kashket, E. R. (1988). Stress- and growth phase-associated proteins of *Clostridium acetobutylicum*. *Appl. Environ. Microbiol.* 54, 1989–1995.

Thiel, T., and Poo, H. (1989). Transformation of a filamentous cyanobacterium by electroporation. *J. Bacteriol.* 171, 5743–5746.

Trevors, J. T. (1990). Electroporation and expression of plasmid pBR322 in *Klebsiella aerogenes* NCTC 418 and plasmid pRK2501 in *Pseudomonas putida* CYM 318. *J. Basic Microbiol.* 30, 57–61.

Trevors, J. T., and Starodub, M. E. (1990a). Electroporation of pKK1 silver-resistance plasmid from *Pseudomonas stutzeri* AG259 into *Pseudomonas putida* CYM318. *Curr. Microbiol.* 21, 103–107.

Trevors, J. T., and Starodub, M. E. (1990b). Electroporation and expression of the broad host-range plasmid pRK2501 in *Azotobacter vinelandii*. *Enzyme Microbiol. Technol.* 12, 653–655.

Trevors, J. T., van Elsas, J. D., Starodub, M. E., and van Overbeek, L. S. (1989). Survival of and plasmid stability in *Pseudomonas* and *Klebsiella* spp. introduced into agricultural drainage water. *Can. J. Microbiol.* 35, 675–680.

van de Guchte, M., van der Vossen, J. M. B. M., Kok, J., and Venema, G. (1989). Construction of a lactococcal expression vector: Expression of hen egg white lysozyme in *Lactococcus lactis* subsp. *lactis*. *Appl. Environ. Microbiol.* 55, 224–228.

van der Lelie, D., van der Vossen, J. M. B. M., and Venema, G. (1988). Effect of plasmid incompatibility of DNA transfer to *Streptococcus cremoris*. *Appl. Environ. Microbiol.* 54, 865–871.

van Elsas, J. D., Trevors, J. T., van Overbeek, L. S., and Starodub, M. E. (1989). Survival of *Pseudomonas fluorescens* containing plasmids RP4 or pRK2501 and plasmid stability after introduction into two soils of different texture. *Can. J. Microbiol.* 35, 951–959.

Vehmaanperä, J. (1989). Transformation of *Bacillus amyloliquefaciens* by electroporation. *FEMS Microbiol Lett.* 61, 165–170.

Wen-jun, S., and Forde, B. G. (1989). Efficient transformation of *Agrobacterium* spp. by high voltage electroporation. *Nucleic Acids Res.* 17, 8385.

Willson, T. A., and Gough, N. M. (1988). High voltage *E. coli* electro-transformation with DNA following ligation. *Nucleic Acids Res.* 16, 11820.

Wirth, R., Fresisenegger, A., and Fiedler, S. (1989). Transformation of various species of gram-negative bacteria belonging to 11 different genera by electroporation. *Mol. Gen. Genet.* 216, 175–177.

Zealey, G., Dion, M., Loosmore, S., Yacoob, R., and Klein, M. (1988). High frequency transformation of *Bordetella* by electroporation. *FEMS Microbiol. Lett.* 56, 123–126.

Zimmerman, U. (1983). Electrofusion of cells: Principles and industrial applications. *Trends Biotechnol.* 1, 149–155.

18

Creating Vast Peptide Expression Libraries: Electroporation as a Tool to Construct Plasmid Libraries of Greater than 10^9 Recombinants

William J. Dower and Steven E. Cwirla

Affymax Research Institute, Palo Alto, California 94304

I. Introduction

Electroporation is now recognized as the most efficient means of transforming many prokaryotes. *Escherichia coli* is especially receptive to electroporation, some strains becoming transformed with efficiencies greater than 10^{10} transformants/μg of plasmid DNA (Dower *et al.*, 1988). This extraordinary level of transformation allows the construction of complex libraries in plasmid vectors when only small amounts of DNA are available.

Highly efficient use of transforming DNA is the best-known attribute of electrotransformation; however, there are several other unique characteristics of the method that are useful for special purposes. These include (1) the relatively small effect of plasmid size on the probability that it will transform a cell (Leonardo and Sedivy, 1990; Dower, 1990), (2) the large proportion of cells competent to take up DNA (Dower *et al.*, 1988), and (3) the remarkably high capacity for DNA (Miller

Guide to Electroporation and Electrofusion
Copyright © 1992 by Academic Press, Inc. All rights of reproduction in any form reserved.

et al., 1988; Dower *et al.*, 1988). We have exploited the advantages of electroporation to construct plasmid libraries of greater than 10^9 independent recombinants.

Our purpose in creating libraries of this size is to produce diverse collections of recombinant peptides as sources of novel ligands for various receptors. We have employed a method of peptide expression whereby an oligonucleotide cloned into the gene for a minor coat protein of filamentous phage is expressed on the outer surface of the virion as a peptide fused to the coat protein. Phage displaying peptides with affinity for an immobilized binding protein can be enriched by an adsorption technique called panning (Smith, 1985; Parmley and Smith, 1988).

To obtain a highly diverse population of peptides, we have synthesized random collections of oligonucleotides and cloned these into the filamentous phage vector fAFF1 (Cwirla et al., 1990), creating very large peptide expression libraries. We have screened one of these libraries for binding to a specific monoclonal antibody. By this approach, we identified new ligands for this antibody, and gained additional information on the binding specificity.

II. Electroporation as a Tool to Create Extremely Large Libraries

The combination of the exceptionally high transformation efficiency and enormous capacity for DNA characteristic of electroporation provides the means to build plasmid libraries of unprecedented size. Our discussion of these enabling characteristics should begin with a description of the terms we are using. The terms transformation efficiency and transformation frequency have occasionally been used interchangeably. Here, we will use transformation efficiency to describe the number of transformants recovered per specified quantity of added DNA. The efficiency value may be expressed either as transformants per microgram of DNA, or transformants per mole of plasmid. Another way of expressing the transforming activity of a plasmid is the probability, P_p, that a plasmid molecule will produce a transformant (P_p = transformants/molecule; Hanahan, 1983). The molar efficiency and the transforming probability are especially useful when comparing the activities of plasmids of different size.

The transformation frequency is the fraction of cells in the sample that become transformed (transformed cells/total cells). During electroporation, a significant number of the cells are damaged and are not recoverable as colony-forming units; therefore, we use frequency to represent the proportion of the survivors that become transformed (transformed cells/surviving cells). The DNA capacity is defined as the point at which the addition of more plasmid DNA fails to yield more transformants (the system is saturated with respect to transforming DNA). Before saturation is reached, however, there is a phase of declining efficiency with increasing DNA. The specialized definition of DNA capacity that we will use here is the point at which addition of more DNA produces a detectable drop in transformation efficiency.

We have reported the effect of increasing DNA concentration on the frequency and efficiency of electrotransformation. Recovery of *E. coli* transformants increases linearly with plasmid concentration over a very large range. This relationship is illustrated in Fig. 1; recovery of transformants shows a first-order dependence on the concentration of pBR329 DNA from 10 pg/ml to 7.5 μg/ml. The transformation efficiency is about the same for every point on this curve, demonstrating the high capacity for intact, supercoiled transforming DNA available with electroporation (Fig. 1 is replotted from data in Dower *et al.*, 1988). This is a very different effect from that seen with chemically mediated transformation, where efficiency declines at DNA concentrations about 100-fold lower than observed with electrotransfor-mation. In contrast to the conditions of the experiment shown in Fig. 1, the DNA used for many cloning applications is not composed solely of intact supercoiled plasmids with high transforming activity; rather, it contains a large proportion of nonligated, noncircularized DNA. DNA taken from ligation reactions for library construction is often 10- to 1000-fold less active in transformation than is the uncut vector. In chemically mediated transformations this inactive DNA may compete for

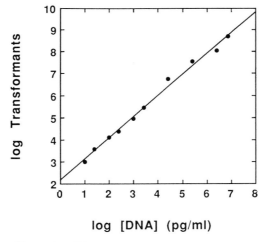

log [DNA] (pg/ml)

Figure 1   pBR329 DNA (0.4 pg to 0.3 μg) was added to 40 μl of LE392 electrocompetent cells and one pulse of 12.5 kV/cm, ~5 ms, was applied to each sample. The cells were suspended in 1 ml of SOC and incubated 1 hr at 37°C before plating on L-agar containing ampicillin (100 μg/ml). Preparation of electrocompetent cells and transformation procedure is described in Dower *et al.* (1988).

putative binding sites on the cells, limiting the effective capacity for actively transforming DNA. Electrically mediated transformation seems not to require specific
binding to the cells (Calvin and Hanawalt, 1988; Dower et al., 1988), a property
that may account for the much higher DNA capacity observed with electroporation.

To examine the effect on plasmid transformation of an excess of nontransforming
DNA, cells were electroporated in the presence of a fixed amount of supercoiled
pUC18 (10 pg) and varying amounts of sheared salmon sperm DNA (0–40 µg).
The result is shown in Fig. 2. A control transformation containing no salmon sperm
DNA produced 2.6×10^{10} transformants per microgram of pUC18 DNA. The
addition of up to 25 µg/ml of nontransforming DNA (up to 10^5-fold mass excess)
has no effect on efficiency. At higher concentrations of salmon sperm DNA (100–
400 µg/ml) the recovery of transformants drops significantly, although even at 1
mg/ml of added DNA (a 4×10^6-fold mass excess), the pUC18 DNA still transforms
with the rather high efficiency of 5×10^9 transformants/µg. This result shows that
even in the presence of a large excess of nontransforming DNA, plasmids can
transform cells with the exceptional efficiencies typical of electrotransformation.

While demonstrating the relative insensitivity of electrical transformation to the
presence of nontransforming DNA, this experiment may not accurately model the
effect on transformation of a large amount of linear vector DNA from a ligation
reaction. Specific plasmid sequences on these linear fragments (the replication origin,

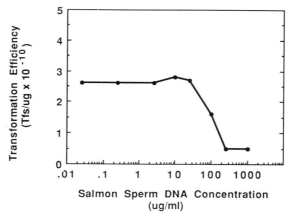

Figure 2   pUC18 DNA (10 pg) and sheared salmon sperm
DNA (0–40 µg) were added to 40 µl of electrocomponent
MC1061 cells and one pulse of 12.5 kV/cm, ~5 ms, was
applied to each sample. The cells were suspended in 1 ml of
SOC (Hanahan, 1983) and incubated for 1 h at 37°C before
plating on L-agar containing ampicillin (100 µg/ml).

for example) could compete for plasmid uptake and maintenance. The most stringent test of this possibility is the measurement of the transforming activity of one plasmid in the presence of an excess of another, incompatible supercoiled plasmid. The results from such an experiment are shown in Fig. 3. Cells were electrotransformed with a small fixed amount of pBR322 (10 pg) and varied amounts of pUR222. Both plasmids contain an identical pMB1 origin and are therefore incompatible, competing strongly with one another for replication. The pBR322 encodes tetracycline resistance, allowing the independent scoring of pBR322 transformants. The control sample containing only pBR322 DNA (10 pg) yielded transformants with an efficiency of $5 \times 10^9/\mu g$. The presence of up to 25 $\mu g/ml$ of pUR222 DNA had no effect on the number of tetracycline-resistant transformants produced with 10 pg of pBR322 DNA. The addition of still higher concentrations of pUR222 did reduce the recovery of pBR322 transformants, inhibiting 40% at 0.1 mg/ml and 90% at 0.25 and 1 mg/ml pUR222. Increasing the concentration of pUR222 DNA had a marked effect on the size of the recovered colonies. At high pUR222 excess, most of the colonies were very small, requiring more than 24 h to reach 1 mm in diameter. A likely explanation for this phenomenon is that during permeabilization, each cell taking up a pBR322 molecule also received a large number of pUR222 molecules. This would result in a low probability at each generation that the pBR322 is selected for replication, and would delay expansion of the tetracycline-

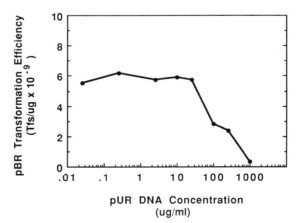

Figure 3   pBR322 DNA (10 pg) and pUR222 (DNA (0–40 μg) were added to 40 μl of electrocompetent MC1061 cells and one pulse of 12.5 kV/cm, ~5 ms, was applied to each sample. The cells were suspended in 1 ml of SOC and incubated 1 h at 37°C before plating on L-agar containing tetracycline (12 $\mu g/ml$) to score for cells transformed by pBR322.

resistant colony. These results demonstrate that the high DNA capacity of electro-transformation extends to a variety of types of DNA, a finding of practical importance for a number of cloning applications.

III. Constructing Peptide Expression Libraries: An Application Exploiting the High DNA Capacity of Electrotransformation

The high efficiency of transformation and capacity for DNA possible with electro-poration provide the means to construct enormous libraries of more than 10^9 recombinants. The need for production of such large libraries is driven by our desire to create a collection of compounds representing vast chemical diversity, and to use this collection as a source of ligands for many receptors. Our strategy is to obtain that diversity as peptides encoded by randomly synthesized oligonucleotides cloned in an appropriate vector. The number of possible peptides increases rapidly with increasing chain length (20^n, for a string of n amino acids). Table 1 provides an example of the enormous diversity available with even very short genetically coded random peptides: There are 64 million possible hexapeptides, 1.3 billion hepta-peptides, 26 billion octapeptides, and so on. To produce a significant fraction of these compounds requires the construction of very large libraries.

Our ability to build such large libraries by electrotransformation is of little use without an effective means to screen for desired properties. For our purposes the property of interest is high affinity for particular receptors, and our means of screening for this property is based on a system devised by G. Smith and colleagues (Smith, 1985; Parmley and Smith, 1988). The screening system allows the expression of peptides fused to a minor coat protein of the filamentous phage fd. The fusion proteins are displayed on the outer surface of the phage particles, where they are

Table 1

Number of Possible Peptides of Length n
Genetically Coded Amino Acids (6- to 12-mers)

	n	20^y
Hexapeptides	6	6.4×10^7
Heptapeptides	7	1.3×10^9
Octapeptides	8	2.6×10^{10}
Nonapeptides	9	5.1×10^{11}
Decapeptides	10	1.0×10^{13}
Undecapeptides	11	2.1×10^{14}
Dodecapeptides	12	4.1×10^{15}

available for interaction with binding proteins. Adsorption and recovery of the phage on immobilized binding proteins permit isolation of only those phage displaying peptides with affinity for the binding protein. If these ligand phage are selected from a library displaying many different peptides, the identity of the ligands can be easily deduced from the sequence of the DNA in the cloning site of the adsorbed phage.

To create libraries of displayed peptides, we constructed the filamentous phage affinity vector fAFF1 (Cwirla *et al.*, 1990). This vector permits the cloning of oligonucleotides in a specific orientation in a double *Bst*XI site in the 5' region of geneIII. A peptide encoded by the cloned oligonucleotide appears on the surface of the virion fused to the N-terminus of the minor coat protein pIII. The C-terminus of pIII is embedded in the phage coat, and the N-terminus, containing the foreign peptide, extends out into the milieu. By cloning a collection of many different oligonucleotides, we can produce a library of phage expressing many different peptides.

Peptide diversity is dependent on the diversity of the cloned oligonucleotides. These are chemically synthesized to contain degenerate codons of the structure $(NNK)_n$, where N is A, C, G, or T, and K is G or T. This motif encodes all 20 amino acids and one (amber) stop codon. We have constructed several libraries with this general approach; three examples are shown in Table 2. Each of these libraries is designed to display the variable peptide on the N-terminus of pIII; thus, the first residue of the peptide is the N-terminal amino acid of the fusion protein. These libraries contain peptides with six, eight, or 12 variable residues. The hexapeptide library is the best characterized of this group and we discuss its characteristics in some detail below. The octapeptide library may be the largest clone library yet constructed, consisting of 1.4×10^9 independent recombinants. About 75% of these yield infective phage; therefore, this library may contain over a billion clones producing phage bearing different octapeptides.

The size of the library is determined by (1) the competence of the cells, (2) the

Table 2
Peptide Expression Libraries Constructed in fAFF1

Peptide		DNA		Transformants	Efficiency
Location	Length	(μg)	Transformations	(Tc^R CFU)	(transformants/μg)
N-Terminal	6-mer	20	$5 \times 80\ \mu l$	3×10^8	1.5×10^7
N-Terminal	8-mer	135	$15 \times 200\ \mu l$	1.4×10^9	1.1×10^7
N-Terminal	12-mer	100	$10 \times 200\ \mu l$	5×10^8	5×10^6

efficiency of ligation of the oligonucleotides to the vector, and (3) the mass of vector DNA used. We usually use as recipients electrocompetent MC1061 cells prepared and transformed as previously described (Dower *et al.*, 1988). These cells yield transformants with an efficiency of about 3×10^{10} per µg of control pUC18 DNA. Intact fAFF1 RF DNA transforms these cells with an efficiency of 3×10^9 per µg DNA. With the oligonucleotide cloning strategy described in Cwirla *et al.* (1990), the vector DNA is utilized much less efficiently, yielding about 10^7 transformants/ µg DNA. Rather large amounts of vector are thus required to build libraries of 10^9 recombinants. The value of electroporation for this cloning application is both the high transformation efficiency it provides and the large amount of DNA that can be used in each transformation. Our current procedure calls for 10µg of ligated DNA in 200 µl of electrocompetent cells (50 µg/ml); therefore, a library of 10^9 recombinants can be constructed with about 100 µg of vector DNA and 10 µg of synthetic oligonucleotide in 10 electrotransformations.

An important limitation in building libraries of this size is the means of amplifying the libraries. In general, we prefer to amplify any library on a solid substrate to reduce competition among the individual clones. This is inconvenient for libraries larger than 10^7 members and impractical for those larger than 10^8. We have amplified our fAFF libraries by allowing limited growth in liquid culture. After a 1-h outgrowth period following transformation, the cells were innoculated at about 3×10^8 transformants per liter in L-broth (with tetracycline) and grown for about 10 doublings at 37°C. This yields up to 10^5 library equivalents of infective phage. Libraries even larger than those described here might be constructed by increasing the scale of the cloning and amplification process.

IV. Analysis of the Libraries

Oligonucleotides synthesized with the NNK codon motif encode 32 codons, one of which is a stop. This stop will occur in the fraction $[1-(31/32)^n]$ of peptides of length *n* amino acids, interrupting translation of the peptide and the pIII sequence downstream. This results in phage that are noninfective and, of course, cannot display a foreign peptide. The expected reduction of peptide diversity by this mechanism is minor for the libraries shown in Table 2. Libraries encoding peptides of 6, 8, and 12 residues would be expected to contain, respectively, 83, 78, and 68% clones producing infective, peptide-bearing phage. There are likely additional biological selections affecting the distribution of peptides expressed that further reduce this number.

The hexapeptide library has been most closely analyzed and serves as an example of the properties and use of these libraries. It contains 3×10^8 independent recombinants, 72% of which make infective phage with the potential to carry peptides. To provide a glimpse of the diversity in the library, the DNA inserts isolated from 52 infective phage in this collection were sequenced. The deduced peptide sequences are shown in Fig. 4. These sequences revealed no marked deviation from a random distribution of residues; most of the amino acids are represented at each of the residue positions. If no biological or chemical bias affected the library, we would predict a library of this size made with the (NNK) codon structure to express 75% of the possible hexapeptides, or almost 50 million different compounds.

In order to test our ability to select ligands from this hexapeptide library, we employed as a receptor the monoclonal antibody 3-E7 that recognizes the sequence Tyr-Gly-Gly-Phe common to peptide opiates (Meo *et al.*, 1983). An unblocked N-terminal tyrosine is important for recognition by the antibody, and the best small peptide ligands known for 3-E7 are Tyr-Gly-Gly-Phe-Met (YGGFM) and Tyr-Gly-Gly-Phe-Leu YGGFL; Met- and Leu-enkephalin). After three rounds of affinity enrichment of the hexapeptide library on mAb 3-E7, the DNA sequences in the cloning site of the recovered phage were determined. The peptides carried by these phage are shown in Fig. 5. Those we analyzed share a remarkable similarity. All have a Tyr (Y) at the N-terminus, a feature known to be of importance for high affinity binding to 3-E7; almost all (94%) have a Gly (G) in the second position. Less similarity is evident in positions 3 and 4, although 93% of the residues in position 4 are occupied by one of the four large hydrophobic amino acids Ile, Leu, Phe, and Trp (I, L, F, and W). The specificity of enrichment was extraordinarily high, no peptides were identified that were not obviously related to the natural

G M L Q R L	C R G N S G	D W V G G A	D L S P K V
M S R K L F	E R K A S V	T A M Q P G	D S E V S L
L W A G H E	W A E V F M	L D L K R L	R A A R D C
V G I T Q L	V A A G L N	V R N S M G	I Y T L H R
K T S Y G G	G P L P L F	R V C N K T	I T A P Y S
Q K R G E D	S N K G W A	R W S W E Q	S N D L S G
T L T K R Q	K H M L R W	G N M A H F	R S L H A G
S V S L Q A	V Q R L G K	F D S F G R	R W T W L G
A G S F E A	S G L Q R G	Q A V L M Q	A T L G F S
A I A A R A	W E K P R R	G K H Y Q W	I P G L L L
K A R L G L	R L V S T H	Q S R K S F	V C L L T V
L A F L A M	C A S L R S	G Y S S V D	G G G F T M
E R C R V D	Y A P S T R	S I G Q S K	V C P Q F C

Figure 4   Amino acid sequences (deduced from DNA sequence) of N-terminal hexapeptides on pIII of infectious phage randomly selected from the library. Sequences begin at the signal peptidase cleavage site. The single-letter code for amino acids is: A, Ala; C, Cys; D, Asp; E, Glu; F, Phe; G, Gly; H, His; I, Ile; K, Lys; L, Leu; M, Met; N, Asn; P, Pro; Q, Gln; R, Arg; S, Ser; T, Thr; V, Val; W, Trp; Y, Tyr. Reprinted from Cwirla *et al.* (1990).

```
YGGLGL  YGSLVL  YGALGG  YGWWGL  YGLWQS
YGGLGI  YGSLVQ  YGALSW  YGWWLT  YGFWGM
YGGLGR  YGSLVR  YGALDT  YGWLAT  YGKWSG
YGGLNV  YGSLAD  YGALEL  YGWANK  YGPFWS
YGGLRA  YGSLLS                  YGEFVL
YGGLEM  YGSLNG  YGAIGF  YGNWTY  YGDFAF
        YGSLYE          YGNFAD
YGGIAS          YGAWTR  YGNFPA  YAWGWG
YGGIAV  YGSWAS*                 YAGFAQ
YGGIRP  YGSWAS*         YGTFIL
        YGSWQA          YGTWST  YSMFKE
YGGWAG
YGGWGP  YGSFLH          YGVWAS
YGGWSS                  YGVWWR

YGGMKV
YGGFPD
```

Figure 5   Amino acid sequences (deduced from DNA sequence) of N-terminal peptide of pIII of 51 phage isolated by three rounds of panning on mAb 3-E7. *Identical nucleotide sequences. Reprinted from Cwirla *et al.* (1990).

ligand (YGGFM), despite the complexity of the peptide mixture (tens of millions) in the starting phage population., The details of these panning experiments are described in Cwirla *et al.* (1990).

Two other examples of the use of this approach in identifying ligands have appeared. Scott and Smith (1990) constructed a library of random hexapeptides inserted several amino acids from the N-terminus of pIII. Screening this library with monoclonal antibodies revealed two groups of ligands, one group closely related to the known epitope, and another quite different group. Devlin *et al.* (1990) created a library of 15-residue peptides located near the N-terminus of pIII and screened this against streptavidin. The screening revealed a group of phage containing a common tripeptide that compete with biotin for binding to streptavidin.

V. Conclusions

Two technical advances, the development of a system for the expression and efficient detection of peptide ligands, and the use of electroporation to construct extremely large libraries in plasmid vectors, have provided a recombinant approach to producing and analyzing collections of hundreds of millions of compounds. These collections are potential sources of ligands for antibodies, enzymes, and receptors of interest for scientific, pharmaceutical, agricultural, and industrial purposes.

References

Calvin, N. M., and Hanawalt, P. C. (1988) High-efficiency transformation of bacterial cells by electroporation. *J. Bacteriol.* **170**, 2796–2801.

Cwirla, S. E., Peters, E. A., Barrett, R. W., and Dower, W. J. (1990). Peptides on phage: A vast library of peptides for identifying ligands. *Proc. Natl. Acad. Sci. USA* **87**, 6378–6382.

Devlin, J. J., Panganiban, L. C., and Devlin, P. E. (1990). Random peptide libraries: A source of specific protein binding molecules. *Science* **249**, 404–406.

Dower, W. J., Miller, J. F., and Ragsdale, C. W. (1988). High efficiency transformation of *E. coli* by high voltage electroporation. *Nucleic Acids Res.* **16**, 6127–6145.

Dower, W. J. (1990). Electroporation of bacteria: A general approach to genetic transformation. In "Genetic Engineering" (J. K. Setlow, ed.), Vol. 12, pp. 275–295. Plenum Press, New York.

Hanahan, D. (1983). Studies on transformation of *Escherichia coli* with plasmids. *J. Mol. Biol.* **166**, 557–580.

Leonardo, E. D., and Sedivy, J. M. (1990). A new vector for cloning large eukaryotic DNA segments in *Escherichia coli*. *Biotechnology* **8**, 841–844.

Miller, J. F., Dower, W. J., and Tompkins, L. S. (1988). High-voltage electroporation of bacteria: Genetic transformation of *Campylobacter jejuni* with plasmid DNA. *Proc. Natl. Acad. USA* **85**, 856–860.

Meo, T., Gramsch, C., Inan, R., Hollt, V., Weber, E., Herz, A., and Rietmuller, G. (1983). Monoclonal antibody to the message sequence Tyr-Gly-Gly-Phe of opiod peptides exhibits the specificity requirements of mammalian opiod receptors. *Proc. Natl. Acad. Sci. USA* **80**, 4084–4088.

Parmley, S. F., and Smith, G. P. (1988). Antibody-selectable filamentous fd phage vectors: Affinity purification of target genes. *Gene* **73**, 305–318.

Scott, J. K., and Smith, G. P. (1990). Searching for peptide ligands with an epitope library. *Science* **249**, 386–390.

Smith, G. P. (1985). Filamentous fusion phage: Novel expression vectors that display cloned antigens on the virion surface. *Science* **228**, 1315–1317.

19

Electroporation and Electrofusion Using a Pulsed Radio-Frequency Electric Field

Donald C. Chang,[1] John R. Hunt, Qiang Zheng, and Pie-Qiang Gao

Department of Molecular Physiology and Biophysics, Baylor College of Medicine, Houston, Texas 77030

I. Introduction

Exposure of cells to a pulsed, high-intensity electric field can induce permeabilization (electroporation) and fusion (electrofusion) of cells. Such techniques have important applications in biological research and biotechnology, including gene transfection

---

[1]Present address: Department of Biology, Hong Kong University of Science and Technology, Kowloon, Hong Kong.

Guide to Electroporation and Electrofusion
      303

and gene transformation, introduction of exogenous molecules into cells, somatic hybridization, formation of hybridomas, and monoclonal antibody production. In comparison to traditional chemical or biological methods, the electrical methods are more efficient, applicable to a wider selection of cell types, and have fewer harmful side effects (Zimmermann, 1986; Chu *et al.*, 1987; Potter, 1988). For example, electroporation-mediated gene transfection can provide a much higher efficiency than the traditional calcium phosphate method (Chu *et al.*, 1987; McNally *et al.*, 1988, Chang *et al.*, 1991); and in plant somatic hybridization, the electro-fusion of plant protoplasts was at least 10 times more efficient than the PEG method (Bates *et al.*, 1987).

As these electrical methods have become widely applied, there is a need to further improve their efficiency and cell viability. Conventional electroporation and elec-trofusion methods typically use a direct current (DC) electric field in the form of either a rectangular or an exponential decay pulse (Fig. 1a and b). Such methods still have a few shortcomings. For example, in many experiments, an excessive amount of cells could be killed. Also, DC elecroporation methods are extremely sensitive to cell size. For a spherical cell, the equation for the membrane potential induced by an external electric field is (Cole, 1968)

$$V_m = 1.5rE \cos \theta \tag{1}$$

where V_m is the voltage drop across the membrane, r is the radius of the cell, E is the field strength of the applied electric field, and θ is the angle between the normal to the membrane and the electric field vector. It is clear from this equation that the induced membrane potential is proportional to the cell diameter. When a population of cells of mixed sizes is electroporated, the large cells would experience a large voltage drop across the membrane, and they are more likely to be irreversibly damaged by the electrical treatment. Conversely, the small cells would experience a small voltage drop across the membrane, and they have less chance to be porated.

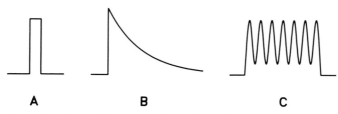

A   **B**   **C**

Figure 1   Three different waveforms of pulsed electric field used for electroporation and electrofusion. (A,B) Typical DC pulses used in conventional methods. (C) DC-shifted RF pulse, which was used in this study.

Hence, it is difficult to optimize the conditions for electroporating a nonhomogeneous cell population.

Recently, we have developed a new method in which a pulsed electric field oscillating at ratio frequency (RF) is used for electroporation and electrofusion of cells (Chang, 1989a, 1989b). The main advantage of using such an oscillating electric field is to counterbalance the cell size-effect with an opposite effect of cell relaxation. Equation (1) is valid only when the applied electric field is varied slowly with time. When the applied field is an oscillating electric field, the equation becomes (Holzapfel *et al.*, 1982)

$$V_m = 1.5 \frac{rE \cos \theta}{[1 + (\omega\tau)^2]^{1/2}} \tag{2}$$

where ω is the angular frequency and τ is the "relaxation" time of the cell, which is given by

$$\frac{1}{\tau} = \frac{1}{R_m C_m} + \frac{1}{rC_m(\rho_i + 0.5 \rho_e)} \tag{3}$$

where R_m and C_m are specific resistance and specific capacitance of the membrane, and ρ_i and ρ_e are the specific resistances of the intracellular medium and the extracellular medium, respectively. For a cell of several microns in diameter, τ is typically on the order of 1 μs.

Since the R_m of most cells is very large, for practical purposes, the above equation can be simplified to

$$\tau = rC_m(\rho_i + 0.5\rho_m) \tag{4}$$

Thus, the relaxation time is approximately proportional to the radius of cell. That means for a larger cell, the potential takes a longer time to build up. When an oscillating field with a high frequency is applied so that $\omega\tau > 1$, the denominator of Eq. (2) approaches $\omega\tau$. Substituting Eq. (4) for τ in Eq. (2), it is easy to see that V_m is no longer dependent on the radius r. Hence, with an oscillating electric field, it is possible to use the same electrical pulse to electroporate (or fuse) cells of different sizes.

Furthermore, an oscillating field is more effective for electroporation or electrofusion. A radio-frequency electric field can produce a vibrating motion of molecules in the cell membrane due to an electromechanical coupling effect. This motion is equivalent to a localized sonication, which can produce a mechanical fatigue in the cell membrane and thus enhance the formation of membrane pores (Chang, 1989a, 1989b).

In this chapter, sample results from various applications of the pulsed RF method to electroporation and electrofusion of a variety of cell systems are presented. These applications include gene transfection of cultured mammalian cells, introduction of antibodies, proteins and small molecules into fibroblasts, and fusion of red blood cells and mammalian culture cells.

II. Materials and Methods

A. Instrumentation

The RF pulses used in electroporation and electrofusion were generated from an instrument designed and built in our laboratory. The basic design has been described in earlier publications (Chang, 1989a, 1989b). Briefly, the apparatus contained an electronic circuit that produced a rectangular DC pulse of the desired pulse width. This pulse was used to gate the output of an RF oscillator. This gated RF signal was later mixed with the DC pulse to produce a DC-shifted RF pulse (see Fig. 1c). Finally, this waveform was amplified using a power amplifier and then applied to the electrodes in the cell chamber. The output waveform at the cell chamber was monitored using a digital oscilloscope (Tektronix, model 2210).

B. Cell Preparation

Several fibroblastlike cell lines, including green monkey COS-M6 and CV-1, hamster CHO, mouse 3T3 and C3H, and human IMR-90, were maintained in culture at 37°C in Dulbecco's modified Eagle's medium (DMEM) containing 4.5 g/l of glucose (GIBCO, Gaithersberg, MD), supplemented with 10% fetal bovine serum (FBS, from GIBCO) and 100 IU/ml each of penicillin and streptomycin (GIBCO).

Human red blood cells (RBC) were isolated from human blood, which was collected with heparin to prevent clotting. The blood cells were washed several times in isotonic phosphate-buffered saline solution (PBS), resuspended in PBS and stored at 4°C.

C. Electroporation

1. Gene Transfection of Suspected Culture Cells

For electroporation of suspended cells, the cultured cells grown to mid-log phase were first detached by trypsin-EDTA (ethylenediamine tetraacetic acid) treatment, washed in poration medium twice, and then concentrated to an approximate density of 10^7 cells/ml. The standard poration media used for gene transfection were (1) HEPES-buffered poration medium (HPM), which consisted of 1 mM MgCl$_2$, 10 mM HEPES buffer, and 270 mM sucrose, and (2) phosphate-buffered poration medium (PPM), which was similar to HPM except it contained 260 mM sucrose and 15 mM of Na phosphate buffer. The pH was adjusted to 7.3.

The fibroblast cell line COS-M6 was initially used to evaluate the efficiency of gene transfection by RF electroporation. Two marker genes were used. The first one was chloramphenicol acetyltransferase (CAT), derived from *Escherichia coli* (Gorman *et al.*, 1982), in the form of the plasmid construct pSV2-CAT. The other one was an *E coli* β-galactosidase gene (*lacZ*) in the form of the plasmid RSV-β-gal (MacGregor *et al.*, 1990).

For transfection of cells with CAT, approximately 100 μl of M6 cell suspension (approximate 2 million cells) in poration medium was pipetted into a well of a 96-well culture plate (Corning 25860). Then about 0.1 μg of supercoiled CAT DNA in Tris-EDTA buffer was added and mixed thoroughly with the cells. An electrode probe consisting of two concentric stainless steel electrodes (Chang, 1989b) was inserted into the well, such that the suspended cells were drawn into the space between the electrodes. After the application of RF pulses, the cells trapped between the electrodes were washed into the well of the culture plate with DMEM.

In the comparative experiments where cells were transfected using the conventional DC pulse method, COS-M6 cells were electroporated in disposable cuvettes with embedded aluminum electrodes at a gap of 0.4 cm (Bio-Rad Laboratories, Richmond, CA). The experiments were conducted using a capacitor-discharge poration device (Gene Pulser, Bio-Rad Laboratories, Richmond, CA) according to the procedures specified by the manufacturer.

Cells transfected with the plasmid pSV2-CAT were allowed to grow in culture for 48 h before they were harvested and a cell extract was prepared. CAT activity was measured by a modification of the method of Gorman *et al.* (1982). Typically, 25 μg of protein extract was mixed with [^{14}C]chloramphenicol and acetyl-coenzyme A for 90 min at room temperature. The reaction products (acetylated chloramphenicol derivatives) were separated by thin-layer chromatography (TLC) and imaged on x-ray film by autoradiography. The TLC plate was then cut to match the spots on the x-ray film to separate the radioactive acetylated products, and the activity was measured in a liquid scintillation counter.

For β-gal transfection, the cells were electroporated with RSV-β-gal DNA following a similar procedure as for CAT transfection. The electroporated cells were allowed to culture in 60-mm plastic culture dishes for 48 h. Then, the expression of β-gal was assayed by a histochemical staining procedure similar to that of Macgregor *et al.* (1990). Basically, the β-galactoside analog 5-bromo-4-chloro-3-indolyl-β-D-galacto-pyranoside (X-Gal) was used as a chromogenic indicator of enzyme activity, and a mixture of ferrocyanide and ferricyanide ions was used as an oxidation catalyst to localize and intensify the color of the reaction. The number of stained cells was counted under an inverted bright-field microscope.

Other mammalian cell lines, including CV-1, CHO, 3T3, and rabbit hepatocytes, were also transfected with CAT or β-gal gene using the methods outlined above.

2. Gene Transfection of Attached Culture Cells Using an *in Situ* Electroporation Method

COS-M6 cells were first plated on 12-mm-square glass coverslips (VWR Scientific, Inc., San Francisco) and allowed to grow for 8 h before electroporation. The coverslip with cells was then washed with a poration medium (PM, 2 mM HEPES, 15 mM K phosphate buffer, 250 mM mannitol, 1 mM MgCl$_2$, pH 7.2) and mounted on a chamber, in which two parallel platinum wires (1.5 mm apart) were placed in the middle of the coverslip (Zheng and Chang, 1990, 1991). Approximately 50 µl of PM mixed with 5 µg RSV-β-gal DNA was added to cover the cells. Then, three trains of RF electric pulses (5 pulses per train, 1 s between pulses, 10 s between trains, 1 ms pulse width, frequency 40 kHz) were applied to the cells via the two wires. Only cells situated between the two parallel electrodes were electroporated. The cells in the regions outside of the electrodes were not exposed to the applied electric field and thus served as controls. After electroporation, the cells were first incubated in a postporation medium (PM plus 10% FBS) for 30 min at 37°C before returning to normal culture.

After 10–48 h in culture, the cells were fixed and assayed for β-gal expression using the histochemical staining procedure described above.

3. Loading of Antibodies and Marker Molecules into Attached Cells

The above *in situ* electroporation method can also be used for loading marker molecules and antibodies into cultured cells. Examples of these applications are discussed here.

a. Loading of Small Molecules. CV-1 cells grown on coverslip were electroporated with either 5 mM Calcein (an analog of fluorescein, Calbiochem Co., San Diego, CA) or 1 mg/ml rhodamine-conjugated dextran (MW 20 kD, Sigma Chemical Co., St. Louis, MO) following a procedure similar to that used for gene transfection. Following electroporation, the cells were incubated in the postporation medium for 30 min and DMEM for another 30 min. Then the cells were washed with DMEM, mounted in HEPES-buffered DMEM (10 mM HEPES in DMEM), and viewed using fluorescence microscopy.

b. Labeling of F-Actin of Living Cells with Rhodamine-Phalloidin. Several cell lines, including CV-1, C3H, and IMR-90, were used for this study. Cells grown on coverslips were electroporated following a procedure similar to (a), except that sucrose was substituted for mannitol in the poration medium.

Rhodamine-phalloidin (Rh-Ph) was prepared by evaporating the solvent from a 3.3 µM stock solution in methanol (obtained from Molecular Probes, Eugene, OR). The Rh-Ph was then dissolved in poration medium to achieve the desired concentration of phalloidin (typically 0.3–3.3 µM). Following electroporation, the cells were incubated for varying amounts of time. Then the cells were viewed using fluorescence microscopy, either as live cells or after fixation. For live cell viewing, the cell coverslips were placed on HEPES-buffered DMEM medium, and the cells were sandwiched into a chamber filled with DMEM. For fixed cells, coverslips were rinsed with PBS, and fixed in 3.7% formaldehyde in PBS for 30 min, washed with PBS, and mounted in Mowiol medium (Osborn and Weber, 1982). Fixed or live cells were viewed in a Zeiss Axiophot microscope, equipped with phase and epi-fluorescence.

   c. Loading of Antibodies.   Antibodies were introduced into C3H mouse fibroblasts under conditions that were virtually identical to those for phalloidin loading. Two antibodies were used: a rhodamine-labeled goat anti-mouse IgG (American Qualex), and a monoclonal antibody directed against the nucleolar protein B23 (obtained from Dr. P. K. Chan). Antibodies were added in PM to a final concentration of approximately 1 mg/ml.

   In the experiment where cells were loaded with rhodamine-IgG, the cells were directly viewed under a fluorescence microscope. In the experiment where cells were loaded with the anti-B23 antibody, the electroporated cells were incubated for 2–16 h at 37°C, and then fixed in 3.7% formaldehyde in PBS for 30 min and washed with PBS. The fixed cells were then stained with a second antibody, fluorescein-labeled goat anti-mouse IgG, and then mounted in the Mowiol medium and viewed by fluorescence microscopy.

D. Electrofusion

1. Electrofusion of Red Blood Cells

Human red blood cells were washed three times in a fusion medium, which consisted of 27 mM Na phosphate buffer and 150 mM sucrose, pH 7.4. A small portion of the washed cells were labeled with the lipophilic fluorescent dye DiI (1,1'-dihexadecyl-3,3,3',3'-tetramethylindocarbacyanine perchlorate, obtained from Molecular Probes, Eugene, OR), following the procedure of Sowers (1984). Labeled cells were then mixed with unlabeled cells in a ratio of about 1 : 10 and were ready for use in the electrofusion experiment.

   The fusion chamber was constructed using two glass coverslips separated by 0.3-mm-thick spacers. Two thin parallel platinum wires, 0.4 mm apart, were sandwiched between the coverslips (see Chang et al., 1989a). The RBC cells suspended in the medium were first brought into close contact to form "pearl chains" by the process of dielectrophoresis (Pohl, 1978) using a low-amplitude AC (alternating current) electric field (100–400 V/cm). Then one or three high-amplitude RF pulses (0.3 ms, 4–5 kV/cm, 100 kHz) were applied to cells to induce fusion. The fusion process was monitored under a Nikon inverted microscope equipped with an epi-fluorescence attachment.

2. Electrofusion of Attached CV-1 Cells

The procedures of electrofusion of attached CV-1 cells were very similar to the procedures of *in situ* electroporation as described in Section C,2, except the fusion medium consisted of 2 mM HEPES, 280 mM mannitol, 2 mM MgCl$_2$, pH 7.2, and only a single RF pulse (100 kHz, 0.2 ms, 1 kV/cm) was applied. In this case, application of a dielectrophoretic field was unnecessary, since some of the attached cells were already in contact during cell growth. (For more details, see Chapter 12 of this book.)

III. Results

A. Gene Transfection of Suspended Cells

To demonstrate the effectiveness of the RF electroporation method, COS-M6 cells were transfected with the marker gene chloramphenicol acetyltransferase (CAT) using RF pulses. A sample record of the CAT assay in this experiment is shown in Fig. 2A, in which chloramphenicol (bottom row) and monoacetylated chloramphenicol (top two rows) are separated by thin-layer chromatography. The ratio between the acetylated chloramphenicol and total chloramphenicol, called "acetylation rate," is a rough measurement of the number of copies of gene expressed in the cells. Each column in Fig. 2A represents the result obtained from using a different amount of plasmid DNA during electroporation. It is found that a high level of CAT activity can be obtained using only 0.1 μg of DNA. By comparison, usually 5–40 μg of plasmid DNA is required to transfect a few million cells with the conventional methods. Results of this study thus demonstrate that the RF electroporation method is highly efficient.

To provide a more detailed comparison, we have conducted a series of transfection experiments using three methods: (1) the RF electroporation method, (2) the conventional electroporation method using a DC field (generated by a capacitor-discharge device), and (3) the Ca phosphate method. All these methods were applied under their optimized conditions. The reporter gene was pSV2-CAT. The results of this comparative study are plotted in Fig. 2B. It is evident that the acetylation rate

Figure 2   (A) Sample record of CAT assay where cells were transfected using RF electroporation. Expression of pSV2-CAT gene in COS-M6 cells was shown to vary as a function of DNA concentration. (B) Comparison of CAT activity of cells transfected by three different transfection methods: ● results form six experiments using RF electroporation, ■ results from four experiments using DC electroporation, ▲ results from four experiments using the Ca phosphate method. (Error bar represents standard error.) (From Chang *et al.*, 1991.)

Expression of pSV2-CAT Gene
in COS-M6 Cells

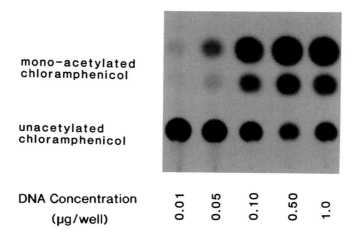

mono-acetylated
chloramphenicol

unacetylated
chloramphenicol

DNA Concentration   0.01   0.05   0.10   0.50   1.0

(μg/well)

A

B

obtained by the RF method is significantly higher than that of the DC method. This difference is particularly apparent at the low DNA concentrations, where the CAT reaction is in the linear region. The transfection efficiency using the Ca phosphate method is much lower than either electroporation method. Using our assay conditions, the average acetylation rate obtained by the calcium phosphate method with 10 μg of CAT DNA is about 25%. With the RF method, a higher acetylation rate can be obtained using 0.1 μg of DNA. Thus, in term of DNA consumption, the transfection efficiency in M6 cells using the RF method is about 100-fold higher than that of the calcium phosphate method.

The improved efficiency of the RF electroporation method is also confirmed in a second experiment using β-gal as the reporter gene. Table 1 shows the results of a series of experiments in which a RSV-β-gal plasmid was introduced into COS-M6 cells using three different methods: RF, DC, and calcium phosphate. Again, the transfection efficiency obtained by the RF method is significantly higher than that obtained by the other two methods.

Table 1
Expression of β-gal Gene in COS-M6 Cells: Comparison of
Three Different Transfection Methods[a]

Transfection method	[DNA] (μg/well)	Relative number of transfected cells	Condition
Electroporation by RF field	20	424	1.0 kV/cm
	20	472	1.0 kV/cm
	20	860	1.2 kV/cm
	20	784	1.2 kV/cm
Electroporation by DC field	20	155	1.0 kV/cm
	20	201	500 μF
	40	250	
	40	266	
Ca phosphate	20	51	Standard protocol
	20	80	followed by
	40	107	DMSO shock
	40	180	

[a]Each sample contained 3×10^6 COS-M6 cells. The plasmid DNA was RSV-β-gal. In the RF field method, 3 trains of pulses (5 pulses per train) were applied, pulse width 2 ms, frequency 40 kHz. In the DC field method, only a single pulse was applied; the time constant of the exponential decay pulse was approximately 4 ms. The relative numbers of transfected cells were determined by adding the number of blue-stained cells in 10 random fields observed under a microscope with $100\times$ magnification.

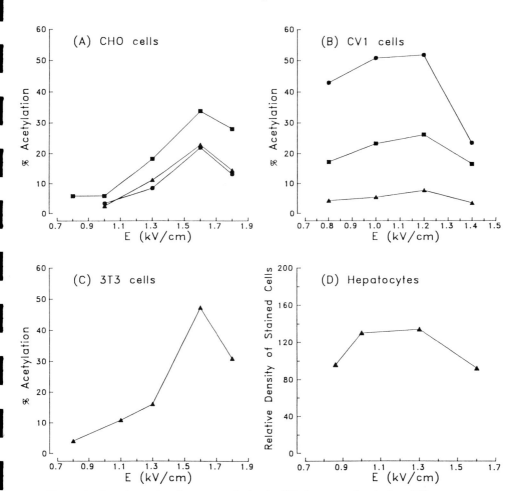

Figure 3   Sample results of gene transfection by RF electroporation in four different types of mammalian cells. (A) Transfection of CHO cells with pSV2-CAT, (B) transfection of CV-1 cells with CAT, (C) transfection of 3T3 cells with CAT, and (D) transfection of rabbit hepatocytes with CMV-β-gal. The families of curves in CHO and CV-1 represent results obtained using buffers of different K phosphate concentrations: ▲ (15mM), ■ (30mM), and ● (50mM). Data were averaged values from six sets (CV-1) and four sets (CHO) of experiments.

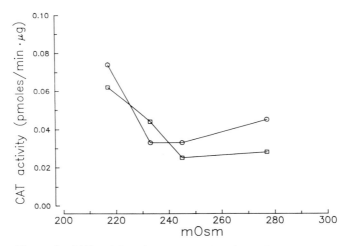

Figure 4   CAT activity of CV-1 cells plotted as a function of the osmolarity of the poration medium. Osmolarity was adjusted by changing the concentration of sucrose. Results are from two sets of experiments.

In addition to COS-M6 cells, the RF electroporation method has been used effectively to introduce cloned genes into a variety of other cell types. We have successfully used the RF method to introduce the CAT gene into CV-1 cells, CHO cells, HeLa cells, and 3T3 cells, and to introduce the β-gal gene into K562 cells and hepatocytes, a primary cell culture isolated from rabbit. Some sample results of these transfection experiments are shown in Fig. 3.

We have also examined a number of factors that may affect the efficiency of gene transfection. Such factors include pH, temperature, osmolarity, ionic strength, concentration of divalent ions, etc. As can be seen in Fig. 3, the concentration of phosphate salt in the poration medium has a significant effect on the transfection efficiency in many cell types (Chang *et al.,* 1991). Figure 4 shows the results of two experiments examining the effects of osmolarity on the transfection of CV-1 cells

---

Figure 5   Expression of the β-gal reporter gene in COS-M6 cells that were electroporated in an attached state. (A) A low-magnification view of the coverslip on which the cells were attached and transfected. Most cells in the region between the two electrodes took up and expressed the β-gal gene and thus were stained blue. They appear as a "dark band" in the middle of the coverslip. (B) A magnified view of the cells in the regions outside the electrodes. (C) A magnified view of the cells inside the "dark band." Solid arrow, stained cell; open arrow, unstained cell.

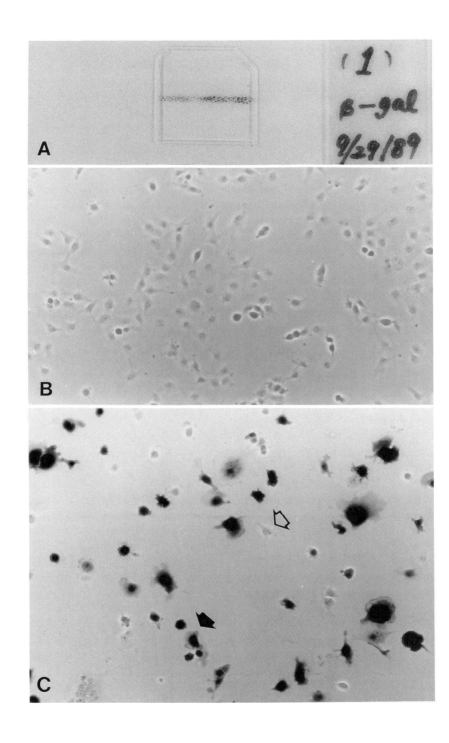

by pSV2-CAT. Although there is a small shift between the two sets of data, the results basically suggest that a reduction of osmolarity in the poration medium may enhance the transfection efficiency.

B. Gene Transfection of Attached Cells

Another factor that can affect the efficiency of gene transfection is cell treatment. In the standard electroporation procedure, cells were usually detached using EDTA-trypsin treatment, which might reduce the viability of the cells. Hence, we have developed an *in situ* electroporation method, in which RF electric pulses were applied directly to cells attached to substrata (Zheng and Chang, 1991). In this case, the EDTA-trypsin treatment was avoided. A sample result of gene transfection using such *in situ* method is shown in Fig. 5.

Panel 5A is a low magnification view of cells grown on a coverslip that were stained to assay for β-gal expression. Only the cells that were situated between the electrodes (in the middle of the coverslip) were exposed to the applied RF electric field. Because most of these electroporated cells took up and expressed the β-gal gene, they appear as blue dots forming a band on the coverslip. The width of this band is the same as the gap between the two parallel electrodes. No blue cells were found in the region outside the electrodes (Fig. 5B.) Panel 5C shows a magnified view of the cells inside the electrode region. It can be seen that up to 80% of the survived cells expressed the β-gal gene (blue stained).

By comparison, the percentage of cells expressing the β-gal gene was much lower

Table 2
Gene Transfection of Attached and Suspended Cells by RF Electroporation

RF pulse strength (kV/cm)	Attached cells[a]		Suspended cells[b]	
	efficiency (%)	viability (%)	efficiency (%)	viability (%)
0.4	0.8 ± 0.6	92.9 ± 5.7	—	—
0.6	3.2 ± 0.7	89.1 ± 4.1	0.1 ± 0.0	100 ± 0.0
0.8	20.3 ± 0.8	81.4 ± 9.1	0.9 ± 0.3	89.2 ± 6.1
1.0	40.9 ± 2.1	66.7 ± 10.7	2.5 ± 0.4	70.3 ± 8.0
1.2	82.4 ± 3.1	54.2 ± 7.6	5.7 ± 1.8	53.7 ± 7.4
1.4	83.8*	31.2*	11.8 ± 0.9	17.3 ± 4.2
1.6	—	—	19.0 ± 2.4	7.1 ± 3.6

[a]Attached cells were allowed to grow for 10 hs before assay; data were from three experiments except the ones marked with asterisk which were from one experiment.
[b]Suspended cells were allowed to grow for 48 hs before assay; data were from four experiments.

if the cells were electroporated in suspension, even though the electrical protocol, poration medium, and the concentration of plasmid DNA were identical. Results of this comparative study are shown in Table 2. In every level of electric field strength, the transfection efficiency of the suspended COS-M6 cells was always lower than that of the attached cells.

C. Loading of Antibodies and Marker Molecules

In the study of various aspects of cellular structure and functions, there is often a need to introduce exogenous molecules into living cells. Such exogenous molecules include pharmacological agents, antibodies, and other proteins. The *in situ* electroporation method is particularly useful for such loading, because one can monitor the same cells before and after the introduction of the exogenous agents. Results of several sample studies are reported here.

In the first example, rhodamine-conjugated phalloidin molecules were introduced into CV-1, IMR-90, and C3H cells to study the structure and dynamics of microfilaments in living cells. The structures of F-actin (for example, the stress fibers) in living cells can be easily labeled with this method. Figure 6A and B show live C3H and CV-1 fibroblasts with F-actin labeled with rhodamine-phalloidin introduced by *in situ* electroporation. At present, we are using such labeled cells as model systems to study the role of actin in cell motility. With the *in situ* RF electroporation, the efficiency for loading small molecules into living cells is extremely high. Such high-efficiency loading is demonstrated in the results shown in Fig. 6C and D. Here, CV-1 cells were loaded with calcein (a fluorescent calcium indicator, 600 Da) using the RF method. By comparing the differential interference contrast (DIC) (Fig. 6C) and fluorescent (Fig. 6D) images, one can see that essentially all cells in the region between electrodes were labeled with the marker molecules, and with no cell damage. Similarly, rhodamine-labeled dextran (20 kD) could also be introduced into CV-1 cells with very high efficiency (see Fig. 6E).

Finally, it has been demonstrated that antibodies introduced by the *in situ* RF electroporation method can retain their ability to bind to specific intracellular antigens within the living cell. In this study, an antibody against nucleolar protein B23 (Chan *et al.*, 1986) was introduced into living C3H mouse fibroblasts using the RF electroporation method. After 2–16 h of incubation, presence of the antibody was assayed using an immunofluorescence method. Bright staining of the nucleolus (indication of antibody binding, Fig. 6F) was detected in some of the electroporated cells (about 2–4% of the cell population). The low percentage of nucleolar staining is not due to the difficulty of introducing antibody into the cell, since paired experiments using rhodamine-conjugated goat IgG indicated that the efficiency of electroporation for introducing IgG could be nearly 100%. The low percentage of nucleolar staining by the anti-B23 antibody may be due to the diffusion barrier of

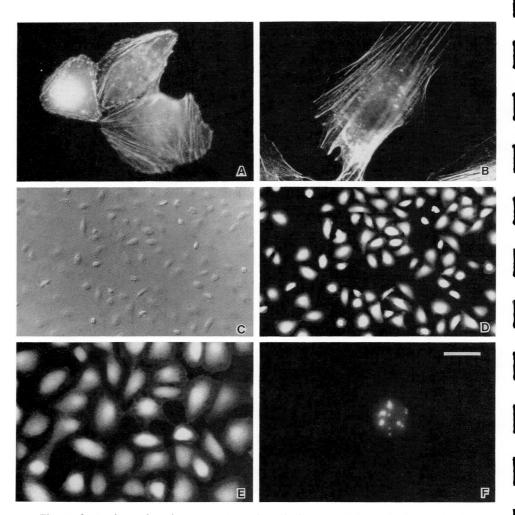

Figure 6 Loading of marker molecules and antibodies into living cells by the *in situ* electroporation method. (A) Live CV-1 cells loaded with rhodamine-phalloidin. (B) A live C3H cells loaded with rhodamine-phalloidin. (C-D) Live CV-1 cells loaded with calcein molecules. (E) Live CV-1 cells loaded with rhodamine-dextran. (20 kD). (F) C3H cells loaded with anti-B23 antibody. All panels are fluorescent micrographs, except (C) which is obtained by DIC optics. Scale bar is 30 μm (A, B, and F), 58 μm (E), and 117 μm (C, D).

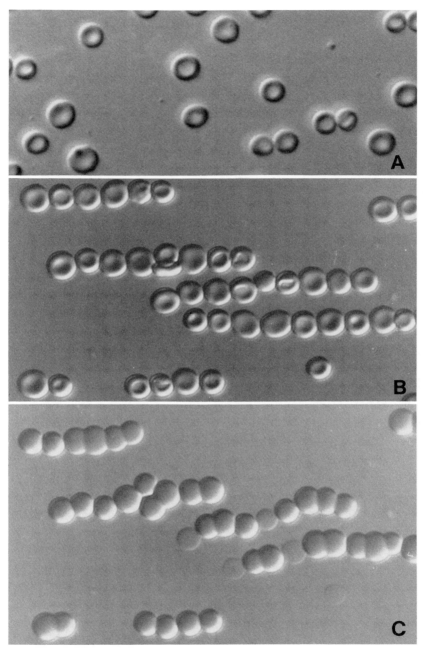

Figure 7 Fusion of human red blood cells (RBC) by the RF field as observed using DIC optics. (A) Before any electric field was applied. (B) After a low-intensity, continuous AC field (200 V/cm, 60 kHz) was applied, RBC were lined up in "pearl chains" by dielectrophoresis. (C) At 1.5 min after applying three RF pulses (3.5 kV/cm, 100 kHz, 0.1 ms). Some of the neighboring RBC began to fuse together (see arrows). Some cells also lysed (indicated by arrowheads) after electrical pulsation.

the nuclear envelope. It is possible that even if a substantial amount of antibody can be introduced into the cytoplasm by electroporation, very few antibody molecules might enter the interphase nucleus to bind to the nucleolar proteins.

D. Electrofusion of Human Red Blood Cells

The oscillating electric field is not only useful for cell poration, it is also highly effective in inducing cell fusion. We have used human red blood cells (RBC) as a cell model to study cell fusion using the RF electrical pulse. Figure 7 shows the event at different stages of the electrofusion process. RBC were suspended between two platinum electrodes and were observed under a light microscope. When there was no electric field applied across the electrodes, the RBC distributed randomly as shown in Fig. 7A. After a low-intensity AC field was applied for 0.5 min, the cells were aligned by a dielectrophoretic process to form "pearl chains" (Fig. 7B). At this point, three pulses of RF electric field (0.1 ms, 3 kV/cm, 100 kHz) were applied. The dielectrophoretic field was then turned on again. In a few seconds, the RBC became rounded in shape, and some neighboring cells began to fuse with each other. At 1.5 min after applying the RF electrical pulses, many pairs of fused cells were formed (Fig. 7C).

Figure 7 shows light micrographs obtained by DIC (differential interference contrast) optics. Fusion events can be clearly detected when the cytoplasms of the fusing cells start to merge. However, sometimes neighboring cells may fuse without merging their cytoplasms. Another method of assaying the fusion yield is to use a fluorescent dye transfer method (Sowers, 1984). Figure 8 shows fluorescence micrographs taken at different times before and after applying the RF pulses. About 10% of the suspended RBC were prelabeled with a lipophilic fluorescent dye (DiI), the rest of the RBC were unlabeled. This dye, which concentrated in the cell membrane and gave a brilliant fluorescent image, normally could not transfer from one cell to another. Thus, before application of the RF pulses, the labeled cells appeared as isolated monomers viewed under a fluorescence microscope (Fig. 8A). After the application of RF pulses, fusion was induced between neighboring cells. Such fusion allowed the dye to diffuse from the membrane of labeled cells to the unlabeled cells (Fig. 8B). With time the dye became evenly distributed between the prelabeled and the newly labeled cells; they appeared as dimers under a fluorescence microscope. As the fusion progressed, many newly labeled cells fused with their next neighbors, and chains of cells stained with the fluorescent dye could be observed (Fig. 8C).

To compare the effectiveness of electrofusion of the RF field method with that of the more conventional DC field method, we have conducted a study in which RBC were divided into two groups; one group was fused using RF electrical pulses and the other group was fused using rectangular DC pulses. The pulse protocol, pulse width, and pulse amplitude were identical in these two cases (0.1 ms, 3.7

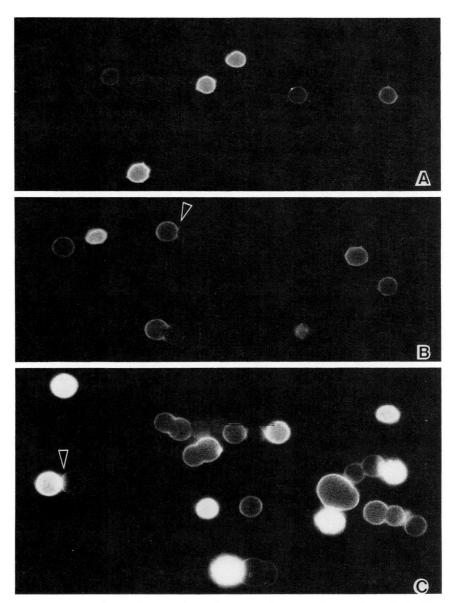

Figure 8 Electrofusion of RBC as observed using the fluorescent dye transfer method. About 10% of the RBC were prelabeled with DiI. The other cells were unlabeled. (A) Before the RF pulses were applied, the labeled cells appeared as isolated cells under a fluorescence microscope, although in reality they were embedded within pearl chains made up of many unlabeled cells. (B) One minute after applying three RF pulses, some of the labeled cells started to fuse with their unlabeled neighbors (arrowhead). The fluorescent dye now could be transferred from the labeled cell to the unlabeled fusing partner, making both cells fluorescent. (C) Four minutes after application of the RF pulses, some of the fused cells had merged their cytoplasms to form larger cells. Some cells fused their membranes but did not merge their cytoplasms (see arrowheads).

kV/cm, 3 pulses applied 1 s apart). The results were summarized in Table 3. Using the fluorescent dye method, the average fusion yield obtained by the DC field is 3.4%, while the average fusion yield obtained by applying the RF field is 18.3%. The difference between the RF field and the DC field becomes even more prominent when the fusion yield was measured among the RBC that remained unlysed after 5 min (note: such measurement was done using the DIC optics). Over four experiments, the average fusion yield obtained using the RF field (at 100 kHz) was 55.6%, while no cytoplasmic fusion was observed using a DC field with the same pulse width and amplitude (see Table 3). Hence, it is clear that the RF field is more effective than the DC field in fusing red blood cells. It has been reported previously that using DC pulses, intact RBC without pronase treatment were very difficult to fuse (Stenger and Hui, 1986). We found that the pronase treatment was unnecessary if the electrofusion was performed using a RF electric field.

E. Electrofusion of CV-1 Cells

The oscillating electric field is also highly effective in fusing mammalian cultured cells. One example is the fusion of CV-1 cells that were grown on a glass coverslip. The method is very similar to *in situ* electroporation. Figure 9 shows a sample result of electrofusion of CV-1 cells. Before any electrical pulse was applied, the attached CV-1 cells appeared as typical mononucleated interphase fibroblast cells (Fig. 9A).

Table 3
Electrofusion of Red Blood Cells Using Multiple Pulses[a]

Experiment number	Fusion yield (%)		Method
	RF pulse	DC pulse	
050388	50.5	3.3	Fluorescent dye transfer
051088	16.0	0.0	
051188	13.5	4.9	
051988	9.0	6.0	
052488	9.0	3.0	
052588	12.0	3.0	
Mean ± SD	18.3 ± 15.9	3.4 ± 2.0	
070788	58.3	0.0	DIC optics
070988	58.6	0.0	
071188	53.8	0.0	
071288	51.6	0.0	
Mean ± SD	55.6 ± 3.4	0.0 ± 0.0	

[a]Fusion was induced by applying 3 pulses (pulse width 0.1 ms, 100 kHz, 3.7 kV/cm).

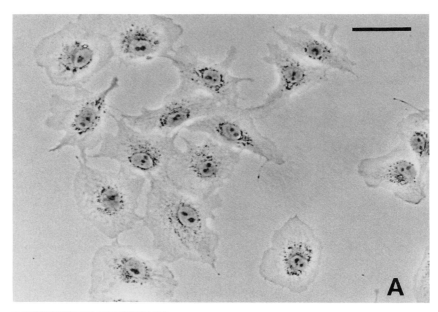

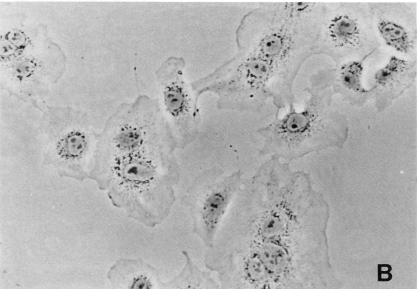

Figure 9   Electrofusion of attached CV-1 cells using RF electric field. (A) Phase optics micrograph of CV-1 cells in control; almost all of the cells were mononucleate. (B) Phase optics micrograph of fused CV-1 cells at 3 h after application of the electric pulse. Scale bar: 50 μm.

After a single pulse of RF field (1 kV/cm, 0.2 ms, 100 kHz) was applied, many neighboring cells began to fuse. The fusion process took a few hours. At 2–3 h, most of the fusing cells had merged their cytoplasms, and their nuclei had aggregated (Fig. 9B). We found that the cytoskeletal system of the fusing cells had also undergone a reorganization during this fusion process. The detailed findings are reported in Chapter 12 of this book.

IV. Discussion

Results of this study indicate that the efficiency of both electroporation and electrofusion can be significantly improved by using an oscillating electric field rather than a DC field to treat the cells. It is evident in our gene transfection experiments that the RF method is more effective than the chemical method or the conventional electroporation method using a DC field. For example, we showed that with the RF method, 2 million of COS-M6 cells can be transfected with high efficiency using only 0.1 μg CAT DNA. The β-gal experiment also showed that up to 80% of cells took up and expressed the reporter gene when the cells were elctroporated *in situ* using the RF field. Regarding cell fusion, we demonstrated that intact red blood cells (without pronase treatment) can be fused in high frequency using an RF field, while the efficiency of fusion using a DC field was found to be much lower.

The RF field method has another advantage in that it is generally less damaging to the cells when compared to the conventional DC field. An earlier study (Zimmermann *et al.*, 1980) has indicated that when a cell was exposed to a DC electric field with a long duration (longer than 20 μs), the porated cell would be easily damaged depending on the intensity of the electric field. However, if the duration of the field was reduced to less than 5 μs, the porated cell was much more resistant to electrical damage. From this observation, it is not difficult to understand why the RF field is more conducive to cell survival than the DC field. With an oscillating electric field, the cell is never continuously exposed to a high-intensity field for an extensive period, and thus would have less chance of being damaged by the electrical treatment.

In addition, since the RF field method is less cell-size dependent than the DC field method, it is particularly useful for electroporating cells with nonhomogenous sizes or irregular morphology. This consideration is important in the application of *in situ* electroporation. Most cultured cells grown on substrata have irregular shapes; their effective diameters are non-uniform and vary depending on the cell orientation. The RF electroporation can work more effectively than a DC field for porating such cells. Indeed, our results indicate that the *in situ* method using an RF electric field is extremely effective for transferring DNA and other exogenous molecules into many types of attached cells.

The lack of cell-size dependence in the RF field method is also desirable in cell fusion for somatic hybridization. For example, suppose one wants to fuse myeloma cells with lymphocytes for production of monoclonal antibodies. Because of the large difference in cell size, it is difficult to optimize the electrical conditions with the conventional DC electrofusion method. However, this difficulty can be overcome by using the RF field. By choosing a proper frequency, one can counterbalance the effect of cell size with the cell relaxation effect. Thus, it is possible to select a field strength that is equally effective for fusing large and small cells. Such advantage should enable one to greatly improve the efficiency of hybridoma production.

References

Bates, G. W., Nea, L. J., and Hasenkampf, H. A. (1987). Electrofusion and plant somatic hybridization. *In* "Cell Fusion" (A. E. Sowers, ed.), Plenum Press, New York, pp. 479–496.

Chan, P. K., Aldrich, M., Cook, R. G., and Busch, H. (1986). Amino acid sequence of B23 protein phosphorylation site. *J. Biol. Chem.* **261**, 1868–1872.

Chang, D. C. (1989a). Cell fusion and cell poration by pulsed radio-frequency electric fields. *In* "Electroporation and Electrofusion in Cell Biology" (E. Neumann, A. E. Sowers, and C. A. Jordan, eds.), Plenum Press, New York, pp. 215–227.

Chang, D. C. (1989b). Cell poration and cell fusion using an oscillating electric field. *Biophys. J.* **56**, 641–652.

Chang, D. C., Gao, P. Q., and Maxwell, B. L. (1991). High efficiency gene transfection by electroporation using a radio-frequency electric field. *Biochim. Biophys. Acta* **1992**, 153–160.

Chu, G., Hayakawa, H., and Berg, P. (1987). Electroporation for the efficient transfection of cells with DNA. *Nucleic Acids Res.* **15**, 1311–1326.

Cole, K. S. (1968). In "Membranes, Ions, and Impulses: A Chapter of Classical Biophysics." University of California Press, Berkeley, pp. 12–18.

Gorman, C. M., Moffat, L. F., and Howard, B. H. (1982). Recombinant genomes which express chloramphenicol acetyltransferase in mammalian cells. *Mol. Cell Biol.* **2**, 1044–1051.

Holzapfel, C., Vieken, J., and Zimmermann, U. (1982). Rotation of cells in an alternating electric field: Theory and experimental proof. *J. Membrane Biol.* **67**, 13–26.

MacGregor, G. R., Nolan, G. P., Fiering, S., Roederer, M., and Herzenberg, L. A. (1990). Use of *E. coli LacZ* (β-galactosidase) as a reporter gene. *In* "Methods in Molecular Biology" (E. J. Murray, and J. M. Walker, eds.), Vol. 7, Humana Press, Clifton, NJ.

McNally, M. A., Lebkowski, J. S., Okarma, T. P., and Lerch, L. B. (1988). Optimizing electroporation parameters for a variety of human hematopoietic cell lines. *BioTechniques* **6**(9), 882–886.

Osborn, M., and Weber, K. (1982). Immuno-fluorescence and immunocytochemical procedures with affinity purified antibodies: Tubulin-containing structures. *In* "Methods in Cell Biology," Vol. 24, Chapter 7, Academic Press, New York.

Pohl, H. A. (1978). "Dielectrophoresis." Cambridge University Press, London.

Potter, H. (1988). Electroporation in biology: Methods, applications, and instrumentation. *Anal. Biochem.* **174**, 361–373.

Sowers, A. E. (1984). Characterization of electric field-induced fusion in erythrocyte ghost membranes. *J. Cell Biol.* **99**, 1989–1996.

Stenger, D. A., and Hui, S. W. (1986). Kinetics of ultrastructural changes during electrically induced fusion of human erythrocytes. *J. Membrane Biol.* **93**, 43–53.

Zheng, Q., and Chang, D. C. (1990). Dynamic changes of microtubule and actin structures in CV-1 cells during electrofusion, *Cell Motil. Cytoskeleton* **17**(4), 345–355.

Zheng, Q., and Chang, D. C. (1991). High-efficiency gene transfection by *in situ* electroporation of cultured cells. *Biochim. Biophys. Acta* **1088**(1), 104–110.

Zimmermann, U. (1986). Electrical breakdown, electropermeabilization and electrofusion. *Rev. Physiol. Biochem. Pharmacol.* **105**, 176–256.

Zimmermann, U., Vienken, J., and Pilwat, G. (1980). Development of drug carrier systems: electric field induced effects in cell membranes. *J. Electroanal. Chem.* **116**, 553–574.

20

Electroinsertion: An Electrical Method for Protein Implantation into Cell Membranes

Youssef Mouneimne, Pierre-François Tosi,[1] Roula Barhoumi, and Claude Nicolau[1]

Texas A&M University, Institute of Biosciences and Technology, Cell Biology Section, College Station, Texas 77843

[1]Present address: Center for Blood Research, Harvard Medical School, Boston, Massachusetts 02115.

Guide to Electroporation and Electrofusion
Copyright © 1992 by Academic Press, Inc. All rights of reproduction in any form reserved.

I. Introduction

The electroporation of membranes has been based on the observation that transient hydrophylic pores are generated in the membrane using electrical field strength equal to or higher than a critical value E_c (Sugar and Neumann, 1984; Serpersu *et al.*, 1985). This phenomenon is used in order to induce cell fusion: electrofusion (Teissié *et al.*, 1982; Sowers, 1984; Zimmermann, 1986), gene transfer: electrotransfection (Neumann *et al.*, 1982; Potter *et al.*, 1984; Hashimoto *et al.*, 1985) or cell loading (Kinosita and Tsong, 1978). In this chapter we present evidence that at field strengths slightly below the critical field E_e, it becomes possible to insert a number of xenoproteins into the plasma membrane of a variety of cells without apparent damage to the cells. Figure 1 illustrates the different effects of a pulsed electric field. The inserted protein exposes its normal epitopes and functions, at least as far as the extracellular sequence of the protein is concerned (Nicolau *et al.*, 1990). We named this novel method of insertion "electroinsertion" (Mouneimne *et al.*, 1989).

The electroinsertion consists of the application of pulsed electrical field of up to a millisecond duration, on a suspension of cells, in the presence of a selected membrane protein having a membrane spanning sequence. It results in the implantation of the protein in the cell's plasma membrane.

A number of methods have been developed for the insertion of proteins in the membrane of both prokaryotic and eukaryotic cells as well as in artificial lipid membranes.

Thus, cytochrome b5 was transferred from large, unilamellar lipid vesicles to small, sonication liposomes (Enoch *et al.*, 1979), bacteriorhodopsin and UDP-glucuronosyltransferase were spontaneously inserted in liposomes bilayers (Scotto and Zakim, 1988), and Band 3, the erythrocyte anion transporter, was transferred in native orientation from red blood cells (RBC) to liposomes (Huestis and Newton, 1986). Many other proteins were inserted in artificial membrane, but some could be inserted in biomembranes. CD4 and glycophorin were associated after low-pH exposure to human red blood cells (Arvinte *et al.*, 1989), with CD4 retaining its gp120-binding activity (Arvinte *et al.*, 1990).

None of the methods using detergents (Zalman *et al.*, 1987; Rigaud *et al.*, 1988) attempted to reuse the cells with inserted protein *in vivo*. The low-pH treatment of RBC, for instance, severely curtails the life span of the RBC-X in circulation (R. Kravtzoff, unpublished observations, 1988). The observation by Kinosita and Tsong (1978) that electroporated erythrocytes retained their normal life span in circulation encouraged us to try to apply a modified form of electroporation to the insertion of membrane receptors (CD4 in this case) into RBC membranes, in order to obtain a long-lived form of CD4 in circulation. Electroinsertion—the technology we developed—has proved simple and reproducible (Mouneimne *et al.*, 1989, 1990, 1991). Moreover, it does not require detergents or other chemicals that might damage the RBC plasma membrane, affecting its life span in circulation. Therefore, the electroinsertion of proteins or specific receptors into intact red blood cell membrane can

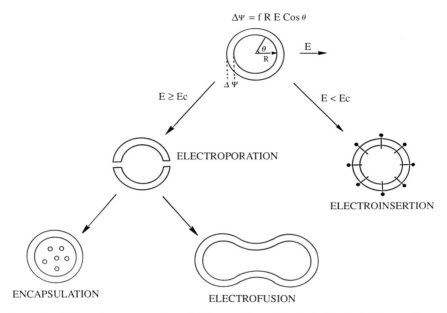

Figure 1 Schematic representation of different effects of pulsed electric fields on cells at strengths higher and lower than the E_c field. The induced membrane potential $\Delta\Psi$ of the cell membrane at a given point is proportional to the external field E, to the cell radius R, to the cell shape defined by a factor f (equal to 1.5 for a spherical cell), and to $\cos\theta$, where θ is the angle between the given point and the field direction. The critical voltage for electroporation $\Delta\Psi_c$ is first reached in the direction of the field for $E = E_c$ (critical field for electroporation), and at other points not in the field direction for $E > E_c$. Electrofusion and encapsulation are observed at $E \geqslant E_c$. Electroinsertion occurs at $E < E_c$. The schematic diagram shows the inserted proteins distributed over the cell membrane due to the lateral diffusion after insertion.

be used to extend the circulatory life span of these receptors when injected into the bloodstream, providing long-term protection or therapy. The insertion of CD4 receptor into erythrocyte membranes might provide a therapy for AIDS (Nicolau *et al.*, 1988, 1990; Zeira *et al.*, 1991).

II. Some Key Factors of Electroinsertion

A. Electric Field

Considering Fig. 2 relative to the electroinsertion of glycophorin into a 2-week-old piglet red blood cell membrane, it appears that the electroinsertion is optimal at

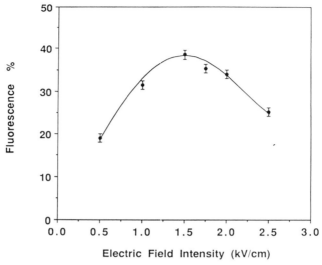

Figure 2 Determination of the optimal electric field strength for electroinsertion of glycophorin in piglet red blood cell membrane. The percent shift of fluorescence intensity with respect to the control RBC (BSA replacing glycophorin) in the flow cytometry histogram is plotted against the applied electric field. Pulse duration is 1 ms. The optimal insertion is obtained at $E \approx 1.5$ kV/cm.

$E \approx 1.5$ kV/cm. Given the radius of the piglet erythrocyte of 2.43 μm as measured by a Coulter counter (ZM Channelyzer 256, Coulter, Hialeah, FL) and using Eq. (1) (Fricke, 1953)

$$V = 1.5 \, Er \qquad (1)$$

where V is the induced membrane potential (V) at the points perpendicular to the electric field E (V/cm) and r is the radius of the spherical cell (cm), the induced membrane potential at 1.5 kV/cm is ~0.55 V. This induced membrane potential is less than 1 V, the critical potential for electroporation, as mentioned by different authors (Glaser *et al.*, 1988, Kinosita and Tsong, 1977; Benz *et al.*, 1979; Neumann, 1989). This example shows that electroinsertion is optimum at a field strength slightly below the critical field for electroporation E_c.

Regardless of the two basic mechanisms describing the effect of electric fields on cell membranes, namely, the electromechanical model (Crowley, 1973) and the statistical model (Powell *et al.* 1986), we can deduce that the electroinsertion occurs due to the presence of hydrophobic pores, or, more generally, due to the perturbation of the membrane structure. The latter is maximum at field strengths, slightly lower than those required for membrane electroporation (Abidor *et al.*, 1979; Crowley, 1973; Sugar and Neumann, 1984; Dimitrov, 1984; Glaser *et al.*, 1988).

The hypothesis that the electric field-induced perturbation is responsible for the electroinsertion of proteins into cell membranes is compatible with the idea that the insertion with acid or detergent occurs due to the perturbations induced by these agents (Traüble and Eibl, 1974; Enoch *et al.*, 1979; Arvinte *et al.*, 1989).

B. Temperature

The electroinsertion of proteins (glycophorin and CD4) into erythrocytes membrane was substantially greater at 37°C than at 24°C and could not be detected at 4°C for mammalian cells. The temperature dependency of the electroinsertion is compatible with a process requiring membrane mobility, which is a characteristic of the phospholipids of the membrane.

C. Hydrophobic Domain of Proteins

The electroinsertion in our hands is limited to membrane proteins presenting a hydrophobic, membrane-spanning domain such as full-length CD4 or native glycophorin. Proteins without hydrophobic domain such as soluble CD4 (sCD4) (courtesy of SmithKline Beecham Laboratory, USA) or human β_2-microglobulin (β_2-m) could not be electroinserted into erythrocyte membrane as shown by immunofluorescence (Mouneimne *et al.*, 1990). β_2-Microglobulin is a soluble protein (M_r 11,800) that was found associated with cell surface antigens of the major histocompatibility complex (Nakamuro *et al.*, 1973; Peterson *et al.*, 1974). On the other hand, preliminary studies showed that the size of the cytoplasmic domain of the electroinserted membrane proteins were without effect on the efficiency of insertion; electroinsertion of full length recombinant CD4 or CD4 with a truncated cytoplasmic domain (31 C-terminal amino acids deleted) was identical. The requirement for the presence of hydrophobic sequence in the protein to be inserted suggests that hydrophobic pores or structure perturbation of the phospholipid bilayer are decisively involved in this phenomenon.

III. Procedures

A. Reagents

1. *Chemical*

The electroinsertion medium consists of 0.14 M NaCl and 10 mM Na$_2$HPO$_4$/NaH$_2$PO$_4$ adjusted to pH 8.8 by NaOH. The washing buffer contains 0.145 M NaCl and 5 mM phosphate at pH 7.4 (PBS). The anti-CD4 monoclonal antibodies including OKT4A, OKT4C, and OKT4D are provided by P. Rao from Ortho Diagnostic Systems, Inc. (Raritan, NJ). Leu 3a is from Becton Dickinson Immuno-Cytometry System (Mountain View, CA), and the BL4/10T4 is from AMAC, Inc. (Westbrook, ME). Phycoerythrin conjugated affinipure F(ab')2 fragment goat anti-mouse IgG (Gam-PE) is obtained from Jackson Immuno Research Laboratories (West Grove, PA). Antiphycoerythrin monoclonal antibody, clone number PE-85 (Anti-PE), fluorescein isothicyanate isomer I (FITC), bovine serum albumin (BSA), human glycophorin type MM, and human β_2-microglobulin are from Sigma (St. Louis, MO). FITC-labeled rabbit anti-human β_2-microglobulin antibodies are from Accurate Chemical and Scientific Corporation (Westbury, NY). The anti-human glycophorin monoclonal antibodies 10F7 are kindly provided by Dr. R. M. Jensen from the Biomedical Science Division, Livermore National Laboratory, University of California, and has been described by Bigbee *et al.* (1984); D2-10 (mouse isotype IgG1) is from AMAC (Westbrook, ME). Anti FITC antibodies and phycoerythrin labeled avidin are from Molecular Probes (Eugene, OR). NHS biotin is from Serva (Heidelberg, NY). Highly purified lyophylized, full-length recombinant CD4 is prepared in our laboratory (Webb *et al.*, 1989).

2. FITC Labeling of Glycophorin and CD4

FITC was covalently bound to human glycophorin or CD4 according to Goldman (1968). Briefly, glycophorin at a concentration of 10 mg/ml in 20 mM Tris-HCl, pH 8.9, or CD4 at a concentration of 4 mg/ml in borax/boric acid, pH 8.0, 0.25 M, is incubated with fluorescein isothicyanate (FITC) at a ratio of 100 μg/mg protein. After 4 h on ice, free fluorescein is removed by chromatography on Sephadex G25 (Pharmacia). The labeled protein suspension is then collected, concentrated (Amicon Micropartition System), and lyophilized.

B. Instrumentation

1. Pulse Generator

The electroinsertion requires the sample to be subjected to several electrical pulses of defined strength and width. Therefore, the square pulse is preferred to the exponential decay pulse because the latter depends mostly on the sample. For this purpose, we use a square pulse generator Cober 606 (Cober Electronics, Stamford, CT), which provides a constant and uniform voltage (within 4%), with a rise time of 100 ns and fall time of 35 ns. Voltage and current are monitored through a voltage divider by a Nicolet 2090 digital storage oscilloscope (Nicolet, Madison, WI).

2. Chamber

The sample is pulsed in a chamber with parallel electrodes to provide uniform electric field. We use a Cober chamber (P/N ST 126-26), which consists of a Teflon cylinder of 1.2 cm diameter with each end formed by 1.2×2.5 cm stainless steel flat electrodes, the electrodes gap being variable (0.2–1 cm). Electroporation curvettes with aluminum electrodes from BRL (Gaitherburg, MD), P/N 1601 AB, are used also, after modification in our laboratory, in order to have parallel electrodes with a 3.5-mm gap (factory electrodes have 24% divergence), and to reduce the volume unexposed to the electric field. To work on a small sample volume (25 μl) we use the BRL disposable microelectroporation chamber (P/N 1608 AJ), after modification to improve sample holding between the electrodes' bosses. Parallel stainless-steel electrodes, cuvette electrodes BTX 471,470,474, with electrode gaps of 3.5, 1.9, and 0.5 mm (BTX San Diego, CA) are also used. These electrodes fit firmly into disposable semimicrocuvettes (VWR Scientific, P/N 58017-847).

During the experiment, the electroinsertion chamber was maintained at 37°C.

C. Electroinsertion

Electroinsertion medium should be isoosmolar. In early experiments we used mannitol (Mouneimne *et al.*, 1989) in order to have low-conductivity medium, since low-conductivity medium is more effective in delivering most of the pulse energy to the cells. However, prolonged incubation of the erythrocytes at 37°C in mannitol induces lysis. Presently, isotonic 10 mM phosphate-buffered saline is used in our laboratory.

Before pulse, the erythrocyte–protein suspension is preincubated on ice in order to establish close contact between the protein and the cells. This contact can be enhanced by adding divalent cations to the insertion medium (0.5 mMCa^{2+}, 0.5 mM Mg^{2+} or 0.1 mM Mn^{2+}). The overall efficiency of electroinsertion is higher when these ions are not used.

Incubation at 37°C after pulse application appears necessary for cell recovery and enhancement of insertion efficiency. All the cells subjected to one electric pulse expose inserted proteins, but the application of several pulses is required to increase the number of inserted molecules per cell.

Considering the previous remarks, the actual procedure of protein electroinsertion into erythrocytes membrane is as follows:

1. Human erythrocytes are separated from fresh whole blood of donors with citrate buffer as anticoagulant. Mouse erythrocytes are separated from heparinized fresh whole blood obtained from BALB/c strain mouse by retro-orbital sinus puncture. Rabbit erythrocytes are separated from fresh whole blood obtained from rabbits

by marginal ear vein puncture in citrate buffer as anticoagulant. Pig erythrocytes are collected in citrate buffer from the jugular vein.

2. Wash erythroyctes three times with the electroinsertion medium and prepare an erythrocyte stock suspension.
3. Dissolve the lyophilized protein in the electroinsertion medium. In the case of CD4, the concentration is 1 mg/ml, and for glycophorin the concentration is 20 mg/ml.
4. Add the protein suspension to the erythrocyte suspension.
5. Incubate erythrocytes and proteins in suspension 20 min on ice.
6. Raise the sample temperature to 37°C and apply four square electrical pulses of 1 ms duration at 15-min intervals. The field intensity depends on the type of erythrocyte: 1.3 kV/m for human, 1.5 kV/cm for piglet, 1.6 kV/cm for rabbit, and 2.1 kV/cm for mouse erythrocyte.
7. Incubate 1 h at 37°C.
8. Wash the erythrocytes 3 times in PBS, pH 7.4.

IV. Detection of Electroinserted Proteins

A. Presence of Immunologically Active Epitopes

The reaction with monoclonal antibodies (mAb) of the electroinserted CD4 in erythrocyte membranes shows that CD4 exposes the different active epitopes: Leu3a, OKT4A, OKT4C, OKT4D, and BL4/10T4. These mAb are specific for different epitopes of the CD4 molecule (Rao *et al.*, 1983; Sattentau *et al.*, 1986). The first two epitopes are located in the region of binding of the HIV-1 envelope protein gp120 (Jameson *et al.*, 1988; Landau *et al.*, 1988).

Knowing that these epitopes are distributed over all the external sequence of CD4 (Sattentau *et al.*, 1986; Jameson *et al.*, 1988; Mizukami *et al.*, 1988; Peterson and Seed, 1988) and that the fluorescence intensity, measured by flow cytometry, upon reaction of each mAb separately with the same RBC-CD4 sample is identical, it appears that electroinserted CD4 exposes the different active epitopes in the proper orientation and ratio. This further supports the view that adsorption of CD4 is unlikely to be the result of subjecting RBC and CD4 to electrical fields. Should the CD4 molecules have been adsorbed, some epitopes would have been less detectable than others. Finally, the presence of the different epitopes indicates that the CD4 molecule is not denatured during the electroinsertion procedure.

Similar results are obtained with the human glycophorin electroinserted into animal erythrocytes (RBC-Glyc) when reacted with D2-10 and 10F7 mAb (Mouneimne *et al.*, 1991).

B. Mobility of Electroinserted Proteins (Patching)

Examination of the RBC-CD4 or RBC-Glyc under the fluorescent microscope upon reaction with mAb show a punctuate fluorescent pattern that is characteristic of the patching of the inserted proteins upon reaction with mAb (Fig. 3). FITC-labeled proteins show uniformly fluorescent cells and show the patching only after reaction with mAb (Mouneimne et al., 1991). This indicates that the patchy distribution does not occur upon insertion of the proteins; it only happens due to antibody cross-linking and is possible because of the lateral mobility of the inserted proteins. Considering that fluorescent probes adsorbed on cell surfaces do not display lateral mobility (Szoka et al., 1980), this is an indication that electroinsertion implants the protein molecules in the cells' lipid membrane. The patching observed with RBC-CD4 upon reaction with anti CD4 mAb is similar to that observed in CEM cells (a line of human T-lymphoblasts) under the same conditions.

C. Quantitative Analysis by Flow Cytometry

Flow cytometry analysis of RBC-CD4 shows a single fluorescence peak with a complete shift toward the higher fluorescence intensity with respect to the control (BSA replaces CD4 or glycophorin) (Fig. 4). This means that the erythrocyte population subjected to the electroinsertion procedure exposes up to 7000 CD4 epitopes inserted per mouse RBC (Nicolau et al., 1990), 8000 per human RBC, and 2000 per rabbit RBC. For human glycophorin, we observed up to 10,000 epitopes per mouse RBC and 4000 per rabbit RBC.

D. Arguments against Adsorption

The strength of the membrane attachment of inserted CD4 was first assayed as the capacity of RBC-CD4 to accumulate successive immune complexes (Guy et al., 1988). Briefly, RBC-CD4 already stained with OKT4D and Gam-PE were reacted with a mouse anti-phycoerythrin (APE) mAb (clone PE-85), and then reacted again with Gam-PE. The enhancement of the fluorescence signal is quantified by flow cytometry for each cycle of Phycoerythrin-anti-phycoerythrin (PEAPE) complex. Figure 5 shows that the inserted CD4 into RBC membrane can withstand four PEAPE complexes up to saturation. This behavior is similar to the native CD4 expressed on the membrane of CEM-CM3 cells (acute lymphoblastic leukemia T-cell line).

This indicates that CD4 molecules are much more tightly attached to the RBC membrane than would be the case with adsorbed molecules. Adsorbed CD4 on the

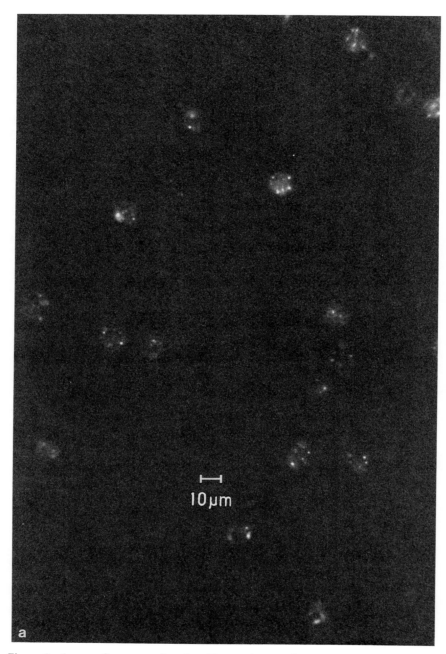

Figure 3 Immunofluorescence imaging: Human RBC-CD4 stained with Leu3a mAb and phycoerythrin conjugated goat anti-mouse (GamPE) mAb. (a) RBC-CD4, excitation wavelength 488 nm; (b) the same RBC-CD4, direct light.

b

Figure 3   (*Continued*)

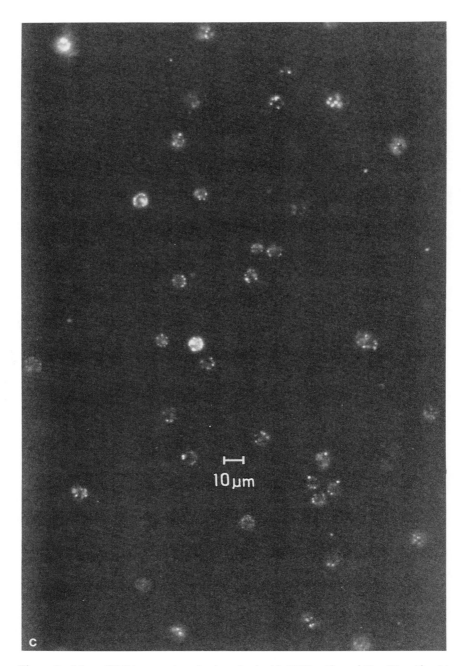

Figure 3 Mouse RBC-human glycophorin stained with 10F7 mAbs and GamPE mAbs. (c) RBC-glycophorin, excitation wave length 488 nm; (d) the same RBC-glycophorin, direct light. Note the patching in (a) and (c) and the normal shape of RBC in (b) and (d).

338

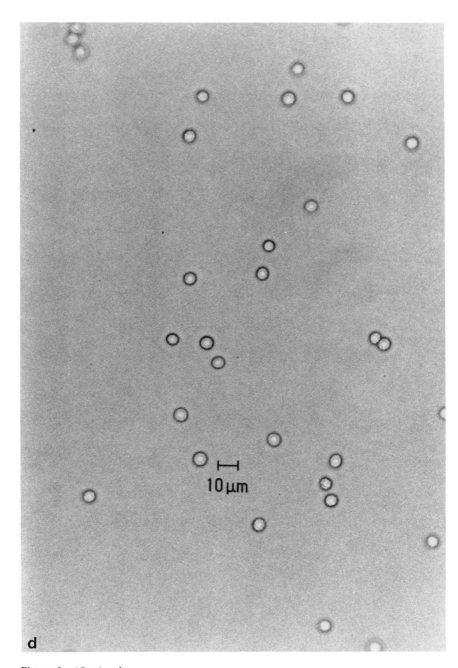

d

Figure 3   (*Continued*)

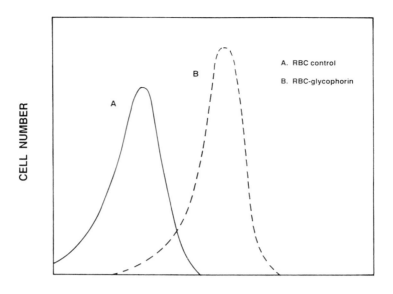

LOG FLUORESCENCE

Figure 4   Flow cytometry histograms (red fluorescence): (A) control RBC; (B) rabbit RBC-glycophorin stained with 10F7 and GamPE mAb. The RBC-glyc had a mean peak channel corresponding to 4000 epitopes per RBC.

surface of RBC would not resist the mechanical stress caused by the four successive immune complexes accumulated and would very probably be desorbed from the RBC surface. Repeated washing at high salt concentration (NaCl, 1 *M*) failed to remove CD4 or glycophorin from RBC-CD4 and RBC-glyc systems. Moreover, in ghosts obtained from mouse RBC-glyc, the inserted proteins are detected by immunoblot analysis. All these arguments make a strong case for insertion of the proteins in the plasma membranes, as opposed to mere adsorption.

V. Orientation of the Electroinserted Proteins

It is important to know the orientation of the electroinserted proteins in the RBC membrane. To this end, FITC-labeled glycophorin was inserted in mouse RBC and FITC-labeled CD4 was inserted into human RBC, and the fluorescence emission intensity was measured by flow cytometry. The RBC-CD4-FITC and RBC-glyc-

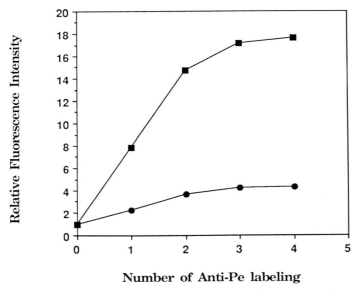

Number of Anti-Pe labeling

Figure 5 Variation of fluorescence intensity with the number of Anti-PE successive labeling of RBC-CD4 (●). Mouse RBC-CD4 and CEM cells (■), labeled with OKT4D and GamPE and successively with Anti-PE and Gam-PE antibodies to saturation. The fluorescence intensity is plotted versus fluorescence intensity of the first GamPE. (From Mouneimne *et al.*, 1990a, with permission.)

FITC were then reacted with anti-FITC antibodies quenchers of the FITC fluorescence exposed, and the fluorescence emission of the cells was measured again.

From the values in Table 1 it appears that 70–74% of the fluorescence of electroinserted FITC-labeled glycophorin and CD4 is quenched by the anti-FITC antibodies, which suggests that this is also the proportion of correctly "exposed" molecules among the electroinserted ones. This result is based on the assumption that only the carbohydrate-containing, extracellular segment of the proteins is labeled with FITC (Rüppel *et al.*, 1982).

The 70–74% "correct" orientation is compatible with the results of Van Zoelen *et al.* (1978), showing that glycophorin incorporated in lipid vesicles is oriented in an asymetric way with 75–80% of its sugar residues directed to the outside of the vesicles.

A tentative explanation of this result may be that, since the hydrophobic pores generated in the membrane by the electric field are probably the sites of electroinsertion, the bulky, heavily hydrated carbohydrate moieties in both CD4 and

Table 1

Quenching of Fluorescence of Electroinserted FITC-
Labeled Proteins by anti-FITC-Antibodies, Assayed
by Flow Cytometry[a]

Sample	F %	Q %
Human RBC-CD4-FITC	77.2	
Human RBC-CD4-FITC-Anti FITC	20.0	74.1
Mouse RBC-Glyc-FITC	33.3	
Mouse RBC-Glyc-FITC-Anti FITC	10.0	70.0

[a]F %: Mean peak fluorescence intensity expressed as percentage
of the shift with respect to the control RBC (where BSA replaces
the glycophorin and CD4). Q %: Quenching %, percent ratio
of quenched intensity to the intensity before reaction with anti-
FITC. This number represents the percentage of the right-side-
out inserted proteins.

glycophorin determine the predominantly outward orientation of the proteins in the
RBC membrane.

VI. Life-Span Study

In order to verify that electroinsertion did not significantly damage the erythrocyte
membrane, their life span *in vivo* was measured. The life-span study is also important
for the assessment of the stability of the inserted proteins in the membrane of
erythrocytes in circulation.

Mouse RBC-glyc were biotinylated according to Suzuki and Dale (1987), and
injected into mice. Samples of blood were assayed for the presence of RBC-glyc by
phycoerythrin-avidin and for the glycophorin by phycoerythrin stained monoclonal
antiglycophorin antibodies, using the flow cytometry.

The results of this study indicated (Mouneimne *et al.*, 1991):

1. The half-life time of the RBC is ~12 days, the normal value for mouse erythrocytes, in agreement with Kinosita and Tsong (1978).
2. Immunologically active, electroinserted human glycophorin exposed on the RBC
   membrane shows the same half-life time as the RBC in mouse

Electroinsertion appears not to affect the life span of RBC, and the inserted
proteins are stable and present in circulation as long as the RBC survive.

VII. Electroinsertion in Nucleated Cell Plasma Membrane

Preliminary results of the insertion of glycophorin into CEM cells (human T-lymphoblast cell line), SF9 insect cells (*Spodoptera frugiperda*), and primary cell G1/G3 (bovine T8) indicate that up to 10^5 glycophorin molecules per cell could be inserted in the plasma membrane, as measured by flow cytometry. (Results not shown.)

It is important to mention that for some cell lines the temperature of electroinsertion is not 37°C but that of cell growth. For example, in the case of the electroinsertion of glycophorin into SF9 cells the electroinsertion is performed at 27°C, the temperature at which these cells grow in airlift fermentors.

VIII. Conclusions

In this chapter we show that subjecting the cells to a pulsed electrical field in the presence of foreign membrane proteins, under certain conditions, results in the insertion of these proteins in the plasma membrane of the cells. We show that electroinsertion is dependent on the electric field strength, and it is optimal at values below the critical electric field for electroporation of cells. The temperature dependency observed and the failure to insert membrane proteins having a hydrophobic spanning domain suggest the involvement of the hydrophobic pores or structure perturbations in the phospholipid bilayer in this phenomenon.

The inserted proteins are immunologically active and present different epitopes. Upon reaction with monoclonal antibodies, patching is observed, indicating lateral mobility of the inserted molecules.

Experiments on quenching the fluorescence of FITC-labeled inserted proteins suggest that 70–74% of the inserted proteins are "correctly" oriented. The relatively large number of epitopes detected per erythrocyte, the normal life span of mouse RBC (half-life = 12 days), and the *in vivo* stability of the inserted glycophorin with the same half-life indicate that red blood cells with specific receptors inserted in their membranes may be used as long-lived, targeted carriers.

The electroinsertion is not limited to red blood cells, as it has been achieved also with nucleated cells at different growing temperature. The large number of inserted glycophorin molecules suggests the possibility of using electroinsertion to study certain physiological properties of these cells and to understand particular aspects of the lipid–protein interactions in the cell plasma membrane.

References

Abidor, I. G., Arakelyan, V. B., Chernomordik, L. V., Chizmadzhev, Y. A., Pastushenko, V. F., and Tarasevich, M. R. (1979). Electric breakdown of bilayer lipid membrane I.

The main experimental facts and their qualitative discussions. *Bioelectrochem. Bioenerg.* 6, 37–52.

Arvinte, T., Schultz, B., Cudd, A., and Nicolau, C. (1989). Low-pH association of proteins with the membranes of intact red blood cells I. Exogenous glycophorin and the CD4 molecule. *Biochim. Biophys. Acta* 981, 51–60.

Arvinte, T., Schulz, B., Madoulet, C., and Nicolau, C. (1990). Red blood cells bearing CD4 bind to gp120 covered plates and aggregate with cells expressing gp120. *J. AIDS* 3/11, 1041–1045.

Benz, R., Beckers, F., and Zimmermann, U. (1979). Reversible electrical breakdown of lipid bilayer membranes: A charge-pulse relaxation study. *J. Membrane Biol.* 48, 181–204.

Bigbee, W. L., Langlais, R. G., Vandelaan, M., and Jensen, R. M. (1984). Binding specificities of eight monoclonal antibodies to human glycophorin A. Studies with McM and MkEn (UK) variant human erythrocytes and M- and MNV-Type chimpanzee erythrocytes. *J. Immunol.* 133, 3149–3155.

Crowley, J. M. (1973). Electrical breakdown of biomolecular lipid membranes as an electromechanical instability. *Biophys. J.* 13, 711–724.

Dimitrov, D. S. (1984). Electric field-induced breakdown of lipid bilayers and cell membranes: A thin viscoelastic film model. *J. Membrane Biol.* 78, 53–60.

Enoch, H. G., Fleming, P. J., and Strittmatter, P. (1979). The binding of cytochrome b5 to phospholipid vesicles and biological membranes. Effect of orientation on intermembrane transfer and digestion by carboxypeptidase Y. *J. Biol. Chem.* 254, 6483–6488.

Fricke, H. (1953). The electric permittivity of a dilute suspension of membrane ellipsoids. *J. Appl. Phys.* 24, 644–646.

Glaser, R. W., Leiken, S. L., Chernomordik, L. V., Pastushenko, V. F., and Sokirko, A. I. (1988). Reversible electrical breakdown of lipid bilayers: Formation and evolution of pores. *Biochim. Biophys. Acta* 940, 275–287.

Goldman, M. (1968). *In* "Fluorescent Antibody Methods," pp. 101–103, 123–124, Academic Press, New York.

Guy, K. Crichton, D. N., and Ross, J. A. (1988). Indirect immunofluorescence labelling with complexes of phycoerythrin and monoclonal anti-phycoerythrin antibodies (PEAPE complexes). *J. Immunol. Methods* 112, 261–265.

Hashimoto, M., Morikawa, H., Yamada, M., and Kimma, A. (1985). A novel method for transformation of intact yeast cells by electroinjection of plasmid DNA. *Appl. Microbiol. Biotechnol.* 21, 336–339.

Huestis, W. H., and Newton, A. C. (1986). Intermembrane protein transfer. Band 3, the erythrocyte anion transporter, transfers in native orientation from human red blood cells into the bilayer of phospholipid vesicles. *J. Biol. Chem.* 261, 16274–16278.

Jameson, B. A., Rao, P. E., Kong, L. I., Hahn, B. H., Shaw, G. M., Hood, L. E., and Kent, S. B. H. (1988). Location and chemical synthesis of a binding site for HIV-1 on the CD4 protein. *Science* 240, 1335–1338.

Kinosita, K., Jr., and Tsong, T. Y. (1977). Voltage induced pore formation and hemolysis of human erythrocytes. *Biochim. Biophys. Acta* 47, 227–242.

Kinosita, K., Jr., and Tsong, T. Y. (1978). Survival of sucrose loaded erythrocytes in circulation. *Nature* 272, 258–260.

Landau, N. R., Warton, M., and Littman, D. R. (1988). The envelope glycoprotein of the human immunodefficiency virus binds to the immunoglobulin-like domain of CD4. *Nature* **334,** 159–162.

Mizukami, T., Fuerst, T. R., Berger, E. A., and Moss, B. (1988). Binding regions for human immunodeficiency virus (HIV) and epitopes for HIV-blocking monoclonal antibodies of the CD4 molecule defined by site directed mutagenesis. *Proc. Natl. Acad. Sci. USA* **85,** 9273–9277.

Mouneimne, Y., Tosi, P. F., Gazitt, Y., and Nicolau, C. (1989). Electro-insertion of xenoglycophorin into the red blood cell membrane. *Biochem. Biophys. Res. Commun.* **159,** 34–40.

Mouneimne, Y., Tosi, P.-F., Barhoumi, R., and Nicolau, C. (1990). Electroinsertion of full length recombinant CD4 into red blood cell membrane. *Biochim. Biophys. Acta* **1027,** 53–58.

Mouneimne, Y., Tosi, P.-F., Barhoumi, R., and Nicolau, C. (1991). Electroinsertion of xeno proteins in red blood cell membranes yields a long lived protein carrier in circulation. *Biochim. Biophys. Acta* **1066,** 83–89.

Nakamuro, K., Tanigaki, N., and Pressman, D. (1973). Multiple common properties of human beta-2-microglobulin and the common portion fragment derived from HL-A antigen molecules. *Proc. Natl. Acad. Sci. USA* **70,** 2863–2865.

Neumann, E, (1989). The relaxation hysteresis of membrane electroporation. *In* "Electroporation and Electrofusion in Cell Biology" (E. Neumann, A. E. Sowers, and C. A. Jordan, eds.), pp. 61–82. Plenum Press, New York.

Neumann, E., Schäfer-Ridder, M., Wang, Y., and Hofschneider, P. H. (1982). Gene transfer into mouse lyoma cells by electroporation in high electric fields. *EMBO J.* **1,** 841–845.

Nicolau, C., Ihler, G. M., Melnick, J. L., Noonan, C., George, S. R., Tosi, F., Arvinte, T., and Cudd, A. (1988). Targeted drug delivery via protein mediated membrane fusion. Specific killing of HIV infected lymphocytes. *J. Cell Biochem.* **12B,** 254.

Nicolau, C., Tosi, P.-F., Arvinte, T., Mouneimne, Y., Cudd, A., Sneed, L., Madoulet, C., Schulz, B., and Barhoumi, R. (1990). CD4 inserted in red blood cell membranes or reconstituted in liposome bilayers as a potential therapeutic agent against AIDS. *In* "Horizons in Membrane Biotechnology" (C. Nicolau and D. Chapman, eds.), pp. 147–177. Wiley-Liss, New York.

Peterson, P. A., Rask, L., and Lindblom, J. B. (1974). Highly purified papain-solubilized HL-A antigens contain beta 2-microglobulin. *Proc. Natl. Acad. Sci. USA* **71,** 35–39.

Peterson, A., and Seed, B. (1988). Genetic analysis of monoclonal antibody of HIV binding sites on the human lymphocyte antigen CD4. *Cell* **54,** 65–72.

Potter, H., Weir, L., and Leder, P. (1984). Enhancer-dependent expression of human K immunoglobulin genes introduced into mouse pre-B lymphocytes by electroporation. *Proc. Natl. Acad. Sci. USA* **81,** 7161–7165.

Powell, K. T., Derrik, E. G., and Weaver, J. C. (1986). A quantitative theory of reversible electrical breakdown in bilayer membranes. *Bioelectrochem. Bioenerg.* **15,** 243–255.

Rao, P. E., Talle, M. A., Kung, P. C., and Goldstein, G. (1983). Five epitopes of a differentiation antigen on human inducer T cells distinguished by monoclonal antibodies. *Cell Immunol.* **80,** 310–319.

Rigaud, J. L., Paternostre, M. T., and Bluzat, A. (1988). Mechanisms of membrane protein insertion into liposomes during reconstitution procedures involving the use of detergent.

2. Incorporation of the light-driven proton pump bacteriorhodopsin. *Biochemistry* **27**, 2677–2688.

Rüppel, D., Kapitza, H. G., Galla, H. J., Sixl, F., and Sackmann, E. (1982). On the microstructure and phase diagram of dimyristoylphosphatidylcholine-glycophorin bilayers. The role of defects and the hydrophilic lipid-protein interaction. *Biochim. Biophys. Acta* **692**, 1–27.

Sattentau, Q. J., Dalgleish, A. G., Weiss, K. A., and Beverley, P. C. L. (1986). Epitopes of the CD4 antigen and HIV infection. *Science* **339**, 1120–1127.

Scotto, A. W., and Zakim, D. (1988). Reconstitution of membrane proteins. Spontaneous incorporation of integral membrane proteins into preformed bilayers of pure phospholipids. *J. Biol. Chem.* **263**, 18500–18506.

Sowers, A. E. (1984). Characterization of electric field-induced fusion in erythrocyte ghost membranes. *J. Cell Biol.* **99**, 1989–1996.

Serpersu, E. H., Kinosita, K., Jr., and Tsong, T. Y. (1985). Reversible and irreversible modification of erythrocyte membrane permeability by electric field. *Biochim. Biophys. Acta* **812**, 779–785.

Sugar, I. P., and Neumann, E. (1984). Stochastic model for electric field-induced membrane pores electroporation. *Biophys. Chem.* **19**, 211–225.

Suzuki, R., and Dale, G. L. (1987). Biotinylated erythrocytes: *In vivo* survival and *in vitro* recovery. *Blood* **70**, 791–795.

Szoka, F., Jacobson, K., Derzko, Z., and Papahadjopoulos, D. (1980). Fluorescence studies on the mechanism of liposome-cell interactions in vitro. *Biochim. Biophys. Acta* **600**, 1–18.

Teissié, J., Knutson, V. P., Tsong, T. Y., and Lane, M. D. (1982). Electric pulse-induced fusion of 3T3 cells in monolayer culture. *Science* **216**, 537–538.

Träuble, H., and Eibl, H. (1974). Electrostatic effects on lipid phase transitions: Membrane structure and ionic environment. *Proc. Natl. Acad. Science USA* **71**, 214–219.

Van Zoelen, E. J. J., Verkleij, A. J., Zwaal, R. F. A., and Van Deenen, L. L. M. (1978). Incorporation and asymetric orientation of glycophorin in reconstituted protein-containing vesicles. *Eur. J. Biochem.* **86**, 539–546.

Webb, N. R., Madoulet, C., Tosi, P. F., Broussard, D. R., Sneed, L., Nicolau, C., and Summers, M. D. (1989). Cell surface expression and purification of human CD4 produced in baculovirus-infected insect cells. *Proc. Natl. Acad. Sci. USA* **86**, 7731–7735.

Zalman, L., Wood, L. M., Frank, M. M., and Müller-Eberhard, H. J. (1987). Deficiency of the homologous restriction factor in paroxysmal nocturnal hemoglobinuria. *J. Exp. Med.* **165**, 572–577.

Zeira, M., Tosi, P.-F., Mouneimne, Y., Lazarte, J., Sneed, L., Volsky, D. J., and Nicolau, C. (1991). Full-length CD4 electroinserted in the red blood cell membrane as a long-lived inhibitor of HIV infection. *Proc. Natl. Acad. Sci. USA* **88**, 4409–4413.

Zimmermann, U. (1986). Electrical breakdown, electropermeabilization and electrofusion. *Rev. Physiol. Biochem. Pharmacol.* **105**, 175–256.

21

Electroporation as a Tool to Study Enzyme Activities *in Situ*

R. R. Swezey and D. Epel

Hopkins Marine Station, Stanford University, Department of Biological Sciences,
Pacific Grove, California 93950

I. Introduction

Electroporation, the breakdown of the plasma membrane's permeability barrier by exposing cells to electrical fields, has been used by biologists mainly to (1) introduce genetic material into cells (reviewed in Potter, 1988), and (2) study the means by which membrane secretory events are regulated (Knight and Scrutton, 1986). These types of studies do not exhaust the possible uses of electrically permeabilized cells; once access to the cell interior is established by this approach, many questions concerning cellular regulation may become amenable to experimental analyses. In this chapter, we discuss how our laboratory has used electroporated cells to study the behavior of enzymes, with particular emphasis on how other cell structural components may exert regulatory influences on enzymatic activities. These types of influences may be lost when cells are homogenized in order to assay activities, but,

Guide to Electroporation and Electrofusion
Copyright © 1992 by Academic Press, Inc. All rights of reproduction in any form reserved.

347

as will be described, they appear to be preserved if the assays are carried out using electroporated cells (Swezey and Epel, 1988).

Our experimental system is the sea urchin egg, a cell that undergoes a marked enhancement in physiological activities upon fertilization by sperm. The biological questions of interest to us are what enzymes are specifically affected to cause this upswing in metabolic activities, and what are the mechanisms by which these enzymes are regulated in the egg. We and others (Suprynowicz and Mazia, 1985) have found electroporation to be very useful for assaying enzyme activities with minimal perturbation to the integrity of the cell. In the most extreme preservation of this integrity, we have found that enzyme activities can be monitored *in vivo* by using a reversible electroporation protocol to load labeled substrates into cells.

The theoretical basis of how electric fields perforate plasma membranes will not be discussed here, as this will receive much attention in other chapters in this volume. However, a few characteristics germane to our discussion should be noted. First, a key advantage of electroporation over other permeabilization regimes (i.e., chemical, enzymatic, or mechanical) is that the effect can be selective for the plasma membrane, leaving intracellular membrane systems unaffected (Knight and Baker, 1982). This allows the experimenter to have access to the cytoplasm, with minimal mixing of intracellular compartments. Second, the effect is localized on the plasma membranes where the electrical field strength experienced by the cell is the greatest, that is, at the cell poles closest to the apparatus electrodes (Knight and Baker, 1982; Kinosita *et al.,* 1988). This allows the introduction of discrete lesions on the cell surface, permeabilizing the cells without causing large-scale disruption that may weaken the cell surface to the point of cellular lysis. Microscopic examination of electrically permeabilized cells shows a uniform population with respect to permeability to small molecules results from this treatment (Swezey and Epel, 1988), and that the ultrastructure of these cells is little compromised compared to untreated cells (Epel and Swezey, 1990). Third, electroporation is a physically based approach. Therefore, no foreign molecules, such as detergents or pore-forming agents, are introduced into the system that could potentially interfere with the events to be monitored within the permeabilized cells.

III. Experimental Considerations

A. Permeabilization Apparatus

Figure 1 (top) shows a schematic of the apparatus we have used to electroporate sea urchin eggs. It utilizes the capacitor discharge approach, which means that the cells in the permeabilization chamber are exposed to electrical fields that decay exponentially upon reaching their peak strength. A standard power supply capable of generating 1500 V is used to charge a capacitor (nonelectrolytic); using a DPDT switch,

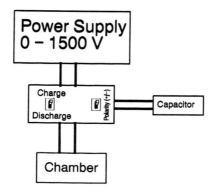

CELL SUSPENSION

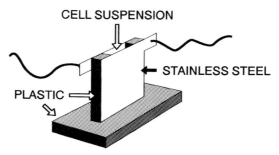

PERMEABILIZATION CHAMBER

Figure 1 Electroporation apparatus. (Top) Schematic diagram of the electrical components and connections in our homemade electroporation device. A more detailed wiring diagram is available upon request to the authors. (Bottom) Design of the electroporation chamber. The stainless-steel plates are held to the plastic spacers with epoxy glue, and the edges can be covered with silicon sealant to prevent leakages.

the capacitor is then disconnected from the power source and connected across the permeabilization chamber electrodes. The duration of the field across the chamber electrodes is a function of (1) the size of the capacitor and (2) the resistance of the system. The resistance in turn is proportional to (1) the ionic composition of the permeabilization medium used and (2) the geometry of the chamber (the greater the distance between the electrodes for any given volume of permeabilization medium, the greater the resistance of the system). The field decays by $1/e$ of its strength over a period defined by the product of the resistance (R) and the capacitance (C):

$t_{1/e} = R \times C$ (Knight and Baker, 1982). Once the capacitor has been discharged it is reconnected to the power source to recharge it for the next pulse. Between pulses we reverse the polarity of the applied field, using another DPDT switch, as this minimizes electrolytic damage to the chamber electrodes.

The design of the permeabilization chamber is given in Fig. 1 (bottom). It is constructed from two stainless steel plates, cut from a conventional pancake flipper, that are glued (epoxy) to plastic spacers that serve as the other walls of the chamber, and this assembly is then glued to a plastic base. Wires then connect the chamber electrodes to the switchbox, and the suspension of cells in the permeabilization medium is added/withdrawn from the top of the chamber. The width of the plastic spacers is an important consideration in the applied field strength (V/cm), and thus high field strengths can be obtained by using narrow spacers. To determine accurately the resistance of the permeabilization chamber containing the suspension of cells in the permeabilization medium, for the purpose of calculating the decay time of the applied field, it is necessary to use an alternating current ohmmeter since with standard ohmmeters the electrodes quickly polarize yielding unreliable values. For those constructing their own such apparatuses, the best DPDT switches available should be used. These are the first components to fail, especially when voltages above 1000 V are used (when inexpensive switches fail they have the annoying tendency to sometimes discharge into the experimenter rather than into the permeabilization chamber).

B. Permeabilization Medium

A key consideration in electroporation is the choice of the extracellular medium as this will affect the lifetime of the electropores and the health status of the cells while they are in the permeabilized condition (Hughes and Crawford, 1989; Rols and Teissie, 1989). Our medium for suspending cells to be permeabilized has been designed to mimic the intracellular milieu of the sea urchin egg with respect to ionic composition and osmotic strength. Thus, the intracellular constitutents will equilibrate with the medium and minimize any perturbations to the cells. The hope is that this will provide the best chance that the results we obtain are physiologically meaningful. The exact compositions of the permeabilization media used have been published elsewhere (Swezey and Epel, 1988; Swezey and Epel, 1989), and they are characterized by having low contents of Na^+, Cl^-, and Mg^{2+}, high contents of K^+ and glycine, and they are devoid of calcium. This last consideration is important since we found that Ca^{2+} at millimolar levels is sufficient to reseal electropores in sea urchin eggs (Swezey and Epel, 1989).

Other workers have used very different media with success in electroporating other cell types. The media tend to be one of two types: (1) low ionic strength media made isotonic with sugars (Mehrle et al., 1985; Stopper et al., 1987; Escande-

Gerand *et al.*, 1988; Lopez *et al.*, 1988; Rols and Teissie, 1989), and (2) physiological saline solutions of high ionic strength (Kinosita and Tsong, 1977; Kinosita and Tsong, 1979) whose ionic compositions are quite dissimilar to that of the cytoplasm of most cells. The higher the ionic strength of the medium, the more rapidly the field will decay in capacitor-discharge-based electroporation systems, since the resistance is lower (conductance higher). Furthermore, at high ionic strength, lower electric field intensities are required to permeabilize the plasma membrane than at low ionic strength (Rols and Teissie, 1989).

III. Permeabilized Cell Assays

There are two types of enzyme assays we carry out using electrically permeabilized cells, and they are fundamentally different in aim. The first utilizes high-voltage discharges to render the cells permanently permeable to molecules, which can be of molecular dimensions up to 40 kD (Epel and Swezey, 1990). The permanence of the permeability here allows continuous access of substrates to the enzymes within the cells, and thus a means of assaying activities while the enzymes remain at their normal cellular addresses. The cells are dead, however, after this treatment. The other type of assay is to render the cells reversibly permeable to substrates, such that the electropores can be resealed at later times (usually 2–5 min after their formation), thereby allowing the loading of labeled, normally impermeant, molecules into cells; since the cells remain viable, the *in vivo* fates of the substrates can be determined. Under these conditions, the size of the substrates that can enter the cells is 1000 Da or less (Swezey and Epel, 1989).

A. Permanently Permeabilized Cell Assays

In many ways the assay of enzyme activities using permanently permeabilized cells (referred to as *in situ* assays) is similar to assays carried out in cell homogenates (see Fig. 2); substrate molecules are added to a suspension of permeabilized cells, and then at various times the amount of product formed is determined. The assays can be carried out continuously (e.g., fluorimetric assays of dehydrogenase enzymes) or discontinuously (e.g., stopping reactions with trichloroacetic acid and subsequent chemical analyses of the amount of product). Some important differences between assays *in vitro* and those *in situ* merit consideration.

First, when performing *in situ* assays using suspensions of permeabilized cells, it is necessary to correct for any enzyme activity in the suspension that is catalyzed by enzyme molecules outside of the cells. This activity can arise from enzymes that have escaped the cells either by leakage through the electropores or by cellular lysis. To control for such extracellular activity, a parallel batch of permeabilized cells is centrifuged to remove the cells, and the supernatant is then assayed to determine the contribution of extracellular catalysis in the *in situ* assays with suspensions of permeabilized cells.

Second, assays of enzymes within cells are analogous to assays of enzymes that have been immobilized onto an insoluble matrix. Since the enzymes are not dispersed

Permeabilized Cell Assay

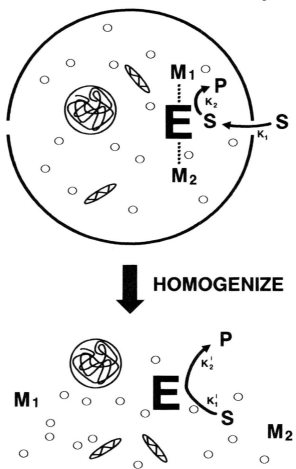

Homogenized Cell Assay

in the medium to the same extent that are substrates are, there can be some effect of the time it takes for substrates to diffuse to the active sites of the intracellular enzymes. This is illustrated in in Fig. 2. Substrate S diffuses to the active site of the enzyme E, with a time constant of k_1. Whether this will be of consequence to the overall rate of product formation, and therefore the enzymatic rate assayed, depends on the relative values of k_1 and k_2 (the rate of product formation from the enzyme–substrate complex). If $k_2 \ll k_1$, then the overall rate is limited by the catalytic steps, and thus the effect of "immobilization" will be negligible. However, if $k_2 \gg k_1$, then the reaction is diffusion limited, and thus the overall rate cannot be equated with the rate of enzyme catalysis, which complicates the analyses of *in situ* activities considerably. The most definitive way to determine whether diffusion is limiting in these assays is to perform a standard kinetic analysis, that is, measure reaction velocities as a function of substrate concentration. If diffusion is limiting, then the concentration of substrate required for half-maximal activity (the apparent k_m) will be greater for assays performed in permeabilized cells than for assays with a solubilized form of the enzyme, as in homogenates. This is because in diffusion-limited reactions the concentration of substrate in the microenvironment of the enzyme (i.e., in the permeabilized cells) will be lower than its concentration in the bulk extracellular phase (Goldstein, 1976). In Fig. 3 we show such a comparison for the enzyme glucose-6-phosphate dehydrogenase (G6PDH) assayed in permeabilized sea urchin eggs (A) and in homogenates of these eggs (B). The same total amount of enzyme was present in both assays. Since the values of the k_m are indistinguishable between the two types of assays, we conclude that diffusion is not limiting for this enzyme in this cell, and therefore the rates observed do correspond to the rates of enzymatic conversion *in situ*.

A third difference between *in situ* and *in vitro* assays is also indicated in Fig. 2. The enzyme in the permeabilized cell exists in a highly structured environment containing a high concentration of proteins, whereas in cell-free homogenates the enzyme is separated from the other cellular constituents to a much greater extent. If these constitutents, labeled M_1 and M_2 in the figure, interact with the enzyme

---

Figure 2   Comparison of assays in permeabilized cells and homogenized cells. This shows the principle of the assay of enzymes (E) within permeabilized cells. Substrate (S) diffuses to the enzyme through the electropores in the membrane with a time constant of k_1, and is converted to product (P) with a time constant of k_2. The enzyme in the permeabilized cells is subject to influences from macromolecular constituents (M_1 and M_2) that may affect k_2. If the interactions between E and M_1/M_2 are inherently weak, the complexes will dissociate upon homogenization, thereby causing the loss of these regulatory effects in the homogenized cell assay. Note that different kinetic constants (k_1' and k_2') are given for the homogenized cell assay. It is probable that $k_1 \le k_1'$, that, substrate diffusion to E is slower for permeabilized cells assays than for homogenized cell assays. The relative values of k_2 and k_2' will depend on whether M_1 and M_2 are activators or inhibitors of E.

PERMEABILIZED CELLS

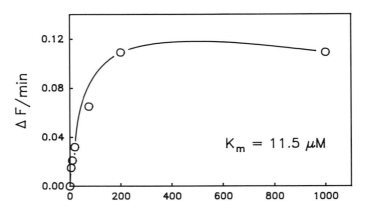

HOMOGENIZED CELLS

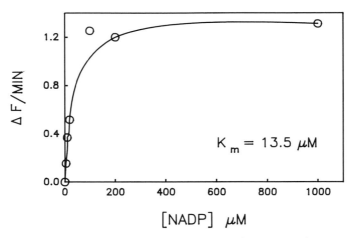

[NADP] μM

Figure 3 Kinetics of G6PDH with respect to cofactor NADP for assays conducted in permeabilized cells (top) and homogenates of permeabilized cells (bottom). Assays were conducted fluorimetrically at 16°C, and contained 400 μM G6P. Note the near identity of the values of K_m, suggesting that the assay in permeabilized cells is not diffusion-limited.

in permeabilized cells to affect its kinetic behavior, then such modulatory interactions may become lost upon homogenization if they are inherently weak in dilute solutions (see Minton, 1983). For the sake of simplicity we have indicated only two cell components (M_1 and M_2) in this figure; of course there could be many more interactions occurring *in vivo*.

Preservation of such interactions that may be an important component of enzymatic regulation is in fact the motivation for developing the *in situ* assay approach for sea urchin eggs. For decades, an understanding of the enzymatic basis for activation of egg metabolism after fertilization had remained obscure because the levels of enzyme activities in homogenates of eggs prepared before and after fertilization were generally identical (Swezey and Epel, 1988). Egg biochemistry was clearly not mirroring egg physiology. However, we found (Table 1) that when enzymes of energy metabolism were assayed in permeabilized eggs, two features were outstanding. First, the activities available in the permeabilized cells (i.e., the V_{max} values) were always much less than the activities present in homogenates prepared from these same permeabilized eggs, suggesting that some modulation (here an inhibition) of activity was occurring in the cell. Second, the activities in permeabilized cells were greater for fertilized eggs than for unfertilized eggs, suggesting that this inhibition was partially lifted after fertilization. This is a clear-cut demonstration of fertilization-dependent increases in the activity of metabolic enzymes in sea urchin eggs. Since these changes are made visible in the permeabilized cells, it now opens

Table 1

Changes in Activities of Sea Urchin Egg Enzymes at Fertilization[a]

Enzyme assay (species)	Permeabilized cell activities (% of total)		Fold increase
	Eggs	Embryos	
Glucose-6-phosphate dehydrogenase (*S. purpuratus*)	1	15	15.0
6-Phosphogluconate dehydrogenase (*S. purpuratus*)	2	23	11.5
Glutathione reductase (*S. purpuratus*)	4	26	6.5
Thymidylate kinase (*L. pictus*)	17	51	3.0
NAD kinase (*S. purpuratus*)	63	71	1.1
Hexokinase (*L. pictus*)	15	53	3.5

[a] All assays contained saturating concentrations of substrates and cofactors (determined empirically or taken from literature values). See Swezey and Epel (1988) for details of the assay protocols.

up a new experimental avenue for determining what the regulatory elements in these eggs are (i.e., what are M_1, M_2, etc.) and how they change after fertilization.

The *in situ* approach therefore has the advantage of permitting activity measurements while the enzyme molecules are subject to the influences of the cellular microenvironment. As noted above, this influence does affect the V_{max} (Table 1), but not the k_m (Fig. 3), of G6PDH. A most striking effect of the microenvironment was seen for G6PDH when the *in situ* and *in vitro* assays were compared as a function of pH (Swezey and Epel, 1988). Here activity in homogenates shows a progressive increase with medium alkalinity such that the rate at pH 9 is fivefold greater than at pH 6. The corresponding pH/activity profile for G6PDH in permeabilized cells shows a near constant rate from pH 6 to 8.5, with a progressive *decrease* in activity at higher pH values. While the cause of this disparity in pH profiles is not known, it is not due to a lessened permeability (i.e., resealing of the pores) in media of higher pH.

B. Reversibly Permeabilized Cell Assays

Since the assays described above involve cells that are no longer alive, it is important to verify that the changes in V_{max} noted in Table 1 do correspond to changes that occur *in vivo*. We have approached this problem by modifying our permeabilization protocol to permit induction of a reversible increase in egg membrane permeability. We use the same apparatus, shown in Fig. 1, for this. The major differences between the two electroporation techniques are that for reversible permeabilization we use lower voltages (125–150 V/cm) than for permanent permeabilization (2000 V/cm), and the permeabilization medium is slightly different in composition (see Swezey and Epel, 1989).

Under these conditions we find that the permeability enhancement is much less pronounced in that the size limit of permeant substances is about 1 kD, and that several minutes are required for full equilibration of labeled molecules between the interior and exterior of the permeabilized cells (Swezey and Epel, 1989). The enhanced permeability persists for at least 20 min after the cells are electroporated, but these cells can be resealed at any time by adding calcium ions to the medium to a concentration of 1 mM. This concentration is optimal in causing rapid closure of the pores without allowing appreciable influx of Ca^{2+} into the eggs, which would alter the physiology considerably. The scheme for *in vivo* measurement of enzyme activity is thus (1) subjecting eggs to reversible electroporation in the presence of labeled substrate, (2) allowing this substrate to diffuse into the egg, (3) sealing these substrate molecules into the eggs by raising the calcium level to 1 mM, and (4) following the rate of conversion of these loaded substrates into product by some appropriate analytical means.

To minimize conversion to product during loading we have found it useful to carry out this step at 0°C, which reduces enzyme activities but has little effect on substrate diffusion. Sea urchin eggs do not permeabilize at 0°C with our standard protocol, but we have found that cells permeabilized at 16°C (the usual temperature used) can be chilled to 0°C and still retain their permeabilized state. Thus, the approach here is to permeabilize at 16°C, transfer cells to a 0°C medium containing

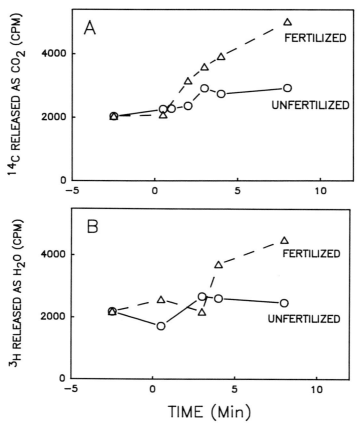

Figure 4   Metabolism of ^{14}C- and ^{3}H-labeled G6P before and after fertilization. Unfertilized sea urchin eggs were loaded with labeled G6P as described previously (Swezey and Epel, 1989), then half of the eggs were fertilized at time zero. At the indicated times, samples were killed with trichloroacetic acid (10% final), and the amount of radiolabel liberated as (A) CO_2 and (B) H_2O was quantified by scintillation counting. Circles represent unfertilized samples, and triangles represent fertilized samples.

FATE OF $(1-{}^{14}C)$ GLUCOSE-6-PHOSPHATE

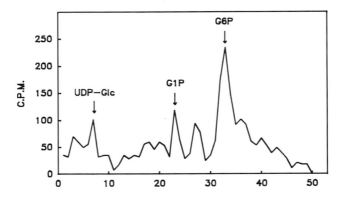

8 MIN AFTER LOADING

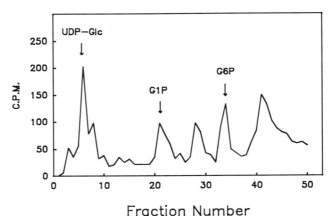

13 MIN AFTER LOADING

Fraction Number

Figure 5 Distribution of radiolabel in acid soluble metabolites in sea urchin eggs loaded with [1-^{14}C]-G6P. Unfertilized sea urchin eggs were loaded with labeled G6P as described previously (Swezey and Epel, 1989), and killed with 5% HClO$_3$ at 8 and 13 min after loading. The aqueous extracts were neutralized with KHCO$_3$, centrifuged to remove KClO$_3$ precipitate, concentrated by evaporation, and subjected to HPLC chromatographic analysis on a weak anion exchange column. Peaks readily identifiable with standards are indicated on the chromatograms. (UDP-Glc = UDP-glucose; G-1-P = glucose-1-phosphate.)

the labeled substrate, then carry out incubation and resealing steps at 0°C before transferring back to a 16°C medium to "start" the assay of enzyme activity.

An example of the type of data obtainable with this approach is shown in Figs. 4 and 5. Glucose 6-phosphate (G6P), some of which was labeled with ^{14}C at the C-1 position and some with ^3H at the C-5 position, was loaded into unfertilized eggs of *Strongylocentrotus purpuratus* at 0°C. The cells were brought back to 16°C, and divided into two batches, one of which was fertilized at time zero. We assayed for the rate of conversion of labeled G6P to ^3H_2O, an index of glycolytic rate, and to $^{14}CO_2$, an index of both glycolysis and pentose shunt, that is, G6PDH and 6-phosphogluconate dehydrogenase (6PGDH) activities, (Fig. 4). Also, the acid-soluble metabolites were analyzed by anion-exchange HPLC (Fig. 5).

The data in Fig. 4 show that fertilization stimulates the oxidation of G6P significantly, and since the increased rate of labeled CO_2 production precedes that of labeled H_2O production, this indicates that the earliest stimulatory events occur in the pentose shunt enzymes. Therefore, the activities of G6PDH and/or 6PGDH do increase after fertilization *in vivo*, which parallels at least qualitatively the increases seen for these enzymes in permanently permeabilized cells (Table 1).

Analysis of the acid-soluble metabolite pool (Fig. 5) showed that G6P depletion is partly due to conversion to substances other than end products, that is, glucose 1-phosphate and UDP-glucose. Accumulation of these metabolites suggests that conversion of sugar phosphates to glycogen is occurring in these eggs, which we have subsequently confirmed (data not shown).

We wanted to dissect whether the increase in pentose shunt activity is due to stimulation of G6PDH or 6PGDH rates. We approached this by trying to load labeled 6-phosphogluconate into unfertilized eggs, and carry out analyses similar to those describe for G6P. As it turned out, the rate of turnover of the 6PG pool is so rapid in the unfertilized eggs, even at 0°C, that we effectively could not load the label into the eggs (it came off as $^{14}CO_2$ during the loading and resealing steps). This prevented us from assaying 6PGDH activity *in vivo* in a straightforward way, but it also challenged us to develop an alternative strategy described below, which we believe will prove to be of more general utility for cell physiologists.

C. Use of "Caged" Compounds

Our approach is to chemically modify the labeled substrate with a blocking group that (1) prevents enzymatic utilization and (2) can be cleaved from the substrate photolytically. The common term for such blocked compounds is "caged" compounds. This cleavage can thus be performed after electroporative loading of the caged substrate by illuminating the cells with an intense flash of light, thereby releasing the labeled substrate within the cell at a discrete time.

We prepared "caged" 6PG using a procedure devised for caging phosphorylated compounds (Walker *et al.*, 1988), and have found that this compound is not utilized by eggs until it has been uncaged by an intense pulse of light (we use a xenon photographic flash source for this). With this modification of the *in vivo* assay technique we find that the turnover of the 6PG pool is complete in less than 7 s in unfertilized eggs. This is at the limit of our experimental temporal resolution, and thus we cannot say at this point whether the turnover rate is in fact greater for fertilized eggs (as would be predicted based on the data in Table 1). However, since the turnover of the G6P pool is on the order of several minutes (Fig. 5), whereas that of the 6PG pool is of seconds, it appears that the rate-limiting enzyme for the increased oxidation of G6P after fertilization (Fig. 4) must be G6PDH.

IV. Conclusions

We have shown that electroporation can be used to probe the activity of enzymes without having to remove them from their cellular environment during the measurements. This permits investigators to detect regulatory interactions that exist in the intracellular milieu, especially interactions that lack sufficient strength to persist through the dilution and disruption inherent to homogenization of cells. While we have stressed the use of electroporated cells in the assay of single enzymes, we mention here that this technique has been successful for measuring *in situ* rates of more complex systems in sea urchin eggs, such as synthesis of protein and DNA (Swezey and Epel, 1987).

With reversibly electroporated cells it should be possible to ascertain the velocity *in vivo* of any given reaction with substrates smaller than 1 kD; by using caged substrates the temporal resolution of the measurements becomes much more precise. This is an exciting breakthrough. Since most metabolites are by nature membrane-impermeant it has not been possible up to now to load them directly into cells to follow the rates of their utilization. With this new approach it should be possible to discern more definitively where the points of metabolic regulation are in cells, and to ask questions about how cell structure may affect the metabolic status of cells.

References

Epel, D., and Swezey, R. R. (1990). A role for structural changes in cell activation at fertilization: Differential inhibition of enzymatic activities. *In* "Structural and Organizational Aspects of Metabolic Regulation" (P. A. Srere, M. E. Jones, and C. K. Mathews, eds.), Wiley-Liss, New York, pp. 13–25.

Escande-Gerand, M. L., Rols, M. P., Dupont, M. A., Gas, N., and Teissie, J. (1988). Reversible plasma membrane ultrastructural changes correlated with electropermeabilization in Chinese hamster ovary cells. *Biochim. Biophys. Acta* **939**, 247–259.

Goldstein, L. (1976). Kinetic behavior of immobilized enzymes systems. *Methods Enzymol.* 44, 397–443.

Hughes, K., and Crawford, N. (1989). Reversible electropermeabilisation of human and rat blood platelets: evaluation of morphological and functional integrity *in vitro* and *in vivo*. *Biochim. Biophys. Acta* 981, 277–287.

Kinosita, K., Jr., and Tsong, T. Y. (1977). Formation and resealing of pores of controlled sizes in human erythrocyte membrane. *Nature (Lond.)* 268, 438–441.

Kinosita, K., Jr., and Tsong, T. Y. (1979). Hemolysis of human erythrocytes by a transient electric field. *Proc. Natl. Acad. Sci. (USA)* 74, 1923–1927.

Kinosita, K., Jr., Ashikawa, I., Saita, N., Yoshimura, H., Itoh, H., Nagayama, K., and Ikegami, A. (1988). Electroporation of cell membrane visualized under a pulsed-laser fluorescence microscope. *Biophys. J.* 53, 1015–1019.

Knight, D. E., and Baker, P. F. (1982). Calcium-dependence of catecholamine release from bovine adrenal medullary cells after exposure to intense electric fields. *J. Membrane Biol.* 68, 107–140.

Knight, D. E., and Scrutton, M. C. (1986). Gaining access to the cytosol: the technique and some applications of electropermeabilization. *Biochem. J.* 234, 497–506.

Lopez, A., Rols, M. P., and Teissie, J. (1988). ^{31}P NMR analysis of membrane phospholipid organization in viable, reversible electropermeabilized Chinese hamster ovary cells. *Biochemistry* 27, 1222–1228.

Mehrle, W., Zimmermann, U., and Hampp, R. (1985). Evidence for asymmetrical uptake of fluorescent dyes through electropermeabilized membranes of *Avena* mesophyll protoplasts. *FEBS Lett.* 185, 89–94.

Minton, A. P. (1983). The effect of volume occupancy upon the thermodynamic activity of proteins: some biochemical consequences. *Mol. Cell. Biochem.* 55, 119–140.

Potter, H. (1988). Electroporation in biology: methods, applications, and instrumentation. *Anal. Biochem.* 174, 361–373.

Rols, M.-P., and Teissie, J. (1989). Ionic-strength modulation of electrically induced permeabilization and associated fusion of mammalian cells. *Eur. J. Biochem.* 179, 109–115.

Stopper, H., Hones, H., and Zimmermann, U. (1987). Large scale transfection of mouse L-cells by electropermeabilization. *Biochim. Biophys. Acta* 900, 38–44.

Suprynowicz, F. A., and Mazia, D. (1985). Fluctuation of the Ca^{2+}-sequestering activity of permeabilized sea urchin embryos during the cell cycle. *Proc. Natl. Acad. Sci. USA* 82, 2389–2393.

Swezey, R. R., and Epel, D. (1987). Regulation of egg metabolism at fertilization. *In* "Molecular Biology of Invertebrate Development" (J. D. O'Connor, ed.), Alan R. Liss, New York, pp. 71–85.

Swezey, R. R., and Epel, D. (1988). Enzyme stimulation upon fertilization is revealed in electrically permeabilized sea urchin eggs. *Proc. Natl. Acad. Sci. USA* 85, 812–816.

Swezey, R. R., and Epel, D. (1989). Stable, resealable pores formed in sea urchin eggs by electric discharge (electroporation) permit substrate loading for assay of enzymes *in vivo*. *Cell Regul.* 1, 65–74.

Walker, J. W., Reid, G. P., McCray, J. A., and Trentham, D. R. (1988). Photolabile 1-(2-nitrophenyl)ethyl phosphate esters of adenine nucleotide analogues. Synthesis and mechanism of photolysis. *J. Am. Chem. Soc.* 110, 7170–7177.

22

Comparison of PEG-Induced and Electric Field-Mediated Cell Fusion in the Generation of Monoclonal Antibodies against a Variety of Soluble and Cellular Antigens

Uwe Karsten,[1] Peter Stolley,[1] and Bertolt Seidel[2]

[1]Central Institute of Molecular Biology, D-(O)-1115 Berlin-Buch, Germany
[2]Institute of Neurobiology and Brain Research, D-(O)-3090 Magdeburg, Germany

I. Introduction
II. Electrofusion as Presently Used in Berlin-Buch
  A. Equipment
  B. Methods
III. Results
IV. Conclusions
  References

I. Introduction

Electric field-mediated cell fusions, although already established in other areas of research for many years [1], have only been more recently employed as a means to generate hybridomas [2–5]. Our first successful experiment was performed in April 1984 [5]. The environment leading to success was, in our case, provided by the availability of an effective instrument, specific experiences of one of us (P.S.) with this instrument in a number of model systems, and our long-term experience with cell cultivation and the hybridoma technology in general. We were surprised by the obvious advantage of electrofusion over polyethylene glycol (PEG) mediated fusions, which was confirmed in later experiments by ourselves [6, 10] as well as by others [7–9]. This chapter is intended to outline our version of electrofusion as presently employed, and to discuss what we consider to be essential and/or critical steps in its application.

II. Electrofusion as Presently Used in Berlin-Buch

A. Equipment

The instrument used was developed and built in limited numbers at the Division of Research Technology of the Central Institute of Molecular Biology, Berlin-Buch. Although designed for manual adjustment, it is still a powerful and versatile instrument. Its technical parameters are:

Dielectrophoresis:	Frequency 0.5, 1.0, or 2.0 MHz
	Amplitude 0–40 V
Pulse (square):	Voltage 0–300 V
	Pulse length 1–15 μs

Several types of fusion chambers were constructed. The simplest was the most successful. It consists of a space of 28 mm in diameter and 0.5 mm in height provided by two flat, stainless-steel electrode plates, which are held apart by an insulating ring with a thickness of 0.5 mm (Fig. 1, left). In the final version (Fig.

Figure 1   Fusion chamber used for electrofusions. Left, laboratory version, opened. Right, final version, outer view.

1, right) this chamber is easily assembled from its parts in the sterile room. The cell suspension is placed as a small drop in the center of the lower electrode before the upper electrode is put in place. The electrode plates and insulating rings are exchangeable and can be replaced after each fusion by a new set of sterile parts, thus allowing a series of fusions to be performed in sequence. This type of chamber provides a homogeneous electric field. Nevertheless, cell alignment does occur in this situation due to field inhomogeneity within the cells themselves, provided that the conductivity of the medium differs from that of the cytoplasm. Our fusion chamber is well suited for batch-type fusions and accepts a higher number of cells than others designed to provide an inhomogeneous field *per se.*

B. Methods

1. Electrofusion

As a general rule, the methods developed by Zimmermann [1, 2] were employed. A detailed protocol is given in Part 3 of this handbook. In brief, spleen cells taken from immunized mice are depleted from erythrocytes and thrombocytes by density gradient centrifugation, washed, mixed at a ratio of 2 : 1 with pronase-treated plasmocytoma cells, transferred to low-ionic-strength medium BUW-8401 [2], and suspended in this medium at high cell density (10^5 cells/μl). This cell mixture is placed in the fusion chamber and treated as follows:

Pretreatment (dielectrophoresis): 2 MHz, 500 V/cm, 10 s
One single pulse is given while the dielectrophoresis is interrupted: 5 kV/cm, 4 μs
Posttreatment: (dielectrophoresis): 2 MHz, 500 V/cm, 5 s

These conditions have been found the most favorable in a number of experiments performed with the fusion chamber as described above. However, we would like to stress that the parameters should be reevaluated and adapted to each type of chamber individually, and, in principle, to each type of cell to be fused. Moreover, the electrical parameters must be controlled by an oscilloscope. Among the variables examined, we found so far in no case an advantage in applying two or more pulses.

2. Cell Culture

This brief chapter is not intended to give specific procedural advice (we refer in this respect to Part 3), but instead meant to stress the importance of considering, in all phases of the fusion experiment, the aspects of cell biology. The total procedure should be kept as short as possible, and pipetting and centrifugal stress reduced to the minimal necessary extent. The principles of good tissue culture practice have to be obeyed throughout. The parental myeloma cell line must be free from mycoplasma, and other mycoplasma-infected cell lines should never be kept alongside with myeloma or hybridoma cells in the same room. After fusion, cells should be transferred without delay to "normal" medium and kept in a CO_2 incubator for some hours before being distributed to microtest plates. An essential point is that most newly generated hybridomas are lymphokine dependent. This is especially important in the case of electrofusion. Therefore, feeder cells or an equivalent have to be employed during hybridoma selection, cloning, and growing up, at least as long as the clones have not yet been frozen.

III. Results

Since 1984, hybridomas producing monoclonal antibodies have been routinely generated by electrofusion in our (two) laboratories. In a considerable number of cases a conventional PEG fusion was performed in parallel. Many types of antigens have been used, from purified proteins to viruses and whole cells. If one compares the results, two advantages of electrofusion are obvious. First, the yield, calculated as the number of growing colonies per 10^5 splenocytes, was consistently higher. In a total of 36 experiments in which both techniques were compared, the mean hybridoma yield amounted to 2.46 ± 2.55 in the case of electrofusions, whereas only 0.29 ± 0.37 hybridoma colonies per 10^5 spleen cells were counted in PEG-induced fusions. The difference is statistically highly significant ($p < 1\%$). This effect was independent of a number of experimental settings such as different batches of PEG, rat or mouse splenocytes, the type of serum used (FCS or horse), or the selection medium employed (azaserine vs. HAT). If the hybridoma yield is compared in each individual experiment, the difference becomes even more pronounced. Furthermore, there exists a significant correlation between both methods in individual experiments [10]. Taken together, this means that, although fusions differ in their outcome depending on the type of antigen and details of immunization, electrofusion leads to better results in terms of quantity. Since the percentage of hybridomas producing specific monoclonal antibodies is not diminished in electric field-mediated fusions, the result is a real increase in useful hybridomas. The second difference between both techniques is a qualitative one. We observed that hybridomas arising from electrofusion appeared earlier and grew faster during the first weeks. Although the reason for this effect is at present unknown, it was very welcome. Otherwise, we found no differences between hybridomas generated by either technique. This applies, for instance, to the patterns of isotypes, antibody titers, the stability of antibody production, or the transplantability. The reproducibility of electrofusions performed in parallel (with aliquots of the same cell suspension) was good (Table 1). In some of the hybridomas generated by electrofusion we analyzed the DNA distribution patterns by flow cytometry employing ethidium bromide and olivomycin

Table 1

Experimental Reproducibility of Electrofusions

Experiment number	Immunogen	Hybridoma yield[a]	
		Individual	Mean
1	Transformed rat fibroblasts	5.7	
		4.9	5.3
2	Human milk fat globule membranes	5.4	
		5.2	
		5.5	5.4
3	Human asialo erythrocytes	5.5	
		6.9	6.2

[a]Number of hybridoma colonies obtained from 10^5 splenocytes. Aliquots of cells were treated under identical conditions.

[5] and the stability of the DNA content per cell during the critical period, the first half year (Fig. 2). As the example in Fig. 2 shows, hybridoma A53-B/A2 had a stable DNA content per cell from the second month onward during 6 months of uninterrupted cultivation without recloning. The DNA content of the cells of this clone amounted to about 1.5 times that of the parental myeloma cells. Taking into consideration the well-known segregation phenomenon in cell hybrids, this indicates

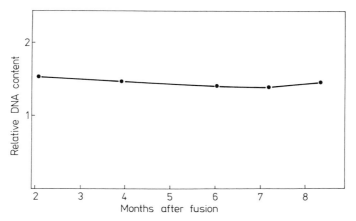

Figure 2   Relative DNA content of hybridoma clone A53-B/A2, generated by electrofusion. Flow cytometry; DNA content of parental myeloma cells (X63-Ag8.653) = 1.

that A53-B/A2 is the result of a two-cell fusion. Another hybridoma obtained from the same experiment, A53-B/A8, had a DNA content corresponding to a three-cell product. This clone grows remarkably well and produces very high titers of antibody.

IV. Conclusions

Electric field-mediated fusions are superior to PEG-mediated fusions for the formation of monoclonal antibody producing hybridomas in terms of hybridoma yield and growth properties in early stages after fusion. Other groups have independently come to similar conclusions [7–9,11]. By using hypo-osmolar solutions the number of hybridomas obtained by electrofusion can be further enhanced [12]. The reason for the better performance of electrofusion is at present not clear. One obvious experiment would be to fuse cells prepared for electrofusion (i.e., suspended in low ionic strength medium) by both electric field pulses and PEG. The more general advantage of electrofusion (i.e., strictly controllable fusion conditions) requires efforts to adapt these conditions very carefully to a given situation. In any case, the aspects of cell biology must be given high priority.

References

1. Zimmermann, U. (1982). Electric field-mediated fusion and related electrical phenomena. *Biochim Biophys. Acta* **694**, 227–277.
2. Zimmermann, U. (1987). Electrofusion of cells. *In* "Methods of Hybridoma Formation" (A. H. Bartal and Y. Hirshaut., eds.), pp. 97–147. Humana Press, Clifton, NJ.
3. Bischoff, R., Eisert, R. M., Schedel, I., Vienken, J., and Zimmermann, U. (1982). Human hybridoma cells produced by electrofusion. *FEBS Lett.* **147**, 64–68.
4. Lo, M. M. S., Tsong, T. Y., Conrad, M. K., Strittmatter, S. M., Hester, L. D., and Snyder, S. H. (1984). Monoclonal antibody production by receptor-mediated electrically induced cell fusion. *Nature (Lond.)* **310**, 792–794.
5. Karsten, U., Papsdorf, G., Roloff, G., Stolley, P., Abel, H., Walther, I., and Weiss, H. (1985). Monoclonal anti-cytokeratin antibody from a hybridoma clone generated by electrofusion. *Eur. J. Cancer Clin. Oncol.*, **21**, 733–740.
6. Karsten, U., Stolley, P., Walther, I., Papsdorf, G., Weber, S., Conrad, K., Pasternak, L., and Kopp, J. (1988). Direct comparison of electric field-mediated and PEG-mediated cell fusion for the generation of antibody producing hybridomas. *Hybridoma* **7**, 627–633.
7. Ohnishi, K., Chiba, J., Goto, Y., and Tokunaga, T. (1987). Improvement in the basic technology of electrofusion for generation of antibody-producing hybridomas. *J. Immunol. Methods* **100**, 181–189.
8. Van Duijn, G., Langedijk, J. P. M., de Boer, M., and Tager, J. M. (1989). High yields of specific hybridomas obtained by electrofusion of murine lymphocytes immunized in vivo or in vitro. *Exp. Cell Res.* **183**, 463–472.

9. Foung, S. K. H., and Perkins, S. (1989). Electric field-induced fusion and human monoclonal antibodies. *J. Immunol. Methods* **116**, 117–122.
10. Karsten,U., Stolley, P., and Seidel, B. (1990). Comparison of poly(ethylene glycol)- and electric field-mediated cell fusion for the formation of hybridomas. *In* "Methods in Enzymology," Academic Press, Orlando, FL (in press).
11. Tomita, M., and Tsong, T. Y. (1990). Selective production of hybridoma cells: Antigenic-based pre-selection of β lymphocytes for electrofusion with myeloma cells. *Biochim. Biophys. Acta* **1055**, 199–206.
12. Schmitt, J. J., and Zimmermann, U. (1989). Enhanced hybridoma production by electrofusion in strongly hypo-osmolar solutions. *Biochim. Biophys. Acta* **983**, 42–50.

23

Production of Genetically Identical Embryos by Electrofusion

Lawrence Charles Smith

Centre de Recherche en Reproduction Animale, Faculté de Médecine Vétérinaire,
Université de Montréal, Saint Hyacinthe, Québec J2S 7C6, Canada

I. Introduction

Recent improvements in micromanipulation techniques have opened new possibilities for experimental research in mammalian embryology. The combined use of the techniques of nuclear transplantation and cell electrofusion have enabled further investigations into the genetic and epigenetic mechanisms by which early embryonic development is controlled. These studies have clarified our understanding of determination and differentiation in mammalian embryonic cells during the preimplantation stages and also of the abilities of oocyte cytoplasm to reprogram these developmental events. Live offspring have been derived from nuclear transplanted embryos in laboratory and domestic species with the purpose of producing genetically identical animals ("clones").

This review aims to provide readers with some background knowledge on factors influencing the development of nuclear transplanted embryos, hoping that this will stimulate further research on nucleocytoplasmic interactions in early embryos. The first section is devoted to the cloning model used for amphibia eggs, since this has been studied in more detail and for a longer time than the mammalian model. Although nuclear transplantation techniques in mammals have been around for at least one decade, cloning procedures have only become viable in the last 5 years when they were first successfully applied to farm animals (Willadsen, 1986). The mammalian section begins by describing the steps leading to the current methodology and then outlines the effects the donor nucleus and the recipient cytoplasm have on the ability of nuclear transplanted embryos to develop normally. The section on the nuclear–cytoplasmic interactions discusses more fully the events taking place at fusion, such as the effects of cell cycle stage, cytoskeleton interactions, and others. The final section concludes by detailing some practical uses for nuclear transplantation and the production of genetically identical mammals.

II. The Amphibian Cloning Model

A. Methodology

Current procedures for cloning mammalian embryos originated from nuclear transplantation experiments in amphibians performed initially by Briggs and King (1952). The transplantation operation in amphibia is carried out in two main steps. First, the recipient frog eggs are activated by pricking with a glass needle; this causes rotation and brings the animal pole to lie uppermost. The chromosomes are visualized as a "black dot" (i.e., pigment granules clustered about the second metaphase meiotic spindle) that can be extirpated surgically by lifting the spindle away from the surface of the egg with a glass needle (Briggs and King, 1953). Effective enucleation can also be attained by ablating the chromosomes with either ultraviolet (Gurdon, 1960) or microbeam ruby laser irradiation (Ellinger et al., 1975). The second step involves the isolation of the donor cells and the nuclear transplantation itself. Donor cells are isolated, either mechanically with a fine-diameter pipette or chemically with enzymatic treatment. A given cell is then drawn into the tip of a sharp micropipette, the inner diameter of which is somewhat smaller than that of the cell. This causes rupture of the plasma membrane, but the cytoplasm contents are not dispersed and remain surrounding the nucleus. The pipette is then inserted through the membrane of the recipient enucleated egg and the contents of the broken cell are ejected, liberating the nucleus into the egg cytoplasm.

B. Development Potential

Amphibian nuclear transplantation experiments using several embryonic cell types have revealed that, as donor nuclei are tested from progressively older stages of embryogenesis, there is a decrease in the number of individuals that develop normally (reviews by King, 1966; DiBerardino and Hoffner, 1970; Gurdon, 1974; Briggs, 1977; McKinell, 1978). Although already containing approximately ten thousand cells with some degree of morphological differentiation, most, if not all, blastula nuclei are thought to be able to support normal development when fused to enucleated eggs. However, only a few early gastrula-stage nuclei have been shown to be totipotent, and a very low percentage of nuclei taken at any later embryonic stage are fully able to support development (Briggs and King, 1960). Studies carried out using nuclei derived from regions of larvae already commited to a specific developmental pathway have shown that only around 0.2% develop to adults and another 4% arrest at the larva stage (pluripotent nuclei). Together, these results suggest that the developmental capacity of nuclei becomes progressively restricted during the process of cell type determination and differentiation.

Several investigators have tested the developmental capacity of amphibian nuclei from adult cells. The most advanced stage of development reported from using nuclei of adult cells has been a feeding larva obtained from spermatogonia cells of *Rana pipens* (DiBerardino and Hoffner, 1971). Among adult somatic nuclei tested in *Xenopus*, there are 33 cases of early larvae obtained in support of nuclear pluripotency (Laskey and Gurdon, 1970; Gurdon *et al.*, 1975; McAvoy *et al.*, 1975; Wabl *et al.*, 1975; Brun, 1978). One of these larvae was reported to be normal and originated from a crest cell nucleus of the intestine, but this apparent normal larva died during an early larval stage (McAvoy *et al.*, 1975). The remaining nuclear transplant larvae were morphologically abnormal and the percent success ranged from 0.3 to 6.0%. No adult frog has yet developed from transplanted adult nucleus.

In order to determine the nature of the developmental restrictions displayed by nuclei from advanced cell types, an extensive series of studies was performed on abnormal embryos and larvae of *Rana* nuclear transplants (reviewed by DiBerardino, 1979). Most abnormal nuclear transplants examined exhibited abnormalities in chromosome number and/or structure that, in most cases, arose during the first cell cycle of the egg (DiBerardino and Hoffner, 1970). These abnormalities are now known to be the cause of developmental arrest in this species. The most severe chromosomal alterations cause developmental arrest at the blastula stage, whereas relatively minor karyotypic alterations permit development to early larval stages. Evidence has been presented that most of these numerical and structural changes in the chromosomes (1) involve chromosomal loss, (2) are a reflection of chromosomal differentiation acquired progressively though embryogenesis, and (3) are not a result, for the most part, of technical damage (DiBerardino, 1979, 1980; DiBerardino and

Hoffner, 1980). It appears that the cytoplasmic cell cycle of the amphibian egg, which is much faster than the nuclear cycle of advanced cell types, induces premature changes in transplanted nuclei resulting in incomplete DNA replication, chromosome breaks and arrangements, and hypoaneuploidy in the most severely affected embryos.

III. The Mammalian Cloning Model

A. Methodology

As the volume of individual mammalian eggs is close to one thousand times smaller than that of frog eggs, it is not surprising that more refined methods of microsurgery were required before the development of techniques for nuclear transplantation could be usefully applied to mammals. The successful microinjection of embryonic nuclei into the cytoplasm of one-cell rabbit (Bromhall, 1975) and mouse embryos (Modlinski, 1978) suggested that nuclei from later embryonic stages were able to participate with the egg's genome in supporting preimplantation development. The first report of mice born from nuclear transplantation came from the work of Illmensee and Hoppe (1981), showing that nuclei derived from the inner cell mass rather than trophectoderm of mouse blastocysts were able to support development to the morula–blastocyst stage in 34% of the transferred embryos, and that 19% of these would develop to term. These reports brought many scientists to speculate that this technique would provide a means of making an infinite number of genetically identical copies from a single embryo. However, it was later reported that attempts of other laboratories using this technique were unsuccessful (Marx, 1983; McGrath and Solter, 1984a; McLaren, 1984). Suggestions were made that Illmensee and Hoppe could not have properly enucleated the recipient zygotes, which would have allowed for their development to full term (McGrath and Solter, 1984a).

In support of these suggestions were the results from a novel method developed for nuclear transplantation by McGrath and Solter (1983a). Their technique avoids the necessity of penetrating the plasma membrane of either the donor or the recipient egg. This is possible by placing the eggs in cytoskeletal inhibitors for a short period before microsurgery, followed by zona pellucida penetration with a sharp micropipette and sucking out the two pronuclei surrounded by a piece of plasma membrane—a karyoplast. The donor karyoplast, which can be obtained from any embryonic or somatic cell, is injected under the zona pellucida of the enucleated recipient egg (Fig. 1). The following step of this noninvasive method for nuclear transplantation relies on an effective means for fusing the nuclear (karyoplast) and cytoplasmic (cytoplast) portions enclosed in the zona pellucida. Contrary to the poor levels of success obtained using Illmensee and Hoppe's invasive technique, McGrath and Solter's method provided virtually 100% of success, and the yield of live mice is very high when fertilized eggs are used as nuclear donors.

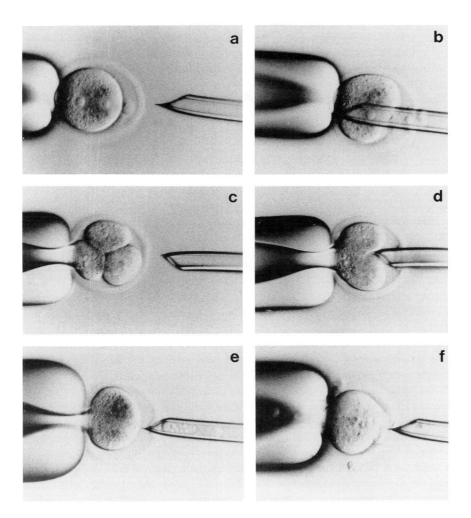

Figure 1   Technical procedure use for the enucleation of pronuclear zygotes followed by the collection and transplantation of karyoplasts. (a) Positioning of pronuclear zygote for enucleation; (b) removal of male and female pronuclei in a karyoplast; (c) nuclear donor four-cell stage embryo positioned for microsurgery; (d) removal of four-cell stage karyoplast; (e) positioning of donor karyoplast into zona pellucida; (f) recipient cytoplast and donor karyoplast ready for fusion.

Several methods are available for fusion, some of which are either unreliable and toxic, such as polyethylene glycol (PEG) and lysolecithin or, as for inactivated Sendai virus solutions, laborious to obtain and dangerous to raise due to its highly infectious nature. The latter also has the inconvenience that it is impossible to ascertain exactly the moment of fusion, since it occurs between 15 to 60 min after manipulation according to the amount and activity of the solution injected into the perivitelline space. However, possibly due to its high levels of success and to the ease with which manipulations are performed, Sendai-mediated fusion has been the most widely used approach in mouse nuclear transplantations. Willadsen (1986) reported the use of electric pulses to fuse the nuclear donor cell to the cytoplasm of the enucleated egg. Although Sendai virus-mediated fusion was successfully achieved in sheep, electro-fusion techniques are more effective and less variable when species other than mice are used.

Electrofusion has been successfully used to fuse blastomeres and for nuclear transplantation experiments in mammalian embryos derived from species as diverse as mice (Kubiak and Tarkowski, 1985; Tsunoda et al., 1987a; Kono and Tsunoda, 1988; Barra and Renard, 1988; Clement et al., 1988), rabbit (Ozil and Modlinski, 1986; Stice and Robl, 1988; Clement et al., 1988), sheep (Willadsen, 1986; Smith and Wilmut, 1989), cattle (Prather et al., 1987; Bondioli et al., 1990), and pigs (Clement et al., 1988; Prather et al., 1989). Embryonic cells (karyoplasts and cytoplasts) to be fused are placed in a chamber with electrofusion medium and positioned between two electrodes connected to a pulse generator (Fig. 2). Successful fusion can be attained using a large range of parameters for the direct current (DC) fusing pulse. Normally, a field intensity of around 1.0 kV/cm with durations between 50 and 100 μs is used. Two or three consecutive pulses separated by a fraction of a second may, in some instances, improve the levels of success. Fusion seems to be caused by the reversible electrical breakdown in the zone of membrane contact between the cytoplasm and nuclear donor cells. This breakdown is followed im-mediately by their repair into a single membrane, leading initially to the formation of small pores between the two cells, which continue growing and joining together until fusion is complete (Zimmerman and Vienken, 1982). Alternating current (AC) pulses may be used to align the cells so as to position their membranes perpendicular to the electrical field where conditions for fusion are most suitable. The preceding AC pulse is particularly important when fusing enucleated oocytes to cells with reduced diameters since the polarization caused by the AC field will aid in bringing their membranes into contact for the DC fusing pulse (Smith and Wilmut, 1989).

Willadsen (1986) reported the development to term of sheep embryos derived from the transplantation of embryonic blastomere nuclei to the cytoplasm of sec-ondary oocytes. The relevance of his findings was not exclusively related to the ability of a single eight-cell blastomere to support development, since this had already been indicated in earlier chimeric studies (Willadsen and Fehilly, 1983), but also that, as for amphibian nuclear transfers, the resulting fused embryo would

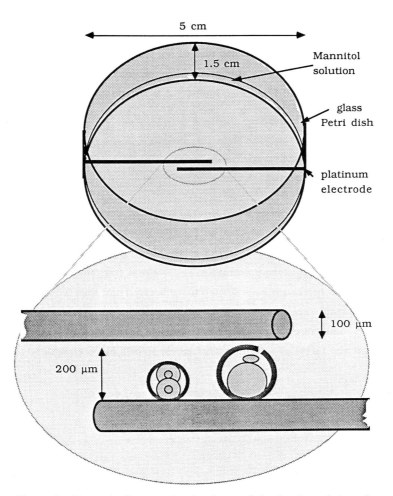

Figure 2 Schematic diagram of a chamber used for the electrofusion of mammalian embryos. Two 100-μm-diameter platinum electrodes are stuck to the bottom of a glass petri dish separated at 200 μm from each other.

develop and differentiate as if commencing from the time of fertilization, indicating an ability of the oocyte's cytoplasm for reprogramming the developmental "clock" of the donor nucleus. Although the technique used by Willadsen is similar to that used with frogs, in that both techniques utilize the secondary oocytes as the recipient cytoplasm, it also uses the noninvasive technique for transplanting the nucleus into the cytoplasm.

B. Development Potential

As for amphibian, many factors are involved in the ability of mammalian nuclear transplanted embryos to develop normally. Because these techniques are still fairly recent in mammals, our understanding of each factor remains limited. The following sections will try to point out some possible reasons why many embryos fail to develop after nuclear transplantation.

1. Nuclear Donor Cell

All cells in an embryo have a nucleus containing a comparable amount of DNA ($2n$), which is derived by continuous replication and segregation at mitosis from a single fusion product of the DNA carried by male (n) and female (n) gametes at fertilization. Assuming that this concept of nuclear equivalence is correct, and there are many reasons to believe so, one can presume that because later embryonic stages possess more cells with genetically identical nuclei, these are more suitable for cloning. There are some factors that limit the extent to which embryos should develop when allocated for use as nuclear donors in the cloning procedure.

The first factor relates to the loss of developmental potential at the expense of cellular determination or differentiation. As described earlier in amphibia, nuclei derived from cells that have committed themselves to a particular pathway seem to lose the ability to be readily reprogrammed and are unable to regain the totipotent status of their ancestral fertilization genome. An early report in mice showed that, although 95% of enucleated zygotes fused with zygote pronuclei will develop to blastocyst, only 13% will develop to blastocyst when 2-cell nuclei are fused and none when four-, eight-, 16-cell, and inner cell mass (ICM) nuclei are fused to enucleated zygotes, indicating a rapid loss in potential already by the 2-cell stage (McGrath and Solter, 1984a). These findings have been confirmed by other laboratories (Robl *et al.*, 1986; Surani *et al.*, 1987; Tsunoda *et al.*, 1987b; Smith *et al.*, 1988), and suggestions have been made that blastomere nuclei lose their potential at the time of maternal-zygotic transition (MZT) due either to genome differentiation or to toxicity of stage-specific factors (Solter *et al.*, 1986; Solter, 1987).

More recent findings in several mammalian species have shown that nuclei derived from stages beyond the MZT are still able to support development (Table 1). Comparisons between results from several mammals indicate that embryos derived from species that experience MZT within a single or two cells after oocyte activation tend to perform poorly after nuclear transplantation. In mice, where MZT occurs at the early two-cell stage (Flach *et al.*, 1982), it has been reported that a small proportion of nuclei from the eight-cell (McGrath and Solter, 1986) and from the inner cell mass (Tsunoda *et al.*, 1988, 1990) will support development to the blastocyst stage when fused to enucleated secondary oocytes. Again in mice, by using karyoplasts derived from embryos at different stages within the two-cell cycle,

Table 1

Pre- and Postimplantation Development Potential of Nuclear Transplanted Embryos Derived from the Fusion of Embryonic Cells to the Cytoplasm of "Enucleated" Secondary Oocytes in Several Mammalian Species[a]

Species	Embryonic cell type	Number fused	Morula or blastocyst	Number transferred	Established pregnancy	Full-term development	Reference
Mouse	2-Cell	86	35%	n.a.	n.a.	n.a.	(1)
	8-Cell	36	None	n.a.	n.a.	n.a.	(1)
	ICM	80	3%	n.a.	n.a.	n.a.	(1)
	ICM	83	4%	n.a.	n.a.	n.a.	(2)
	PGC	135	10%	59	10%	None	(1)
Pig	2-Cell	11	9%	33	n.a.	None	(3)
	4-Cell	83	8%	34	n.a.	3%	(3)
	8-Cell	57	19%	21	n.a.	None	(3)
Rabbit	8-Cell	70	68%	85	2%	1%	(4)
	16-Cell	n.a.	n.a.	110	n.a.	21%	(5)
	32-Cell	67	76%	n.a.	n.a.	n.a.	(5)
	32-Cell	67	55%	67	n.a.	n.a.	(6)
	ICM	52	37%	n.a.	n.a.	n.a.	(6)
	TE	27	None	n.a.	n.a.	n.a.	(6)
Sheep	8-Cell	76%	42%	4	n.a.	75%	(7)
	16-Cell	29	48%	6	50%	n.a.	(7)
	16-Cell	49	35%	14	21%	14%	(8)
	ICM	16	56%	8	13%	13%	(8)
Cattle	<8-Cell	111	12%	12	n.a.	None	(9)
	<16-Cell	50	16%	7	n.a.	29%	(9)
	<32-Cell	24	8%	n.a.	n.a.	n.a.	(9)
	Day 5 morula	604	32%	n.a.	n.a.	n.a.	(10)
	Day 5.5 morula	139	40%	n.a.	n.a.	n.a.	(10)
	Day 6 morula	54	53%	n.a.	n.a.	n.a.	(10)
	Morula	n.a.	n.a.	463	23%	22%	(10)

[a]References: (1) Tsunoda et al., 1990; (2) Tsunoda et al., 1988; (3) Prather et al., 1989; (4) Tsunoda et al., 1989; (5) Collas and Robl, 1990; (6) Collas and Robl, 1991; (7) Willadsen, 1986; (8) Smith and Wilmut, 1989; (9) Prather et al., 1987; (10) Bondioli et al., 1990; n.a. = not available; ICM = inner cell mass; PGC = primordial germ cell; TE = trophectoderm cells.

nuclei derived from embryos just before the MZT are slightly less able to support blastocyst development than those derived from embryos after the MZT (Smith et al., 1988). In pigs, where MZT occurs at the four-cell stage (Norberg, 1973), no significant difference could be detected between using nuclei from two- (9%), four- (8%) or eight-cell (19%) blastomeres in nuclear transplantations to enucleated secondary oocytes (Prather et al., 1989).

Species that experience MZT at or after the eight-cell stage (three or more cell cycles after oocyte activation) tend to perform better in nuclear transplantation trials. In sheep, where MZT occurs at the eight-cell stage (Crosby *et al.*, 1988), it has been reported that eight-cell (42%) and 16-cell (48%) nuclei are able to support development to blastocyst at similar proportions (Willadsen, 1986) and also that nuclei from 16-cell (35%) and the ICM (56%) stages of early blastocysts are similarly competent in supporting development both to blastocyst and to term (Smith and Wilmut, 1989). In cattle, where MZT occurs around the eight-cell stage (Camous *et al.*, 1986), reports have indicated similar development potentials are obtained from nuclei derived from embryos collected at day 5 (32%), day 5.5 (40%), and day 6 (53%) (Bondioli *et al.*, 1990). Therefore, it seems that, at least up to the embryonic stages tested in farm animals, there is no detectable loss of developmental potential of embryonic nuclei during the preimplantation stages utilized to this data. However, reports using rabbit nuclear transplanted embryos have indicated a slight decrease in potential at later embryonic stages. Although MZT occurs at the eight-cell stage in rabbits (Manes, 1977), the potential of nuclei derived from eight-cell (68%), 32-cell (55%), or from the inner cell mass cells (37%) seems to decrease slightly during development (Tsunoda *et al.*, 1989; Collas and Robl, 1990, 1991). Another important finding made in this species is the striking fall in potential (0%) when using committed blastocyst trophectoderm cells in nuclear transplantations (Collas and Robl, 1991).

Technical aspects should also be considered when deciding on the best embryonic stage from which to derive viable cells for nuclear transplantation. At compaction, cells from the outer layer of the embryo develop tight junctions making disaggregation into single cells increasingly more difficult. Although calcium-free medium and/or trypsinization can be used to aid in separating cells, such treatments may be detrimental to the viability of their nuclei even after short periods of exposure. The compact outer layers of late morula and blastocysts can be destroyed through an immunosurgical technique (Solter and Knowles, 1975), permitting an easier disaggregation of their inner cells as reported for the use of cells from the inner cell mass of blastocysts in sheep nuclear transplantations (Smith and Wilmut, 1989). Moreover, cell diameter decreases considerably during the cleavage stages, and their reduced size causes problems during the fusion procedure.

2. Recipient Cytoplasm

Recipient cytoplasm for mammalian nuclear transplantation was derived initially from enucleated zygotes at early stages after fertilization (McGrath and Solter, 1983a). As described in the previous section, nuclear transplantation experiments in mice using cytoplasm from pronuclear zygotes provided very limited support from nuclei beyond the two-cell stage in mice (McGrath and Solter, 1984a). Nuclear transplantations to zygotic cytoplasm have also been attempted in other mammals with very poor results. In rats, although 22% of the embryos derived from pronuclear transfers developed to term, no pregnancies could be attained when using nuclei derived from two-, four-, or eight-cell blastomeres (Kono *et al.*, 1988). In cattle, pronuclear zygotes were centrifuged to reveal the pronuclei for removal and later fused to karyoplasts from early stage blastomeres (Robl *et al.*, 1987). Bovine results were similarly discouraging, with only 17% of the pronuclear transplantations developing to the morula-blastocyst stage after 5 days culture in a sheep's oviduct and not a single normal development after transplantations of two-, four-, and eight-cell blastomere nuclei.

When using cytoplasm from two-cell enucleated blastomeres, Robl *et al.* (1986) were able to obtain morulae and blastocysts after fusion with karyoplasts derived from eight-cell mouse blastomeres. However, they also reported that these apparently normal blastocysts were unable to support development beyond mid-gestation after transfer to synchronized females. With a slight modification of the technique for using two-cell recipient cytoplasm, Tsunoda *et al.* (1987) were able to produce a small number of offspring from nuclei derived from eight-cell embryos but concluded that this technique would not prove useful in cloning mammalian embryos. Generally, recipient cytoplasm from secondary oocytes utilized in mouse nuclear transplantations has provided very little improvement when compared to the postfertilization cytoplasm sources (Fig. 3). However, some development to blastocyst was achieved with eight-cell (5%) and ICM (3%) stages after fusion to enucleated oocytes (McGrath and Solter, 1986; Tsunoda *et al.*, 1990). Moreover, primordial germ cell (PGC) nuclei have also been reported to support development to blastocyst (10%) and to mid-gestation (10%) after fusion to secondary oocyte cytoplasm (Tsunoda *et al.*, 1990).

Secondary oocytes have been the only cytoplasm recipient stage to support full-term development after nuclear transplantation in mammals. Liveborn have been produced in several species including sheep (Willadsen, 1986; Smith and Wilmut, 1989), cattle (Prather *et al.*, 1987; Marx, 1988; Bondioli *et al.*, 1990), rabbit (Stice and Robl, 1988), and pigs (Prather *et al.*, 1989). The optimal stage for utilizing secondary oocytes for nuclear transplantations (timing after ovulation or maturation) has not yet been determined, but there seems to be an increased potential after a short period of aging (Ware *et al.*, 1989). However, this beneficial effect is likely to be related to improved levels of success in causing activation by artificial stimuli after aging. The beneficial effect of aging the oocyte has also been observed in rabbit, where further improvement can be obtained by increasing the number of pulses at or after electrofusion (Collas and Robl, 1990).

In general, secondary oocytes have been obtained directly from the oviduct either during surgical intervention or at slaughter. With the current efficiency of the cloning procedure, the collection of *in vivo* matured oocytes comprises a substantial proportion of the total cost for obtaining a nuclear transplanted offspring. Although Prather *et al.* (1987) reported limited development using *in vitro* matured oocytes,

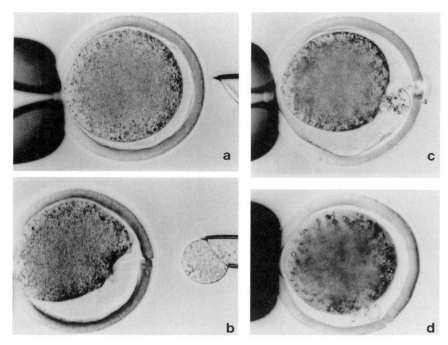

Figure 3 Microsurgical methods used in the preparation of sheep secondary oocytes and embryos for nuclear transplantation by electrofusion. (a) Secondary oocyte positioned for enucleation with the first polar body (1PB) facing the beveled microsurgical pipette. (b) Removal of approximately one-quarter of the cytoplasm surrounding the 1PB in a membrane-bound cell fragment. A single cell derived from either (c) 16- or 32-cell embryos or (d) inner cell mass of early blastocysts has been positioned within the zona pellucida of an enucleated secondary oocyte for the electrofusion procedure.

recent improvements in the methods for *in vitro* maturation (Gordon and Lu, 1990) are likely to make this the most cost-effective route for obtaining recipient oocytes for nuclear transplantation. This slaughterhouse source of cytoplasm has been successfully used for obtaining live offspring in cattle and sheep nuclear transplantation.

One limitation for the use of secondary oocytes as recipient cytoplasm concerns the enucleation procedure. With the exception of the rabbit (Stice and Robl, 1988), metaphase chromosomes cannot be readily visualized in secondary oocytes from other farm species due to the presence of large lipid vesicles in the cytoplasm. This leads to the need for using the position of the first polar body as an indicator of the position of the chromosomal plate followed by the aspiration of cytoplasm from the adjacent area (Fig. 4). However, possibly due to displacement or degeneration of

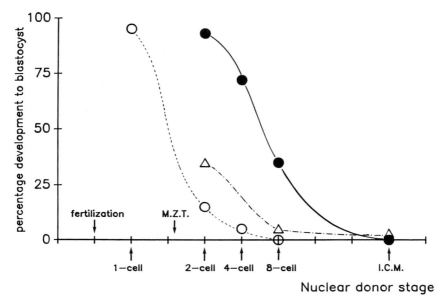

Figure 4   Comparison between the development potential of different recipient cytoplasm stages with regard to their ability in supporting the development to the morula or blastocyst stage after fusion to karyoplasts derived from preimplantation stage embryos. MZT = timing of the maternal–zygotic transition. ○, 1-cell cytoplasm; ●, 2-cell cytoplasm; △, oocyte cytoplasm.

the first polar body, an improper removal of chromosomes still occurs in a substantial proportion of manipulated secondary oocytes. Unsuccessful enucleation has been reported to occur in approximately one-third of the attempts, which, possibly due to aneuploidy, lead to abnormal cleavage and developmental arrest during the early preimplantation stage (Willadsen, 1986; Smith and Wilmut, 1989). Although the success of enucleation can be ascertained using DNA vital stains, further research is required to assess the effect this may have on development due to the possible damaging of mitochondrial DNA and other cytoplasmic components.

IV. Nuclear–Cytoplasmic Interactions

When fusion is complete, donor nucleus and recipient cytoplasm will initiate interactions, which, when compatible, will lead to the normal development of a nuclear transplanted embryo. This section discusses the cellular and molecular interactions that take place at and after fusion.

A. Cell Cycle Stage Effects

Experiments with mouse pronuclear zygotes have indicated that cell cycle stage synchrony between nucleus and cytoplasm is beneficial for further development *in vitro* (Smith *et al.*, 1988). This observation has been further extended to transplantations between two-cell embryos where asynchronous exchanges were highly deleterious to further development (unpublished observation). This effect may be explained either by the disruption of the cell cycle oscillatory mechanisms or by incompatible nuclear–cytoplasmic interactions in controlling critical developmental steps (Smith *et al.*, 1990). Possibly resulting from incompatible nuclear–cytoplasmic interactions, sheep embryos recovered from ligated oviducts 5 days after nuclear transplantation either do not cleave at all (and may fragment after a period of aging) or cleave a few times but cannot compact and blastulate as normal embryos (Smith and Wilmut, 1989). Because observations have been performed mostly after a long period of *in vivo* culture, it is unclear exactly at what stage these embryos arrest development. Better methods for culturing livestock embryos *in vitro* (i.e., coculture with oviductal cells) coupled with detailed biochemical studies on their transcriptional and translational activities will certainly enable a clearer understanding of these nuclear–cytoplasmic incompatibilities.

B. Cytoplasmic Effects

Nucleo–cytoplasmic compatibility has been shown to improve when nuclear transplanted embryos are placed into a medium containing cytochalasin B (CB) for a short period immediately after fusion. In sheep, nuclei derived from both 16-cell stages (35% vs. 11%) and ICM cells (56% vs. 0%) of early blastocysts were significantly better able to support development to morulae and blastocysts after nuclear transplantation when placed in medium containing 7.5 μg CB for 1 hr after the electric stimulus as compared with placing them into medium alone (Smith and Wilmut, 1989). A similar treatment was also effective in improving development to blastocyst (44% vs. 11%) in rabbit 32-cell nuclear transplantations (Collas and Robl, 1990). Together, these results indicate that fused nuclei are affected by cytoskeletal mechanisms operating during the activation of the recipient cytoplasm, possibly leading to chromosomal anomalies by extrusion into pseudo first polar bodies. However, an abstract report has shown that the effect of cytochalasin B is not apparent in cattle nuclear transplantation, which may indicate species variations in nucleo–cytoplasmic interactions after fusion (Levanduski and Westhusin, 1990).

Czolowska *et al.* (1984) observed the swelling of thymocyte nuclei in activated ovum cytoplasm to equal that of pronuclei when nuclear introduction coincided with activation, but diminished with increasing time between activation and nuclear introduction. Nuclear swelling has also been observed in mouse and rabbit nuclear transplantations (Robl *et al.*, 1986; Stice and Robl, 1988). Moreover, Szollosi *et al.* (1988) showed that nuclear membrane breakdown and further reassembly of a new nuclear envelop occurred only when fusions were performed immediately before or up to 30 min after oocyte activation. Possibly, the nuclear membrane functions as a barrier to a few specific factors in the oocyte cytoplasm required to reprogram the genome's developmental pathway. Another explanation for the poor performance of nuclear transplantations to enucleated zygotes as compared with secondary oocytes is that developmentally important factors are sequestered within or around the nuclear membrane and removed during pronuclear enucleation.

Peri- and postimplantational studies on nuclear transplanted embryos are limited in mammals (Table 1). It has been reported that although apparently normal in morphology, many of the morula–blastocyst stage embryos derived from nuclear transplantation fail to produce pregnancy after transplantation to the uteri of synchronized recipients (Prather *et al.*, 1987; Smith and Wilmut, 1989; Bondioli *et al.*, 1990). Although many seem to die soon after implantation, causing an extended estrous cycle, others may develop much further, leading to abortions at later stages of gestation. It has also been noted that fetuses derived from nuclear transplanted embryos are likely to be larger than usual and/or to extend slightly beyond the normal gestation time (Smith and Wilmut, 1989; Dr. K. Bondioli, personal communication).

V. Applications and Conclusions

The technique of nuclear transplantation in mammals is the most rigorous test to ascertain whether embryonic or somatic nuclei retain all of the genetic information found in a zygote nucleus. However, other biological applications of this technique have included (1) studies on the role of paternally and maternally derived genes in development ("imprinting") (McGrath and Solter, 1984b; Surani *et al.*, 1984), (2) studies to test whether events that take place during embryonic development are inherited through the nucleus or cytoplasm (McGrath and Solter, 1983b; McGrath and Solter, 1984c; Mann, 1986), and (3) the control of cleavage in early embryos (Smith *et al.*, 1990).

In practice, the technique of nuclear transplantation already provides a means for cloning embryos in many mammalian species (Fig. 5). Therefore, it is likely that this technique will soon become more widely available for producing genetically identical animals either for research or for commercial application. In research, clones could be used to eliminate the variation caused by the genotype allowing for better assessment of nongenetic factors such as environment, disease, experimental treatments, etc. For animal breeders, cloning will bring the advantage of speeding the annual genetic gain of selection programs and enable a much shorter interval to disseminate the gain to the commercial herds. Commercial organizations using these

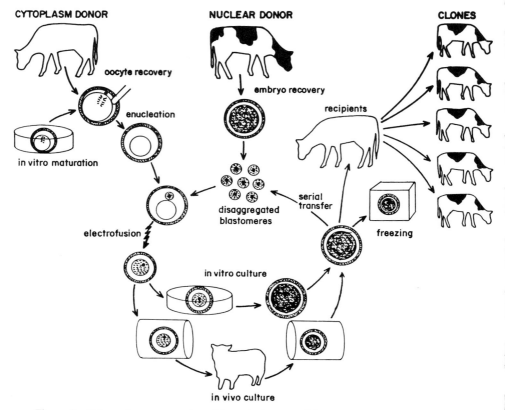

Figure 5   Schematic representation of the techniques used for cloning mammalian embryos. Pathways used for obtaining donor nuclei and recipient cytoplasm are presented, followed by the cyclic procedure for deriving cloned embryos. The cloned embryos can be either recycled, frozen to be used at a later stage, or transferred to recipients to provide cloned offspring.

techniques have reported the successful use of nuclear transplanted embryos as nuclear donors for further transplantations (serial transplantations). Considering that at the present level of success one can derive approximately five embryos from every initial nuclear donor embryo and that the same process could be repeated for each of the resulting nuclear transplanted embryos, after four generations of asexual reproduction (or 4 weeks), a total of 625 genetically identical cloned embryos would be available for transplantation into synchronized recipients (Wilmut and Smith, 1989). It is likely that as many clones would be required for any reason, but this example serves to illustrate the potential of the nuclear transplantation procedure.

Recent reports have indicated progress in deriving embryonic stem (ES) cells from farm animals, which may become a valuable source of genetically identical nuclei for cloning (Evans *et al.*, 1990). The ease of selecting for integration and expression of transfected ES cells *in vitro* will also allow for a more efficient route for producing transgenic livestock than the current methodologies of pronuclear injection and chimeras produced from transfected ES cells. At this point, it is also important to note some of the possible disadvantages in the cloning procedure. One aspect concerns the decrease in the genetic variability by inbreeding. Large foundation populations should be obtained when using cloning in conjunction with MOET (multiple ovulation and embryo transfer) selection schemes. It is also important to verify the degree to which cytoplasmic inheritance can influence animal production, since clones will be exposed not only to different uterine and neonatal environments but also to a different ooplasm. The maternal inheritances of mitochondrial genes are examples of differences that may arise between nuclear transplantation-derived clones. Other potential factors of variation to be considered are the random inactivation of X chromosomes in females and the possibility of differential expression or mutation after nuclear transplantation in cloned embryos. These final comments help to strengthen the notion that many aspects of the interactions between nucleus and cytoplasm after transplantation remain unknown and that it is only by both fundamental and applied research that further improvements may be achieved with the cloning of mammalian species.

Acknowledgments

The author thanks Dr. A. K. Goff for helpful comments and NSERC, CAAB, and FCAR for their financial support during the preparation of this manuscript.

References

Barra, J., and Renard, J.-.P. (1988). Diploid mouse embryos constructed at the late 2-cell stage from haploid parthenotes and androgenotes can develop to term. *Development* 102, 773–779.

Bondioli, K. R., Westhusin, M. E., and Looney, C. R. (1990). Production of identical bovine offspring by nuclear transfer. *Theriogenology* 33, 165–174.

Briggs, R. (1977). Genetics of cell type determination. *In* "Cell Interactions in Differentiation" (L. Saxen and L. Weiss, eds.), pp. 23–44. Academic Press, New York.

Briggs, R., and King, T. J. (1952). Transplantation of living nuclei from blastula cells into enucleated frog eggs. *Proc. Natl Acad. Sci. USA* 38, 455–467.

Briggs, R., and King, T. J. (1953). Factors affecting the transplantability of nuclei of frog embryonic cells. *J. Exp. Zool.* 122, 485–505.

Briggs, R., and King, T. J. (1960). Nuclear transplantation studies on the early gastrula (*Rana pipiens*). *Dev. Biol.* 2, 252–270.

Bromhall, J. D. (1975). Nuclear transplantation in the rabbit egg. *Nature (Lond.)* **258**, 719–722.

Brun, R. B. (1978). Developmental capacities of Xenopus eggs, provided with erythrocyte or erythroblast nuclei from adults. *Dev. Biol.* **65**, 271–289.

Camous, S., Kopecny, V., and Flechon, J.-E. (1986). Autoradiographic detection of the earliest stage of (3H)-uridine incorporation. *Biol. Cell* **58**, 195–200.

Clement, A., Meyer, J., and Brem, G. (1988). Electrofusion of early mammalian embryos. *In* "New Developments in Biosciences; Their Implications for Laboratory Animal Sciences" (A. C. Beynen and H. A. Solleveld, eds.), pp. 343–347. Martins Nijhoff, Dorrecht, Netherlands.

Collas, P., and Robl, J. M. (1990). Factors affecting the efficiency of nuclear transplantation in the rabbit embryo. *Biol. Reprod.* **43** (in press).

Collas, P., and Robl, J. M. (1991). Development of rabbit nuclear transplant embryos from morula and blastocyst stage donor nuclei. *Theriogenology* (in press).

Crosby, I. M., Gandolfi, F., and Moor, R. M. (1988). Control of protein synthesis during early cleavage of sheep embryos. *J. Reprod. Fertil.* **82**, 769–775.

DiBerardino, M. A. (1979). Nuclear and chromosomal behavior in amphibian nuclear transplants. *Inter. Rev. Cytol. Suppl.* **9**, 129–160.

DiBerardino, M. A. (1980). Genetic stability and modulation of metazoa nuclei transplanted into eggs and oocytes. *Differentiation* **17**, 17–30.

DiBerardino, M. A., and Hoffner, N. J. (1970). Origin of chromosomal abnormalities in nuclear transplants—A reevaluation of nuclear differentiation and nuclear equivalence in amphibians. *Dev. Biol.* **23**, 185–209.

DiBerardino, M. A., and Hoffner, N. J. (1971). Development and chromosomal constitution of nuclear transplants derived from male germ cells. *J. Exp.Zool.* **176**, 61–78.

DiBerardino, M. A., and Hoffner, W. J. (1980). The current status of cloning and nuclear reprogramming in amphibian eggs. *In* "Differentiation and Neoplasia," Vol. 2 (R. G. McKinnel *et al.,* eds.), pp. 53–64. Springer-Verlag, Berlin.

Ellinger, M. S. (1978). The cell cycle and transplantation of Blastula Nuclei in *Bombina orientalis*. *Dev. Biol.* **65**, 81–89.

Evans, M. J., Notarianni, E., Laurie, S., and Moor, R. M. (1990). Derivation and preliminary characterization of pluripotent cell lines from porcine and bovine blastocysts. *Theriogenology* **33**, 125–128.

Flach, G., Johnson, M. H., Braude, P. R., Taylor, R. A. S., and Bolton, V. N. (1982). The transition from maternal to embryonic control in the 2-cell mouse embryo. *EMBO J.* **1**, 681–686.

Gordon, I., and Lu, K. H. (1990). Production of embryos in vitro and its impact on animal production. *Theriogenology* **33**, 77–88.

Gurdon, J. B. (1960). The effects of ultra-violet irradiation on uncleaved eggs of *Xenopus laevis*. *Q. J. Microsc. Soc.* **101**, 299–311.

Gurdon, J. B. (1974). The genome in specialized cells as revealed by nuclear transplantation in amphibia. *In* "The Cell Nucleus," Vol. 1 (H. Bush, ed.), pp. 471–488. Academic Press, New York.

Gurdon, J. B., Laskey, R. A., and Reeves, O. R. (1975). The developmental capacity of nuclei transplanted from keratinized skin cells of adult frogs. *J. Embryol. Exp. Morphol.* **34**, 93–112.

Illmensee, K., and Hoppe, P. C. (1981). Nuclear transplantation in *Mus musculus:* Developmental potential of nuclei from preimplantation embryos. *Cell* 23, 9–18.

King, T. J. (1966). Nuclear transplantation in amphibia. *Methods Cell Biol.* 2, 1.

Kono, T., and Tsunoda, Y. (1988). Effects of induction current and other factors on large scale electrofusion for pronuclear transplantation of mouse eggs. *Gamete Res.* 19, 349–357.

Kono, T., Shioda, Y., and Tsunoda, Y. (1988). Nuclear transplantation of rat embryos. *J. Exp. Zool.* 248, 303–305.

Kubiak, J. Z., and Tarkowski, A. K. (1985). Electrofusion of mouse blastomeres. *Exp. Cell Res.* 157, 561–566.

Laskey, R. A., and Gurdon, J. B. (1970). Genetic content of adult somatic cells tested by nuclear transplantation from cultured cells. *Nature (Lond.)* 228, 332–358.

Levanduski, M. J., and Westhusin, M. E. (1990). Effect of cytoskeletal inhibitors on fusion and development of bovine nuclear transfer embryos. *Theriogenology* 33, 273 (abstr.).

Manes, C. (1977). Nucleic acid synthesis in preimplantation rabbit embryo. III. A "dark period" immediately following fertilization and the early predominance of low molecular weight RNA synthesis. *J. Exp. Zool.* 201, 247–258.

Mann, J. L. (1986). DDK-egg foreign sperm incompatibility in mice is not between the pronuclei. *J. Reprod. Fertil.* 76, 779–781.

Marx, J. L. (1983). Swiss research questioned. *Science* 220, 1023.

Marx, J. L. (1988). Cloning sheep and cattle embryos. *Science* 239, 463–464.

McAvoy, J. W., Dixon, R. E., and Marshall, J. A. (1975). Effects of differences in mitotic activity, stage of cell cycle and degree of specialization of donor cells of nuclear transplantation in *Xenopus laevis*. *Dev. Biol.* 45, 330–352.

McGrath, J., and Solter, D. (1983a). Nuclear transplantation in the mouse embryo by microsurgery and cell fusion. *Science* 220, 1300–1302.

McGrath, J., and Solter, D. (1983b). Nuclear transplantation in mouse embryos. *J. Exp. Zool.* 288, 355–362.

McGrath, J., and Solter, D. (1984a). Inability of mouse blastomere nuclei transferred to enucleated zygotes to support development *in vitro*. *Science* 226, 1317–1319.

McGrath, J., and Solter, D. (1984b). Completion of mouse embryogenesis requires both the maternal and paternal genomes. *Cell* 37, 179–183.

McGrath, J., and Solter, D. (1984c). Maternal Thp lethality in the mouse is a nuclear, not cytoplasmic defect. *Nature (Lond.)* 308, 550–551.

McGrath, J., and Solter, D. (1986). Nucleocytoplasmic interactions in the mouse embryo. *J. Embryol. Exp. Morphol.* 97, suppl., 277–289.

McKinnel, R. G. (1978). "Cloning, Nuclear Transplantation in Amphibia." University of Minnesota, Minneapolis.

McLaren, A. (1984). Methods and success of nuclear transplantation in mammals. *Nature (Lond.)* 309, 671–672.

Modlinski, J. A. (1978). Transfer of embryonic nuclei to fertilized mouse eggs and development of tetraploid blastocysts. *Nature (Lond.)* 273, 466–467.

Norberg, H. S. (1973). Ultrastructural aspects of the preattached pig embryo: cleavage and early blastocyst stages. *A. Anat. EnstGesch* 143, 45–114.

Ozil, J.-P., and Modlinski, J. A. (1986). Effect of electric field on fusion rate and survival of 2-cell rabbit embryos. *J. Embryol. Exp. Morphol.* 96, 211–228.

Prather, R. S., Barnes, F. L., Sims, M. M., Robl, J. M., Eyestone, M. H., and First, N. L. (1987). Nuclear transplantation in the bovine embryo: Assessment of donor nuclei and recipient oocyte. *Biol. Reprod.* 37, 859–866.

Prather, R. S., Sims, M. L., and First, N. L. (1989). Nuclear transplantation in early pig embryos. *Biol. Reprod.* 41, 414–418.

Robl, J. M., Gilligan, B., Critser, E. S., and First, N. L. (1986). Nuclear transplantation in mouse embryos: Assessment of recipient cell stage. *Biol. Reprod.* 34, 733–739.

Robl, J. M., Prather, R. S., Barnes, F., Eyestone, W., Northey, D., Gilligan, B., and First, N. L. (1987). Nuclear transplantation in bovine embryos. *J. Anim. Sci.* 64, 642–647.

Smith, L. C., and Wilmut, I. (1989). Influence of nuclear and cytoplasmic activity on the development in vivo of sheep embryos after nuclear transplantation. *Biol. Reprod.* 40, 1027–1035.

Smith, L. C., Wilmut, I., and Hunter, R. H. F. (1988). Influence of cell cycle stage at nuclear transplantation on the development in vitro of mouse embryos. *J. Reprod. Fertil.* 84, 619–624.

Smith, L. C., Wilmut, I., and West, J. D. (1990). Cleavage control in single-cell reconstituted mouse embryos. *J. Reprod. Fertil.* 88, 655–663.

Solter, D. (1987). Inertia of the embryonic genome in mammals. *Trends Genet.* 3, 23–27.

Solter, D., and Knowels, B. B. (1975). Immunosurgery of mouse blastocysts. *Proc. Natl. Acad. Sci. USA* 72, 5099–5102.

Solter, D., Aronson, J., Gilbert, S. F., and McGrath, J. (1986). Nuclear transfer in mouse embryos: activation of the embryonic genome. *In* "Molecular Biology of Development," 50th Symp. Quant. Biol. Vol. L, pp. 45–50. Cold Spring Harbour, New York.

Stice, S. L., and Robl, J. M. (1988). Nuclear reprogramming in nuclear transplant rabbit embryos. *Biol. Reprod.* 39, 657–664.

Surani, M. A. H., Barton, S. C., and Norris, M. L. (1984). Development of reconstituted eggs suggested imprinting of the genome during gametogenesis. *Nature (Lond.)* 308, 548–550.

Surani, M. A. H., Barton, S. C., and Norris, M. L. (1987). Experimental reconstruction of mouse eggs and embryos: An analysis of mammalian development. *Biol. Reprod.* 36, 1–16.

Szollosi, D., Czolowska, R., Szollosi, M. S., and Tarkowski, A. K. (1988). Remodeling of mouse thymocyte nuclei depends on the time of their transfer into activated homologous oocytes. *J. Cell Sci.* 91, 603–613.

Tsunoda, Y., Kato, Y., and Shioda, Y. (1987a). Electrofusion for the pronuclear transplantation of mouse eggs. *Gamete Res.* 17, 15–20.

Tsunoda, Y., Yasui, T., Shioda, Y., Nakamura, K., Uchida, T., and Sugie, T. (1987b). Full term development of mouse blastomere transplanted into enucleated two-cell embryos. *J. Exp. Zool.* 242, 147–151.

Tsunoda, Y., Shioda, Y., Onodera, M., Nakamura, K., and Uchida, T. (1988). Differential sensitivity of mouse pronuclei and zygote cytoplasm to Hoeschst staining and ultraviolet irradiation. *J. Reprod. Fertil.* 82, 173–178.

Tsunoda, Y., Maruyama, Y., and Kila, M. (1989). Nuclear transplantation of 8- to 16-cell embryos into enucleated oocytes in the rabbit. *Jpn. J. Zootech. Sci.* 60, 846–851.

Tsunoda, Y., Tokunaga, T., Imai, H., and Uchida, T. (1990). Nuclear transplantation of male primordial germ cells in the mouse. *Development* **107**, 407–411.

Wabl, M. R., Brun, R. B., and Pasquier, L. (1975). Lymphocytes of the toad *Xenopus laevis* have the gene set for promoting tadpole development. *Science* **190**, 1310.

Ware, C. B., Barnes, F. L., Maija-Lauiro, M., and First, N. L. (1989). Age dependence of bovine oocyte activation. *Gamete Res.* **22**, 265–275.

Willadsen, S. M. (1986). Nuclear transplantation in sheep embryos. *Nature (Lond.)* **320**, 63–65.

Willadsen, S. M., and Fehilly, C. B. (1983). The developmental potential and regulatory capacity of blastomeres from 2-, 4- and 8-cell sheep embryos. *In* "Fertilization of the Human Egg in Vitro; Biological and Clinical Applications" (H. M. Beier and H. R. Lindner, ed.), pp. 27–52. Springer-Verlag, Berlin.

Wilmut, I., and Smith, L. C. (1989). Biotechnology and the bovine embryo: At present and in the future. *In* "Colloque Scientifique", 4th ed., pp. 19–31. A. E. T. E., Lyons, France.

Zimmermann, U., and Vienken, J. (1982). Electric field induced cell-to-cell fusion. *J. Membrane Biol.* **67**, 165–182.

24

Development of Cell–Tissue Electrofusion for Biological Applications

Richard Heller[1] and Richard Gilbert[2]

[1]Department of Surgery, College of Medicine, University of South Florida, Tampa, Florida 33612

[2]Department of Chemical Engineering, College of Engineering, University of South Florida, Tampa, Florida 33620

I. Introduction

The relatively new and expanding field of cell–cell electrofusion formed the basis for the discovery of cell–tissue electrofusion (CTE). Senda *et al.* (1979) were the first to describe the fusion of two cells by electrofusion. Plant protoplasts were pushed together with micropipettes and fused with a single electrical pulse. Subsequently, other investigators reported similar results with animal (Neumann *et al.*, 1980; Richter *et al.*, 1981) and yeast cells (Weber *et al.*, 1981a, 1981b). Over the past 10 years, the literature in cell–cell electrofusion has been expanding rapidly (Bates *et al.*, 1987; Chang *et al.*, 1989; Powell *et al.*, 1989; Rols *et al.*, 1989, 1990a, 1990b; Sowers *et al.*, 1989, 1990; Zimmermann *et al.*, 1981; Zimmermann, 1982, 1986). Cell–tissue electrofusion expands this new technology even further.

During the past several years many protocols have been developed that facilitate the incorporation of individual cells into intact tissue. Cell–tissue electrofusion technology employs electromechanical processes by which individual separated animal cells can be electrofused directly to histologically intact tissues *in vitro, in situ,* or *in vivo* in anesthetized animals (see Fig. 1A). The formation of somatic cell hybrids *in vivo* is accomplished by applying induced electrical fields that result in the coalescence of juxtaposed plasma membranes of the individual cells and the cells within the tissues. These procedures have been carried out routinely in the laboratory. The further development of this technology should result in varied biological applications. For example, some applications might include:

1. Novel site-specific delivery systems for cancer chemotherapy and gene therapy.
2. Improved surgical procedures.
3. The establishment of novel bioengineered animal models for the study of specific human pathogens.

II. Biological Applications

Further development of CTE can lead to several applications in many areas of biomedical research. These applications include a site-specific drug delivery system, *in vivo* gene therapy, surgical applications, and tissue–tissue electrofusion.

A. Potential Applications

It has been demonstrated that liposomes can be fused to cells by the process of electrofusion *in vitro* (Chernomordik *et al.*, 1990; Tsong, 1987). By developing liposome–tissue electrofusion, liposomes containing drugs, genes, or other agents could be delivered to specific sites *in vivo*. For example, electrofusion of such drug-

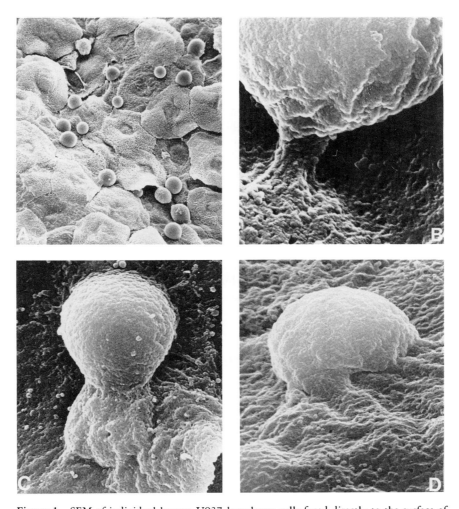

Figure 1   SEM of individual human U937 lymphoma cells fused directly to the surface of rabbit corneal epithelium. (A) Corneal surface (1300×) showing a representative distribution of 14 smooth, spherically shaped human cells electrofused to the rabbit epithelium. (B) Plasma membrane "stalk" that forms very shortly after electrofusion formed between the upper smoother human cell and the lower microvilli covered rabbit cell surface (15,000×). (C, 6000×, and D, 8600×) Coalescing human–rabbit plasma membranes of somatic cell hybrids beyond the initial stages of their formation in the intact corneal epithelium. (From Grasso *et al.*, 1989; reproduced with permission.)

encapsulated particles directly into tumor tissue should result in high concentrations of the agent where it is needed most and simultaneously should reduce undesirable systemic side effects associated with currently employed routes of drug administration.

Cell–tissue electrofusion may also be a useful tool in several surgical applications including arthroscopic surgery. For example, when bone spurs are removed from knee joints, chondrocytes must proliferate to replace cartilage in the areas denuded by the surgery. To accomplish this, the patient's own chondrocytes could be collected from adjacent areas and electrofused to the surgical areas. This procedure might accelerate the healing process, allowing cartilage to form faster.

Although tissue–tissue electrofusion has yet to be performed in the laboratory, it has many potential applications that could speed up the healing process in burn patients, skin grafts, and other topical procedures. Furthermore, tissue–tissue electrofusion techniques could be employed to perform epikeratophakia surgery. Today, this ophthalmic surgical procedure requires suturing, which distorts the topology of the transplanted corneal tissue and is time-consuming. Instead, electrofusion could be employed with an electrode unit shaped to the desired curvature of the corneal tissue. This procedure should eliminate undesirable distortion. Thus, this ocular surgical procedure should be more beneficial to the patient and should be able to be performed in a fraction of the time required presently.

B. Development of an Animal Model

The creation of model systems for the *in vitro* and *in vivo* study of species-specific infectious diseases will lead to a better understanding of the pathogenic mechanisms of a variety of receptor-mediated processes. CTE could be utilized to produce novel animal models that will facilitate the transfer of distinctive features from a susceptible animal or human to a resistant animal. These animals would possess biological properties that would differ fundamentally from those displayed naturally by the unaltered animal species.

The first step in obtaining a model system is the establishment of a method that allows for the transfer of membrane surface components from susceptible cells to intact tissue of laboratory animals. Individual cells, which contain the appropriate receptors, can be incorporated into excised tissue. The selected tissue is naturally resistant to attachment by host or tissue specific microbial pathogens. The details of the transfer procedure and the usefulness of the system could be tested and characterized *in vitro* before attempting to establish an animal model.

An *in vitro* and an *in vivo* model system for examining the pathogenic mechanisms of the obligate human pathogen *Neisseria gonorrhoeae* are presented to illustrate the procedure for developing an animal model. This particular model was chosen because there have been numerous unsuccessful attempts at developing a receptor-mediated animal model for this infectious disease for more than 100 years (Arko, 1989). The basic idea for this animal model was to transfer gonococcal receptors from the surface of susceptible human cells to intact corneal epithelium of rabbits. The cornea was selected to be the tissue to have human cells fused to its surface for a variety of technical and scientific reasons. For example:

1. This new biotechnology can be utilized to establish an ocular model much simpler than either a vaginal, cervical, or urethral model.
2. The eye is a natural site of gonococcal infections in humans [e.g., ophthalmia neonatorum and keratoconjunctivitis in adults (Ghosh, 1987; Ullman *et al.*, 1987; Wan *et al.*, 1986)].
3. The pathogen would colonize the tissue in the animal model as it does naturally on human epithelial surfaces.
4. Gonococcal ocular lesions could be elicited in a controlled manner with relative ease because the infection site is readily accessible.
5. Direct visual observations of clinical symptoms of infection could be made and evaluated with a noninvasive grading scheme.
6. The initiation of pathogenesis of ocular gonorrhea in the *in vivo* model would be mediated by the incorporation of functional human gonococcal attachment receptors into non human epithelium.

For these reasons, the rabbit cornea was selected as the anatomical model site for testing the ability to utilize CTE technology to create model systems for the study of species-specific infectious diseases.

C. Verification of Established Model

For CTE to be utilized in the creation of any novel model systems, or any other biological applications, it is necessary to accomplish the following:

1. Demonstrate that individual cells can be electrofused to intact tissue.
2. Demonstrate that membrane surface components present on the individual cells are present and remain functional on the newly formed somatic cell hybrids.
3. Demonstrate that cell tissue electrofusion techniques are safe and cause no ill effects to an animal during *in vivo* procedures.
4. Demonstrate that a human specific pathogen can cause an infection in a naturally resistant animal.

The procedures discussed in this chapter demonstrate how the system to study *Neisseria gonorrhoeae* met this list of requirements. The *in vitro* model will be presented first, followed by the *in situ* and *in vivo* procedure.

III. *In Vitro* Cell–Tissue Electrofusion

A. Apparatus

Cell–tissue electrofusion can be accomplished *in vitro* with commercially available cell fusion instruments; however, the chambers and other accessories that accompany these instruments may prove to be inadequate for some applications. Specially designed chambers and electrodes may be necessary to facilitate the fusion of individual cells to intact tissue. A Plexiglas well slide with an attached coverslip and metal post can serve as a chamber for cell–tissue electrofusion experiments. It should also be outfitted with a removable grid electrode designed to fit into the hole of the well slide over the coverslip. This grid electrode consists of four 24-gauge platinum wires flatly spaced 2–3 mm apart in a regular parallel and perpendicular planar array over a circular area 10 mm in diameter. The second platinum electrode, 3 mm in

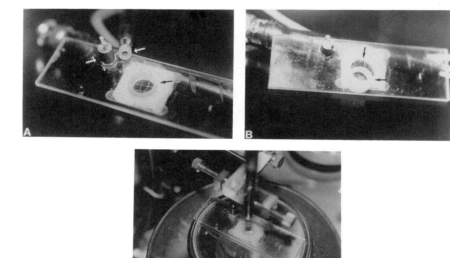

Figure 2 *In vitro* cell–tissue electrofusion chamber and electrodes, showing grid electrode chamber with electrode holder complete with recessed platinum electrode. (A) Well slide with grid electrode in place (black arrow); 3-mm disc electrode is recessed into the electrode holder (white arrow on right); post (white arrow on left) is electrically connected to the grid electrode and is where the return wire (back to pulse generator) is connected. (B) Grid electrode (lower black arrow) can be removed from well (upper black arrow). (C) Entire apparatus configured for use. Electrode holder is held by a micromanipulator and can be lowered down to the well slide to make contact with the cells and tissue in the well.

diameter and disc shaped, is contained in an insulated holder and is recessed 2–3 mm from the tip (Fig. 2).

B. Procedure

A typical *in vitro* cell-tissue electrofusion experiment can be performed as described by Grasso *et al.* (1989). Briefly, an 8-mm corneal button is excised, placed in the well of the Plexiglas slide, and then covered with a low-ionic-strength solution of 0.3 M D-glucose, 0.5 mM MgCl$_2$, 0.5 mM CaCl$_2$, and 6 mM histidine. A suspension ($>1.0 \times 10^7$ cells/ml) of human U937 lymphoma cells is placed dropwise into the well to cover the cornea. An additional drop of cell suspension is placed by capillary action into the recessed portion of the disc electrode. The electrode is lowered (a micromanipulator can be used to facilitate this) until the suspension in the disc electrode makes contact with the suspension in the well. The cells are exposed to 30 s of AC at 2 V with a frequency of 2 MHz to generate a dielectrophoretic field to align the cells prior to fusion. Following the alignment phase the cells and cornea are exposed to three 20-μs, 0.06-kV/cm square pulses 1 s apart to initiate the fusion process. The disc electrode is raised and the cornea removed from well. A vigorous wash and fixation in 2.5% glutaraldehyde follows. After the fixation step, the fusion results can be examined by scanning electron microscopy (SEM) as shown in Fig. 1.

C. Complications

Although Fig. 1 demonstrates that this initial setup could be utilized to successfully electrofuse individual cells to intact tissue, there are several experimental observations that indicate modifications of the original CTE procedures would be necessary if this new technology is to be used either *in situ* or *in vivo*. For example, successful CTEs conducted by the procedures outlined above produce fused cells that form a pattern very similar to that of the grid electrode at the bottom of the fusion chamber. The location of most of the fused cell clusters correlated with the location of the crosshairs of the grid. In addition, due to the variety of locations and shapes of tissues, it is not possible to obtain an evenly distributed layer of cells on a selected tissue site for fusion. One way to eliminate this problem is to remove cells from suspension before placement onto the tissue. Unfortunately, this modification may result in excessive heating of the cells and tissue during extended exposure of the tissue to AC fields. In addition, this extended exposure may cause discomfort to animals or humans during *in vivo* electrofusion procedures. An alternate procedure that addresses this concern is discussed below.

D. Alternate *in Vitro* Procedure

1. *Apparatus*

Mechanical pressure is a suitable substitute for AC dielectrophoresis to obtain contact between the individual cells and the cells of intact tissue. Figure 3 illustrates new electrodes and a new *in vitro* cell–tissue electrofusion chamber designed for the electrofusion of pressure-aligned cells. The *in vitro* chamber consists of six wells. Each well has a flat, round, 10-mm-diameter platinum electrode at the bottom. Each well is insulated and the chamber can be attached to a water bath to maintain the desired thermal environment. The pulse application electrode includes a collection of six individual electrode heads of various shapes and sizes that screw into an insulated handle (insert). The head is selected to match the cell–tissue contact required of the application. The electrode shapes included a concave electrode, which was machined to reflect the curvature of the rabbit cornea.

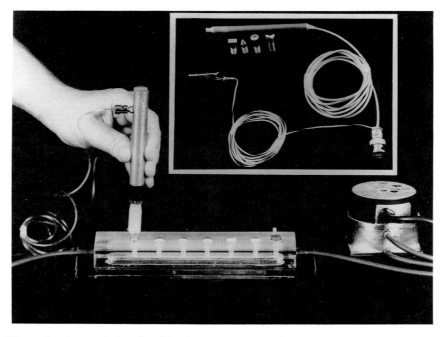

Figure 3 Custom designed and built cell–tissue electrofusion accessories. Six-well fusion chamber is shown in lower portion of figure. Attached air pump circulates temperature regulating water. Insert shows prototype electrode holder with a variety of shapes and sizes of electrodes.

2. *Cell Placement*

A new method for placing the cells in contact with the corneal tissue would be needed if pressure is to replace the AC alignment field. This method's principle attribute must be to increase the cells' exposure to the corneal surface area and to localize and maintain the cells in a specific area when performing CTE procedures both *in situ* and *in vivo*. One method is to layer the cells onto Millipore filters by centrifugation and then place the filter, with the cell side down, onto the tissue. The mechanical pressure is then applied with the custom-designed electrodes (appropriate shape to fit tissue) to obtain juxtaposition between the individual cells and tissue immobilized cells.

3. *Procedure*

Excised corneas, 10 mm in diameter, are placed in the wells of the six-well chamber. Human U937 or HL60 lymphoma cells are centrifuged onto Millipore filters. Filters are placed cell side down onto the corneas. The disc electrode is lowered with a micromanipulator until contact is obtained between the electrode and filter. Pressure is exerted through the electrode until the desired force is reached (600–700 g/cm^2). The force can be estimated by placing the entire apparatus on a torsion beam balance (or similar scale). The pressure is maintained for 3 min and then three 20-μs, 0.20-kV/cm square pulses are delivered 1 s apart to initiate the fusion process.

IV. *In Situ* and *in Vivo* Cell-Tissue Electrofusion

A. Procedure

The procedures for *in situ* and *in vivo* electrofusion are similar. (The animal is anesthetized for the *in vivo* procedure and humanely sacrificed for *in situ* procedures.) A complete description of the procedures is given in Grasso *et al.* (1989) and Heller and Grasso (1990). Briefly, human and nonhuman cells are washed by centrifugation in PBS and layered onto Millipore filters by centrifugation. Supernatant fluids are discarded and the filters containing about 10^7 cells are removed with forceps and placed cell side down on the PBS-washed corneal surfaces of rabbits. Mechanical pressure plus three 20-μs, 20-V square pulses at a 1-s repetition rate were applied simultaneously to a concave titanium electrode shaped to the curvature of the rabbit eye. For initial *in situ* and *in vivo* fusion studies, providing a field value (kV/cm) was not practical because the ground potential was referenced to the rabbit's anatomy. Depending on the instrument available, a prefusion step is performed. If the CGA/Precision (Chicago, IL) Zimmermann cell fusion system is used, a less than 0.1 V, <1 Hz AC dielectrophoretic field is applied for less than 1 s before the three fusion pulses are delivered. (Due to the electronics of the CGA instrument, this very weak and very brief AC electrical field is necessary to activate the circuitry that delivers the fusion pulses.) If the BTX (San Diego, CA) Transfector 800 instrument is used, this AC step is not required. (The authors do not recommend using AC fields for *in vivo* procedures. *In vivo* CTE's conducted at the University of South Florida are accomplished with an instrument designed at the university. This instrument, the VIVOFUSER, does not use any AC fields for cell–tissue alignment.) Following the administration of the electrical pulses all eyes are washed with PBS to remove unfused cells.

B. Transfer of Membrane Surface Receptors

One key issue with CTE generated animal models is whether membrane surface receptors will also be transferred with the cell. To illustrate how to determine if membrane surface components do transfer successfully, two examples are provided. The first presents an *in situ* transfer experiment while the second discusses an *in vivo* situation.

1. Demonstration of in Situ Transfer

Neisseria gonorrhoeae strain Pgh 3-2 was utilized to determine if gonococcal receptors present on human cells were present on the surface of these newly formed phenotypically altered human–rabbit somatic cell hybrids. With the newly devised electrodes and the demonstration that mechanical pressure could substitute for AC alignment, the procedures were performed *in situ* on a freshly killed rabbit.

The *in situ* fusion experiments, in combination with bacterial adherence assays, were performed with an electrode that was machined to match the curvature of the rabbit eye. This electrode was used to exert mechanical pressure to the Millipore filter and rabbit cornea during the application of the square pulses. The electrical parameters and conditions used to electrofuse individual cells to intact tissue were the same as described above except additional control experiments were performed. These experiments included fusing nonhuman cells as well as including corneas not subjected to electrofusion. After electrofusing human HL60 or U937 lymphoma cells, human buccal cells obtained by scraping the inside of the cheek, rabbit skin cells, or monkey kidney Vero cells directly to the rabbit tissue, the corneas were excised and employed in qualitative gonococcal adherence assays and then examined by SEM. The gonococci attached only to those corneas that were fused with human cells.

It is evident from these observations that the administration of an electrical field is not sufficient to modify the rabbit corneal epithelium to allow the attachment of this obligate human pathogen to the rabbit tissue. In addition, the incorporation

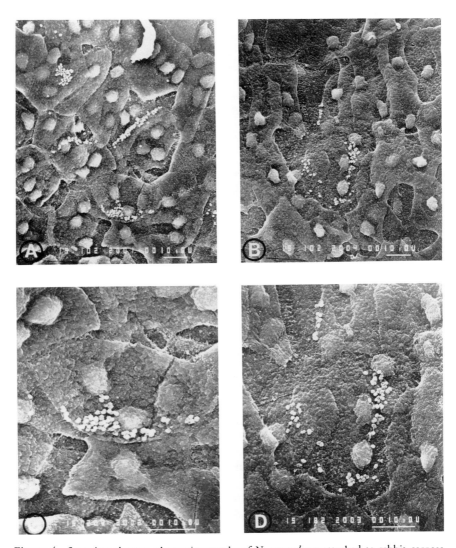

Figure 4   Scanning electron photomicrographs of *N. gonorrhoeae* attached to rabbit corneas histologically modified with electrofused human HL60 cells: receptor-mediated attachment of *N. gonorrhoeae* Pgh 3-2 to somatic human–rabbit cell hybrids formed within rabbit corneal epithelium by electrofusion *in situ*. After human HL60 cells were electrofused to corneal epithelium, the corneas were excised, employed in gonococcal adherence assays, and examined by SEM. (A and B) 1000×, (C) 2000×, and (D) 1800×.

of nonhuman cells into the intact tissue did not allow attachment of the organism. Attachment was observed only when human cells were electrofused to the rabbit corneas (Fig. 4). The human and rabbit plasma membranes appeared to have coalesced completely due to the extended incubation times required to perform the adherence assays (Fig. 4) in contrast to the early fusion stages (Fig. 2). From these observations it is evident that the process of cell–tissue electrofusion does not modify or damage cell surface membrane components.

2. Demonstration of In Vivo Transfer

Human HL60 cells were electrofused to the corneas of anesthetized rabbits to determine if CTE could be performed without harm to the animals. The electrofusion procedure utilized was the same as described above. A series of four experiments was set up; each experiment utilized four rabbits. The experimental design utilized was described previously (Heller and Grasso 1990). Briefly, rabbit number 1 was set up as the electrofusion and human cell control. Fusion pulses were administered to the left eye in the absence of cells, while the right eye was exposed to human cells in the absence of fusion pulses. In all experiments, there were no observable signs of ocular inflammation in these eyes after exposure to the pathogen. Rabbit, murine, and human cells were electrofused to both eyes of rabbits 2, 3, and 4, respectively. The left eye of each animal was not exposed to gonococci and served as the inflammation control. There were no visible signs of inflammation detected in the left eyes of rabbits 2, 3, and 4. When the right eyes of rabbits 2 and 3 bearing electrofused nonhuman cells were exposed to the bacteria, neither inflammatory reactions nor ocular lesions developed. In striking contrast, a purulent keratoconjunctivitis developed in the right eye of rabbit 4 bearing electrofused human cells. Clinical symptoms of these ocular lesions appeared approximately 2 h after infection and peaked between 8 and 12 h (Fig. 5). Inflammation was completely resolved after 24 h, and no additional inflammation was observed during an extended 7-day observation period in any of the rabbits. This experimental design was repeated four times with similar results. The only time inflammation was observed was if the rabbit eye had human cells fused to its surface and had been exposed to N. gonorrhoeae.

Figure 5   Purulent gonococcal conjunctivitis in rabbit cornea following transfer of gonococcal receptors by cell–tissue electrofusion. (A) Photograph of a mild gonococcal lesion that occurred in the right eye of rabbit 4 only 4 h after initiating the bacterial infection. Human–rabbit somatic cell hybrids were formed 30 min prior to the addition of the bacteria, by electrofusing human HL60 cells to the rabbit corneal epithelium. (B) Photograph of the right eye of rabbit 3, which had been similarly infected 4 h earlier. Murine–rabbit somatic cell hybrids were formed 30 min prior to the addition of the bacteria, by electrofusing murine WEHI-3 cells to the rabbit corneal epithelium. (From Heller and Grasso, 1990, reproduced with permission.)

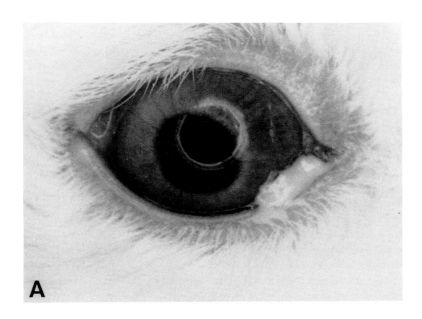

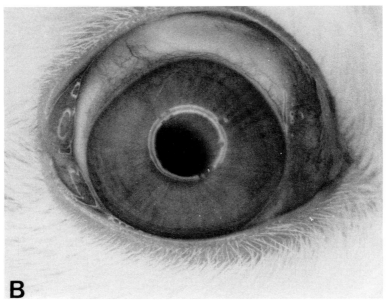

C. Complications

A variation in the severity of the observed inflammation from mild to moderately severe was noted when comparing the results found in the right eye of rabbit 4 from each of the four *in vivo* experiments. This variation is attributed to undesirable variables that need to be identified and controlled. One possible explanation for these results is that the severity may be directly related to the number of cells fused. Although cell–tissue electrofusion was successfully accomplished in each of the experiments, there may have been a variation in the number of cells fused. Therefore, further assessment of this problem is dependent upon the development of a method to quantitate the number of cells fused. A new assay procedure is presented in Section V of this chapter.

The variation in the number of cells fused may be attributed to the way the cells are placed in contact with the tissue prior to fusion. Placing a suspension of cells in a defined volume onto the tissue and allowing the cells to settle by gravity did not yield an evenly distributed layer of cells on the tissue. Employing Millipore filters to place cells onto the tissues improved the situation but ultimately proved to be inadequate. Cells had a tendency to fall off or shift position when the filter was turned upside down to be placed onto the tissue. Therefore, a modified cell placement method was examined.

D. Modified Cell Placement Method

Electrostatically charged membranes, (Zeta-Probe discs) can be utilized in a modified cell placement procedure. When individual cells are centrifuged onto the surface of Zeta-Probe discs, as described in Heller and Grasso (1991), the cells remain loosely adhered to the discs. This is attributed to the electrostatic attraction between the negatively charged cells and positively charged disc. This attraction is sufficient to retain the cells during the removal of the disc from the centrifuged plates and alignment with the surface of the tissue. Furthermore, the cells remain adhered to the disc at ambient temperatures for several hours. The disc can be picked up and then inverted so the cells can be placed on the tissue without any cells being lost. After the electrofusion pulses are applied, all of the cells are released from the disc.

A reproducible number of cells can be layered onto these disks by centrifuging for 5 min at 180 × g. A vital fluorescent dye, hydroethidine (HE), available from Polyscience Inc. (Warrington, PA), can be used as part of a quantitation method to determine the number of cells on the disc (Heller and Grasso, 1991). In addition, there are no differences in any of the characteristics of non-HE-stained cells when subjected to these procedures. This is critically important to establish because quantitation is based upon HE stained cells and *in vivo* experiments are carried out with non-HE-stained cells.

V. Quantitative Assay for Cell–Tissue Electrofusion

The majority of procedures developed and information gathered about the process of CTE have been qualitative. Quantitative information about the approximate number of cells that are electrofused to tissues under various biological and electromechanical conditions is not easy to obtain. A new spectrofluorometric assay to quantify CTE has been devised. This method, together with a series of experiments designed to determine optimal CTE operating conditions, will provide the information about electromechanical conditions needed to assure good yields of fused cells. *In vitro* cell–tissue electrofusion studies were performed with excised rabbit corneas, in order to work out the method's technical details. *In vitro* instead of *in vivo* studies were conducted because these experiments were much less expensive, were much more practical, and were much easier to conduct under controlled conditions.

The procedure is outlined as follows:

1. Cells in suspension are stained with hydroethidine, a vital fluorescent dye.
2. Stained cells are washed by centrifugation to remove extracellular dye.
3. Corneal buttons (10 mm diameter) are dissected from excised tissue.
4. Stained cells layered on Zetaprobe discs are electrofused to rabbit corneal epithelium on the excised tissue.
5. Unfused stained cells are removed by washing the tissue.
6. Stained fused cells on the corneal buttons are lysed by exposure to 0.2% sodium dodecyl sulfate (SDS), which releases the dye.
7. The remaining corneal tissue is removed and fluorescent intensities in the supernatant fluids are measured with a spectrofluorometer.
8. The number of cells that had been electrofused is estimated from standard curves.
9. These curves are generated from fluorescence measured in dilutions of lysed cells prepared from the same stained cell preparation that is used for electrofusion.

VI. Standardized Reproducible Cell–Tissue Electrofusion

A standard reproducible *in vitro* cell–tissue electrofusion procedure as demonstrated by the previously described quantitative assay is as follows:

1. Corneal buttons (10 mm) are excised and placed in the fusion chamber (Fig. 3).
2. Individual cells are centrifuged onto Zeta-Probe discs and placed cell side down onto corneas.
3. Flat platinum electrode is placed in holder (Fig. 3) and lowered by a micromanipulator to apply pressure to juxtapose individual cells and cells in tissue.

4. A single square-wave pulse of 0.6 kV/cm is applied via a DC generator to initiate the fusion process.
5. The electrode is raised and the cornea is removed and washed.
6. Histologically modified cornea can now be fixed and examined by SEM or the number of cells fused can be quantitated by the previously described method.

The quantitative assay can be utilized to more fully characterize the process of cell–tissue electrofusion.

VII. Summary

Cell–tissue electrofusion has evolved through several different phases to reach a point where the procedure has been standardized. CTE has been used to demonstrate that:

1. Individual cells can be electrofused to intact tissue *in vitro, in situ,* and *in vivo.*
2. *In vivo* electrofusion can be performed without causing lethality or ocular inflammation to anesthetized animals.
3. Human gonococcal membrane receptors remain functional after their incorporation into human–rabbit somatic cell hybrids.
4. An obligate human pathogen can produce a purulent inflammatory response mediated by transferred human microbial attachment receptors in a common laboratory animal.

When these results are examined it is evident that a model system for studying the infectivity of species-specific diseases can be developed by fusing susceptible cells to the tissue of naturally resistant animals by the process of cell–tissue electrofusion.

Unfortunately, the ability to perform CTE is hampered by the lack of commercially available instruments and accessories. CTE experiments can be performed with existing commercially available pulse generators but specially designed electrodes must be utilized. In addition, further development of this technology will be dependent on optimizing the number of cells fused. Furthermore, the examination and development of the various biological applications, such as the *N. gonorrhoeae* model discussed in this chapter are possible with current combinations of commercially available instrumentation using modified electrodes. Continued development in *in vivo* cell–tissue electrofusion applications will undoubtably result in enhanced instrumentation designed for this purpose.

References

Arko, R. J. (1989). Animal models for pathogenic *Neisseria* species. *Clin. Microbiol. Rev.* 2(Suppl), S56–S59.

Bates, G. W., Saunders, J. A., and Sowers, A. E. (1987). Principles and applications of electrofusion. *In* "Cell Fusion" (A. E. Sowers, ed.), Plenum Press, New York, pp. 367–395.

Chang, D. C., Hunt, J. R., and Gao, P. Q. (1989). Effects of pH on cell fusion induced by electric fields. *Cell Biophys.* 14(3), 231–243.

Chernomordik, L. V., Sokolov, A. V., and Budker, V. G. (1990). Electrostimulated uptake of DNA by liposomes. *Biochim. Biophys. Acta* 1024(1), 179–183.

Ghosh, H. K. (1987). Gonococcal eye infection in the elderly. *Med. J. Austral.* 146, 330.

Grasso, R. J., Heller, R., Cooley, J. C., and Haller, E. M. (1989). Electrofusion of individual animal cells directly to intact corneal epithelial tissue. *Biochim. Biophys. Acta* 980, 9–14.

Heller, R., and Grasso, R. J. (1990). Transfer of human membrane surface components by incorporating human cells into intact animal tissue by cell-tissue electrofusion *in vivo*. *Biochim. Biophys. Acta* 1024, 185–188.

Heller, R., and Grasso, R. J. (1991). Reproducible layering of tissue culture cells onto electrostatically charged membranes. *J. Tissue Culture Methods* 13, 25–30.

Neumann, E., Gerisch, G., and Opatz, K. (1980). Cell fusion induced by high electric impulses applied to *Dictyostelium*. *Naturwissenschaffen* 67, 414–415.

Powell, K. T., Morgenthaler, A. W., and Weaver, J. C. (1989). Tissue electroporation. Observation of reversible electrical breakdown in viable frog skin. *Biophys. J.* 56(6), 1163–71.

Rols, M. P., and Teissie, J. (1989). Ionic-strength modulation of electrically induced permeabilization and associated fusion of mammalian cells. *Eur. J. Biochem.* 179(1), 109–115.

Rols, M. P., Dahhou, F., Mishra, K. P., and Teissie, J. (1990a). Control of electric field induced cell membrane permeabilization by membrane order. *Biochemistry* 29(12), 2960–2966.

Rols, M. P., and Teissie, J. (1990b). Modulation of electrically induced permeabilization and fusion of Chinese hamster ovary cells by osmotic pressure. *Biochemistry* 29, 4561–4567.

Senda, M., Takeda, J., Abe, S., and Nakamura, T. (1979). Induction of cell fusion of plant protoplasts by electrical stimulation. *Plant Cell Physiol.* 20, 1441–1443.

Sowers, A. E. (1989). Evidence that electrofusion yield is controlled by biologically relevant membrane factors. *Biochim. Biophys. Acta* 985(3), 334–338.

Sowers, A. E. (1990). Low concentrations of macromolecular solutes significantly affect electrofusion yield in erythrocyte ghosts. *Biochim. Biophys. Acta* 1025(2), 247–251.

Tsong, T. Y. (1987). Electric modification of membrane permeability for drug loading into living cells. *Methods Enzymol.* 149, 248–259.

Ullman, S., Roussel, T. J., Culbertson, W. W., Forster, R. K., Alfonso, E., Mendelsohn, A. D., Heidemann, D. G., and Holland, S. P. (1987). *N. gonorrhoeae* keratoconjunctivitis. *Opthamology* 94, 525–531.

Wan, W. L., Farkas, G. C., May, W. N., and Pobin, J. B. (1986). The clinical characteristics and course of adult gonococcal conjunctivities. *Am. J. Opthalmol.* 102, 575–583.

Weber, H., Forster, W., Berg, H., and Jacob, H.-E. (1981a). Parasexual hybridization of yeasts by electric field stimulated fusion of protoplasts. *Curr. Genet.* 4, 165–166.

Weber, H., Forster, W., Jacob, H.-E., and Berg, H. (1981b). Microbiological implications of electric fields. III. Stimulation of yeast protoplast fusion by electric field pulses. *Allg. Mikrobiol.* 21, 555–562.

Zimmermann, V., Pilwat, G., and Richter, H.-P. (1981). Electric field stimulated fusion: Increased field stability of cells induced by pronase. *Naturwissenschaffen* **68**, 557–579.

Zimmermann, U. (1982). Electric field-mediated fusion and related electrical phenomena. *Biochim. Biophys. Acta* **694**, 227–277.

Zimmermann, U. (186). Electrical breakdown, electropermeabilzation and electrofusion. *Rev. Physiol. Pharmacol.* **105**, 176–256.

25

Novel Applications of Electroporation

Sean R. Gallagher[1] and Donald C. Chang[2,3]

[1]Hoefer Scientific Instruments, San Francisco, California 94107

[2]Department of Molecular Physiology and Biophysics, Baylor College of Medicine, Houston, Texas 77030

I. Introduction
II. Protein Loading
   A. Restriction Enzymes
   B. Antibodies
   C. Phalloidin
III. Virus
   A. Plant
   B. Protozoa
   C. Insect Cells
IV. Whole Tissue
   A. Plants
   B. Animals
V. Transgenic Fish
VI. Cell Monolayers
VII. Improved Cell Viability and Transformation Efficiency
   A. Cell Synchronization
   B. Additives
   References

I. Introduction

The intent of this chapter is to present areas of research that, while not directly addressed by the other chapters of this book, are of great importance. Thus, the following discussion is not intended to be exhaustive, but will be topical in nature.

[3]Present address: Department of Biology, Hong Kong University of Science and Technology, Kowloon, Hong Kong.

The wide range of research covered illustrates one of the most fascinating aspects of electroporation. Initially, as a tool for molecular biology, it moved rapidly to the forefront as a method of choice for transformation in eukaryotic and then prokaryotic systems. However, because of the way electroporation works—by reversibly breaking down the lipid bilayer surrounding the cell—a number of different compounds and indeed whole organisms can be introduced to the cytosol, opening up new and novel ways to explore cellular physiology.

Restriction enzymes and antibodies represent recent examples of this new application, and are discussed in detail in Section II. Simultaneous infection of a large population of cells with whole virion or viral nucleic acid is also possible with electroporation, as presented in Section III. This simplifies the study of viral infection sequence, as well as speeding the development of vectors. Section IV covers electroporation of whole tissue. This should help the development of genetically engineered crop plants as well as the study of transformed cells in the more natural state of intact tissue. Applications in Section V are directed at improving cell viability and transformation efficiency by coelectroporation with restriction enzymes, and by the addition of serum or PEG to the electroporation media.

Other chapters in this book covering the areas of general applications of electroporation (Part II), gene transfer in plants (Chapters 15, 16, and 29), mammalian cells (Chapters 27 and 28), yeasts (Chapter 31), bacteria (Chapters 17, 18, and 30), and loading of macromolecules (Chapters 20 and 21) should be consulted for more detailed information about specific systems.

II. Protein Loading

Several recent reports indicate that electroporation is an effective technique for introducing a large variety of proteins into cells (Table 1). Under optimal conditions, electroporation causes minimal damage to the cells and permits rapid loading of antibodies (Kwee *et al.*, 1990; Chakrabarti *et al.*, 1989; Berglund and Starkey, 1991), cytotoxic proteins (Mir *et al.*, 1988), and active enzymes (Winegar *et al.*, 1989; Morgan *et al.*, 1990a; Morgan and Winegar, 1989; Moses *et al.*, 1990). However, the cell cycle can be adversely affected (Goldstein *et al.*, 1989; Morgan *et al.*, 1990b). The proteins as large as 230 kD have been electroporated (Lambert *et al.*, 1990), and the cells can either be in suspension (e.g., Winegar *et al.*, 1989) or in monolayers (Zheng and Chang, 1991; Raptis and Firth, 1990; Kwee *et al.*, 1990). The conditions for electroporating proteins are similar to those for DNA.

A. Restriction Enzymes

Electroporating active restriction enzymes into Chinese hamster ovary (CHO) cells illustrates the power of this technique for protein work. Carcinogenic substances

Table 1

Electroporation Conditions for Protein Loading

System	[Protein]	Field strength	Pulse type	Reference
HeLa, L5178Y cells (antibodies)	2 mg/ml	750 V/cm	0.7–0.9 ms Pulse, capacitor	Chakrabarti et al. (1989)
Mouse melanoma (antibodies)	1/1000 dilution	688 V/cm	Capacitor	Berglund and Starky (1991)
Rat fibroblasts (antibodies)	2 mg/ml	50–150 V	32 μF Capacitor	Raptis and Firth (1990)
Human monolayer (monoclonal antibodies)	—	300–400 V/cm	0.1–1 μF Capacitor	Kwee et al. (1990)
NIH 3T3 cells (pokeweed antiviral protein, PAP)	10^{-11} M PAP	1400 V/cm	Square wave, 8 × 100 μs	Mir et al. (1988)
CHO cells (various proteins)	10 μg/ml	750 V/cm	960 μF Capacitor	Lambert et al. (1990)
Xeroderma pigmentosum (endogenous endonuclease)	0.8–5 μg/ml	1.7 kv/cm	Capacitor	Tsongalis et al. (1990)
Ltk⁻ (Hind III and XbaI restriction enzymes)	0–200 U/ml	1750 V initial	Capacitor	Yorifuji and Mikawa (1990)
CHO cells (various restriction enzymes)	12.5–125 U/ml	650–750 V/cm	1600 μF Capacitor	Winegar et al. (1989), Morgan et al. (1990b)
CHO cells (various restriction enzymes)	5–40 U/ml	400–850 V/cm	1600 μF Capacitor	Moses et al. (1990)
Polymorphonuclear leukocytes (PMNs) (actin labeling with phalloidin)	80 μM	500 V/cm	3 × 3 μs Pulses	Hashimoto et al. (1989)
Carrot suspension (actin labeling with phalloidin)	100 μM	1 kV/cm	5 μF capacitor	Traas et al. (1987)

and UV and ionizing radiation cause many different types of damage in DNA (Morgan et al., 1988). Effects vary depending on the agent, but DNA damage is usually measured by quantitating the number of chromosome aberrations and sister chromatid exchanges. Restriction enzymes, because they cause double-strand breaks at specific DNA sequences, provide a unique tool to study the effects of defined lesions in the chromatin. Previously, various techniques using inactivated Sendai virus, pelleting, osmolytic shock, and expression vectors were required to introduce restriction enzymes into cells (Morgan and Winegar, 1989). These procedures suffered from low yields, high variability, a requirement for a large amount of enzyme, or a complicated and lengthy protocol.

Optimal conditions for electroporation of restriction enzymes into Chinese hamster ovary cells have been discussed in detail (Winegar et al., 1989; Morgan et al., 1990a, 1990b; Moses et al., 1990). Exponentially growing cells are trypsinized, washed, and suspended into HEPES-buffered saline (Winegar et al., 1989). Moses et al. (1990) preferred Hanks balanced salts due to the cytotoxicity of HEPES in their system. Up to 100 U of restriction enzyme was added to 0.8 ml of the cell suspension. Although electroporation was at room temperature, Winegar et al. (1989) placed the cells on ice for 5 min immediately afterward. At this point, the authors stress, the cells are very fragile. Electroporation under these conditions causes a delay in the cell cycle, necessitating culturing the cells for 22 h to select for cells electroporated during G1 (Morgan et al., 1990b). Furthermore, blunt-end-producing restriction enzymes caused significantly more cytogenetic damage than those producing cohesive ends. This may be related to the efficiency of cutting within the cell. Pulsed field electrophoresis can be used to monitor the extent of DNA breakage and thus the activity of various restriction enzymes following electroporation (Morgan et al., 1990a, 1990b). Restriction enzymes that normally produce blunt and staggered end breaks in DNA are remarkably effective in vivo and induced chromosomal aberrations in greater than 90% of the electroporated cells (Winegar et al., 1989). One general conclusion from this work was that double-stranded breaks in DNA lead to chromosome aberrations but not to sister chromatid exchange.

Endonucleases also have potential for improving transformation efficiency and, in an application using an endogenous endonuclease, correcting genetic defects. Yorifuji and Mikawa (1990) showed that mixing the restriction enzymes Hind III and Xba I with DNA just prior to electroporation increased the transformation efficiency of the thymidine kinase gene. Tsongalis et al. (1990) reported that introducing by electroporation an endonuclease isolated from normal lymphoblastoid cells into repair-deficient xeroderma pigmentosum cells will restore DNA repair of UV radiation damage via unscheduled DNA synthesis (UDS) and correct the primary defect of the complementation group A (XPA) xeroderma pigmentosum cells.

B. Antibodies

Electroporation provides a simple and very effective way to introduce antibodies into cells while maintaining high viability. Using monoclonal antibodies directed against bovine asparagine synthase, Chakrabarti *et al.* (1989) demonstrated that an antibody that inhibits the tumor form of asparagine synthase *in vitro* caused the tumor cells to have a greater dependence for exogenous asparagine, presumably due to enzyme inhibition *in vivo* following electroporation. Optimal conditions for transferring the antibodies to HeLa, human fibroblasts, and murine lymphoma cells included suspending the cells at a concentration of 1×10^7 in Opti-MEM (GIBCO) with 5% FCS, and electroporating with a short pulse length (0.7–0.9 ms) at relatively low capacitance (50–60 μF) and moderate voltage (750 V/cm). The procedure worked best with exponentially growing cells. Cell viability was 80–90%, with 90% of the cells retaining antibody.

In a study using antibodies to the *ras* oncogenes, Berglund and Starky (1991) have identified several important parameters for the introduction of anti-ras antibodies into B16BL6 mouse melanoma cells. Survival was improved by storing the cells in media containing serum, complete RPMI 1640, prior to and immediately after electroporation, while suspending the cells in Hanks balanced saline solution just for the electroporation. Although the cells remained permeable to the low-molecular-weight DNA intercalating fluorochrome propidium iodide (PI) for several hours, the antibody had to be present during the electroporation for maximum antibody labeling. Due to the length of time that the cells are permeable to trypan blue and PI, the use of these reagents for viability testing is questioned.

Antibodies, other proteins, and DNA have also been electroporated into cell monolayers, thus eliminating the need to remove the cells into suspension first (Kwee *et al.*, 1990; Zheng and Chang, 1991; Raptis and Firth, 1990). This is termed *in situ* electroporation and is discussed in detail below.

C. Phalloidin

Fluorescent phalloidin, a cyclic peptide that binds to actin, can be introduced to the cytosol of cells by electroporation (Traas *et al.*, 1987; Hashimoto *et al.*, 1989). Traas *et al.* (1987) electroporated cell suspensions of carrot in the presence of rhodaminyl lysine phallotoxin without prior removal of the cell wall. By avoiding aldehyde fixation, a more complete visualization of the actin network was possible. Similarly, Hashimoto *et al.* (1989) described optimal conditions for labeling of motile polymorphonuclear leukocytes (PMNs) with fluorescein isothiocyanate-tagged phalloidin to observe the distribution of actin during cell movement.

III. Virus

Electroporation of either the whole virion or the naked nucleic acid into plant cells (Hibi, 1989; Osbourn and Wilson, 1992; Osbourn et al., 1989; Watts et al., 1987; Register and Beachy, 1988; Watanabe et al., 1987; Luciano et al., 1987; Nishiguchi et al., 1986), protozoa (Furfine and Wang, 1990), and insect cells (Mann and King, 1989) illustrates the varied applications of electroporation to the use and study of viruses. The advantage of electroporating viral nucleic acid directly into cells or protoplasts are many. There can be a simple and dramatic increase of efficiency of infection over a traditional procedure, such as is the case with producing recombinant baculovirus by infection of insect cells (Mann and King, 1989). With plants, applications include more easily discerning a viral infection sequence (Hibi, 1989, and references within) and determining the mechanism of plant resistance to viruses (Register and Beachy, 1988; Osbourn and Wilson, 1992; Osbourn et al., 1989). Rapidly and reproducibly infecting cells with viral nucleic acid simplifies the production of recombinant virus and thus improves the chances of developing new vectors. This is important in situations where no genetic vector exists and information about the genetics of potential candidate vectors is limited. This approach has been used with the double-stranded RNA virus (GLV) of the intestinal parasite *Girardia lamblia* (Furfine and Wang, 1990).

A. Plant

Hibi (1989) has summarized results from several laboratories working with plants. Of the several parameters that must be considered—electroporation media, cell and nucleic acid concentration, temperature, electrical conditions—the physiological state of the protoplasts is among the most important. However, there is a wide range of conditions employed by various laboratories (Table 2), and both square-wave and exponential-decay electroporation pulses are used.

Electroporation is an important tool for determining how viruses infect plants. One example is cross protection, where an infection of a mild strain of virus prevents subsequent infection by a more virulent strain. Expression of the viral coat protein in transgenic plants gives some degree of resistance to infection by a virus with an identical or closely related coat protein. Viral RNA, whole virion, and pseudovirus particles—viruslike particles assembled *in vitro* through the interaction with coat protein and mRNA—have all proven very useful in the study of genetically engineered protection against virus infection in plants (e.g., Osbourn and Wilson, 1992; Register and Beachy, 1988). When transgenic plants express a low level of coat protein from tobacco mosaic virus (TMV), for example, they are resistant to infections by whole but not partially uncoated TMV or naked RNA. It is hoped this general strategy of expression of coat protein will produce commercially available crops with

Table 2

Electroporation Conditions for RNA and DNA Viruses

System		V/cm	Pulse type	Reference
RNA Virus	[RNA]			
Giardia lamblia	0.5 mg/ml	12.5 kV/cm	25 μF Capacitor, 400-Ω shunt	Furfine and Wang (1990)
Plant cells	1–10 μg/ml	0.5–2 kV/cm	Square wave, 1–5 × 50–100 μs	Hibi (1989)
Tobacco	0.13–2.5 μg/ml	750 V/cm	100 μF Capacitor	Watanabe et al. (1987)
Tobacco	80 μg/ml	300 V/cm	1000 μF Capacitor	Luciano et al. (1987)
Tobacco	10 μg/ml	10 kV/cm	Square wave, 9 × 90 μs pulse	Nishiguchi et al. (1986)
Tobacco	5–30 μg/ml	2.5 kV/cm	50 nF Capacitor	Watts et al. (1987)
Pseudovirus				
Tobacco	50 μg/ml pseudovirus 2.5 μg/ml mRNA	200 V/cm	Square wave, 200 ms	Osbourn et al. (1989)
DNA virus	[DNA]			
Insect cells (baculovirus)	1 μg/ml		Capacitor, 2.8– 7.7 ms pulse length	Mann and King (1989)

some degree of resistance to infection from several viruses including TMV, cucumber mosaic virus (CMV), potato virus X (PVX), alfalfa mosaic virus (A1MV), tobacco rattle virus (TRV), and tobacco streak virus (TSV).

Pseudovirus particles provide a way to observe the effects of coat protein expression both on the uncoating step of infection and on gene translation of the encapsidated mRNA (Osbourn et al., 1989; Osbourn and Wilson, 1992). To determine at what point in the infection sequence endogenous coat protein expression exerts its effect, protoplasts are electroporated in the presence of a pseudovirus consisting of an mRNA for the β-glucuronidase (GUS) reporter gene. The level of expression of GUS is then a measure of the extent of uncoating that takes place. Factors other than uncoating can also be investigated in this way.

B. Protozoa

In many systems, the lack of a genetic vector is a major roadblock to further study. The recent discovery of double-stranded RNA viruses in a number of protozoan parasites provides researchers with a starting point from which to develop useful vectors. This includes *Giardia lamblia*, a protozoa that causes severe diarrhea throughout the world (Furfine and Wang, 1990). In addition to the double-stranded RNA *Giardia lamblia* virus (GLV), cells infected with GLV also contain a single-stranded full-length RNA copy of the genome (Furfine *et al.*, 1989). Using electroporation of gel-purified and electroeluted RNA, Furfine and Wang (1990) demonstrated that the single-stranded, but not the double-stranded, RNA species is the only nucleic acid needed for infection, and that the infection did not occur without electroporation.

C. Insect Cells

Baculoviruses are important tools for eukaryotic expression of foreign genes. The virus infects insect cells, and the infected cells are not only capable of producing large quantities of soluble recombinant protein, but can also effectively process, modify, and package proteins from higher organisms (Piwnica-Worms, 1990). Mann and King (1989) used electroporation to transfect insect cells (*Spodoptera frugiperda*) with baculovirus (*Autographa californica*) DNA alone, or in combination with a transfer plasmid carrying the β-galactosidase gene to produce recombinant viruses through homologous recombination. Although electroporation was much more effective at transfecting cells with baculovirus DNA when compared to the traditional procedures using calcium phosphate, both procedures gave comparable amounts of recombinant virus. Maximum recovery of recombinants (2.9%) occurs 2 days after electroporation.

IV. Whole Tissue

A. Plants

Typically, electroporation of plant cells is performed with suspensions of individual protoplasts (Table 2). However, the expression of foreign DNA in isolated protoplasts

Table 3
Electroporation Conditions for Intact Tissue

Tissue	[DNA]	Field strength	Pulse type	Reference
Rice	100 μg/ml	375 V/cm	900 μF Capacitor	Dekeyser *et al.* (1990)
Retina	30 μg/ml	400 V/cm	960 μF Capacitor	Pu and Young (1990)
Fish eggs	100 μg/ml	750 V/cm	5×50 μs	Inoue *et al.* (1990)

can be quite different from intact tissue (Dekeyser *et al.*, 1990). To obtain a more realistic view of gene expression in transient gene assays, Dekeyser *et al.* (1990) developed a method for electroporating intact rice, wheat, maize, and barley (Table 3). Compared to procedures with protoplasts, electroporation of intact tissue required an incubation and wash step prior to adding the DNA in order to remove nucleases from the explant. Other modifications included using relatively high concentrations of DNA, preincubating the DNA with the tissue, and electroporating the tissue in a small volume of solution with a relatively high capacitance and long pulse length.

B. Animals

Transfection of intact animal tissue is also possible with electroporation, as was demonstrated with chicken retinal explants (Pu and Young, 1990). Whole complex tissue is ideally suited for the study of transcriptional control using transient assays, and of how cell–cell interactions might affect this. Three DNA transformation procedures—calcium phosphate, cationic lipids, and electroporation—were used. Under their conditions, the signal from the CAT assay was highest in extracts from electroporated tissue. The data obtained from the transient assay of electroporated retinas indicate that cis-active elements of the glutamine synthetase gene are located between 1.3 and 2.5 kb upstream from the gene.

The biophysics of electroporation in whole tissue have been studied by Powell *et al.* (1989). They demonstrated that reversible membrane breakdown is possible without apparent damage to the tissue.

V. Transgenic Fish

The production of stable, germ line-transformed plants and animals is required to answer questions about gene regulation and function in the more complex environment of the whole organism. Many of the techniques required to generate stable transformants in other systems are discussed elsewhere in this book (e.g., Chapters 14 and 27). However, obtaining transgenic fish through electroporation is new and will be discussed briefly below. Because of the difficulties associated with microinjection, Inoue *et al.* (1990) developed a protocol for transforming fertilized fish (medaka) eggs: 3109 eggs were electroporated with a plasmid containing the gene for rainbow trout growth hormone. Of the 783 that hatched, 4% were transgenic. The F1 and F2 offspring also contained the gene, indicating that germ-line transformants are possible with this technique (Table 3).

VI. Cell Monolayers

In addition to eliminating the typical trypsin/ethylenediamine tetraacetic acid (EDTA) treatment to put cells into suspension, reversibly permeabilizing attached cells by *in situ* electroporation has several other advantages (Kwee *et al.,* 1990; Zheng and Chang, 1991; Raptis and Firth, 1990). Under optimal conditions, viability and efficiency are both very high, and a large variety of macromolecules can be introduced to both the cytoplasm and the nucleus. If fluorescent probes are used, the cells are visualized more easily when left attached.

Results between different laboratories are difficult to compare because of the varied methods used (Table 4). Cells are grown and electroporated either on coverslips (Kwee *et al.,* 1990; Zheng and Chang, 1991) or, in a new application, conductive indium-tin-coated glass in conjunction with a unique electrode design (Raptis and Firth, 1990). Potter *et al.* (Chapter 13) demonstrated that cells grown on microcarrier beads can be electroporated with relative ease, and this is discussed in detail elsewhere in this book. Specialized electrodes that fit into 16-mm tissue culture wells are also available for monolayer electroporation (Hoefer Scientific Instruments). Zheng and Chang (1991) showed that with radiofrequency electroporation of the β-gal gene, COS-M6 and CV-1 cell monolayers had higher transformation efficiencies than the identical cells electroporated in suspension. The difference appears to be caused by the trypsin detachment treatment rather than by electroporation *per se*. At the optimal radiofrequency field strength of 1.2 kV/cm, survival of the COS-M6 cells was approximately 50%, with over 80% of the cells expressing β-galactosidase.

In contrast, Raptis and Firth (1990) reported much lower efficiencies with the gene for hygromycin resistance when transforming Fisher rat fibroblasts grown on conductive glass. Conditions were varied from 50 to 150 V, and the capacitance was 32 μF. However, at voltages lower than that required for transfection, antibodies could be loaded into essentially 100% of the cells with minimal cell death. Furthermore, electroporation of monolayers on conductive glass produced no change in the cell cycle length (see, however, Morgan *et al.,* 1990b, and Section II,A above).

Table 4
Electroporation Conditions for Cell Monolayers

Cells	[DNA]	Field strength	Pulse type	Reference
Rat fibroblasts	50 μg/ml	50–150 V	32 μF Capacitor	Raptis and Firth (1990).
Human monolayer	—	300–400 V/cm	0.1–1 μF Capacitor	Kwee *et al.* (1990)
COS-M6 CV-1	100 μg/ml	1.2 kV/cm	RF 40 KHz, 3 pulses/train, 5 pulses/train	Zheng and Chang (1991)

Table 5

Electroporation Conditions for Cell Synchronization, PEG, and Butyrate Treatments

Cells	[DNA]	Field strength	Pulse type	Reference
Cell synchronization				
Mouse Ltk⁻ cells	6 μg/ml	1500 V/cm	$\tau = 1.7$ ms	Yorifuji *et al.* (1989)
Human fibroblasts (combined with postelectroporation butyrate treatment)	120 μg/ml	to 450 V/cm	960 μF Capacitor	Goldstein *et al.* (1989)
Polyethylene glycol				
S. *cerevisiae*	3–19 μg/ml	854 V/cm	500 μF Capacitor	Rech *et al.* (1990)
S. *pombe*	10 μg/ml	8.5 kV/cm	1 μF Capacitor	Hood and Stachow (1991)
N. *tabacum* protoplast	20 μg/ml	1–1.25 kV/cm	1 nF Capacitor ($\tau = 10$ μs)	Shillito *et al.* (1985)
Serum				
Hematopoietic stem WEHI-3B cells	50 μg/ml	550–850 V/cm	7–8 ms Pulse, capacitor	Bahnson and Boggs (1990)

VII. Improved Cell Viability and Transformation Efficiency

A. Cell Synchronization

Through use of synchronized human fibroblast cells and a postelectroporation treatment with butyrate, Goldstein *et al.* (1989) were able to electroporate cells with reduced amounts of DNA and at lower voltages (Table 5). These less deleterious conditions were critical for performing experiments that not only produced reasonable cell survival, but enhanced expression of the transfected DNA while maintaining normal cell cycling. Similarly, Yorifuji *et al.* (1989) found that by first synchronizing the mouse Ltk⁻ cells with hydroxyurea or aphidicolin and then transfecting with the herpes simplex thymidine kinase gene, the transformation efficiency was up to eight times higher at the G2/M phase of cell growth.

B. Additives

Various defined and undefined additives have been mixed with cells prior to electroporation or after electrical treatment in order to improve efficiency.

1. Serum

Bahnson and Boggs (1990) showed that the addition of 20% serum to murine myelomonocytic leukemia WEHI-3B cells before or after electroporation not only modestly improved cell viability, but dramatically increased transformation efficiency. Serum addition caused electroporated cells to reseal much more quickly, as indicated by lucifer yellow permeation. Sealing the membrane more quickly might limit the leakage of cell components, contributing to cell stability. Furthermore, the optimum net transformation efficiency was shifted to 650 from 550 V/cm in the presence of serum. Although it is unclear what the active compounds in the serum area, both high- and low-molecular-weight components are indicated.

2. PEG

Polyethylene glycol (PEG) is required for efficient transformation of both *Saccharomyces cerevisiae* (Rech *et al.*, 1990) and *Schizosaccharomyces pombe* (Hood and Stachow, 1990, 1991). Earlier, Shillito *et al.* (1985) also found that PEG can enhance the efficiency of gene transfer to plants. The effects of PEG on transformation are complex. Hood and Stachow (1991) show that the addition of PEG not only extends the length of time the electroporated cells remain permeable but further increases their permeability. During incubation with PEG, the pores created during electroporation also apparently grow in size. The combined effect of PEG on the size of the pores and their lifetime may enhance the uptake of DNA and thus result in the observed sixfold improvement in transformation efficiency. A level of 30% PEG gives optimal levels of transformants per microgram of DNA. Heat shock, a nonelectrical method of transformation, also uses PEG, but by another mechanism that probably facilitates DNA uptake without creating pores, and this may reflect a second mechanism occurring during electroporation.

3. Butyrate

Butyrate (Goldstein *et al.*, 1989; see Section VII,A above), an inhibitor of the histone deacetylase, acted synergistically with synchronized cells to increase the transformation efficiency. The mechanism by which butyrate may enhance the efficiency of electroporation is not yet clear.

References

Bahnson, A. B., and Boggs, S. S. (1990). Addition of serum to electroporated cells enhances survival and transfection efficiency. *Biochem. Biophys. Res. Commun.* 171, 752–757.
Berglund, D. L., and Starkey, J. R. (1991). Introduction of antibody into viable cells using electroporation. *Cytometry* 12, 64–67.

Chakrabarti, R., Wylie, D. E., and Schuster, S. M. (1989). Transfer of monoclonal antibodies into mammalian cells by electroporation. *J. Biol. Chem.* **264,** 15494–15500.

Dekeyser, R. A., Claes, B., De Rycke, R. M. U., Habets, M. E., Van Montagu, M. C., and Caplan, A. B. (1990). Transient gene expression in intact and organized rice tissues. *Plant Cell* **2,** 591–602.

Furfine, E. S., and Wang, C. C. (1990). Transfection of the *Giardia lamblia* double-stranded RNA virus into *Giardia lamblia* by electroporation of a single-stranded RNA copy of the viral genome. *Molec. Cell. Biol.* **10,** 3659–3662.

Furfine, E. S., White, T. C., Wang, A. L., and Wang, C. C. (1989). A single-stranded RNA copy of the *Giardia lamblia* virus double-stranded RNA genome is present in the infected *Giardia lamblia*. *Nucleic Acids Res.* **17,** 7453–7467.

Goldstein, S., Fordis, C. M., and Howard, B. H. (1989). Enhanced transfection efficiency and improved cell survival after electroporation of G2/M-synchronized cells and treatment with sodium butyrate. *Nucleic Acids Res.* **17,** 3959–3971.

Hashimoto, K., Tatsumi, N., and Okuda, K. (1989). Introduction of phalloidin labelled with fluorescein isothiocyanate into living polymorphonuclear leukocytes by electroporation. *J. Biochem. Biophys. Methods* **19,** 143–154.

Hibi, T. (1989). Electrotransfection of plant protoplasts with viral nucleic acids. *In* "Advances in Virus Research" (K. Maramorosh, F. A. Murphy and A. J. Shatkin, eds.), Vol. 37, pp. 329–342. Academic Press, San Diego.

Hood, M. T., and Stachow, C. (1990). Transformation of *Schizosaccharomyces pombe* by electroporation. *Nucleic Acids Res.* **18,** 688.

Hood, M. T., and Stachow, C. (1991). The influence of polyethylene glycol on electropore size of *Schizosaccharomyces pombe*. *Appl. Environ. Micro.* Submitted.

Inoue, K., Yamashita, S., Hata, J., Kabeno, S., Asada, S., Nagahisa, E., and Fujita, T. (1990). Electroporation as a new technique for producing transgenic fish. *Cell Differentiation Dev.* **29,** 123–128.

Kwee, S., Nielsen, H. V., and Celis, J. E. (1990). Electropermeabilization of human cultured cells grown in monolayers. Incorporation of monoclonal antibodies. *Bioelectrochem. Bioenerg.* **23,** 65–80.

Lambert, H., Pankov, R., Gauthier, J., and Hancock, R. (1990). Electroporation-mediated uptake of proteins into mammalian cells. *Biochem. Cell Biol.* **68,** 729–734.

Luciano, C. S., Rhoads, R. E., and Shaw, J. G. (1987). Synthesis of potyviral RNA and proteins in tobacco mesophyll protoplasts inoculated by electroporation. *Plant Sci.* **51,** 295–303.

Mann, S. G., and King, L. A. (1989). Efficient transfection of insect cells with baculovirus DNA using electroporation. *J. Gen. Virol.* **70,** 3501–3505.

Mir, L. M., Banoun, H., and Paoletti, C. (1988). Introduction of definite amounts of nonpermeant molecules into living cells after electropermeabilization: Direct access to the cytosol. *Exp. Cell Res.* **175,** 15–25.

Morgan, W. F., Fero, M. L., Land, M. C., and Wagner, R. A. (1988). Inducible expression and cytogenetic effects of the EcoRI restriction endonuclease in chinese hamster ovary cells. *Mol. Cell Biol.* **8,** 4204–4211.

Morgan, W.F., and Winegar, R. A. (1989). The use of restriction endonucleases to study the mechanisms of chromosome damage. *In* "Chromosomal Aberrations. Basic and Applied Aspects" (G. Obe and A. T. Natarajan, eds.), pp. 70–78, Springer-Verlag, Berlin.

Morgan, W. F., Phillips, J. W., Chung, H. W., Ager, D. D., and Winegar, R. A. (1990a). The use of restriction endonucleases to mimic the cytogenetic damage induced by ionizing radiations. *In* "Ionizing Radiation Damage to DNA: Molecular Aspects" (S. S. Wallace and R. P. Painter, eds.) pp. 191–199. Wiley-Liss, New York.

Morgan, W. F., Ager, D., Chung, H. W., Ortiz, T., Phillips, J. W., and Winegar, R. A. (1990b). The cytogenic effects of restriction endonucleases following their introduction into cells by electroporation. *In* "Mutation and the Environment, Part B" (M. L. Mendelsohn and R. V. Albertini, eds.), pp. 355–361. Wiley-Liss, New York.

Moses, S. A. M., Christie, A. F., and Bryant, P. E. (1990). Clastogenicity of PvuII and EcoRI in electroporated CHO cells assayed by metaphase chromosomal aberrations and by micronuclei using the cytokinesis-block technique. *Mutagenesis* 5, 599–603.

Nishiguchi, M., Langridge, W. H. R., Szalay, A. A., and Zaitlin, M. (1986). Electroporation-mediated infection of tobacco leaf protoplasts with tobacco mosaic virus RNA and cucumber mosaic virus RNA. *Plant Cell Rep.* 5, 57–60.

Osbourn, J. K. and Wilson, T. M. A. (1992). Applications of GUS to molecular plant virology. In "GUS Protocols: Using the GUS Gene as a Reporter of Gene Expression" (S. Gallagher, ed.). Academic Press, Orlando, FL.

Osbourn, J. K., Watts, J. W., Beachy, R. N., and Wilson, T. M. A. (1989). Evidence that nucleocapsid disassembly and a later step in virus replication are inhibited in transgenic tobacco protoplasts expressing TMV coat protein. *Virology* 172, 370–373.

Piwnica-Worms, H. (1990). Overview of the baculoviral expression system. *In* "Current Protocols in Molecular Biology" (F. A. Ausubel, R. Brent, R. E. Kingston, D. D. Moore, J. G. Seidman, J. A. Smith, and K. Struhl, eds.), pp. 16.8.1–16.8.5. Wiley-Interscience, New York.

Powell, K. T., Morgenthaler, A. W., and Weaver, J. C. (1989). Tissue electroporation. Observation of reversible electrical breakdown in viable frog skin. *Biophys. J.* 56, 1163–1171.

Pu, H., and Young, A. P. (1990). Glucocorticoid-inducible expression of a glutamine synthetase-CAT-encoding fusion plasmid after transfection of intact chicken retinal explant cultures. *Gene* 89, 259–263.

Raptis, L., and Firth, K. L. (1990). Laboratory methods. Electroporation of adherent cells *in situ*. *DNA Cell Biol.* 9, 615–621.

Rech, E. L., Dobson, M. J., Davey, M. R., and Mulligan, B. J. (1990). Introduction of a yeast artificial chromosome vector into *Saccharomyces cerevisiae* cells by electroporation. *Nucleic Acids Res.* 18, 1313.

Register, J. C. III, and Beachy, R. N. (1988). Resistance to TMV in transgenic plants results from interference with an early event in infection. *Virology* 166, 524–532.

Shillito, R. D., Saul, M. W., Paegkowski, J., Müller, M., and Potrykus, I. (1985). High efficiency direct gene transfer to plants. *Biotechnology* 3, 1099–1103.

Traas, J. A., Doonan, J. H., Rawlins, D. J., Shaw, P. J., Watts, J., and Lloyd, C. W. (1987). An actin network is present in the cytoplasm throughout the cell cycle of carrot cells and associates with the dividing nucleus. *J. Cell. Biol.* 105, 387–395.

Tsongalis, G. J., Lambert, W. C., and Lambert, M. W. (1990). Correction of the ultraviolet light induced DNA-repair defect in xeroderma pigmentosum cells by electroporation of a normal human endonuclease. *Mutat. Res.* 244, 257–263.

Watanabe, Y., Meshi, T., and Okada, Y. (1987). Infection of tobacco protoplasts with *in vitro* transcribed tobacco mosaic virus RNA using an improved electroporation method. *FEBS Lett.* **219**, 65–69.

Watts, J. W., King, J. M., and Stacey, N. J. (1987). Inoculation of protoplasts with viruses by electroporation. *Virology* **157**, 40–46.

Winegar, R. A., Phillips, J. W., Youngblom, J. H., and Morgan, W. F. (1989). Cell electroporation is a highly efficient method for introducing restriction endonucleases into cells. *Mutat. Res.* **225**, 49–53.

Yorifuji, T., and Mikawa, H. (1990). Co-transfer of restriction endonucleases and plasmid DNA into mammalian cells by electroporation: Effects on stable transformation. *Mutat. Res.* **243**, 121–126.

Yorifuji, T., Tsuruta, S., and Mikawa, H. (1989). The effect of cell synchronization on the efficiency of stable gene transfer by electroporation. *FEBS Lett.* **245**, 201–203.

Zheng, Q., and Chang, D. C. (1991). High-efficiency gene transfection by *in situ* electroporation of cultured cells. *Biochim. Biophys. Acta* **1088**, 104–110.

[3]Present address: Department of Biology, Hong Kong University of Science and Technology, Kowloon, Hong Kong.

Part III

Practical Protocols for Electroporation and Electrofusion

26

Design of Protocols for Electroporation and Electrofusion: Selection of Electrical Parameters

Donald C. Chang[1]

Department of Molecular Physiology and Biophysics,
Baylor College of Medicine
Houston, Texas 77030

[1]Present address: Department of Biology, Hong Kong University of Science and Technology, Kowloon, Hong Kong.

I. Introduction

A. Waveforms

In both electroporation and electrofusion, a high-intensity electric field is used to induce membrane breakdown in the targeted cells. This electric field is always applied in a pulsed form to prevent irreversible cell damage. The most commonly used waveforms are the rectangular pulse (sometimes called "square pulse") and the exponential decay pulse (Fig. 1). The rectangular pulse is usually generated by gating the output of a high-voltage power supply, while the exponential decay pulse is generated by discharging a capacitor that has been precharged at a high voltage (Fig. 2). In the latter case, the pulse width is characterized by the decay constant τ (see legend of Fig. 1).

These two types of pulses are called DC pulses, because they represent the pulsed form of a DC (direct current) electric field. At this time, most of the commercially available equipment for electroporation and electrofusion are generators of DC pulses. Recently, a new method has been developed in which a pulsed electric field oscillating at a radio frequency (RF) is used to porate or fuse cells (see Chapter 19). There are some advantages to using the RF method, but the equipment required to generate such RF pulses is more complicated.

In this chapter, we will concentrate mainly on discussing the optimization of the DC pulse methods. Only a brief discussion of the RF pulse method will be given at the end of the chapter.

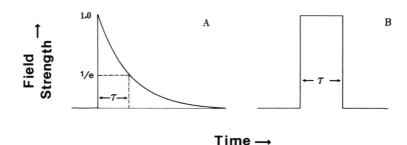

Time →

Figure 1   The most commonly used waveforms of DC pulses. (A) Exponential decay pulse, which is generated by discharging a capacitor. The electric field is a function of time such that $E = E_0 \exp(-t/\tau)$, where τ is the time constant, defined as the length of time at which the field strength is reduced to $1/e$ of the initial value ($e = 2.718$ is the base of natural logarithm). (B) Rectangular pulse (or square pulse).

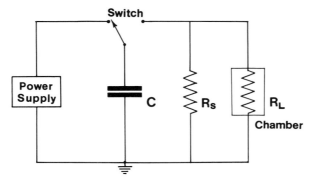

Figure 2   Basic design of a circuit that is commonly used to generate a high-intensity electric pulse for electroporation or electrofusion. The output waveform is an exponential decay pulse. C, capacitance; R_L, load resistance (i.e., resistance of the sample); R_S, shunt resistance.

B. Induced Membrane Potential

In order to understand the principles of selecting electrical parameters for cell poration or cell fusion, one needs to know some simple physics about the electrical treatment of cells. When a spherical cell is exposed to an electric field, a potential is induced at the cell membrane, which is given by

$$V_m = 1.5rE \cos \theta \qquad (1)$$

where V_m is the induced membrane potential, r is the cell radius, E is the electric field strength, and θ is the angle between the normal vector of the membrane and the electric field. From this simple equation, several important points can be seen:

1. The induced membrane potential is proportional to the strength of the applied field.
2. The induced membrane potential is proportional to the radius of the cell.
3. The induced membrane potential is not uniform: V_m is maximum at the cell poles (i.e., the parts of cell facing the electrodes) and is decreased towards the cell equator.

In order to break down the cell membrane (for either cell poration or cell fusion), V_m must reach a critical value, V_c (which is on the order of 1 V). From points 1 and 3, it is apparent that an applied field of high intensity is more effective in breaking down the membrane than a low-intensity field. However, if the applied field is so high that V_m is significantly larger than V_c, the cell will be irreversibly damaged by lysis. Thus, it is important to choose a proper field strength so that cells can be effectively porated (or fused) but without excessive damage.

It is also implied from point 2 that a very-high-intensity field is required to porate cells of small sizes (such as prokaryotic cells), while a less intense field is required for porating large cells (such as most eukaryotic cells).

C. Compensating Relationship between the Field Strength and the Pulse Width

Because cells are not spheres of uniform diameter, Eq. (1) should be used mainly as a qualitative rule of guidance. In the real situation, the field strength can be set within a relatively large range. This is also partially due to the fact that the value of the critical potential V_c is not fixed but can vary depending on a number of parameters, including temperature, membrane composition, and the pulse width (τ). It has been observed that when an electrical pulse of very narrow pulse width was used, the field strength required to give effective poration was usually quite high. Conversely, when the pulse width was large, the required field strength was much lower (Chu et al., 1987). Thus, it appears that, up to a certain extent, the field strength and the pulse width can compensate each other. Some investigators have even suggested that, for a given cell type, the transfection efficiency may be proportional to the power of the applied pulse, that is, the product of $E^2\tau$ (Kubiniec et al., 1990). If this hypothesis is correct, the pulse width may vary inversely with the square of E.

II. Electroporation Using a DC Field

A. Effects of Field Strength (E) and Pulse Width (τ)

For gene transfection or loading of exogenous molecules, the efficiency is generally most sensitive to the setting of the field strength, which is defined as

$$E = V/d \tag{2}$$

where V is the voltage of the applied pulse, and d is the distance between the electrodes in the chamber. The cell survival rate (or cell viability) is also strongly dependent on E.

The typical effects of E on the transfection efficiency and cell viability are demonstrated in a example shown in Fig. 3. The cell viability starts at 100% for low values of E; it then progressively decreases as E increases. The transfection efficiency,

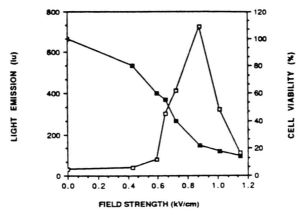

Figure 3   Transfection of B50 rat neuroblastoma cells with 10 μg of plasmid DNA containing the firefly luciferase gene by electroporation using a CD pulse ($\tau = 2$ ms). Light emission was proportional to the expression of the luciferase gene. Data from Andreason and Evans (1988).

on the other hand, is not a monotonic function of E. The efficiency is very low until E reaches a certain threshold level, and then increases sharply with the increase of E. After reaching a maximum value, the transfection efficiency decreases rapidly with further increase of E.

From this example, it is clear that in order to obtain an efficient transfection, one must carefully select a proper setting of E. Also, there seems to be some sort of dilemma here. The required field strength for maximum transfection efficiency is so high that the cell viability becomes unacceptably low. Thus, the best choice of E is probably a compromise setting such that the needs for both transfection efficiency and cell viability can be reasonably satisfied.

Besides the field strength, the efficiency of electroporation and cell viability is also affected by the pulse width τ, although the sensitivity is somewhat less. In most experiments, it is a common practice to first select a trial value for τ based on previous publications, and then vary the setting of E to optimize the transfection (or molecule loading) efficiency and cell viability.

As a practical reference, we have tabulated the experimental conditions of many previous studies that used electroporation to introduce genes (or other exogenous molecules) into various cell types (Table 1). The settings given in this table may be used as a starting point for the newly planned experiments.

Table 1

Electroporation of Animal Cells by DC Pulses

Cell types	Molecules injected	Pulse conditions[a]	References
Bovine adrenal medullary cells	^{86}Rb, ^{51}Cr-EDTA Ca-EDTA	CD: 2 kV/cm 200 μs, 1–100 pulses	Baker and Knight (1978)
Pancreatic islets	^{86}Rb, Ca^{2+}, Ca⁻EGTA	CD: 2.5 kV/cm 250 μs, 5 pulses	Pace et al. (1980)
Islets of Langerhans	Ca^{2+}, ATP, EGTA	CD: 3.4 kV/cm, 200 μs, 3–5 pulses	Yaseen et al. (1982)
Mouse L tk⁻ Cell line	pTk2 gene	CD: 8 kV/cm, 3 μs, 1–10 pulses	Wong and Neumann (1982)
Mouse L tk⁻ fibroblast cells	pAGO with tk gene from herpes virus (HAT selection)	CD: 8 kV/cm, 5 μs, 3 pulses	Neumann et al. (1982)
Mouse B and T lymphocytes and fibroblasts	Mouse and human Ig gene	Power supply directly discharged through cuvette, estimated: 320 kV/cm, 17 ms (CD also tried)	Potter et al. (1984)
Mouse lymphoid cell lines	Plasmids with Ig gene	CD: 8 kV/cm, 5 μs, 3 pulses	Falkner et al. (1984)
Rat neuronal cells (B50 neuroblastoma)	Thy-1 glycoprotein encoding gene	CD: 8 kV/cm, 5 μs, 3 pulses	Evans et al. (1984)
Hamster CHO tk⁻ cells in suspension or monolayer	pAGO with tk gene from herpes virus	CD: 3 × 6 kV/cm, 10 μs SP: 3 × 1.5 kV/cm, 50 μs	Zerbib et al. (1985)
Mouse/human hybrid cell line HU 11	Plasmid with human β-globulin (neo resistance)	CD: 2 kV/cm, τ not given, estimated 700 μs	Smithies et al. (1985)
Mouse L cells	pSV2-neo	CD: 10 kV/cm, 5 μs, 1–10 pulses	Stopper et al. (1985)
Human erythrocytes	^{86}Rb$^+$, ^{22}Na$^+$, ^{14}C Carbohydrates	SP: 0.5–10 μs, 2–4 kV/cm	Serpersu et al. (1985)
Mouse mammary carcinoma cells	pSV2-neo	CD pulses (cutoff, times 2–50 μs): typically 3 kV/cm, 50 μs	Hama-Inaba et al. (1987)

Table 1

(Continued)

Cell types	Molecules injected	Pulse conditions[a]	References
Human fibroblasts, HeLa cells, CV-1 monkey kidney cells, mouse embryo fibroblasts (NIH3T3)	pRSV-gpt, pRSV-neo, pRSV-cat, pMT-hGH-SV2 plasmids	CD: optimal 500–700 V/cm, 7 ms (3.5 ms tested)	Chu *et al.* (1987)
Mouse Friend, human k562 erythroleukemic cells	Plasmid Homer 6 with aph gene (stable transformation), pLW4-CAT	CD: up to 4 kV/cm; no definite τ	Spandidos (1987)
Human BJAB cells	pHIV-CAT	CD: 1500V, no detailed data	Davis *et al.* (1987)
Mouse T-cells	CD8 gene	CD: 7–8 kV/cm, 3 pulses 20 nF capacitor	Dembic *et al.* (1987)
Human T-cells	Epstein-Barr virus episomal replicon	CD: 200 V, 0.7 s total, 1 pulse,	Hambor *et al.* (1988)
Murine 10T 1/2 fibroblasts	FITC-Dextrans (9,41,72,154 kD)	SP: 1–4.2 kV/cm, 40–500 μs, 1 pulse	Liang *et al.* (1988)
Mouse myeloma cells, P3X63Ag8, Hela, CV-1	FITC-Dextran SFV-C-protein	CD: 4 kV/cm, 13 μs, 5 pulses	Michel *et al.* (1988)
Mouse L Tk⁻ cells	H-2D^P mutant gene	CD: 2 kV/cm, 0.7 ms, 1 pulse	Murray *et al.* (1988)
B-EBV, lymphoblastoid, T-lymphocytes	Lucifer yellow, LP/pUC13 plasmid, pEMBL9 plasmid	SP: 1.2–1.4 kV/cm, 100 μs, 8–10 pulses	Press *et al.* (1988)
Human fibroblast	pSV-neo	CD: 1.5 kV/cm, 20–25 ms, 1 pulse	Stevens *et al.* (1988)
Murine hybridoma cells	pSV-neo	CD: 10 kV/cm, 5 μs, 5 pulses	Stopper *et al.* (1988)
Murine BW lymphocytes	Decay-accelerating	CD: 200 V, 1100 μF, 0.7 s total	Tykocinski *et al.* (1988)

(continued)

Table 1

(Continued)

Cell types	Molecules injected	Pulse conditions[a]	References
CHO	factor (DAF) gene		
Spodotera trugiperda cells (insect)	Virus DNA (AcNPV) + pAcRP23-*LacZ*	CD: 500–700 V/cm, 2.8–7.7 ms, 1 pulse	Mann and King (1989)
HeLa cells murine lymphoma cells	Monoclonal antibodies	CD: 750 V/cm, 0.7–0.9 ms, 1 pulse	Chakrabarti *et al.* (1989)
Polymorphonuclear leukocytes	FITC-phalloidin	SP: 5–10 kV/cm, 1–5 µs, 3–5 pulses	Hashimoto *et al.* (1989)
Chicken retinal explant cultures	CAT gene	CD: 400 V, 960 µF	Pu and Young (1990)
RBCF-1 cells from goldfish	pSV2-neo Neo + human C-Ha-ms oncogene	CD: 1.0 kV/cm, 22.5 µF capacitor, 3 pulses	Hayasaka *et al.* (1990)
Medaka fertilized eggs (fish)	pMV-GH (rainbow trout growth hormone cDNA)	SP: 750 V/cm, 50 µs, 5 pulses	Inoue *et al.* (1990)

[a]CD, capacitor discharge; SP, square pulse.

B. Factors That Affect the Pulse Width

When one uses a device that generates a rectangular pulse, the pulse width is set by the circuitry of the device. Hence, under a normal condition, the pulse width is not affected by the configuration of the chamber or the condition of the sample. The same is generally not true if one uses a capacitor discharge (CD) device to generate the electrical pulse. As we have shown in Fig. 1, the so-called "pulse width" in a CD pulse is the "time constant" for the exponential decay function, which is equal to

$$\tau = R\,C \tag{3}$$

where R is the resistance and C is the capacitance. In many commercial devices, R is determined directly by the load resistance R_L, which is the resistance of the sample. The value of R_L is related to the ionic strength of the poration medium, the volume of the cell suspension, and the distance between the electrodes in the chamber. When a large amount of ions is contained in the medium (such as phosphate-buffered saline PBS), the resistance of the sample is very low, and the time constant τ would be very short. If one wants to obtain a pulse with long τ, one must increase the value of C. For example, when one uses a low-ionic-strength medium, such as a HEPES-buffered sucrose solution, a capacitance of 5 μF may be sufficient to generate a pulse with τ in the range of 5 ms. However, if one uses a high-salt medium such as PBS, the resulting pulse width may be only 0.1 ms. In this case, a capacitance of 500 μF might be required in order to obtain a pulse width of several milliseconds.

In the most common types of electroporation chambers, (e.g., commercially available cuvettes), the electrodes are basically a pair of parallel metal plates. In such case, the sample resistance is given as

$$R_L = \rho d/A \tag{4}$$

where ρ is the resistivity of the medium, d is the distance between electrodes, and A is the cross-sectional area of the sample, which is equal to the sample volume divided by d. Thus, the sample resistance is proportional to the gap of the electrodes, and inversely proportional to the volume of the sample. Varying any of these factors will change the resultant pulse width. Thus, if one wants to reproduce certain experiments with precisely the same pulse width, one must carefully control the type of chamber used, the buffer medium, and the volume of sample.

In some commercial devices, an internal shunt resistance (R_S) is provided so that R (and thus the time constant, τ) can be determined by the circuit rather than by the condition of sample (see Fig. 2). However, such an arrangement is useful only when a low-salt medium is used. Since in this case R is determined by

$$1/R = 1/R_L + 1/R_S \tag{5}$$

if R_L is much larger than R_S, R will approach the value of R_S. However, if R_L is very small (as in the case of a high-salt medium), R will be dominated by R_L rather than R_S, and one cannot avoid the effects of the medium on the pulse width.

Even in devices that generate rectangular pulses, the output waveform can be significantly distorted when a high-salt medium is used. This is because the low resistance of the high-salt medium can pose a severe current drain on the pulse generator. For example, a cuvette (with a 0.4-cm electrode gap) containing 0.8 ml of PBS has a load resistance less than 20 Ω.* To general a pulse of 1000 V it will require an output current of more than 50 A. Most pulse generators cannot handle such a large amount of current. As a result, the actual voltage of the output pulse could be far less than what is specified in the settings of the machine.

For these reasons, a storage oscilloscope is recommended to measure the output waveform during an electroporation experiment. Such measurement may be particularly important when one wants to know the actual E and τ so that the results can be reproduced in future studies. A portable digital storage scope (such as Tektronix

*Note: The resistance of the medium is not constant: R_L actually decreases with an increase of the applied voltage. Thus, R_L will be far less than 20 Ω at 1 kV.

model 2211) would be quite suitable for this task, but one must use a 10 × probe in such measurements to avoid damaging the oscilloscope.

C. Procedures for Selecting E and τ

The recommended procedures for selecting the electrical parameters in electroporating different cell types are given as follows.

1. Animal Cells

1. Choose a proper pulse width based on the list given in Table 1. For example, for CHO cells, start with $\tau = 5$ ms. (If τ cannot be set directly, use the following guideline: For low-salt medium, set $C = 25$ μF; for high salt medium, set $C = 500$ μF).
2. Electroporate cells with E varying from 0.5 to 3 kV/cm (in steps of 0.5 kV/cm). Measure cell viability.
3. Select the value of E in which the cell viability is closest to 50%. This value (say, $E = 1$ kV/cm) now becomes the first approximation of the optimized setting, E_0.
4. Repeat electroporation experiments with E varying from $0.6E_0$ to $1.4E_0$, in steps of $0.1E_0$ or $0.2E_0$. Measure the efficiency of gene transfection (or molecule loading) and cell viability. Select the value of E that gives the most desired results as the second approximation of E_0.
5. Once an estimated E_0 is obtained, one can also vary the value of τ to obtain its optimal setting. But after a new setting of τ is chosen, E must be optimized again using the procedures of steps 3 and 4.
6. Finally, to obtain the most reliable choice of E_0, step (4) may be repeated several times.

For those readers who are interested in using electroporation to introduce genes into mammalian cells, refer to Chapters 13, 14, and 28 in this book for more details.

Since the electric field strength and the pulse width can compensate each other, there are many possible combinations of E and τ that one can choose. In the earliest work on electroporation, in which mainly rectangular pulses were used, it was common to apply very short pulses (5–50 μs) with high field strength (2–10 kV/cm). However, in later studies that used capacitor discharge (CD) pulses to porate cells, it was found that higher transfection efficiency can be obtained by using longer pulse width (5–10 ms) and a weaker field (about 0.5 kV/cm) (Chu et al., 1987). At present, this long-pulse/low-field approach is frequently recommended if CD pulses are used to electroporate mammalian cells.

Also, in studies that used rectangular pulses for electroporation, it was common to apply a train of multiple pulses. However, in those studies that used CD pulse, usually only a single pulse was applied. Part of this technical difference is probably due to the fact that most of the capacitor discharge devices can only generate a single pulse at a time. In addition, it has been suggested that application of multiple CD pulses did not enhance the efficiency of transfection in some mammalian cells (McNally *et al.*, 1988). This issue, however, still needs to be thoroughly examined.

2. Plant Cells

The procedures for selecting the electrical parameters for plant cells are basically similar to that for animal cells. However, unlike animal cells, there seems to be no advantage for using long pulses to porate plant cells (J. Saunders, personal communication). Also, it is not desirable to use multiple pulses to porate plant cells, since this would usually result in significant reduction of cell viability (J. Saunders, personal communication). Some of the commonly used settings of E and τ for plants are given in Table 2. For more details, see also Chapter 29 of this book.

Table 2
Electroporation of Plant Cells by DC Pulses

Cell type	Molecules	Pulse conditions	References
Protoplasts from carrot, tobacco, maize (transient expression) mono- and dicots	pNOS-CAT	CD: 875 V/cm, 15 ms	Fromm *et al.* (1985)
Tobacco leaf mesophyll protoplasts	pABDI conferring kn resistance	CD: 1.5 kV/cm, 10 µs, 3 pulses	Shillito *et al.* (1985)
Oat mesophyll protoplasts	EB	SP: 1–2 kV/cm, 20 µs, 1 pulse	Mehrle *et al.* (1985)
Carrot protoplasts	pTiC58 enabling hormone-independent protoblast regeneration	SP: 0.5–3.8 kV/cm, 6 µs (0.2-s intervals), 6 pulses	Langride *et al.* (1985)
Maize protoplasts	pCaMV-NEO conferring kn resistance	CD: 500 V/cm, 2 ms	Fromm *et al.* (1986)
Carrot protoplasts (transient expression)	Plasmids with CAT gene in sense or antisense orientation	CD: 875 V/cm, 10 ms	Ecker and Davis (1986)
Tobacco leaf mesophyll protoplasts	pMON200 conferring kn resistance	CD: 2 kV/cm, 250 µs	Riggs and Bates (1986)

(continued)

Table 2

(*Continued*)

Cell type	Molecules	Pulse conditions	References
Tobacco protoplasts (infection detected by fluorescent antibody to virus proteins)	TMV and CMV RNA TMV and CMV particles	CD: 750 V/cm, 6 ms	Okada *et al.* (1986)
Intact tobacco mesophyll cells (infection detected by immunofluorescence)	TMV RNA	CD: 650 V/cm, τ not given	Morikawa *et al.* (1986a)
Tobacco leaf protoplasts	TMV and CMV RNA	SP: 0.5–11 kV/cm, 1–90 μs (0.1-s intervals), 9 pulses	Nishiguchi *et al.* (1986)
Mung bean protoplasts	Quin 2 (Ca$^+$ indicator)	CD: 3.8 kV/cm, 20 μs, 2 pulses	Gilroy *et al.* (1986)
Rice, wheat, sorghum protoplasts from leaves and suspension cultures (monocots, transient expression)	CAT gene with promoters from CMV and *Drosophila:* p35S-CAT and pCopia-CAT	SP: 2 × 2.5 kV/cm, 100 μs or 20 pulses, each 62 μs, 10 kV in "noncontact mode BAEKON 2000 receptacle" (field strength not given)	Ou-Lee *et al.* (1986)
Tobacco leaf mesophyll protoplasts (infection detected by fluorescent anti-TMV serum)	TMV RNA	SP: 500–900 V/cm, 50 μs, 5–10 pulses (1 s intervals)	Hibi *et al.* (1986)
Nicotiana tabacum, N. plumbaginifolia protoplasts, *Beta vulgaris* protoplasts	CCMV RNA, BMV and CCMV viruses	CD: optimum 1.5/2.5 kV/cm, 1 μs	Watts *et al.* (1987)
Soybean protoplasts	APHII gene coding resistance to kanamycin and G418	CD: 375 V/cm, halftime 45 ms, 1 pulse	Christou *et al.* (1987)
Maize and carrot protoplasts	CAT mRNA	CD: 625 V/cm, 490 μF, 1 pulse	Callis *et al.* (1987)
Tobacco protoplasts	pTET7 (CaMV35S promotor tetR gene)	CD: 340 V, 3 ms, 1 pulse	Gatz and Quail (1988)
Maize protoplasts	NPTII gene	CD: 500–750 V/cm, 1200 μF, 1 pulse	Rhodes *et al.* (1988)
Rice protoplasts	APH (3′) II gene	CD: 500 V/cm, 20 ms, or 750 V/cm, 4 ms	Toriyama *et al.* (1988)

3. Microbials

The procedures for selecting the electrical parameters are similar to those for animal and plant cells, with the exception that the size of microbials is much smaller than most of the eukaryotic cells; therefore the electric field strength required is much higher [see Eq. (1)]. In the electroporation of bacteria, such as *Escherichia coli,* the field strength is usually in the range of 5–10 kV/cm. Some other bacteria may require lower field strength, but some others may require higher (see Table 3). Also,

Table 3
Electroporation by DC Pulses (Other Cell Types)

Cell types	Molecules	Pulse conditions	References
A. Bacteria			
Bacillus cereus protoplasts	pUBIO (kn resistance)	CD: 4 kV/cm, 5 μs, 3 pulses	Shivarova *et al.* (1983)
E. coli	pUC18 pBR329	CD: 7 kV/cm, 20 ms, or 11 kV/cm, 5 ms	Dower *et al.* (1988)
Lactobacillus casei	β-gal gene PL-1 phage DNA	CD: 5 kV/cm, 5 ms, 1 pulse	Chassy and Flickinger (1987)
Campylobacter jejuni, C. coli, E. coli	pILL512 DNA	CD: 16.7 kV/cm, 2 ms	Miller *et al.* (1988)
Enterococcus faecalis protoplasts, *Psuedomones putida, E. coli*	PBR322 pKT230 pAM401	CD: 6.25 kV/cm, 5–7 ms, 1–3 pulses	Fiedler and Wirth (1988)
B. Other cells			
Intact yeast cells (*S. cerevisiae* mutants)	YEp13 (complementing Leu deficiency)	CD: 5 kV/cm, 10 ms, 3 pulses	Hashimoto *et al.* (1985)
Yeast spheroplasts (*S. cerevisiae* mutant)	YRp7 shuttle vector complementing tryptophan deficiency	CD: 10 kV/cm, 50 μs	Karube *et al.* (1985)
Dictyostelium discoideum	Firefly luciferase gene	CD: 1.7–2.5 kV/cm, 0.6 ms, 1 pulse	Howard *et al.* (1988)
Protozoan, *Leptomonas seymouri*	CAT gene	CD: 450 V, 99 ms	Bellofatto and Cross (1989)
Protozoan, *Leishmania enriettii*	CAT and α-tubulin gene	CD: 3 kV/cm, 25 μF capacitor	Laban and Wirth (1989)
Parasite (*Giardia lamblia*)	Virus single-stranded RNA	CD: 1.25 kV/cm, 0.6 ms, 1 pulse	Furtine and Wang (1990)

the required field strength is closely related to the selection of pulse width. (For more details, see Chapters 17, 18, 30, and 31.)

III. Electrofusion Using a DC Field

A. Cell Alignment by Dielectrophoresis

The procedures for electrofusion are very similar to that for electroporation, except that the fusion partners must be brought into stable contact with each other. Cells can be brought into contact by several means, including (1) mechanical manipulations, (2) chemical or biological treatments, and (3) dielectrophoresis. The last method has become the method of choice for most research groups; its advantages include that it is simple to use, highly effective, and results in close contact of membranes in the direction most favorable for electrofusion.

The principle of dielectrophoresis is well known (Pohl, 1978; Zimmermann and Pilwat, 1982). Very briefly, when a cell is exposed to an electric field, its positive and negative charges (at the cell surface) move in opposite directions, causing the cell to become an electric dipole. These induced dipoles can attract each other by their electrical force. As a result, cells are aligned along the field lines of the external field.

Unlike electrophoresis, in which the charged particle moves toward a fixed direction of the field, dielectrophoresis occurs independent of the direction of the field. Hence, dielectrophoresis can be accomplished using either an AC or a DC field. The AC field has an added advantage in that polarization of the electrodes can be avoided. Furthermore, the problem of electrohydrolysis can be prevented by using a low-amplitude AC field. Thus, dielectrophoresis using an AC field has become the standard method of cell contact in electrofusion.

The two important parameters in dielectrophoresis are frequency and amplitude of the AC field. The cells will line up in "pearl chains" only at certain frequencies. If frequency of the AC field is not set properly, the cells will undergo a rotating motion. Obviously, electrofusion cannot be achieved under such conditions. The frequency at which stable dielectrophoresis occurs varies from cell type to cell type, and therefore must be determined empirically. The common method is to construct a small fusion chamber on a glass slide, which can be placed under the objective of a microscope. Then, by directly observing the movement of cells, one varies the frequency of the AC field until cells are aligned in a pearl-chain formation.

The frequency of the dielectrophoretic field is quite high, generally on the order of 1 MHz. The amplitude of this field usually is in the range of 100–400 V/cm. A list of typical values for the dielectrophoretic field used in some of the earlier electrofusion studies is given in Table 4.

Table 4
Electrofusion by DC Pulses

Cell type	Cell alignment	Fusion pulse	References
	A. Animal cells		
Friend erythroleukemia cells	**AC: 100 V/cm, 100 kHz–2 MHz**	**SP: 4 kV/cm**	**Pilwat** *et al.* (1981)
Human erythrocytes	AC: 500 V/cm, 2 MHz	SP: 2 kV/cm, 3 μs	Scheurich and Zimmermann (1981)
NIH 3T3 cells	AC: 400–700 V/cm, 1 MHz	SP: 7 kV/cm, 50 μs	Zimmermann and Pilwat (1982)
Mouse myeloma cells × mouse lymphoctyes	AC: 100 V/cm, 1 MHz	SP: 4 kV/cm, 20 μs	Vienken and Zimmermann (1982)
Human myeloma L363 cells × human B lymphocytes	AC: 100 V/cm, 2 MHz	SP: 3.5 kV/cm	Bischoff *et al.* (1982)
Swiss 3T3 cells	Natural contact in monolayer culture	SP: 1.6 kV/cm, 100 μs, 5 pulses	Teissie *et al.* (1982)
Unilamellar liposome	AC: 200–400 V/cm, 100 kHz	SP: 3–9 kV/cm, 20–50 μs	Buschl *et al.* (1982)
Mouse myeloma cells SP2 and X63	AC: 200 V/cm, 800 kHz	SP: 2.5 kV/cm, 20 μs	Vienken *et al.* (1983)
Mouse myeloma cells × splenic B lymphocytes	Biotin-avidin antigen-antibody cross-link	SP: 4 kV/cm, 5 μs, 4 pulses	Lo *et al.* (1984)
Mouse LM fibroblasts × hamster CHO cells	Natural contact in monolayer culture	SP: 1.5 kV/cm, 50 μs, 5 pulses	Finaz *et al.* (1984)
L5178Y mouse leukemic lymphoblasts	AC: 800 V/cm, 100 kHz	SP: 5–8 kV/cm, 20 μs, 4 pulses	Ohno-Shosaku and Okada (1984)
Human erythrocytes ghosts	AC: 70–150 V/cm, 60 Hz	CD: 500–700 V/cm, 0.2–1.2 ms	Sowers (1984)
L5178Y mouse lymphoma cells	AC: 800 V/cm, 100 kHz	SP: 5–8 kV/cm, 20 μs, 4 pulses	Okada *et al.* (1984)
Primary Leydig cells × primary adrenocortical cells	AC: 200–250 V/cm, 650 kHz	SP: 3.75 V/cm, 15 μs, 9 pulses	Podesta *et al.* (1984)

(continued)

Table 4
(Continued)

Cell type	Cell alignment	Fusion pulse	References
Human lymphoblasts × mouse lymphoblasts	AC: 800 V/cm, 100 kHz	SP: 3.3 kV/cm or 5 kV/cm, 20 μs, 2 pulses	Ohno-Shosaku et al. (1984)
Mouse myeloma cells × mouse B lymphocytes for mAbs to cytokeratin	AC: 1.2 kV/cm, 2 MHz	SP: 4.2 kV/cm, 3 μs, 1 pulse	Karsten et al. (1985)
Mouse myeloma cells × mouse B lymphoctyes	AC: 250 V/cm, 1.5 MHz	SP: 3.5 kV/cm, 20 μs, 3 pulses	Vienken and Zimmermann (1985)
Mouse blastomeres and embryos	Natural contact	SP: 1 kV/cm, 250 μs, 2 pulses	Kubiak and Tarkowski (1985)
Human UC729-6 cells × human lymphocytes	AC: 430 V/cm, 1 MHz	SP: 7.7 kV/cm, 20 μs	Glassy and Hofmann (1985)
Human lymphoma × rabbit corneal epithelium	Mechanical pressure	SP: 20 V, 20 μs, 3 pulses	Grasso et al. (1989)
B. Plant protoplasts			
Vicia faba protoplasts	AC: 250–500 V/cm, 500 kHz	SP: 750 V/cm, 50 μs	Zimmermann and Scheurich (1981)
Nicotiana tabacum, Hordeum vulgare, Lycopersion exculertum	AC: 150 V/cm, 1.2 MHz	SP: 1.7 kV/cm, 25 μs	Jacob et al. (1983)
Nicotiana tabacum	AC: 120 V/cm, 1 MHz	SP: 1.5 kV/cm, 50 μs	Koop et al. (1983)
Oat, corn, vigna, petunia, and amaranthus protoplasts	AC: 200 V/cm, 500 kHz	SP: 700 V/cm, 10–50 μs, 2 pulses	Bates et al. (1983)
Nicotiana tabacum, Acer pseudoplantanus, Catharanthus roseus	Spermine-mediated agglutation	SP: 1.5 kV/cm, 100 μs, 4 pulses	Chapel et al. (1984)
Nicotiana tabacum × N. plumbaginitolia	AC: 20–80 V/cm, 500 kHz	CD: 500 V/cm, 1 ms	Watts and King (1984)
Nicotiana tabacum × N. plumbaginitolia	AC: 150 V/cm, 600 kHz	SP: 1 kV/cm, 50 μs, 2 pulses	Bates and Hasenkampt (1985)

Table 4

(*Continued*)

Cell type	Cell alignment	Fusion pulse	References
Pyscomitrella patens	AC: 20 V/cm, 500 kHz	CD: 800 V/cm, 1 ms	Watts *et al.* (1985)
Nicotiana tabacum	AC: 200 V/cm, 0.9 MHz	SP: 1.2–1.5 kV/cm, 50 μs	Kohn *et al.* (1985)
Nicotiana glauca × *N. langsdorffi*	High-density gravity	CD: 1 kV/cm, 100–200 μs	Morikawa *et al.* (1986b)
Brassica napus L. protoplasts and subprotoplasts	AC: 60–80 V/cm, 1 MHz	CD: 0.9–1.8 kV/cm, 50 μs, 1–5 pulses	Spangenberg and Schweiger (1986)
Brassica oleracea C. × *Brassica compestris* p.	AC: 100–150 V/cm, 1 MHz	CD: 1.5 kV/cm, 500 μs, 3 pulses	Zheng *et al.* (1988)
C. Microbials			
E. coli, Salmonella typhimurium spheroplasts	AC: 1 kV/cm, 1 MHz	SP: 4 kV/cm, 15 μs, 5 pulses	Ruthe and Adler (1985)
Bacillus thuringiensis protoplasts	PEG-induced agglutination	CD: 14–20 kV/cm, 5 μs, 3 pulses	Shivarova and Grigorava (1983)
Saccharomyces cerevisiae protoplasts	AC: 1 kV/cm, 2 MHz	SP: 11 kV/cm, 7 μs	Halfmann *et al.* (1983)
Saccharomyces cerevisiae	AC: 275 V/cm, 800 kHz	SP: 10 kV/cm, 10 μs	Schnettler *et al.* (1984)
D. Others			
Dictyostelium cells	High density	CD: 4–.6 kV/cm, 40 μs, 3 pulses	Neumann *et al.* (1980)
Sea urchin eggs	AC: 100–220 V/cm, 0.5–2 MHz	SP: 400 V/cm, 50–400 μs	Richter *et al.* (1981)

B. Selection of E and τ for the Fusing Pulse

After cells are brought into close contact by the continuous low-amplitude AC field, a pulse (or pulses) of high-intensity field can be applied to induce cell fusion. The strength and pulse width of this high-intensity fusing pulse is basically similar to the electric pulse used for electroporation. Hence, one can follow essentially the same procedures to select the electrical parameters.

However, one may observe that the electrical power required to fuse cells can be significantly less than that required for electroporation. For example, in an experiment using an RF field, we found that a single pulse of 1 kV/cm (0.2 ms wide) was sufficient to induce fusion of CV-1 cells, while at least five pulses were required in order to transfect the same cells by electroporation (see Chapter 19).

In most of the earlier electrofusion experiments, mainly rectangular pulses were used to fuse cells. However, CD pulses were also found to be effective in using certain cell types. In Table 4, we have provided a list of field strength and pulse width used in some of the previous electrofusion studies. For those readers who are interested in electrofusing plant cells, refer to Chapter 29 of this book for more details. For those who are interested in making hybridomas, refer to Chapters 33 and 32.

Most fusion experiments were done using a low-salt medium, mainly for reducing the heating effects during the dielectrophoresis steps. Also, the pulse width required for electrofusion is in general shorter than those used for electroporation. Hence, if one uses a capacitor discharge device to generate the fusing pulse, a small capacitance (e.g., 25 μF) is usually sufficient.

IV. Electroporation and Electrofusion Using a Radiofrequency Field

Although it is conventional to use a pulsed DC field to porate or fuse cells, a new method has recently been developed in which the applied field is oscillating at a radio frequency (RF). The equipment needed for generating such RF pulse is more complicated and has not yet been available commercially. But, it has been demonstrated that this new method has certain advantages over the conventional methods. (For more details, please see Chapter 19 of this book.)

A. Advantages of Using an Oscillating Electric Field

One common problem of electroporation using a DC field is that too many cells could be damaged. This problem can be partially solved by using an oscillating electric field. It was reported in an early study by Zimmermann et al. (1980) that cells can be damaged easily by the intense field if the pulse width was 20 μs or longer. However, when the pulse width was reduced to 5 μs or less, much of the cell damage was avoided. Hence, it seemed to be important to prevent the cells from remaining in a highly polarized state for an extensive period. By using an oscillating electric field, the cells are exposed to a high field only for a very brief duration, and thus should have less chance to be irreversibly damaged.

For example, Fig. 4 shows the results of a gene transfection experiment in which COS-M6 cells were electroporated using a DC-shifted RF field (Chang, 1989b). Even at a field strength where the transfection efficiency was maximum, almost 70% of electroporated cells could still survive. Such a survival rate is apparently better than that obtained by electroporation using a conventional DC field (compare Figs. 3 and 4).

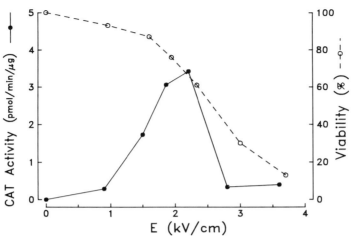

Figure 4   Transfection of COS-M6 cells with 0.1 μg pSV2-CAT DNA by electroporation using a RF field. CAT activity and cell viability are plotted as a function of the peak strength of the applied RF field. Data from Chang (1989).

Another advantage of the RF field method is that, unlike the DC field, one can simultaneously electroporate or electrofuse cells of different sizes (see Chapter 19 of this book). Such ability may be particularly useful for hybridoma research, since the fusing partners in general are in different sizes. In addition, it has been demonstrated that an RF field of proper frequency was far more effective in inducing fusion than a DC field (Chang, 1989b). This improved efficiency may be partially caused by a localized sonication effect (Chang, 1989a) or by a strong dipolar interaction between the fusing partners (see Chapter 11).

B. Selection of Parameters for RF Pulses

Conceivably, a number of waveforms can be used for the RF pulse, including ordinary sinosoidal wave, DC-shifted sinosoidal wave, square wave, etc. So far most of the published studies were done using a pulsed DC-shifted sinosoidal wave (Chang, 1989a, 1989b; Zheng and Chang, 1991). The parameters involved are mainly the field strength, pulse width, frequency of the oscillating field, and the number of pulse used. A list of these parameters used in electroporation and electrofusion experiments on various cell types are summarized in Tables 5 and 6.

For electroporation of most mammalian cells, the optimal field strength was

Table 5
Electroporation by RF Pulses

Cell type	Molecules injected	Pulse conditions	References
COS-M6	pSV2-CAT	1.2–2.6 kV/cm, 40 kHz, 2 ms, 5 pulses	Chang (1989b)
CV-1	pSV2-CAT RSV-β-gal	0.8–1.2 kV/cm, 40 kHz, 2 ms, 3 × 5 pulses	Chang et al. (1991)
CHO 3T3	pSV2-CAT	1.3–1.8 kV/cm, 40 kHz, 2 ms, 3 × 5 pulses	Chang et al. (1991)
Rabbit hepatocytes	CMV-β-gal	0.9–1.3 kV/cm, 40 kHz, 1 ms, 5 pulses	Chang et al. (1991)
HeLa	pSV2-CAT	2.5 kV/cm, 20 kHz, 1 ms, 3 × 5 pulses	Unpublished data
COS-M6 attached cells	RSV-β-gal	1.2 kV/cm, 40 kHz, 1 ms, 3 × 5 pulses	Zheng and Chang (1991)
CV-1 attached cells	RSV-β-gal, calcein, rhodamine-labeled dextran	0.6 kV/cm, 50 kHz, 1 ms, 3 × 5 pulses	Zheng and Chang (1991)

between 1 and 2 kV/cm. The pulse width may vary from 0.2 to 2 ms. The frequency can also be chosen from a relatively broad range (from 10 to 100 kHz). The condition for electrofusion, however, can be quite different. When an RF field was used to fuse human red blood cells, the fusion yield was highly sensitive to the choice of frequency; the optimal value was about 100 kHz. On the other hand, the fusion of

Table 6
Electrofusion by RF Pulses

Cell types	Cell alignment	Fusing pulse	References
Human erythrocytes	AC: 200 V/cm, 60 kHz	3.5–4 kV/cm, 0.1 ms, 100 kHz, 3 pulses	Chang (1989a, 1989b)
Goldfish xanthophore cells × tumor cells	AC: 200 V/cm, 60 kHz	3.3 kV/cm, 40 μs, 400 kHz, 3 pulses	Chang (1989b)
CV-1 cells	Natural contact in monolayer culture	1 kV/cm, 200 μs, 1 pulse	Zheng and Chang (1990)
Cabbage mesophyll protoplasts	100 V/cm, 1 MHz	1.0–1.5 kV/cm, 0.2 ms, 1–3 pulses	Zheng and Chang (1989)

attached mammalian cells or plant protoplasts seemed to be less sensitive to the frequency of the RF field.

For electroporation using an RF field, it was found that multiple pulses are far more efficient than a single pulse. Thus, we always use a train (or trains) of multiple pulses to porate cells. The number of pulses usually varied from 3 to 15. However, the advantage of using multiple pulses was not as pronounced in electrofusion. For example, a single RF pulse was sufficient to induce fusion in human red blood cells or in mammalian fibroblast cells, with satisfactory fusion yield (see Chapter 19).

Acknowledgments

I thank Qiang Zheng for his assistance in preparing the tables of this chapter and Diana Serna for her secretarial assistance.

References

Andreason, G. L., and Evans, G. A. (1988). Introduction and expression of DNA molecules in eukaryotic cells by electroporation. *Biotechniques* 6, 650–660.

Baker, P. F., and Knight, D. E. (1978). Calcium-dependent exocytosis in bovine adrenal medullary cells with leaky plasma membrane. *Nature* 276, 620.

Bates, G. W., and Hasenkampf, C. A. (1985). Culture of plant somatic hybrids following electrical fusion. *Theor. Appl. Genet.* 70, 227–233.

Bates, G. W., Gaynor, J. J., and Shekhawat, N. S. (1983). The fusion of plant protoplasts by electric fields. *Plant Physiol.* 72, 1110–1113.

Bellofatto, V., and Cross, G. A. M. (1989). Expression of a bacterial gene in a trypanosomatid protozoan. *Science* 244, 1167–1169.

Bischoff, R., Eisert, R. M., Schedel, I., Vienken, J., and Zimmermann, U. (1982). Human hybridoma cells produced by electrofusion. *FEBS Lett.* 147, 64–68.

Buschl, R., Ringsdorf, H., and Zimmermann, U. (1982). Electric field-induced fusion of large liposomes from natural and polymerizable lipids. *FEBS Lett.* 150, 38–42.

Callis, J., Fromm, M., and Walbot, V. (1987). Expression of mRNA electroporated into plant and animal cells. *Nucleic Acids Res.* 15, 342.

Chakrabarti, R., Wylie, D. E., and Schuster, S. M. (1989). Transfer of monoclonal antibodies into mammalian cells by electroporation. *J. Biol. Chem.* 264, 15494–15500.

Chang, D. C. (1989a). Cell poration and cell fusion using a pulsed radio-frequency electric field. *In* "Electroporation and Electrofusion in Cell Biology" (E. Neumann, A. E. Sowers, and C. A. Jordan, eds.) pp. 215–227, Plenum Publishing, New York.

Chang, D. C. (1989b). Cell poration and cell fusion using an oscillating electric field. *Biophys. J.* 58, 641–652.

Chang, D. C., Gao, P. Q., and Maxwell, B. L. (1991). High efficiency gene transfection by electroporation using a radio-frequency field. *Biochim. Biophys. Acta* 1992, 153–160.

Chapel, M., Teissie, J., and Alibert, G. (1984). Electrofusion of spermine-treated plant protoplasts. *FEBS Lett.*, 173, 331–336.

450   Donald C. Chang

Chassy, B. M., and Flickinger, J. L. (1987). Transformation of *Lactobacillus* casei by electroporation. *FEMS Microbiol Lett.* 44, 173–177.

Christou, P., Murphy, J. E., and Swain, W. F. (1987). Stable transformation of soybean by electroporation and root formation from transformed callus.*Proc. Natl. Acad. Sci. USA* 84, 3962.

Chu, G., Hayakawa, H., and Berg, P. (1987). Electroporation for the efficient transfection of mammalian cells with DNA. *Nucleic Acids Res.* 15, 1311.

Davis, M. G., Kenney, S. C., Kamine, J., Pagano, J. S., and Huang, E. S. (1987). Immediate-early gene region of human cytomegalovirus trans-activates the promoter of human immunodeficiency virus. *Proc. Natl. Acad. Sci. USA* 84, 8642.

Dembic, Z., Haas, W., Zamoyska, R., Parnes, J., Steinmetz, M., and von Boehmer, H. (1987). Transfection of the CD8 gene enhances T-cell recognition. *Nature* 326, 510.

Dower, W. J., Miller, J. F., and Ragsdale, C. W. (1988). High efficiency transformation of *E. coli* by high voltage electroporation. *Nucleic Acids Res.* 16, 6127–6145.

Ecker, J. R., and Davis, R. W. (1986). Inhibition of gene expression in plant cells by expression of antisense RNA. *Proc. Natl. Acad. Sci. USA* 83, 5372–5376.

Evans, G. A., Ingraham, H. A., Lewis, K., Cunningham, K., Seki, T. Moriuchi, T., Chang, H. C., Silver, J., and Hyman, R. (1984). Expression of the Thy-1 glycoprotein gene by DNA-mediated gene transfer. *Proc. Natl. Acad. Sci. USA* 81, 5532.

Falkner, F. G., Neumann, E., and Zachau, H. G. (1984). Tissue specificity of the initiation of immunoglobulin kappa gene transcription. *Physiol. Chem.* 365, 1331–1343.

Fiedler, S., and Wirth, R. (1988). Transformation of bacteria with plasmid DNA by electroporation. *Anal. Biochem.* 170, 38–44.

Finaz, C., Lefevre, A., and Teissie, J. (1984). Electrofusion: a new, highly efficient technique for generating somatic cell hybrids. *Exp. Cell Res.* 150, 477–482.

Fromm, M., Taylor, L. P., and Walbot, V. (1985). Expression of genes transferred into monocot and dicot plant cells by electroporation. *Proc. Natl. Acad. Sci. USA* 82, 5824–5828.

Fromm, M., Taylor, L. P., and Walbor, V. (1986). Stable transformation of maize after gene transfer by electroporation. *Nature* 319, 791–793.

Furtine, E. S., and Wang, C. C. (1990). Transfection of the *Giardia lamblia* double-stranded RNA virus into *Giardia lamblia* by electroporation of a single-stranded RNA copy of the viral genome. *Mol. Cell. Biol.* 10, 3659–3663.

Gatz, C., and Quail, P. H. (1988). Tn 10-encoded *tet* repressor can regulate an operator-containing plant promoter. *Proc. Natl. Acad. Sci. USA* 85, 1394.

Gilroy, S., Hughes, W. A., and Trewavas, A. J. (1986). The measurement of intracellular calcium levels in protoplasts from higher plant cells. *FEBS Lett.* 199, 217.

Glassy, M. C., and Hofmann, G. (1985). Optimization of electro-cell fusion parameters in generating human-human hybrids with UC729-6. *Hybridoma* 4, 61.

Grasso, R. J., Heller, R., Cooley, J. C., and Haller, E. M. (1989). Electrofusion of individual animal cells directly to intact corneal epithelial tissue. *Biochim. Biophys. Acta* 980, 9–14.

Halfmann, H. J., Emeis, C. C., and Zimmermann, U. (1983). Electrofusion of haploid *Saccharomyces* yeast cells of identical mating type. *Arch. Microbiol.* 134, 1–4.

Hama-Inaba, H., Takahashi, M., Kasai, M., Shiomi, T., Ito, A., Hanaoka, F., and Sato, K. (1987). Optimum conditions for electric pulse-mediated gene transfer to mammalian cells in suspension. *Cell Structure and Function* 12, 173.

Hambor, J. E., Hauer, C. A., Shu, H.K., Groger, R. K., Kaplan, D. R., and Tykocinski, M. L. (1988). Use of an Epstein-Barr virus episomal replicon for anti-sense RNA-mediated gene inhibition in a human cytotoxic T-cell clone. *Proc. Natl. Acad. Sci. USA* **85**, 4010.

Hashimoto, H., Morikawa, H., Yamada, Y., and Kimura, A. (1985). A novel method for transformation of intact yeast cells by electroinjection of plasmid DNA. *Appl. Microbiol. Biotechnol.* **21**, 336.

Hashimoto, K., Tatsumi, N., and Okuda, K. (1989). Introduction of phalloidin labeled with fluorescein isothiocyanate into living polymorphonuclear leukocytes by electroporation. *J. Biochem. Biophys. Methods* **19**, 143–154.

Hayasaka, K., Sato, M., Mitani, H., and Shima, A. (1990). Transfection of cultured fish cells RBCF-1 with exogenous oncogene and their resistance to malignant transformation. *Comp. Biochem. Physiol.* **96**, 349–354.

Hibi, T., Kano, H., Sugiura, M., Kazami, T., and Kimura, S. (1986). High-efficiency electro-transfection of tobacco mesophyll protoplasts with TMV RNA. *J. Gen. Virol.* **67**, 2037–2042.

Howard, P. K., Ahern, K. G., and Firtel, R. A. (1988): Establishment of a transient expression system for *Dictyostelium discoideum. Nucleic Acids Res.* **16**, 2613.

Inoue, K., Yamashita, S., Hata, J., Kabeno, S., Asada, S., Nagahisa, E., and Fujita, T. (1990). Electroporation as a new technique for producing transgenic fish. *Cell Differ. Dev.* **29**, 123–128.

Jacob, H. E., Siegemund, F., Bauer, E., Muhlig, P., and Berg, H. (1983). Fusion of plant protoplasts by dielectrophoresis and electro-field-pulse technique. *Stud. Biophys.* **94**, 99–100.

Karube, I., Tamiya, E., and Matsuoka, H. (1985). Transformation of *Saccharomyces cerevisiae* spheroplasts by high electric pulse. *FEBS Lett.* **182**, 90.

Karsten, U., Papsdorf, G., Roloff, G., Stolley, P., Abel, H., Walther, I., and Weiss, H. (1985). Monoclonal anti-cytokeratin antibody from a hybridoma clone generated by electrofusion. *Eur. J. Cancer Clin. Oncol.* **21**, 733–740.

Kohn, H., Schieder, R., and Schieder, O. (1985). Somatic hybrids in tobacco mediated by electrofusion. *Plant Sci.* **38**, 121–128.

Koop, H. U., Dirk, J., Wolff, D., and Schweiger, H. G. (1983). Somatic hybridization of two selected single cells. *Cell Biol. Int. Rep.* **7**, 1123–1127.

Kubiak, J. Z., and Tarkowski, A. K. (1985). Electrofusion of mouse blastomeres. *Exp. Cell Res.* **157**, 561–566.

Kubiniec, R. T., Liang, H., and Hui, S. W. (1990). Effects of pulse length and pulse strength on transfection by electroporation. *BioTechniques* **81**(1), 16–20.

Laban, A., and Wirth, D. (1989). Transfection of *Leishmania enriettii* and expression of chloramphenical acetyltransferase gene. *Proc. Natl. Acad. Sci USA* **86**, 9119–9123.

Langridge, W. H. R., Li, B. J., and Szalay, A. A. (1985). Electric field mediated stable transformation of carrot protoplasts with naked DNA. *Plant Cell Rep.* **4**, 355–359.

Liang, H., Purucker, W. J., Stenger, D. A., Kubiniec, R. T., and Hui, S. W. (1988). Uptake of fluorescence-labeled dextrans by 10T 1/2 fibroblasts following permeation by rectangular and exponential-decay electric field pulses. *BioTechniques* **6**, 550.

Lo, M. M. S., Tsong, T. Y., Conrad, M. K., Strittmatter, S. M., Hester, L. H., and Snyder, S. H. (1984). Monoclonal antibody production by receptor-mediated electrically induced cell fusion. *Nature* **310**, 792–794.

Mann, S. G., and King, L. A. (1989). Efficient transfection of insect cells with baculovirus DNA using electroporation. *J. Gen. Virol.* **70,** 3501–3505.

McNally, M. A., Lebkowski, J. S., Okarma, T. B., and Lerch, L. B. (1988). Optimizing electroporation parameters for a variety of human hepatopoietic cell lines. *BioTechniques* **6(9),** 882–885.

Mehrle, W., Zimmermann, U., and Hampp, R. (1985). Evidence for asymmetrical uptake of fluorescent dyes through electropermeabilized membranes of *Avena* mesophyll protoplasts. *FEBS Lett.* **185,** 89–94.

Michel, M. R., Elgizoli, M., Koblet, H., and Kempf, C. (1988). Diffusion loading conditions determine recovery of protein synthesis in electroporated P3X63Ag8 cells. *Experientia* **44,** 199.

Miller, J. G., Dower, W. J., and Tompkins, L. S. (1988). High-voltage electroporation of bacteria: Genetic transformation of *Campylobacter jejuni* with plasmid DNA. *Proc. Natl. Acad. Sci. USA* **85,** 856–860.

Morikawa, H., Iida, A., Matsui, C., Ikegami, M., and Yamada, Y. (1986a). Gene transfer into intact plant cells by electroinjection through cell walls and membranes. *Gene* **41,** 121–124.

Morikawa, H., Sugino, K., Hayashi, Y., Takeda, J., Senda, M., Hirai, A., and Yamada, Y. (1986b). Interspecific plant hybridization by electrofusion in *Nicotiana*. *Bio/Technology* **4,** 57–60.

Murray, R., Hutchinson, C. A., III, and Frelinger, J. A. (1988). Saturation mutagenesis of a major histocompatibility complex protein domain: Identification of a single conserved amino acid important for allorecognition. *Proc. Natl. Acad. Sci. USA* **85,** 3535.

Neumann, E., Gerisch, G., and Opatz, K. (1980). Cell fusion induced by high electric impulses applied to *Dictyostelium*. *Naturwissenschaften* **67,** 414–415.

Neumann, E., Schaefer-Ridder, M., Wang, Y., and Hofschneider, P. H. (1982). Gene transfer into mouse lyoma cells by electroporation in high electric fields. *EMBO J.* **1,** 841.

Nishiguchi, M., Langridge, W. H. R., Szalay, A. A., and Zaitlin, M. (1986). Electroporation-mediated infection of tobacco leaf protoplasts with tobacco mosaic virus RNA and cucumber mosaic virus RNA. *Plant Cell Rep.* **5,** 57–60.

Okada, Y., Ohno-Shosaku, T., and Oiki, S. (1984). Ca^{2+} is prerequisited for cell fusion induced by electric pulses. *Biomed. Res.* **6,** 511–516.

Okada, K., Nagata, T., and Takebe, I. (1986). Introduction of functional RNA into plant protoplasts by electroporation. *Plant Cell Physiol.* **27,** 619–626.

Ohno-Shosaku, T., and Okada, Y. (1984). Facilitation of electrofusion of mouse lymphoma cells by the proteolytic action of protease. *Biochem. Biophys. Res. Commun.* **120,** 138–143.

Ohno-Shosaku, T., Hama-Inaba, H., and Okada, Y. (1984). Somatic hybridization between human and mouse lymphoblast cells produced by an electric pulse-induced fusion technique. *Cell Structure Function* **9,** 193–196.

Ou-Le, T. M., Turgeon, R., and Wu, R. (1986). Expression of a foreign gene linked to either a plant-virus or a *Drosophila* promoter after electroporation of protoplasts of rice, wheat, and sorghum. *Proc. Natl. Acad. Sci. USA* **83,** 6815–6819.

Pace, C. S., Tarvin, J. T., Neighbors, A. S., Pirkle, J. A., and Greider, M. H. (1980). Use of a high voltage technique to determine the molecular requirements for exocytosis in islet cells. *Diabetes* **29,** 911.

Pilwat, G., Richter, H. P., nad Zimmermann, U. (1981). Giant culture cells by electric field-induced fusion. *FEBS Lett.* **133**, 169–174.

Podesta, E. J., Solano, A. R., Molina Y., Vedia, L., Paladini, A., Jr., and Sanchez, M. L. (1984). Production of steroid hormone and cyclic AMP in hybrids of adrenal and Leydig cells generated by electrofusion. *Eur. J. Biochem.* **145**, 329–332.

Pohl, H. A. (1978). "Dielectrophoresis." Cambridge University Press, London.

Potter, H., Weir, L., and Leder, P. (1984). Enhancer-dependent expression of human κ immunoglobulin genes introduced into mouse pre-B lymphocytes by electroporation. *Proc. Natl. Acad. Sci. USA* **81**, 7161.

Press, F., Quilet, A., Mir, L., Marchio-Fournigault, C., Feunteun, J., and Fradelizi, D. (1988). An improved electro-transfection method using square shaped electric impulsions. *Biochem. Biophys. Res. Commun.* **151**, 982.

Pu, H. F., and Young, A. P. (1990). Glucocorticoid-inducible expression of a glutamine synthetase-CAT-encoding fusion plasmid after transfection of intact chicken retinal explant cultures. *Gene* **89**, 259–263.

Rhodes, C. A., Pierce, D. A., Mettler, I. J., Mascarenhas, D., and Detmer, J. J. (1988). Genetically transformed maize plants from protoplasts. *Science* **240**, 204.

Richter, H. P., Scheurich, P., and Zimmermann, U. (1981). Electric field-induced fusion of sea urchin eggs. *Dev. Growth Differentiation* **23**, 479–486.

Riggs, C. D., and Bates, G. W. (1986). Stable transformation of tobacco by electroporation. *Proc. Natl. Acad. Sci. USA* **83**, 5602–5606.

Ruthe, H. J., and Adler, J. (1985). Fusion of bacterial spheroplasts by electric fields. *Biochim. Biophys. Acta* **819**, 105–113.

Scheurich, P., and Zimmermann, U. (1981). Giant human erythrocytes by electric field-induced cell-to-cell fusion. *Naturwissenschaften* **68**, 45–46.

Schnettler, R., Zimmermann, U., and Emeis, C. C. (1984). Large-scale production of yeast Saccharomyces cerevisiae hybrids by electrofusion. *FEMS Lett.* **24**, 81–85.

Serpersu, E. H., Kinosita, K., Jr., and Tsong, T. Y. (1985). Reversible and irreversible modification of erythrocyte membrane permeability by electric field. *Biochim. Biophys. Acta* **812**, 779.

Shillito, R. D., Saul, M. W., Paszkowski, J., Muller, M., and Potrykus, I. (1985). High-efficiency direct gene transfer to plants. *Bio/Technology* **3**, 1099–1103.

Shivarova, N., Forster, W., Jacob, H. E., and Grigorova, R. (1983). Microbiological implications of electric field effects. VII. Stimulation of plasmid transformation of *Bacillus cereus* protoplasts by electric field pulses. *Z. Allg. Mikrobiol.* **23**, 595–599.

Shivarova, N., and Grigorova, R. (1983). Microbiological implications of electric field effects VIII. Fusion of *Bacillus thuringiensis* protoplasts by high electric field pulses. *Bioelectrochem. Bioenerg* **11**, 181–185.

Smithies, O., Gregg, R. G., Boggs, S. S. Koralewski, M. A., and Kucherlapati, R. S. (1985). Insertion of DNA sequences into the human chromosomal β-globin locus by homologous recombination. *Nature* **317**, 230.

Sowers, A. E. (1984). Characterization of electric field-induced fusion in erythrocyte ghost membranes. *J. Cell Biol.* **99**, 1989–1996.

Spandidos, D. A. (1987). Electric field-mediated gene transfer (electroporation) into mouse Friend and human K562 erythroleukemic cells. *Gene Anal. Tech.* **4**, 50.

Spangenberg, G., and Schweiger, H. G. (1986). Controlled electrofusion of different types of protoplasts and subprotoplasts including reconstitution in *Brassica napus* L. *Eur. J. Cell Biol.* **41**, 51–56.

Stevens, C. W., Brondyk, W. H., Burgess, J. A., Manoharan, T. H., Hane, B. G., and Fahl, W. E. (1988). Partially transformed, anchorage-independent human diploid fibroblasts result from overexpression of the *c-sis* oncogene: mitogenic activity of an apparent monomeric platelet-derived growth factor 2 species. *Mol. Cell. Biol.* **8**, 2089.

Stopper, H., Zimmermann, U., and Wecker, E. (1985). High yields of DNA transfer into mouse L-cells by electropermeabilization. *Z. Naturforsch.* **40c**, 929–932.

Stopper, H., Zimmermann, U., and Neil, G. A. (1988). Increased efficiency of transfection of murine hybridoma cells with DNA by electropermeabilization. *J. Immunol. Methods* **109**, 145.

Teissie, J., Hnutson, V. P., Tsong, T. Y., and Lane, M. D. (1982). Electric pulse-induced fusion of 3T3 cells in monolayer culture. *Science* **216**, 537–538.

Toriyama, K., Arimoto, Y., Uchimiya, H., and Hinata, K. (1988). Transgenic rice plants after direct gene transfer into protoplasts. *Bio/Technology* **6**, 1072–1074.

Tykocinski, M. L., Shu, H. K., Ayers, D. J., Walter, E. I., Getty, R. R., Groger, R. K., Hauer, C. A., and Medof, M. E. (1988). Glycolipid reanchoring of T-lymphocyte surface antigen CD8 using the 3' end sequence of decay-accelerating factor's mRNA. *Proc. Natl. Acad. Sci. USA* **85**, 3555.

Vienken, J., and Zimmermann, U. (1982). Electric field-induced fusion: electrohydraulic procedure for production of heterokaryon cells in high yield. *FEBS Lett.* **137**, 11–13.

Vienken, J., and Zimmermann, U. (1985). An improved electrofusion technique for production of mouse hybridoma cells. *FEBS Lett.* **182**, 278–280.

Vienken, J., Zimmermann, U., Fouchard, M., and Zagury, D. (1983). Electrofusion of myeloma cells on the single cell level: Fusion under sterile conditions without proteolytic enzyme treatment. *FEBS Lett.* **163**, 54–56.

Watts, J. W., and King, J. M. (1984). A simple method for large-scale electrofusion and culture of plant protoplasts. *Biosci. Rep.* **4**, 335–342.

Watts, J.W., Doonan, J. H., Cove, D. J., and King, J. M. (1985). Production of somatic hybrids of moss by electrofusion. *Mol. Gen. Genet.* **199**, 349–351.

Watts, J. W., King, J. M., and Stacey, N. J. (1987). Inoculation of protoplasts with viruses by electroporation. *Virology* **157**, 40–46.

Winegar, R. A., Philips, J. W., Youngblom, J. H., and Morgan, W. F. (1989). Cell electroporation is highly efficient method for introducing restriction endonucleases into cells. *Mutat. Res.* **225**, 49–53.

Wong, T. K., and Neumann, E. (1982). Electric field mediated gene transfer, *Biochem. Biophys. Res. Commun.* **107**, 584–587.

Yaseen, M. A., Pedley, K. C., and Howell, S. L. (1982). Regulation of insulin secretion from islets of Langerhans rendered permeable by electric discharge. *Biochem. J.* **206**, 81.

Zerbib, D., Amalric, F., and Teissie, J. (1985). Electric field-mediated transformation: Isolation and characterization of a TK + subclone. *Biochem. Biophys. Res. Commun.* **129**, 611.

Zheng, Q., and Chang, D. C. (1989). Fusion of cabbage mesophyll protoplasts using a radio-frequency electric field. *J. Cell Biol.* **109**, 4(2): 309a.

Zheng, Q., and Chang, D. C. (1990). Dynamic changes of microtubule and actin structures in CV-1 cells during electrofusion. *Cell Motil. Cytoskeleton* **17**, 345–355.

Zheng, Q., and Chang, D. C. (1991). High-efficiency gene transfection by *in situ* electroporation of cultured cells. *Biochim. Biophys. Acta* **1088,** 104–110.

Zheng, Q., Wu, Y., Cai, G. P., and Zhao, N. M. (1988). Electrofusion between protoplasts of cabbage and Chinese cabbage. *Acta Biophys. Sinica* **4,** 134–139.

Zimmermann, U., and Pilwat, G. (1982). Electric field-mediated cell fusion. *J. Biol. Phys.* **10,** 43–50.

Zimmermann, U., and Scheurich, P. (1981). High-frequency fusion of plant protoplasts by electric fields. *Planta* **151,** 26–32.

Zimmermann, U., Vienken, J., and Pilwat, G. (1980). Development of drug carrier systems: Electric field induced effects in cell membranes. *J. Electroanal. Chem.* **116,** 553–574.

27

Protocols for Using Electroporation to Stably or Transiently Transfect Mammalian Cells

Huntington Potter

Department of Neurobiology, Harvard Medical School, Boston, Massachusetts 02115

I. Introduction

Over the last 10 years, electroporation has become increasingly the method of choice for introducing macromolecules into various cell types. Because it is a physical rather than a biochemical technique, electroporation has wide applicability. It has been successfully used in essentially all cell types—animal, plant, and microbial—and causes less perturbation of the target cells and electroporated molecules than alternative approaches. In addition to yielding a high frequency of permanent or transient transfectants, electroporation is also substantially easier to carry out than alternative techniques and is highly reproducible (Neumann *et al.*, 1982; Wong *et al.*, 1982; Potter *et al.*, 1984; Chu *et al.*, 1987; for review see Potter, 1988; Potter and Cooke, Chapter 13, this volume).

Although electroporation is effective in a wide variety of cell types, each requires

Guide to Electroporation and Electrofusion
Copyright © 1992 by Academic Press, Inc. All rights of reproduction in any form reserved.

slightly different conditions. Thus a comprehensive description of all the methods and applications of electroporation is now far beyond the scope of one article. Here I will concentrate on electroporation of mammalian cells. The use of the technique in other types of target cells (bacterial or plant, for instance) is discussed elsewhere in this volume. Also, the protocols presented here are intended for use with the most common type of electroporator, which uses a capacitor discharge method to generate the high-intensity electrical pulse (see instrumentation section below). The electrical parameters for other instruments are adjusted according to the manufacturers' instructions.

II. Materials

A. Reagents (Electroporation Buffers)

1. Dulbecco phosphate-buffered saline (without Ca^{2+} or Mg^{2+}) (PBS) (8 g NaCl; 0.2 g KCl; 0.2 g KH_2PO_4; 2.16 g $Na_2HPO_4 \cdot 7H_2O$ per liter, pH 7.3).
2. HEPES-buffered saline (HeBS) (4.76 g HEPES; 8 g NaCl; 0.4 g KCl; 0.18 g Na_2HPO_4; 1.08 g glucose per liter, pH 7.05).
3. Phosphate-buffered sucrose (272 mM sucrose, 7 mM sodium phosphate, pH 7.4, 1 mM $MgCl_2$).
4. Tissue culture medium (without fetal calf serum).

B. Instrumentation

The wide use of electroporation has been made possible in part by the availability of commercial apparatus that are easy to use, extremely reproducible, and safe. Most of our experience has been with the Bio-Rad Gene Pulser, a capacitor discharge device. Commonly used capacitor discharge devices are also available from BRL, BTX, Hoeffer, and IBI. These machines, either in one unit or by add-ons, can deliver a variety of electroporation conditions suitable for most applications. In addition, there are square-wave generators (for instance from BTX or Baekon), which offer greater control over pulse width and repetition and can be more effective for cells that are very sensitive or otherwise difficult to transfect. These machines also carry a higher price tag. Very recently, it has become apparent that alternating current pulses at ~100 kHz can be the most effective waveform for electroporation, and possibly electrofusion (Chang, 1989, 1991, and this volume; Tekle *et al.*, 1991). However, dedicated electroporation devices utilizing such waves are not yet commercially available and must be constructed from components. The experimental protocols outlined below are designed for use with the Bio-Rad Gene Pulser, but will be directly applicable to other capacitor discharge devices, and, with some fiddling, to square-wave generators. Inasmuch as all cell lines will need to be optimized for the particular machine, these protocols are meant primarily as starting guides to be adapted as needed according to the manufacturers' instructions.

III. Electroporation Parameters

The voltage and capacitance settings must be optimized for each cell type. More important, the resistance of the electroporation buffer is critical for choosing the setting. That is, for the low-resistance buffers (high salt) such as PBS and HeBS, start with a capacitor setting of 25 μF and a voltage of 1200 V for 0.4-cm cuvettes, then increase or decrease the voltage until optimal transfection is obtained (usually at about 40–70% cell viability). Optimal permanent and transient transfection occur at about the same instrument settings, so transient expression can be used to optimize conditions for a new cell type.

Some cells are very easily killed and thus electroporate poorly at the high voltages needed for PBS or HeBS electroporation buffers. Phosphate-buffered sucrose can be used at slightly lower voltages. Alternatively, Chu et al. (1987) found many sensitive cells were electroporated more effectively in tissue culture medium with a low-voltage, high-capacitance setting that results in a longer pulse duration. For these conditions, start at 250 V, 960 μF, and change the voltage up to 350 V or down to 100 V in steps to determine optimal settings. While keeping cells at 0°C may improve cell viability, Chu et al. (1987) also found that some cell lines electroporate with higher efficiency at room temperature. Therefore, steps 4–8 should be carried out separately at both temperatures to determine the optimum conditions for a new cell line.

IV. Procedures

A. Suspension Cells

1. Grow cells to be transfected to mid or late log phase in complete medium. Each permanent transfection will usually require 5×10^6 cells to yield a reasonable number of transfectants. Each transient expression may require 1–4×10^7 cells, depending on the promoter.

2. Harvest cells by centrifuging 5 min (600–1000 × g), 4°C, and resuspend the cell pellet in half the original volume of ice-cold electroporation buffer (see discussion of temperature in the section on electroporation parameters, above).

3. Harvest the cells by centrifuging 5 min as in step 2, and resuspend the cells at 1×10^7/ml in electroporation buffer at 0°C for permanent transfection. Higher concentrations of cells (up to 8×10^7) may be used for transient expression. The volume for each electroporation should be 0.5 ml.

4. Transfer aliquots of 0.5 ml of the cell suspension into the desired number of electroporation cuvettes at 0°C.

5. Add DNA to the cell suspension in the cuvettes at 0°C. For permanent transfection, the DNA should be linearized by cleavage with a restriction enzyme that cuts in a nonessential region, and purified by phenol extraction and ethanol precipitation. For transient expression, the DNA may be left supercoiled. In both cases, the DNA should have been purified through two preparative CsCl/ethidium bromide equilibrium gradients followed by phenol extraction and ethanol precipitation. For transient expression, 10–40 μg is optimal. For permanent transfection, 1–10 μg is sufficient. For cotransfection (which we generally avoid), 1 μg of the selectable marker containing DNA and 10 μg of the DNA containing the gene of interest are usually adequate.

6. Mix the DNA/cell suspension by holding the cuvette on the two "window sides" and flicking the bottom.

7. After 5 min at 0°C, place the cuvette in the holder in the electroporation apparatus (at room temperature) and shock one or more times at the desired voltage and capacitance settings, according to the manufacturer's instructions for your instrument. The voltage and capacitance settings that should be used will vary, depending on the cell type and need to be optimized (see below).

8. After electroporation, return the cuvette containing cells and DNA to ice for 10 min.

9. Dilute the transfected cells 20-fold in the nonselective complete medium and rinse the cuvette with the same medium to rinse out all the transfected cells.

10. Grow the cells for 48 h (or about two generations) prior to selection with medium containing the appropriate antibiotic for permanent transformants or for 50–60 h prior to harvesting for transient expression assays.

11a. Place cells to be selected for permanent transfectants in the antibiotic-containing medium. For example, neo selection generally requires ~400 mg/ml G418 in the media. Eco-gpt selection requires 1 μg/ml micophenolic acid, 250 μg/ml xanthine, and 15 μg/ml hypoxanthine in the media. Selection conditions will vary with cell type. For permanent transfection, it is often convenient to plate out the cells at limiting dilution immediately following the shock if they are adherent cells or at the time of antibiotic addition if they are suspension cells.

11b. For transient expression, harvest the cells and assay for expression according to standard protocols.

B. Adherent Cells

Adherent cells may be first detached using a trypsin–ethylenediamine tetraacetic acid (EDTA) treatment and then transfected essentially as described above.

1. Trypsinize the cells from the plate surface (1 ml 0.05% trypsin, 0.53 mM EDTA per 100-mm plate).
2. Inactivate the trypsin with serum-containing medium, and wash the cells by several centrifugations and resuspensions in electroporation buffer at 0°C.
3. Electroporate at setting appropriate for the electroporation buffer and optimized for each cell type.
4. After 5 min at 0°C, plate cells at various dilutions for permanent or transient expression assays as described above for suspension assays.

C. Electroporation of Adherent Cells on Microbead Carriers

Many situations exist where it would be advantageous to be able to introduce macromolecules into adherent cells still attached to their substrate. For instance, many adherent cells, such as neurons or endothelial cells and even fibroblasts, have a unique morphology when attached to their substrates. In addition, they may show importantly different morphologies and behavior when attached to different kinds of substrates. For instance, neurons "growing" (sending out processes) on collagen or tissue culture plastic behave quite differently from neurons growing on laminin.

There are several ways to electroporate adherent cells directly in their attached state (see chapters by Chang *et al.* and Gallagher and Chang, Chapter 25, this volume, and Raptis and Firth, 1990). We have developed an approach to electroporating adherent cells attached to their substrates, which makes use of available methodology and instrumentation in a new way that will be easily adapted in laboratories already carrying our electroporation in the traditional manner. Specifically, we have been able to electroporate DNA into cells attached to the surface of microbeads in suspension (see Potter and Cooke, Chapter 13, this volume, for further discussion). The indication is that the electroporation efficiency is as at least as high as for the same cells in suspension. Because plastic microbeads are very easy to manipulate, come in various types applicable to almost all adherent cells, and can be kept in suspension for short periods of time, it is straightforward to wash a sample of microbeads carrying cells in appropriate electroporation buffer, introduce the beads at any desired concentration into the electroporation chamber in suspension, and carry out the electroporation in the same manner as for suspended cells alone. As long as the concentration of microbeads is not huge so that their contribution to the total volume remains small, drastic changes in electroporation parameters seem, by our preliminary experiments, not to be necessary. The procedure should in principle be applicable to any adherent cell type for both transient and stable expression. As with all electroporation experiments, however, optimization for each cell type would be advisable for obtaining high transfection efficiencies.

To carry out electroporation of cells on microbeads:

1. Prepare the beads (for example, Cytodex 1 from Pharmacia) according to the manufacturer's instructions. Briefly, define the beads by swelling them in Ca^{2+}, Mg^{2+}-free phosphate-buffered saline, allowing them to settle, removing the supernatant, and resuspending them twice. Autoclave the washed microbeads in a 30-fold to 50-fold volume of buffer.
2. Just prior to use, place the desired volume of beads (approximately 1–5 mg per 10^5 cells) into appropriate growth medium [for instance, Dulbecco's Modified Eagle's Medium (DME) with 10% fetal calf serum, glutamate and pen/strep added] and wash once by allowing them to settle in the same medium.
3. Harvest cells growing in tissue culture dishes by standard procedures, add them to the resuspended microbeads, and place beads and cells together in a bacterial or tissue culture dish. Microscopic examination after several hours should indicate good adherence and appropriate morphology of the cells on the microbeads.
4. The following day, remove the microbeads together with their medium from the petri dish and transfer them into electroporation buffer by repeatedly allowing the beads to settle and then resuspending them.
5. After removing the final wash, resuspend the beads with their cells in an appropriate volume of electroporation buffer (0.5 ml) and place them on ice for 3–5 min in the electroporation cuvette.
6. Just prior to electroporation, add 2–5 μg of DNA to the cuvette, resuspend the by-now-settled microbeads containing the cells, and carry out the electroporation as described above for suspension cells. Return the cuvette to ice for 3–5 min.
7. Remove the microbeads from the cuvette and place them in medium for recovery and growth.

Transfected cells can be visualized by standard procedures. For instance, if the transfected DNA contains a β-galactosidase gene,

8. First fix the cells on the microbeads for 10 min in 4% paraformaldehyde, 0.2% glutaraldehyde, wash in PBS, and expose to X-gal solution containing 5-mM KFeCn (ferric), 5 mM KFeCn (ferro), 2 mM $MgCl_2$, 0.1% X-gal for 3–6 h. Blue cells expressing β-galactosidase are clearly visible on the surface of the microbeads, indicating successful transfection (see protocol by MacGregor, Chapter 28, this volume).

Acknowledgments

The work in this laboratory is supported by grants from the National Institutes of Health and the Alzheimer's Disease and Related Disorders Association. Stefan Cooke assisted in the preparation of the manuscript and in the experiments with transfection of adherent cells on microbeads.

References

Chang, D. C. (1989). Cell poration and cell fusion using an oscillating electric field. *Biophys. J.* **56,** 641–652.

Chang, D. C., Gao, P. Q., and Maxwell, B. L. (1991). High efficiency gene transfection by electroporation using a radio-frequency electric field. *Biochim. Biophys. Acta* (in press).

Chu, G., Hayakawa, H., and Berg, P. (1987). Electroporation for the efficient transfection of mammalian cells with DNA. *Nucleic Acids Res.* **15,** 1311–1326.

Neumann, E., Schaefer-Ridder, M., Wang, Y., and Hofschneider, P. H. (1982). Gene transfer into mouse lyoma cells by electroporation in high electric fields. *EMBO J.* **1,** 841–845.

Potter, H. (1988). Electroporation in biology: Methods, applications, and instrumentation. *Anal. Biochem.* **175,** 361–373.

Potter, H., Weir, L., and Leder, P. (1984). Expression of human Ig genes introduced into mouse B-lymphocytes by electroporation: Enhancer effects. *Proc. Natl. Acad. Sci. USA* **81,** 7161–7165.

Raptis, L., and Firth, K. L. (1990). Laboratory methods: Electroporation of adherent cells *in situ. DNA Cell Biol.* **9,** 615–621.

Tekle, E., Astumian, R. D., and Chock, P. B. (1991). Electroporation by using bipolar oscillating electric field: An improved method for DNA transfection of NIH 3T3 cells. *Proc. Natl. Acad. Sci. U. S. A.* **88,** 4230–4234.

Wong, T. K., and Neumann, E. (1982). Electric field mediated gene transfer. *Biochem. Biophys. Res. Commun.* **107,** 584–587.

28

Optimization of Electroporation Using Reporter Genes

Grant R. MacGregor

Department of Cell Biology, Howard Hughes Medical Institute, Baylor College of Medicine, Houston, Texas 77030

I. Introduction

During the past 5 years, electroporation has gained wide acceptance as a powerful method with which to introduce DNA into a variety of prokaryotic and eukaryotic cells in culture. However, a major obstacle to the successful implementation of this technique is the prerequisite determination of conditions that are optimum for the particular cell type being used. Voltage, capacitance, medium composition, cell density, state of cell growth, reaction temperature, and DNA concentration are but a few of the parameters that require consideration.

Of assistance in performing this calibration are plasmids that can express a reporter gene product following their introduction into cells in culture. Cells may be transfected under varying conditions and the relative efficiency of each transfection determined by monitoring the reporter gene activity. Ideally, such a reporter gene product should be stable, innocuous to the cell in which it is being expressed, and readily detectable even in minute quantities. For these reasons, genes that encode enzymes that can be detected using chromogenic or radiolabeled substrates have been widely recruited. Although there exists a variety of reporter genes, the two

Guide to Electroporation and Electrofusion

most commonly used are the bacterial genes that encode chloramphenicol acetyl transferase (CAT; Gorman *et al.*, 1982) and beta-galactosidase (β-Gal; Hall *et al.*, 1983).

Arguably, of the two, *Escherichia coli LacZ* (β-galactosidase; EC 3.2.1.23; Wallenfels and Weil, 1971) is the more versatile. Extensive characterization of both the gene and its product has led to the development of several assays for β-Gal activity that are simple to perform and use relatively inexpensive, nonradioisotopic reagents. In addition, plasmid constructs are available that can be used to generate β-Gal expression in a wide range of mammalian cell types (MacGregor and Caskey, 1989). For the sake of brevity only a histochemical assay will be considered here. For a detailed description of alternative biochemical and biological assays the reader is referred to MacGregor *et al.* (1990).

Of the available assays for β-Gal activity, the *in situ* histochemical assay (Bondi *et al.*, 1982) is the most appropriate for use in performing a determination of relative efficiencies of cellular transfection. Briefly, cells are grown to a mid-log state of growth, trypsinized if adherent, or pelleted if in suspension, resuspended in the medium of choice and electroporated with plasmid that encodes *E. coli* β-Gal under the control of a suitable promoter. After a period of time (24–48 h) to allow recovery and expression of the reporter gene, the cells are fixed and overlaid with a stain containing X-Gal, a chromogenic indicator of β-Gal activity. The X-Gal (5-chloro-4-bromo-3-indolyl-β-D-galactoside) is hydrolyzed by β-galactosidase to generate galactose and soluble indoxyl, which is subsequently converted into insoluble indigo. The indigo is deep blue in color and facilitates determination of the relative proportion of cells that have taken up and expressed the plasmid construct (Fig. 1). The other essential components of the stain are (1) sodium phosphate, which buffers the pH of the system to favor bacterial enzyme activity over endogenous mammalian β-galactosidases (which have more acidic pH optima) and provides sodium ions, an activator of β-Gal, (2) magnesium ions, a cofactor for the enzyme, and (3) potassium ferro- and ferricyanide, which together act as an oxidation catalyst, increasing the rate of conversion of the soluble indoxyl molecules to insoluble indigo, thereby enhancing cellular localization of the enzyme activity. This assay has the advantage that it can detect a single cell expressing β-Gal within a population of nonexpressing cells.

Figure 1   Evaluation of electroporation efficiency using a β-Gal expression vector. NIH 3T3 cells were electroporated with the β-Gal expression vector pCMVβ (MacGregor and Caskey, 1989) that expresses β-Gal under the control of a human cytomegalovirus immediate early promoter. Equal numbers of cells were electroporated with varying voltage. Cells expressing *E. coli* β-Gal appear as different shades of grey. In (b), although the efficiency of transfection (number of expressing cells as a percentage of the total survivors) is greater than in (a), the number of surviving cells is considerably lower. This illustrates the ease with which relative frequencies of transfection and cell survival can be estimated using this technique.

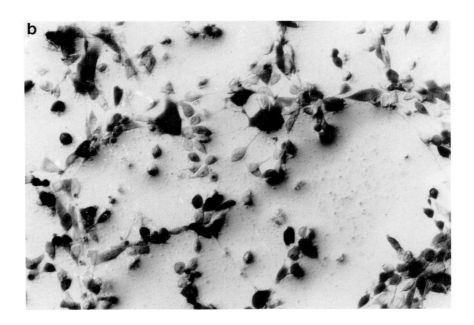

II. Materials

1. Stock solutions of Na_2HPO_4, NaH_2PO_4, and $MgCl_2$, each 1 M, prepared in double distilled (dd) or Millipore (milliQ) grade water. Autoclave to sterilize and store at room temperature. Stable indefinitely.
2. Stock solutions of potassium ferrocyanide $[K_4Fe(CN)_6]$ and potassium ferricyanide $[K_3Fe(CN)_6]$, each 50 mM. Prepare in dd or milliQ water, filter sterilize using a 0.45-μM disposable filtration unit, and store in foil-wrapped glassware (or in dark) at 4°C. Stable for at least 3 months.
3. Stock solution of X-Gal; 20 mg/ml. Dissolve in N,N-dimethylformamide and store in a foil-wrapped glass container (NOT polystyrene) at -20°C. Stable for at least 1 year.
4. Paraformaldehyde; 4% (CAUTION: wear a mask and gloves when handling paraformaldehyde and prepare in a fume cupboard). Dissolve 8 g of powder in 150 ml of 0.1 M sodium phosphate, pH 7.3 (20 mM NaH_2PO_4, 80 mM Na_2HPO_4) by stirring and heating to around 60°C. If necessary, add 10 N NaOH at the rate of 1 drop every 10 s or so until the solution clears. Raise the volume to 200 ml with additional 0.1 M sodium phosphate, pH 7.3, and sterilize by filtration. Store at 4°C for up to 1 month.
5. Gluteraldehyde (Fischer) is purchased as a 25% solution.
6. Phosphate-buffered saline (PBS); composition is 15 mM sodium phosphate, pH 7.3, 150 mM NaCl.
7. To prepare the working fixative (0.2% gluteraldehyde/2% paraformaldehyde), combine 49.2 ml of 0.1 M sodium phosphate, pH 7.3, 0.8 ml of gluteraldehyde, and 50 ml of 4% paraformaldehyde. Store at 4°C for up to 2 weeks.
8. To prepare the X-Gal stain, mix stocks to give final concentrations of 100 mM sodium phosphate, pH 7.3 (80 mM Na_2HPO_4, 20 mM NaH_2PO_4), 1.3 mM $MgCl_2$, 3 mM potassium ferrocyanide, 3 mM potassium ferricyanide, and 1 mg/

Table 1

Quick Calculation Table for X-Gal Stain Ingredients

Compound	Stock concentration	Final concentration	5 ml	10 ml	25 ml	50 ml	100 ml
Na_2HPO_4	1 M	80 mM	400 μl	800 μl	2 ml	4 ml	8 ml
NaH_2PO_4	1 M	20 mM	100 μl	200 μl	500 μl	1 ml	2 ml
$MgCl_2$	1 M	1.3 mM	6.5 μl	13 μl	32.5 μl	65 μl	130 μl
X-Gal	20 mg/ml	1 mg/ml	250 μl	500 μl	1.25 ml	2.5 ml	5.0 ml
$K_3Fe(CN)_6$	50 mM	3 mM	300 μl	600 μl	1.5 ml	3.0 ml	6.0 ml
$K_4Fe(CN)_6$	50 mM	3 mM	300 μl	600 μl	1.5 ml	3.0 ml	6.0 ml
H_2O			3.65 ml	7.3 ml	18.2 ml	36.4 ml	72.8 ml

ml X-Gal. Filter through a 0.45-μ*M* disposable filtration unit prior to use. (See Table 1 for a quick guide to preparation.) Make fresh for each occasion.

III. Methods

A. Adherent Cells

1. Aspirate media from tissue culture dishes containing cells to be assayed.
2. Rinse cell monolayers twice with PBS at room temperature.
3. Overlay cells with fixative and incubate at 4°C for 10 min.
4. Aspirate the fixative and rinse twice with PBS.
5. Overlay cells with X-Gal stain and incubate anywhere from 15 min to overnight at 37°C in a humidified incubator. Examine cells under microscope for β-Gal activity.
6. Aspirate X-Gal stain and rinse plates thoroughly, three times with 70% alcohol. This rinses out the stain and inactivates β-Gal.

B. Suspension Cells

1. Pellet the cells (from 10^4 to 10^7) using a centrifuge.
2. Wash once with 5 ml of PBS.
3. Pellet cells and aspirate PBS.
4. Agitate the tube to disrupt the cell pellet.
5. Add two ml of fixative and incubate at 4°C for 10 min.
6. Pellet the cells, aspirate fixative, wash cells in 5 ml PBS.
7. Pellet cells, resuspend in 2 ml X-Gal stain.
8. Transfer cells to a 24-well tissue culture plate.
9. Incubate anywhere from 15 min to overnight at 37°C in a humidified incubator.
10. Examine cells under microscope for β-Gal activity.
11. If storage is desired, aspirate X-Gal stain and rinse with 70% ethanol as for adherent cells.

C. Notes

1. Heterogeneity of staining within cell populations has often been observed (e.g., Fig. 1). This does not appear to be due to variation of cell permeability to the X-Gal stain, but probably reflects differences in the level of expression of β-Gal from cell to cell.
2. Commercially available antibodies have been shown to have greater sensitivity

in detecting β-Gal (MacGregor *et al.*, 1987). However, after staining with X-Gal, cells cannot subsequently be stained immunocytochemically.

Acknowledgment

The author is a research associate of the Howard Hughes Medical Institute.

References

Bondi, A., Chieregatti, G., Eusebi, V., Fulcheri, E., and Bussolati, G. (1982). The use of β-galactosidase as a tracer in immunocytochemistry. *Histochemistry* 76, 153–158.

Gorman, C. M., Moffat, L. F., and Howard, B. H. (1982). Recombinant genomes which express chloramphenicol acetyl transferase in mammalian cells. *Mol. Cell Biol.* 2, 1044–1051.

Hall, C. V., Jacob, P. E., Ringold, G. M., and Lee, F. (1983). Expression and regulation of *E. coli LacZ* gene fusions in mammalian cells. *J. Mol. Appl. Genet.* 2, 101–109.

MacGregor, G. R., Mogg, A. E., Burke, J. F., and Caskey, C. T. (1987). Histochemical staining of clonal mammalian cell lines expressing *E. coli* β-Galactosidase indicates heterogeneous expression of the bacterial gene. *Somat. Cell Mol. Genet.* 13, 253–265.

MacGregor, G. R., and Caskey, C. T. (1989). Construction of plasmids that express *E. coli* β-Galactosidase in mammalian cells. *Nucleic Acid Res.* 17, 2365.

MacGregor, G. R., Nolan, G. P., Fiering, S., Roederer, M., and Herzenberg, L. A. (1991). Use of *E. coli* LacZ (β-galactosidase) as a reporter gene. *In* "Methods in Molecular Biology" (E. J. Murray, ed.), Vol 7, pp. 217–235. Humana Press, Clifton, NJ.

Wallenfels, K., and Weil, R. (1971). β-Galactosidase. *In* "The Enzymes" (P. Boyer, ed.), 3rd ed., pp. 617–663, Academic Press, New York.

29

Genetic Manipulation of Plant Cells by Means of Electroporation and Electrofusion

James A. Saunders[1] and George W. Bates[2]

[1]Plant Sciences Institute, Beltsville Agricultural Research Center, United States Department of Agriculture, Beltsville, Maryland 20705

[2]Department of Biological Science, Institute of Molecular Biophysics, Florida State University, Tallahassee, Florida 32306

I. Introduction

Genetic manipulation of plants by electroporation and electrofusion has primarily involved isolated protoplasts employing the same basic principle: the application of a brief, high-voltage electrical shock to the cells, which results in a reversible permeabilization of the plasma membrane. Thus, protoplasts suspended in a solution of plasmid DNA can be induced to take up and incorporate some of that DNA, and cells in physical contact at the time of the electrical shock may undergo membrane fusion. Despite the basic simplicity of the technique, a large number of different approaches have been utilized for the electroporation and electrofusion of plant cells. In this chapter, we discuss critical parameters for the successful electroporation and electrofusion of plant cells, and provide several specific protocols that have worked well in different laboratories. It is our hope that this discussion, and these protocols, will provide a useful starting point for other investigators wishing to use electroporation and/or electrofusion in their own research.

A. Choice of Experimental Material

Although electroporation and electrofusion techniques have been successfully applied to plant, animal, bacterial, and fungal cells, there are some important differences in both the procedures and the equipment used for different species and cell types. In plants, electroporation and electrofusion have primarily involved the use of plant protoplasts due to the barrier created by the cell wall. Plant protoplasts are isolated by enzymatic digestion of various tissues to remove the cell walls. Procedures for protoplast isolation are described in numerous reviews (Gamborg *et al.*, 1981; Reinert and Yeoman, 1982; Saunders and Bates, 1987). Protoplasts isolated from leaf mesophyll tissue, suspension cultures, or callus cultures are usually used for electroporation and electrofusion because of the ease of their isolation; however, protoplasts from roots and hypocotyls can also be used. An attribute of plant protoplasts, which clearly differentiates them from animal cells with regard to gene transfer procedures, is the ability of a single plant cell to regenerate into an entire mature organism in tissue culture. This key feature has made the electroporation and electrofusion of plant protoplasts an important research tool for the production of genetically modified germplasm.

The ploidy of the source tissue can also be an important criterion in experiments designed to produce genetically transformed or hybrid plants. Mesophyll cells are usually euploid, whereas suspension-culture cells are frequently polyploid or aneuploid. Thus, protoplasts isolated from suspension cultures often result in polyploid or aneuploid plants. On the other hand, suspension-culture protoplasts often grow more rapidly after electroporation or electrofusion than do mesophyll protoplasts. The durability of suspension-cultured protoplasts can be valuable in experiments where transient expression of promoter constructs is sought. Of course, in experiments where the goal is the regeneration of plants, the regenerability and genetic stability of the protoplasts may be the overriding concern.

Although the cell wall represents a formidable barrier to the uptake of DNA and RNA by electroporation, there are examples of experiments on plant cells of successful electroporation on plant cells with cell walls. For example, germinating pollen grains have been successfully electroporated with the β-glucuronidase gene in a process termed pollen electrotransformation (Matthews *et al.*, 1990; Abdul-Baki *et al.*, 1990; and Chapter 15 by Saunders, Matthews, and Van Wert in this volume). In this case, although the pollen grain itself has a substantial pollen cell wall, the germinating pollen tube has a diminutive cell wall that is known to be porous. This condition is unusual in plant tissues and represents a transitory plant cell essentially without a cell wall.

Other reports have suggested that electroporation can proceed directly through the cell wall of intact tissues (Morikawa *et al.*, 1986; Dekeyser *et al.*, 1990). For example, Dekeyser *et al.* (1990) recently reported that leaf bases of etiolated rice seedlings showed transient expression of β-glucuronidase when electroporated. This development in electroporation technology may play a significant role in the induction of chimeric genetic modifications of plants. Due to the nature of the experiments, however, it is unlikely to produce genetically uniform tissue with stably transformed progeny. It does show, however, that the cell wall of plant cells, even in highly organized and differentiated plant structures such as the leaf base of rice, is not an absolute permeability barrier for electroporation enhanced DNA uptake.

DNA is not the only genetic material to have been introduced via electroporation. There have also been several investigations which examined the introduction of RNA into plant protoplasts. For example, T. Hibi *et al.* (1986) established transfection conditions in which high cell viability and infection rates were obtained. They examined varying RNA and protoplast concentrations, ionic conditions of the media, and pulse durations and amplitudes, in an effort to monitor the effects of these parameters on electro-mediated transfection.

While there are a few additional examples of successful electroporation of other nonprotoplast tissues such as suspension culture cells (Lindsey and Jones, 1987), protoplasts are absolutely required for electrofusion because of the need for contact between the plasma membranes of the two cells. It is helpful when attempting electrofusion between protoplast from different sources to select cells with approximately the same diameter. This is due to the fact that the field strengths necessary for fusion thresholds to be reached are inversely proportional to the diameter of the protoplast. Therefore it is more difficult to fuse protoplasts of differing sizes and recover viable hybrids.

B. Electrical Parameters: Square Waves versus Capacitive Discharges

As one examines the details of the electroporation parameters reported in the literature some discrepancies appear concerning the optimal field strength, pulse duration, and pulse shape. These discrepancies may be due to the use of pulses of differing shape, differences between the size of the protoplasts from different species, or other variable factors. Two basically different types of electrical pulses, square-wave pulses and exponentially decaying capacitive discharges, have been used effectively for both electroporation and electrofusion. An examination of the shape of the electroporation pulse on the effectiveness of the uptake of RNA has been examined in detail using tobacco protoplasts (Saunders *et al.*, 1989a). They found that the shape of the electroporation wave has a significant effect on the transformation efficiency. When square waves are used, optimal pulses are short (10–100 microseconds) and of high voltage (1–3 kV/cm). For exponential pulses, longer (1–20 ms), and lower-voltage (0.25–0.75 kV/cm) discharges are most effective. In general, square-wave generators can provide transformation efficiencies over a broader range of electroporation conditions than exponential decaying pulses. The effect seems to be due, not to an intrinsically more efficient wave form, but to the higher percentage of viable cells that survive the electroporation treatment when square waves are used.

The choice of which pulse type to use is often made on economic grounds; the equipment needed to produce short, high-voltage square waves is more expensive than that for exponential decaying capacitance discharges. However, electrofusion is probably more effective when the less damaging, short, high-voltage square-wave pulses are utilized, whereas higher levels of transient gene expression may be obtained using long-duration exponential discharges. This is because fusion only requires poration of the plasma membranes at the poles of the cells, where neighboring cells are brought into contact. In contrast, long pulses cause pores to form over a larger fraction of the cell surface, and this enhances electroporation by increasing the effective surface area for DNA uptake. In transient expression experiments, where the long-term survival of the protoplasts is not as crucial, the use of longer exponential discharges can lead to a higher overall level of gene expression, as compared with the use of square-wave pulses. In some cases, it may be an advantage to use an exponential pulse even if fewer cells survive the electroporation treatment. For example, in situations where no screening criteria are available for the transformants it may be desirable to leave a high proportion of the surviving protoplasts transformed, even though relatively few protoplasts survive. Although the range of effective field strength conditions producing this effect is quite limited, it significantly decreases the postelectroporation screening procedures on the surviving cells. In stable transformation, where both DNA uptake and cell viability must be maximized, both waveforms can be utilized with positive results.

Even though the equipment used for electrofusion can also be used for electroporation, there are important differences between electrofusion and electroporation. In electrofusion, cell-to-cell contacts are formed by application of a high-frequency AC field. Thus, an AC-field generator is required for electrofusion that is not needed for electroporation. Several investigators have shown effects on the viability of the protoplasts induced by the AC alignment sine wave as well as the fusion pulse in electrofusion procedures (Saunders *et al.*, 1986; Boston *et al.*, 1987).

It seems apparent that both the amplitude of the electroporation pulse and the duration of the pulse are important criteria in the formation of electroporation "pores" in the protoplast membrane. Optimal conditions for cell viability, DNA uptake, and hybrid selection must be determined empirically for each new experimental system by using a series of preliminary experiments to determine the optimal pulse parameters. Clearly, difficulties would be encountered in attempting to duplicate electrical parameters of a previous report if two laboratories were using pulse generators that produced different waveforms.

C. Chamber Design

Many different chambers and electrode configurations have been described for electroporation and electrofusion. Generally, either flat plates (Watts and King, 1984) or parallel wires (Bates, 1985) are used as electrodes. Chambers like these can be homemade or purchased commercially. The overall size of the chamber is important because for most species successful culture requires large numbers of protoplasts. However, as chambers are made larger, their electrical conductivity, when filled with medium, also increases. As a result, more powerful electrical equipment is needed to create the electrical fields required for electroporation and electrofusion. Consult the chapter by Chassy, Saunders, and Sowers in Section IV of this volume for detailed information on both homemade equipment and that available commercially.

II. Transformation by Electroporation

A number of protocols for electroporation have been published in the literature, and many of them work quite well. Two separate electroporation protocols are included here, one of which uses exponential wave pulses and the other uses square-wave pulses. Aspects of these protocols illustrate both the overlapping and divergent experimental options available in transformation by electroporation.

A. Exponential-Wave Pulse Electroporation Procedures

1. Freshly isolated protoplasts are washed once in HBS (HEPES-buffered saline: 150 mM KCl, 4 mM $CaCl_2$, 10 mM HEPES, pH 7.2, plus enough mannitol to osmotically stabilize the protoplasts), and are resuspended in the same medium at a density of 4×10^6 protoplasts/ml.
2. Mix an equal volume of protoplasts with a solution of plasmid DNA (40 μg/ml in HBS + mannitol) and introduce 0.5 ml of the mixture into a sterile electroporation chamber.
3. Apply an 8- to 10-ms exponential discharge (pulse length defined by the RC

time constant; a 325-μF capacitor will deliver a pulse of this length when the sample is 0.5 ml of HBS and the electrodes are 0.4 cm apart). The voltage required will depend on the distance between the electrodes and the protoplast source. For electrodes 0.4 cm apart, optimal electroporation of carrot suspension culture protoplasts is obtained by a 300-V pulse (750 V/cm).

4. After 3 min, remove the protoplasts from the chamber and dilute them 10× with culture medium.
5. Transient gene expression is optimal 24 h after electroporation. Protocols for the selection of stable transformants are given below.

Notes

1. High-quality protoplast preparations are essential.
2. Finding those pulse parameters that kill approximately 50% of the protoplasts can be used as a starting point for determining the optimal pulse settings for both transient expression and stable integration. (Protoplast viability should be determined 1–24 h after the pulse. The exclusion of Evan's blue or the uptake of fluorescein diacetate are suitable viability tests.)
3. The plasmid DNA may be either linear or supercoiled [some studies have found linear DNA to be more effective for stable transformation (Negrutiu *et al.*, 1987)]. Higher DNA concentrations and the inclusion of carrier DNA (i.e., heat-denatured salmon sperm DNA) may be beneficial (Shillito *et al.*, 1985).
4. Chilling the protoplasts during electroporation is useful only if very long pulses (>30 ms) are used. Holding the protoplasts in the electroporation medium more than a few minutes after the pulse does not seem to increase the amount of DNA uptake (Bates *et al.*, 1988).

B. Square-Wave Pulse Electroporation Procedures

1. A 500-μl aliquot of the protoplast/DNA suspension is placed in an electroporation chamber consisting of a sterile cuvette that is capped with two stainless steel flat electrodes (BTX, San Diego, CA, model 470). The protoplasts are suspended in media consisting of 0.6 M mannitol without buffer at a concentration of 1.5 × 10^5 protoplasts/ml and the DNA and/or RNA is added to a final concentration of 10–30 μg/ml. A single direct current (DC) pulse is delivered across the 1.9-mm electrode gap supplied by a square-wave pulse generator (BTX model 200 cell manipulator). Both amplitude and duration can be measured for the square-wave pulse from zero voltage immediately preceding the pulse to the beginning of the pulse decay. The duration of the square wave pulse is maintained at 80 μs.
2. Leave the electroporation chamber undisturbed for 3 min. Take an aliquot of

the protoplasts for determination of the number of intact protoplasts and for protoplast viability after the pulse. The number of protoplasts before and after each treatment is determined by five replicate counts on a Spencer AO Bright-Line hemacytometer using standard cell counting techniques. Cell viability can be determined 1 h after the electroporation treatment on a Nikon fluorescent microscope fitted with a B filter using fluorescein diacetate following methods described previously (Saunders *et al.*, 1986). the remaining protoplasts can be cultured for 48 h at 25°C and assayed for transient gene expression. For tobacco protoplasts Takebe's culture media, pH 5.4, has worked well (Takebe *et al.*, 1968).

Notes

1. To account for protoplasts that were exploded during the electroporation treatment, the total number of protoplasts before and after each treatment should be recorded. Although the number of viable cells remaining in a population after a particular electroporation treatment is an important criterion, the essence of the experiment is the ability of the protoplasts to incorporate and express the foreign genetic material. Therefore, quantitatively assay the uptake or expression of your particular DNA/RNA over a variety of pulse field strengths.
2. Multiple pulses are generally discouraged due to the increase in detrimental effects on the viability of the protoplasts.
3. The use of mannitol by itself lowers the ionic strength of the media and allows a lower field strength pulse to be effective in "porating" the cell membrane. The protoplasts are not stable in mannitol without buffer for more than a couple of hours, and care should be taken to return them to culture media without extended delay following the electroporation pulse. Higher ionic strength electroporation media can be used, but this may require the use of higher field strength pulses, which reduce cell viability.
4. Due to the intrinsic differences between the square wave and the exponential wave pulses, it is not necessary to kill 50% of the protoplasts in order to achieve maximal electroporation when using the square-wave generator. Protoplast viabilities as high as 80% may be optimal for uptake of DNA or RNA.

III. Electrofusion

A. Procedure A

This protocol is based on the use of a Zimmermann cell-fusion power supply (GCA Precision Corp.) and the fusion chamber described by Bates (1985). This fusion chamber consists of a series of parallel wires mounted on a Plexiglas slide. The wires are 0.5 mm apart and every other wire is connected to an electrical lead of opposite polarity. A Plexiglas collar is mounted over the wires to create a sample well that holds up to 0.5 ml of protoplast suspension.

1. Freshly isolated protoplasts are washed twice with 0.5 M mannitol and resuspended in 0.5 M mannitol + 0.5 mM $CaCl_2$ at a density of 5×10^5 protoplasts/ml.
2. The two types of protoplasts to be fused are mixed together and 0.25 ml of protoplast suspension is pipetted into the fusion chamber. An AC alignment field of 600 kHz, 100 V/cm, is applied immediately.
3. After allowing 3–4 min for the protoplasts to become aligned, the AC voltage is reduced to 20 V/cm and two DC square-wave pulses (1000 V/cm, 50 μs) are applied 1 s apart.
4. The AC alignment voltage is gradually reduced to 0 V over the next 60 s.
5. The chamber is rocked back and forth a few times to dislodge the protoplasts. Then the protoplasts are carefully removed from the chamber with a Pasteur pipet, transferred to a petri dish, and diluted with 3 volumes of culture medium.

Notes

1. The low-ionic-strength fusion medium is necessary to obtain protoplast alignment in the AC field. If the protoplasts do not align when the AC field is applied, wash them again in 0.5 M mannitol. Preparations of damaged protoplasts often do not align well because of the leakage of salts from the cells into the medium. Inclusion of a small amount of calcium in the fusion medium helps reduce cell lysis; however, calcium concentrations above 1 mM usually prevent protoplast alignment by the AC field.
2. The DC pulse length, voltage, and number of pulses used are determined empirically, and may vary slightly for each type of protoplast to be fused. If pulse-induced protoplast lysis is a problem, using shorter pulses should help (reducing the pulse voltage will reduce lysis but will also reduce fusion yield). Most commercially available electrofusion machines, including the GCA Precision machine, include a circuit that shuts off the AC field during the time that the DC pulse is being applied. This feature also reduces protoplast lysis.
3. Slowly removing the AC field after applying the DC pulse(s) gives the fused protoplasts time to coalesce and stabilize before they are pipetted out of the fusion chamber. Some workers leave the protoplasts in the fusion chamber for as long as 10 min before attempting to remove them.
4. The procedure described here results in the formation of long, thick chains of protoplasts during the AC alignment phase. Application of DC pulses to such pearl chains results in the formation of many multinucleate as well as binucleate

fusion bodies. This approach usually maximizes the absolute number of binucleate heterokaryons that are recovered, and multinucleate fusion products are generally not a problem because most of them die during cell culture. However, if desired, the fraction of fusion products that are binucleate can be increased by reducing the cell density used for fusion (i.e., number of protoplasts/ml fusion medium), by shortening the AC alignment phase, or by reducing the voltage and duration of the DC pulse(s).

B. Procedure B

This protocol is based on the use of potato protoplasts with a BTX model 200 electro cell manipulator and a fusion chamber consisting of two serrated parallel plates 0.8 mm apart (BTX model 473, San Diego, CA).

1. Fusion conditions: protoplasts were isolated from leaves of axenically propagated plants of *Solanum tuberosum* and *Solanum chacoense*. The digestion solution was filter sterilized through a 0.2-μm membrane filter and the leaves were digested for 16 h at 30°C. The protoplasts were filtered through sterile cheesecloth and diluted with an equal volume of fresh digestion media without hydrolytic enzymes. The protoplasts were collected by floating centrifugation in a Babcock bottle at 50 × g for 10 min. The protoplasts were washed with a rinse solution consisting of the digestion media lacking the hydrolytic enzymes and resuspended to a concentration of 1.5×10^5 protoplasts/ml in 0.6 M mannitol.
2. Isolated mesophyll protoplasts were fused using a single 40-μs DC square pulse of 2.2 kV/cm supplied by a BTX model 200 electro cell manipulator. The protoplasts were aligned with a 20-s AC sine wave of 100 V/cm. After a 5-min waiting period the protoplasts were collected by centrifugation at 50 × g for 5 min and resuspended in rinse solution described above. The resuspended protoplasts were plated into cell-layer media and the subsequent tissue culture regeneration steps followed those described by Shepard (1980).

Notes

Fusion hybrids were selected at the microcallus stage 2–4 weeks after the electro-fusion treatment, based on the vigor of their growth in tissue culture. Under the tissue culture conditions utilized in this experiment, protoplasts of *Solanum chacoense* grow and regenerate very slowly, if at all. Protoplasts of *Solanum tuberosum* grow and regenerate readily; however, cells derived from the fused hybrids demonstrated a noticeable hybrid vigor in growth that could be seen in the size of the microcalli. This hybrid vigor allowed for an effective selection criterion (Saunders *et al.*, 1989b).

IV. Use of Oscilloscope for Pulse Analysis

The use of an oscilloscope is often helpful in exploring the optimal electrical parameters for both electroporation and electrofusion. The duration of the square-wave pulse is definable based on the initiation and the termination of the pulse voltage. The square-wave pulse duration can be easily modified on most square-wave generators by simply changing a calibrated dial. In the case of the exponential pulse generator the decay of the pulse can be modified by changing the capacitors used to generate the pulse. Due to the nature of the exponential pulse it is difficult to measure the absolute termination of the pulse duration; therefore, defining this waveform must be done by some other convention. In most "homemade" electroporation devices there is limited potential for changing the duration of the pulse, and therefore limited efforts have been expended in defining the duration of the pulse. Typically exponential pulse durations are specified to be in the 10–100 ms range, although the definition of the termination of the pulse is often neglected. Two main ways are used for defining capacitance discharges. The first procedure is the RC time constant, which is the time required for the voltage to fall to $1/e$ of its initial value, and the other definition of the pulse is the decay half-time ($t_{1/2}$). The RC time constant accurately defines the pulse because it contains information on the resistance and capacitance of the circuit; however, the $t_{1/2}$ is easier to measure from an oscilloscope screen. The $t_{1/2}$ is based on the premise that a time-dependent change in the voltage from a discharging capacitor follows a defined path such that for any arbitrary point on the discharge path (X) there can be determined a subsequent point on the path that is half as high in amplitude ($\frac{1}{2}X$). If X equals the maximum amplitude of the peak, then $\frac{1}{2}X$ is a point on the curve that equals the decay half-time for the peak and defines the shape of the curve. Thus, each curve can be defined by the peak voltage and the time required to reach one-half of the peak voltage after the pulse. This convention can be used in concert with other methods of identifying the duration of the wave and represents a convenient procedure to accurately describe the wave form.

V. Selection

A. Kanamycin Resistance Using the Agarose Bead Technique

This procedure can be used to recover stable transformants following the introduction of plasmids containing a functional neomycin phosphotransferase gene (kanamycin-resistance gene) into protoplasts by electroporation. A modification of this same technique can be used to recover somatic hybrids when one of the parental lines used in fusion carries a stably integrated kanamycin-resistance gene (Bates *et al.,*

1987). Agarose bead culture of protoplasts was originally described by Shillito *et al.* (1983).

1. Culture the electroporated protoplasts in 30 × 10 mm petri plates containing 2 ml of protoplast culture medium.
2. Day 4 (third day of culture): begin reducing the osmolarity of the culture medium by adding 0.2 ml of medium containing only 3–4% sucrose (or glucose) with no additional osmotic stabilizer. The callus culture medium (CM) used for the species at hand is often suitable.
3. Day 6: add 0.32 ml CM.
4. Day 7: add 0.48 ml CM.
5. Day 8:

   a. Gently scrape the cell clusters off the bottom of the petri plate with the tip of a Pasteur pipet and transfer the culture to a 60 × 15 mm petri plate.
   b. Add an equal volume of CM containing 200 μg/ml kanamycin and 1.8% SeaPlaque agarose. (Before it is mixed with the protoplasts, the agar containing medium should be cooled until it is no longer warm to the touch.) The kanamycin should be prepared as a filter-sterilized stock and added after autoclaving the medium.
   c. Mix the contents of the petri plate thoroughly by swirling; let the media harden for 15 min at 4°C.
   d. Using the broad end of a small weighing spatula, divide the solidified media into eight wedge-shaped segments and transfer the segments to a 100 × 15 mm petri plate. Add 5 ml of liquid CM + 100 μg/ml kanamycin to the culture.

6. Day 10: add an additional 5 ml of CM + 100 μg/ml kanamycin to the culture.
7. Day 15: remove 5 ml of medium from the culture and replace it with 5 ml of fresh medium. Repeat this medium change weekly.
8. When calli are 2–3 mm in diameter transfer them to solid CM + 100 μg/ml kanamycin.
9. Regenerate shoots and plants as usual but maintain the selection (100 μg/ml kanamycin) until the shoots have rooted.

Notes

The schedule for diluting the protoplasts will depend on the protoplast system and the quality of the protoplast preparation. Slow-growing or poor-quality protoplasts must be diluted more slowly in the beginning. Dilutions are usually begun when most of the protoplasts have completed their first division. Browning of the culture and the presence of excessive numbers of shrunken and dead protoplasts indicates to rapid dilution.

Two approaches can be taken for determining transformation frequency.

1. Use 5-ml subsamples, transferred to 30 × 10 mm plates and solidified by addition of 1.5 ml of agar media either containing or lacking kanamycin.
2. Selection with kanamycin can be delayed until calli are transferred to solid callus medium.

VI. Conclusion

The information in this chapter has shown successful example protocols for both electrofusion and electroporation of plant protoplasts. Since protoplasts from all sources are not identical these protocols should serve as a functional guide for researchers involved in genetic manipulation of protoplasts from other plants. Researchers and encouraged to optimize the pulse field strength, pulse duration, and fusion media for optimal cell viability as well as maximizing the DNA uptake during electroporation or fusion yield during electrofusion.

References

Abdul-Baki, A. A., Saunders, J. A., Matthews, B. F., and Pittarelli, G. W. (1990). DNA uptake during electroporation of germinating pollen grains. *Plant Sci.* 70, 181–190.
Bates, G. W. (1985). Electrical fusion for the optimal formation of protoplast heterokaryons in *Nicotiana*. *Planta* 165, 217–224.
Bates, G. W., Hasenkampf, C. A., Contolini, C. L., and Piastuch, W. C. (1987). Asymmetric hybridization in *Nicotiana* by fusion of irradiated protoplasts. *Theor. Appl. Genet.* 74, 718–726.
Bates, G. W., Piastuch, W., Riggs, C. D., and Rabussay, D. (1988). Electroporation for DNA delivery to plant protoplasts. *Plant Cell Tissue Org. Cult.* 12, 213–218.
Boston, R. S., Becwar, M. R., Ryan, R. D., Goldsbrough, P. B., Larkins, B. A., and Hodges, T. K. (1987). Expression from heterologous promoters in electroporated carrot protoplasts. *Plant Physiol.* 78, 742–746.
Dekeyser, R. A., Claes, B., De Rycke, R. M. U., Habets, M. E., Van Montague, M. C., and Caplan, A. B. (1990). Transient gene expression intact organized rice tissues. *Plant Cell* 2, 591–602.
Gamborg, O. L., Shyluk, J. P., and Shahin, E. A. (1981). Isolation, fusion and culture of plant protoplasts. *In* "Plant Tissue Culture" (T. A. Thorpe, ed.), Academic Press, New York, pp. 115–154.
Hibi, T., Kano, H., Sugiura, M., Kazami, T., and Kimura, S. (1986). High efficiency electro-transfection of tobacco mesophyll protoplasts with tobacco mosaic virus RNA. *J. Gen. Virol.* 67, 2037–2042.
Lindsey, K., and Jones, M. G. K. (1987). Transient gene expression in electroporated protoplasts and intact cells of sugar beet. *Plant. Mol. Biol.* 10, 43–52.
Matthews, B. F., Abdul-Baki, A. A., and Saunders, J. A. (1990). Expression of a foreign gene in electroporated pollen grains of tobacco. *Sex. Plant Reprod.* 3, 147–151.

Morikawa, H., Iida, A., Matsui, C., Ikegami, M., and Yamada, Y. (1986). Gene transfer into intact plant cells by electroinjection through cell walls and membranes. *Gene* **41**, 121–124.

Negrutiu, I., Shillito, R., Potrykus, I., Biasini, G., and Sala, F. (1987). Hybrid genes in the analysis of transformation conditions. I. Setting up a simple method for direct gene transfer in plant protoplasts. *Plant Mol. Biol.* **8**, 363–373.

Reinert, J., and Yeoman, M. M. (1982). "Plant Cell and Tissue Culture, A Laboratory Manual." Springer-Verlag, New York, pp. 39–45.

Saunders, J. A., and Bates, G. W. (1987). Chemically induced fusion of plant protoplasts. *In* "Cell Fusion" (A. E. Sowers, ed.), Plenum Press, New York, pp. 497–520.

Saunders, J. A., Smith, C. R., and Kaper, J. M. (1989a). Effects of electroporation pulse wave on the incorporation of viral RNA into tobacco protoplasts. *BioTechniques* **7**(10), 1124–1131.

Saunders, J. A., Matthews, B. F., and Miller, P. D. (1989b). Plant gene transfer using electrofusion and electroporation. *In* "Electroporation and Electrofusion in Cell Biology" (E. Neumann, A. E. Sowers, and C. Jordan, eds.), Plenum Press, New York, pp. 343–354.

Saunders, J. A., Roskos, L. A., Mischke, B. S., Aly, M., and Owens, L. D. (1986). Behavior and viability of tobacco protoplasts in response to electrofusion parameters. *Plant Physiol.* **80**, 117–121.

Shepard, J. F. (1980). Mutant selection and plant regeneration from potato mesophyll protoplasts. *In* "Genetic Improvement of Crops" (I. Rubenstein, B. Gengenbach, R. L. Phillips, and C. E. Green, eds.), University of Minnesota Press, Minneapolis, pp. 185–219.

Shillito, R. D., Paszkowski, J., and Potrykus, I. (1983). Agarose plating and a bead type culture technique enable and stimulate development of protoplast-derived colonies in a number of plant species. *Plant Cell Rep.* **2**, 244–247.

Shillito, R. D., Saul, M. W., Paszkowski, J., and Potrykus, I. (1985). High efficiency direct gene transfer to plants. *Biotechnology* **3**, 1099–1103.

Takebe, I., Otsuki, Y., and Aoki, S. (1968). Isolation of tobacco mesophyll cells in intact and active state. *Plant Cell Physiol.* **9**, 115–124.

Watts, J. W., and King, J. M. (1984). A simple method for large-scale electrofusion and culture of plant protoplasts, *Biosci. Rep.* **4**, 335–342.

30

Protocols for the Transformation of Bacteria by Electroporation

William J. Dower,[1] Bruce M. Chassy,[2] Jack T. Trevors,[3]
and Hans P. Blaschek[2]

[1]Affymax Research Institute, Palo Alto, California 94304

[2]Department of Food Science, University of Illinois at Urbana-Champaign,
Urbana, Illinois 61801

[3]Department of Environmental Biology, University of Guelph, Guelph,
Ontario, Canada N1G 2W1

I. Introduction

This chapter gives detailed protocols for electro-transforming bacteria. The protocols for *Escherichia coli* can be applied to most strains to yield very high frequencies and efficiencies (10^6 to 10^{10} transformants/μg DNA) of transformation. Methods for cell culture and electroporation are provided. Additionally, detailed comments on the method should assist the reader in troubleshooting difficulties should they arise. Most strains of gram-negative bacteria can be transformed by these methods, or

Guide to Electroporation and Electrofusion

simple adaptations of them. As a group, gram-negative bacteria have proven more tractable to electroporation than have gram-positive bacteria. Further insight into the factors controlling the electrotransformation of gram-negative bacteria can be found in Chapter 18 (Dower and Cwirla) and Chapter 17 (Trevors *et al.*).

The gram-positive bacteria are somewhat less amenable to transformation by electrical means than are gram-negative bacteria. Some strains do yield high frequencies and efficiencies of transformation; however, the results do not compare favorably with those obtained using gram negative bacteria. Seldom will more than 1–10% of the population be transformed; concomitant efficiencies of 10^7 transformants/μg DNA are considered excellent with a gram-positive bacterium. Some strains defy attempts at transformation by electrical means. The reasons for the difficulties in electrotransforming certain gram-positives are not clear. It may well be that their thick, dense cell walls render the cells less vulnerable to electric fields or prevent the passage of macromolecules once membranes are permeabilized; that the high internal pressures of these cells may prevent influx of DNA once the envelope is permeable; or that there are fundamental limitations in the ability of this group to establish and maintain plasmid function. When working with strains for which there exists no established alternative transformation system, it is indeed difficult to assess the nature of the barriers to transformation in a particular case (Mercenier and Chassy, 1988). Barriers to transformation could lie outside the realm of electroporation; nonspecific nucleases, restriction endonucleases, plasmid incompatibility, failure to replicate or express genes, and possible toxicity of products, among others, must all be considered.

Fortunately, many gram-positive bacterial strains can be transformed by electroporation sufficiently well to permit useful genetic analysis and manipulation of species that are intractable to other methods of transformation. Electroporation is simple enough that a variety of conditions can be quickly assessed. This chapter provides the general principles for transformation of gram-positive bacteria and describes a generalized protocol. Application of the protocols for gram-positives described here may not necessarily yield positive results. The researcher must be prepared to evaluate a variety of experimental conditions as noted in the protocol. Specific methods and comments for the bacilli, lactic acid bacteria, and the clostridia are given. Chapter 17 (Trevors *et al.*) gives a list and references to most of the examples of bacterial electroporation reported in the literature.

II. General Protocol for the Electrotransformation of Gram-Negative Species

Although many prokaryotes can be transformed by electroporation, the method seems to be more generally applicable to gram-negative species. A great many of the members of this group may be prepared and electrotransformed at high efficiencies using very similar protocols (see Chapter 17). This is in contrast to the many gram-positive species that have required special adaptations in the preparation of the cells and the application of the electrical pulses, often with disappointing transformation frequencies (see Chapter 17). Many of the important early demonstrations of elec-trotransformation of gram negative cells employed electrical fields that we would now consider to be of modest intensity (up to ~6 kV/cm). This was due to the limitations of the equipment then available. In almost every case, the subsequent use of higher fields (10–20 kV/cm) has resulted in increases in transformation efficiencies and frequencies of several orders of magnitude.

The general protocol described below [based on a method described for *E. coli* (Dower *et al.*, 1988)] is offered as a starting point for establishing electroporation conditions for gram-negative bacteria. This approach has often provided amazingly high levels of transformation in the initial trials with a new species. Occasionally, only moderate recoveries of transformants are obtained, and some simple refinements are required to produce the high efficiencies desired for certain applications. In the notes following each protocol, some suggestions for finding the optimal conditions are provided. Inevitably, some species will be resistant to this general approach, and the reader is directed to a more detailed discussion of the variables and problems of electrotransformation found in Dower (1990).

A. Preparation of Electrocompetent Cells

1. Inoculate 1 l (or less, as required) of a rich broth appropriate for rapid cell growth with 1/10 to 1/100 volume of fresh overnight culture.
2. Grow cells at the growth temperature and conditions (± shaking) optimal for the strain. Harvest when cells reach early to mid log phase growth (usually around $OD_{600 \text{ nm}}$ of ~0.5).
3. To harvest, chill the growth flask on ice for 15–30 min, and centrifuge in a cold rotor at 4000 × g for 15 min.
4. Resuspend the pellets in a total of 1 l of cold water. Centrifuge as above.
5. Resuspend in 0.5 l of cold water. Centrifuge as above.
6. If cells are to be stored frozen, resuspend in ~20 ml of the cryoprotectant (usually 10% glycerol). Cells to be used fresh may be resuspended in water at this point. Centrifuge as above.
7. Resuspend to a final volume of 2–3 ml in cryoprotectant or water. The cell concentration should be high. With some species a concentration of up to 5 × 10^{10} cells/ml is desirable.
8. At this point, many species may be frozen in 10% glycerol and stored at −70°C for up to a year with little loss of electrocompetence.

B. Factors Affecting the Preparation of Electrocompetent Cells

1. A rich broth allows the cells to reach high density while still in exponential growth. An example is the use of "superbroth" (3.2% tryptone, 2.0% yeast extract, 0.5% NaCl) for *E. coli*. If the cells carry an antibiotic marker, the cells may usually be grown in the presence of the antibiotic with no effect on the competence of the cells.

2. The optimal growth conditions depend on the particular species and strain. A simple rule of thumb is to employ the conditions normally used with a given cell type to obtain the maximum rate of growth to a fairly high density. In general, it is very important to harvest when the cells are still in the exponential growth phase. For some species, *E. coli* for example, the competence of the cells falls off precipitously as the growth of the population begins to slow (Dower, 1990).

3. The importance of chilling the cells during preparation has not been established by systematic analysis. Almost all protocols call for chilling at this step and it seems wise to emulate this approach except when dealing with cold labile cells. In the latter case, preparation at room temperature may be acceptable. Cell types that are difficult to prepare by centrifugation (filamentous species, for example) can be washed and concentrated by filtration.

4. The extensive washing is designed to reduce the electrical conductivity of the cell preparation, consequently the wash solutions are of the lowest ionic strength that can be tolerated by the cells. In some cases, very dilute buffering agents have been used (i.e., 1 mM HEPES), though it seems unlikely that this low buffer concentration provides much pH stabilization. Cells that are osmotically unstable may be stabilized in a nonionic, high-osmolarity solution such as sucrose. The best osmotic agent may require empirical determination, as some cells may be sensitive to the influx of specific osmoprotectants during electropermeabilization.

5. The concentration of cells can have a profound effect on the efficiency of transformation. The cells should be resuspended at 10^{10} to 10^{11} cells/ml. Initial trials with a species should be done with freshly prepared cells; however, many gram-negative species are highly competent after storage in glycerol at $-70°C$. The rate at which the cells are frozen can have a significant effect on the preservation of activity. The competence of some strains of *E. coli* is significantly reduced when frozen in liquid nitrogen rather than powdered dry ice. Some species may require a bit of experimentation with different cryoprotectants and freezing rates to find the optimal protocol, but the convenience of frozen competent cells is well worth the effort. Aliquots of *E. coli* can be thawed and refrozen several times with little effect on their competence.

C. General Protocol for Electrotransformation of Gram-Negative Bacteria

1. Chill the electrode cuvettes and the sliding cuvette holder on ice.
2. Set the pulse generator to deliver a pulse of 12.5 kV/cm with ~5 ms duration.
3. Hold fresh cells on ice before use. Frozen cells should be thawed on ice immediately before used.
4. To a cold 0.5 or 1.5 ml polypropylene tube, add 40 μl of the cell suspension and 1–5 μl of DNA in a low-ionic-strength buffer such as TE (10 mM Tris-HCl, 1 mM EDTA, pH 8). Mix and let sit on ice for ~30–60 s.
5. Transfer the mixture of cells and DNA to a *well chilled* 0.2-cm electroporation cuvette, and quickly shake the solution to the bottom of the cuvette.
6. Apply one pulse at the settings described above.
7. *Immediately* add 1 ml of outgrowth medium (at room temperature) to the cuvette and gently, but quickly, resuspend the cells with a pasteur pipette.
8. Transfer the suspension to an appropriate vessel (usually a 4- or 12-ml polypropylene tube) for the outgrowth period.
9. Incubate at the optimal temperature for growth of the particular cells for a period equivalent to 1 or 2 cell divisions.
10. Plate appropriate aliquots on selective medium. For initial attempts to transform a strain, it is useful to also plate dilutions (pre- and postpulse) on nonselective medium to assess survival rate.

D. Comments on the Gram-Negative Electrotransformation Protocol

1. For some strains, the temperature of the sample when the pulse is applied has a pronounced effect on the recovery of transformants (Dower, 1990). For these cells, chilling the electrodes and the sample of cells on ice is required in order to achieve the highest electroporation efficiencies.
2. The optimum field strength is strain dependent. Fortunately, many gram-negative species transform well under conditions similar to those applied to *E. coli:* therefore, starting electroporation trials with the conditions described above for *E. coli* very often brings success in the initial trials. Further optimization of the electrical conditions for the strain of interest is accomplished by varying the field strength from 10 to 20 kV/cm in 1- to 2-kV increments. Additional adjustment of the time constant from 1 to 20 ms can also be helpful in attaining the highest levels of transformation for a given strain. Note that the field strength is a function of the voltage applied to the electrodes and the distance between the electrodes. The 12.5 kV/cm starting conditions recommended here are obtained by applying 2.5 kV to electrodes 0.2 cm apart, or 1.25 kV to a 0.1-cm interelectrode gap. Depending on the pulse generator available, pulse length is usually controlled by the selection of a specific capacitor and a chosen resistor in parallel

with the sample. Consult the manufacturers' literature and operating manual for the correct settings. Detailed discussions of the effects of varying the electrical parameters are found in Dower *et al.* (1988), Dower (1990), Miller *et al.* (1988), and Hülsheger *et al.* (1981).

3. Sample volumes of up to 400 μl may be used in a typical 0.2-cm cuvette (up to 80 μl in a 0.1-cm cuvette). DNA samples of up to 10% of the cell volume may be used if the DNA is contained in a buffer of low ionic strength (such as TE). The DNA concentration may be quite high. A useful characteristic of electroporation is the high capacity for DNA. In some cases final DNA concentrations of greater than 10 μg/ml may be used with no decline in the efficiency of transformation. The proportion of cells transformed is dependent on the DNA concentration over a very wide range (Dower *et al.*, 1988; Miller *et al.*, 1988; see also Chapter 18, this volume). After mixing the cells and DNA, the sample is left on ice for a short period to allow the sample to drain off the sides of the tube; this practice affords better recovery of the sample.

4. The process of transferring the sample to the cold cuvette, shaking the suspension to the bottom of the tube, and applying the pulse is done as quickly as possible to prevent warming of the electrodes and cell suspension. The importance of well-chilled samples and electrodes deserves special emphasis.

5. With *E. coli,* the immediate addition of outgrowth medium is important to obtain the highest efficiencies. Some other species seem not to be sensitive to a delay at this point; however, there is no evidence that any benefit is derived from delaying the addition of medium.

6. The optimal period of outgrowth will vary with the strain and the nature of the selection. A good starting point is a period equal to one or two cell doublings.

7. The cell survival data obtained by plating cell dilutions on nonselective media is useful if no transformation is obtained. The electroporation pulses should kill a significant portion of the cells. Generally 50–90% of the cells are killed by pulses that produce the highest recovery of transformants. However, there are reported examples of cells that transform well under conditions that cause little cell death (Miller *et al.*, 1988). Often, if no transformation is obtained and few cells are killed, the remedy is to increase the field strength, the time constant, or both.

III. The Electrotransformation of Gram-Positive Species

A. A Generalized Method for Gram-Positive Bacteria

1. *Culture cells* in a growth medium that gives satisfactory results for the chosen strain. Choice of medium can affect results; several different media should be

used to prepare cells for electroporation. Typically, cells prepared in rich, limiting, defined, or minimal media should be evaluated. Growth substrate and cultivation temperature can be varied as well. Additions to the medium that inhibit, weaken, or alter cell wall synthesis may increase the transformation frequency (i.e., penicillin, glycine, threonine).

2. *Harvest cells* at early to mid-exponential growth phase or late-exponential/ stationary phase. It is usually more convenient to use stationary-phase cells (overnight cultures); however, some strains can be transformed best when harvested during the active growth phase. Harvest by centrifugation. Cell are usually kept chilled from the point of harvest.

3. *Wash cells* in the chilled solution to be used for electroporation. Cells are washed thoroughly (usually three to four times) to completely remove contaminating growth medium and electrolytes that leach from the cells. The wash and electroporation medium must meet two basic needs: (a) that the cells remain viable to repeated washes in the medium, and (b) that the medium be relatively nonconductive. Buffers containing 1–7 mM HEPES, MOPS, or phosphate around neutral pH have been reported to work well for many strains. High-quality (high-resistance) water works equally well in many applications and has the advantage of being less susceptible to arcing at higher voltages. The electroporation medium can also contain 0.27–1.0 M sucrose, 10–15% glycerol, 1 M sorbitol, polyethylene glycol (PEG), or other similar agents. Usually the agent chosen is not metabolized by the strain. The exact role of these additives is not clear. Although they are frequently referred to in the literature as "osmotic stabilizers," no evidence that they function in this manner during electroporation has been presented. In some cases the addition of such compounds greatly enhances electroporation; in others, it interferes.

4. *Resuspend* the cells in chilled electroporation medium to a final concentration of 5 × 10^9 to 5 × 10^{10} cells/ml. Keep on ice until use; many strains retain full activity for hours. Most strains can be frozen for future use at this point with little loss of activity, provided a suitable cryoprotectant is added and the freezing rate is controlled to maximize survival.

5. *Selection of pulse parameters* is made by consulting specific references for the genus and strain being transformed. If no report can be found, evaluate parameters reported for closely related species. For generators that deliver capacitive discharge or exponential decay waveforms, the small spherical cocci will require higher field strengths (7–18 kV/cm) than the rodlike bacilli (4–8 kV/cm). Time constants for gram-positive bacteria should be varied between 2.5 and 7.5 ms. Choice of capacitance and resistance settings will vary with the commercial apparatus used and the cuvette chosen. The controlling parameters are field strength (kV/cm) and time constant (τ, ms). Square-wave generators may allow higher voltages and shorter time constants to be applied. Most literature reports have employed pulse generators that deliver exponentially decaying waveforms;

however, square-wave generators have given excellent results in a number of cases. It may be advisable to assess in a preliminary experiment the effect of field strength on bacterial survival in order to determine the maximum voltage the strain will tolerate.

6. *Electrotransform* by placing a sample of cells in a chilled electroporation cuvette. Sterile disposable cuvettes with 1-, 2-, or 4-mm electrode gaps are recommended. Electrode gap is important only in determination of the field strength applied. Sample size will vary from 25 μl to 0.8 ml, depending on the cuvette chosen. Add DNA (0.1–1 μg in 1–5 μl of TE) and mix quickly. Deliver the pulse.

7. *Dilute and plate* the bacteria directly on selective media. Medium is added directly to the electroporation cuvette. A postincubation in growth medium may facilitate recovery of transformants. Although not generally required for transformation, postincubation allows the expression of antibiotic resistance markers and permits time for the cells to recover from the effects of the electric pulse. If a significant postincubation period is employed, quantitative assessment of transformation frequency can only be achieved if the total number of cells present at the *beginning* and *end* of the postincubation period are enumerated. If outgrowth occurs during postinoculation, some of the transformant colonies scored will be siblings and not initial transformants; measurement of outgrowth will allow a correction to be applied.

B. Lactic Acid Bacteria

The electrotransformations of a number of strains of lactic acid bacteria are referenced in Chapter 17 (see also Chassy and Flickinger, 1987; Harlander, 1987; Chassy *et al.*, 1988; Luchansky *et al.*, 1988b; McIntyre and Harlander, 1989; van der Lelie *et al.*, 1988; Aukrust and Nes, 1989). An examination of the literature reveals that a confusing array of growth media and conditions, electroporation media, and electroporation parameters has been applied to these bacteria. It is clear that many lactic acid bacteria can be transformed by electroporation; how best to approach an untried strain is somewhat less clear. The procedure outlined below works sufficiently well with the majority of lactic acid bacteria to allow a process of optimization to begin; it is drawn primarily from experience with the lactobacilli and streptococci, but can be applied to most lactic acid bacteria. Failure to observe transformants following first attempts at electrotransformation will require a comprehensive evaluation of growth media, electroporation medium, transforming DNA, selection system, and electrical parameters.

1. Culture the cells in an appropriate medium: MRS for lactobacilli and BHI or M17 for streptococci are commonly employed. Supplement the medium with 10 m*M* L-threonine or 20 m*M* DL-threonine. Cells may also be prepared by growth in medium containing 1–4% glycine.

2. Harvest the cells at mid-exponential phase of growth or late-exponential/stationary-phase growth.

3. Wash cells four times in chilled deionized water–0.5 M sucrose (optional).

4. Resuspend the cells to a final concentration of 5×10^9 to 5×10^{10} cell/ml. Place a portion (usually 40–50 μl; $> 5 \times 10^8$ cells) in a sterile disposable cuvette. Add 0.1–1 μg DNA in 1–5 μl of TE and mix briskly.

5. Deliver the electroporation pulse. For streptococci evaluate six to eight field strengths between 7 and 14 kV/cm and three time constants between 2.5 and 7.5 ms. For lactobacilli, evaluate five to six field strengths between 4 and 8 kV/cm and two time constants between 2 and 5 ms.

6. Dilute the cells in outgrowth medium and postincubate for 1–3 h; also attempt to plate directly on selective plates prior to outgrowth.

C. The Bacilli

Table 1 summarizes some examples of electrotransformation of *Bacillus* spp. It is noteworthy that strain-to-strain variability is very common when electrotransforming *Bacillus* spp. Electrotransformation protocols for the bacilli have been published by Belliveau and Trevors (1989), Bone and Ellar (1989), Luchansky *et al.* (1988a, 1988b), Brigidi *et al.* (1990), Schurter *et al.* (1989), and Vehmaanperä (1989). Although different *Bacillus* spp. were used as recipients with a variety of plasmids of varying sizes, the protocols are fundamentally the same. Various buffers, voltages, capacitance values, and different plasmids have been employed successfully by various investigators; these are summarized in Table 1.

The protocol of Belliveau and Trevors (1989) is as follows:

1. *Bacillus cereus* 569 is grown in 10 ml LB broth at 37°C for 16 h with shaking at 100 rpm.

2. Two milliliters of a 16-h culture is transferred to 10 ml fresh LB broth and incubated as before for 2.5 h until O.D.$_{600nm}$ = 1.0 (equivalent to 2.2×10^8 CFU/ml).

3. Harvest cells at $4000 \times g$ for 10 min at 4°C and wash three times in 10 mM HEPES buffer (N-2-hydroxymethylpiperazine-N-2-ethanesulfonic acid) and resuspend in 4 ml of the same buffer (yields 5.5×10^8 CFU/ml).

4. Hold cells in ice; 0.8 ml of cells and 30 μl of DNA (0.5 μg) are mixed in an electroporation cuvette.

5. A single electrotransformation pulse is delivered. Using *B. cereus* 569 and plasmid pC194, recovery of cells was in 7.2 ml of LB broth at 30°C for 1 h with shaking at 60 rpm.

6. Serial dilutions are made in sterile peptone water (1 g peptone/l) and aliquots are plated on selective LB agar (containing 10 μg/ml chloramphenicol for selection of transformants containing pC194).

Table 1
Examples of Electrotransformation Parameters for Bacillus spp.

Organism	Transfer DNA	Selection markers	Voltage (kV/cm)	Capacitance (μF)	Time (ms)	Buffer	Transformation		Reference[b]
							Frequency	Efficiency	
B. cereus	pC194	Cm	3750	3	2.7	HEPES	2.0×10^{-5}		1
B. cereus	pGB130	Hg^{2+}	3750	3	1.2	HEPES	1.0×10^{-5}	9.5×10^{3}	2
B. thuringiensis (7 strains)	7 Different plasmids		4520–6250	25	4.1–13.3	SPB		$0–2 \times 10^{5}$	3
B. thuringiensis (21 strains)	14 Different plasmids		7000	25	7–9	Water or TE		$0–1 \times 10^{5}$	4
B. cereus	pGK12	Cm Em	6250	25	—	PEB (2.5× HEB)	—	6.5×10^{1}	5

[a]Abbreviations: Cm, chloramphenicol; Em, erythromycin; Hg^{2+}, mercuric chloride; HEPES, hydroxymethylpiperazine N-2-ethanesulfonic acid, pH 7.0; SPB, 272 mM sucrose, 7 mM sodium phosphate, 1 mM MgCl$_2$, pH 7.4; PEB, 272 mM sucrose, 7 mM phospate, 1 mM MgCl$_2$, pH 7.3; HEB, 272 mM sucrose, 7 mM HEPES, 1 mM MgCl$_2$, pH 7.3.

[b]References: 1, Belliveau and Trevors (1989), 2, Belliveau and Trevors (1990), 3, Bone and Ellar (1989), 4, Mahillon et al. (1989), 5, Luchansky et al. (1988).

D. The Clostridia

The successful application of electroporation for the transformation of *Clostridium* spp. with plasmid DNA has been reported for biomedically important *C. perfringens* (Allen and Blaschek, 1988, 1990; Kim and Blaschek, 1989), and industrially significant *C. acetobutylicum* (Oultram *et al.*, 1988; Kim *et al.*, 1990). Research with these two species of clostridia underscores the need to carefully optimize experimental parameters if the highest frequency is sought. Allen and Blaschek (1990) recently examined specific parameters involved in improving the efficiency and versatility of the *C. perfringens* electroporation-induced transformation system. Two protocols that demonstrate the dependence of results on subtle changes in experimental conditions are described below.

1. *Clostridium perfringens Sucrose-Based Electroporation Protocol*

1. Culture the cells in TGY medium (trypticase, 30 g; glucose, 20 g; yeast extract, 10 g; L-cysteine, 1 g; per liter of distilled water).
2. Harvest late-exponential phase culture (O.D.$_{600 \text{ nm}}$ of 1.3) by centrifugation at 5000 × g for 10 min.
3. Wash the cells in an equal volume of cold E-buffer (0.27 M sucrose, 1 mM MgCl$_2$, 5 mM Na$_2$HPO$_4$; pH 7.4) and resuspend in one-third volume of cold E-buffer.
4. Mix the cell suspension with plasmid DNA (0.8 ml cell suspension; final DNA concentration, 1 μg/ml) in the prechilled electroporation cuvette (4 mm electrode gap) and incubate on ice for 5 min.
5. Deliver the electroporation pulse (6.25 kV/cm and 25 μF), and incubate the cuvette containing sample for 5 min on ice.
6. Dilute the cells into 5 volumes of TGY expression medium (TGY containing 25 mM CaCl$_2$, 25 mM MgCl$_2$, and 0.075% agar, pH 6.4) and incubate for 3 h at 37°C.
7. Plate the cell suspension on TGY-selective agar (pH 6.4) and incubate anaerobically for 48 h at 37°C.

   a. Comments on the *C. perfringens* Sucrose-Based Electroporation Protocol. The following factors were found to improve the efficiency of transformation of *C. perfringens* by electroporation: (1) a reduction in cuvette sample volume (DNA and cell suspension) to 0.8 ml, (2) use of a 1 μg/ml concentration of transforming DNA, (3) use of late-exponential phase cells, (4) three-fold increase in the cell density to 3.0 × 10^8 CFU/ml), and (5) a reduction in the pH of the expression and selective plating medium to 6.4. Application of these conditions resulted in transformation efficiencies for *C. perfringens* 3624A ranging from 7.1 transformants/μg DNA for plasmid pIP401 to 9.2 × 10^4 transformants/μg DNA for plasmid pAK201. In

Table 2

Intact-Cell Electroporation-Induced Transformation of Various
Clostridium perfringens Strains with Plasmid pAK201[a]

Strain	Type	Transformation efficiency[b]	Source/reference
3624	A	5.5×10^4	ATCC
NTG-4	A	3.0×10^1	Blaschek and Klacik (1984)
10543	A	6.2×10^3	ATCC
13	A	9.8×10^6	Julian Rood
14810	B	0	ATCC
3626	B	0	ATCC
12917	B	0	ATCC
3628	C	1.4×10^2	ATCC
15[c]	D	7.5×10^0	Vijay Singh
3629[c]	D	1.5×10^1	ATCC
27324[c]	E	1.5×10^1	ATCC

[a]A final concentration of 1 μg/ml of pAK201 was utilized; selection of transformants was performed on TGY agar containing 20 μg/ml Cm.
[b]Calculated as transformants/μg DNA.
[c]Electrotransformation of these *C. perfringens* strains was performed by Teresa Campos as part of an Undergraduate Senior Research Project directed by Steven Allen.

addition, *C. perfringens* 3624A, 13A, 10543A, 3628C, NTG-4A, 15D, 3629D, and 27324E were successfully transformed with plasmid pAK201 (Table 2). Although all types A, C, D, and E strains were transformable, the three type B strains tested were not transformed under these conditions.

An additional protocol useful with both *C. perfringens* (Kim and Blaschek, 1989) and *C. acetobutylicum* (Kim *et al.*, 1990) is presented next.

2. *Clostridium perfringens*/*C. acetobutylicum* Glycerol or PEG-based Electroporation Protocol

1. Inoculate (1 : 10) and culture cells in TGY as above.
2. Harvest stationary-phase (O.D.$_{600nm}$ = 1.8) *C. perfringens* or late-exponential phase *C. acetobutylicum* (O.D.$_{600nm}$ = 1.2) cells by centrifugation at 5000 × g for 10 min.
3. Wash the cells with ½ volume of a solution of 15% v/v glycerol in distilled water and suspend in one-twentieth of the original volume to give ~10^{12} to 10^{14} cells/ml for *C. perfringens* and 10^{11} for *C. acetobutylicum*. Alternatively, 10% PEG may be substituted for glycerol.

4. Mix the cell suspension (0.8 ml) with 10–20 µl of DNA (~1 µg) in the electroporation cuvette (4 mm electrode gap) and incubate for 10 min on ice.
5. Deliver the electroporation pulse (6.25 kV/cm and 25 µF), and incubate the cuvette containing sample for 10 min on ice.
6. Dilute the cells into 9 volumes of TGY medium and incubate for 1 h (*C. perfringens*) or 9 h (*C. acetobutylicum*) at 37°C.
7. Plate the cell suspension on TGY-selective agar and incubate anaerobically for 36 h at 37°C.

a. Comments on the Combined Electroporation Protocol: Glycerol-based Technique.   Application of the glycerol-based protocol to *C. perfringens* results in transformation efficiencies comparable to that reported for the sucrose-based protocol. However, this approach has the advantage of being applicable to *C. acetobutylicum*. With the exception of cell harvesting and electro-shocking, all manipulations of *C. acetobutylicum* were carried out under anaerobic conditions. 10^3 transformants/µg DNA were obtained with *C. acetobutylicum* ATCC 824 when using the pAMβ1 plasmid deletion derivative pVA677 or the shuttle vector pKG201. The transformation of *C. acetobutylicum* was successful only when late exponential phase cells were used. Following application of the pulse, *C. acetobutylicum* requires an 8-h longer expression period than does *C. perfringens* if transformants are to be seen.

b. PEG-Based Technique.   Recent experiments demonstrate that electroporation in the presence of 10% PEG increases the transformation efficiency of both *C. perfringens* and *C. acetobutylicum* up to 100-fold (A. Y. Kim, personal communication). In the case of *C. acetobutylicum,* the use of PEG in combination with higher concentrations of added DNA (10 µg) has resulted in transformation efficiencies up to 10^4 transformants/µg DNA. Furthermore, it was observed that the rather lengthy postshock expression period required to obtain transformants of *C. acetobutylicum* could be reduced to 1 h, and only 16 h (overnight) was required to produce visible colonies. Use of cuvettes with a 1-mm electrode gap results in production of up to 10^5 transformants/µg DNA (S. Broussard, personal communication).

References

Allen, S. P., and Blaschek, H. P. (1988). Electroporation-induced transformation of intact cells of *Clostridium perfringens. Appl. Environ. Microbiol.* 54, 2322–2324.
Allen, S. P., and Blaschek, H. P. (1990). Factors involved in the electroporation-induced transformation of *Clostridium perfringens. FEMS Microbiol. Lett.* 70, 217–220.
Aukrust, T., and Nes, I. F. (1988). Transformation of *Lactobacillus plantarum* with the plasmid pTV1 by electroporation. *FEMS Microbiol. Lett.* 52, 127–131.

Belliveau, B. H., and Trevors, J. T. (1989). Transformation of *Bacillus cereus* vegetative cells by electroporation. *Appl. Environ. Microbiol.* **55**, 1649–1652.

Belliveau, B. H., and Trevors, J. T. Mercury resistance determined by a self-transmissible plasmid in *Bacillus cereus* 5. *Biol. Metals* (in press).

Bone, E. J., and Ellar, D. J. (1989). Transformation of *Bacillus thuringiensis* by electroporation. *FEMS Microbiol. Lett.* **58**, 171–178.

Brigidi, P., De Rossi, E., Bertarini, M. L., Riccardi, G., and Matteuzzi, D. (1990). Genetic transformation of intact cells of *Bacillus subtilis* by electroporation. *FEMS Microbiol. Lett.* **67**, 135–138.

Chassy, B. M., and Flickinger, J. L. (1987). Transformation of *Lactobacillus casei* by electroporation. *FEMS Microbiol. Lett.* **44**, 173–177.

Chassy, B. M., Mercenier, A., and Flickinser, J. (1988). Transformation of bacteria by electroporation. *Trends Biotechnol.* **6**, 303–309.

Dower, W. J. (1990). Electroporation of bacteria: A general approach to genetic transformation. In "Genetic Engineering—Principles and Methods," (J. K. Setlow, ed.) vol. 12, pp. 275–296, Plenum Publishing, New York.

Dower, W. J., Miller, J. F., and Ragsdale, C. W. (1988). High efficiency transformation of *E. coli* by high voltage electroporation. *Nucleic Acids Res.* **16**, 6127–6145.

Harlander, S. K. (1987). Transformation of *Streptococcus lactis* by electroporation. *In* "Streptococcal Genetics" (J.J. Ferretti and R. C. Curtiss III, eds.), American Society for Microbiology, Washington, D.C., pp. 229–233.

Hülsheger, H., Potel, J., and Niemann, E.-G. (1981). Killing of bacteria with electric pulses of high field strength. *Radiat. Environ. Biophys.* **20**, 53.

Kim, A. Y., and Blaschek, H. P. (1989). Construction of an *Escherichia coli—Clostridium perfringens* shuttle vector and plasmid transformation of *Clostridium perfringens. Appl. Environ. Microbiol.* **55**, 360–365.

Kim, A. Y., Allen, S. A., and Blaschek, H. P. (1990). Electroporation-induced transformation and recovery of plasmid DNA in *Clostridia* spp. *Abstr. Int. Conf. Electroporation and Electrofusion,* Woods Hole, MA, p. 45.

Luchansky, J. B., Muriana, P. M., and Klaenhammer, T. R. (1988a). Electrotransformation of gram-positive bacteria. *Bio-Rad Lab. Bull.* **1350**, 1–3.

Luchansky, J. B., Muriana, P. M., and Klaenhammer, T. R. (1988b). Application of electroporation for transfer of plasmid DNA to *Lactobacillus, Lactococcus, Leuconostoc, Listeria, Pediococcus, Bacillus, Staphylococcus, Enterococcus* and *Propionibacterium. Mol. Microbiol.* **2**, 637–646.

Mahillon, J., Chungjatupornchai, W., Docodk, J., Dierickx, S., Michiels, F., Perferoen, M., and Joos, J. (1989). Transformation of *Bacillus therigiensis* by electroporation. *FEMS Microbiol. Lett.* **60**, 205–210.

McIntyre, D. A., and Harlander, S. K. (1989). Improved electroporation efficiency of intact *Lactococcus lactis* subsp. *lactis* cells grown in defined media. *Appl. Environ. Microbiol.* **55**, 2621–2626.

Mercenier, A., and Chassy, B. M. (1988). Strategies for the development of bacterial transformation systems. *Biochimie,* **70**, 503–517.

Miller, J. F., Dower, W. J., and Tompkins, L. S. (1988). High-voltage electroporation of bacteria: genetic transformation of *Campylobacter jejuni* with plasmid DNA. *Proc. Natl. Acad. Sci. USA* **85**, 856–860.

Oultram, J. D., Loughlin, M., Swinfield, T.-J., Brehm, J. K., Thompson, D. E., and Minton, N. P. (1988). Introduction of plasmids into whole cells of *Clostridium acetobutylicum* by electroporation. *FEMS Microbiol. Lett.* **56,** 83–88.

Schurter, W., Geiser, M., and Mathé, D. (1989). Efficient transformation of *Bacillus thuringiensis* and *Bacillus cereus* via electroporation: Transformation of acrystalliferous strains with a cloned delta-endotoxin gene. *Mol. Gen. Genet.* **218,** 177–181.

van der Lelie, D., van der Vossen, J. M. B. M., and Venema, G. (1988). Effect of plasmid incompatibility on DNA transfer to *Streptococcus cremoris*. *Appl. Environ. Microbiol.* **54,** 865–871.

Vehmaanperä, J. (1989). Transformation of *Bacillus amyloliquefaciens* by electroporation. *FEMS Microbiol. Lett.* **61,** 165–170.

31

Protocol for High-Efficiency Yeast Transformation

Daniel M. Becker and Leonard Guarente

Department of Biology, Massachusetts Institute of Technology,
Cambridge, Massachusetts 02139

I. Introduction
II. Materials
   A. Reagents
   B. Instrumentation
III. Procedures
IV. Modifications
   A. *Schizosaccharomyces pombe*
   B. Life Technologies/BRL Cell-Porator
   C. Freezing
   References
   Appendix A

I. Introduction

A number of recent papers describe protocols for electrotransformation of the budding yeast *Saccharomyces cerevisiae* and the fission yeast *Schizosaccharomyces pombe* (Becker and Guarente, 1991; Delorme, 1989; Meilhoc *et al.*, 1990; Hill, 1989; Rech *et al.*, 1990; Hood and Stachow, 1990; Weaver *et al.*, 1988). In general, these procedures require fewer steps than chemical (Ito *et al.*, 1983) or spheroplast transformation protocols (Hinnen *et al.*, 1978), and provide comparable or superior efficiency. The protocol described here provides efficiencies as high as 5×10^4 transformants/100 ng plasmid with a minimum of manipulation. Like other protocols for electrotransformation of yeast, however, the process saturates at low levels of input DNA: optimal efficiency (transformants/μg) is seen at or below 10 ng plasmid/transformation. As a result, this protocol is least appropriate for applications that

Guide to Electroporation and Electrofusion
501

that call for maximizing the total number of transformants, such as screening expression libraries in yeast (Fikes *et al.*, 1990). It is best used for the routine introduction of plasmid clones into individual yeast strains, and should be especially useful when the DNA is limiting. For transformation of a large number of different strains in a single experiment, the more rapid (albeit lower efficiency) procedure of Delorme (1989; Appendix A) may be preferred.

Although the efficiency of transformation depends in part upon the strain, the protocol as originally presented (Becker and Guarente, 1991) has been used successfully for episomal transformation of a wide variety of *S. cerevisiae* strains—including petites (H. Edwards, personal communication)—and for integrative transformation of various strains of *S. cerevisiae* (H. Edwards, personal communication) and *Schizosaccharomyces pombe* (Olesen *et al.*, 1991). While we have not investigated in detail the effect of plasmid size on transformation efficiency, the original procedure permits transformation with yeast artificial chromosomes (M. Bell, personal communication).

II. Materials

A. Reagents

1. YPD

> 1% (w/v) Yeast extract (Difco)
> 2% (w/v) Bactopeptone (Difco)
> 2% (w/v) Glucose
> Sterilize by autoclaving.

2. 1 *M* sorbitol

> 182.2 g/l in Milli-Q (Millipore; or other high grade) H_2O
> Sterilize by ultrafiltration.

3. Selection plates

> 0.67% (w/v) Yeast nitrogen base without amino acids (Difco).
> 2% (w/v) Carbon source (usually glucose).
> 1 M Sorbitol (added as an osmotic protectant; not appreciably metabolized).
> Amino acids (40 mg/l) as needed for strain auxotrophies; omit amino acid to be used for plasmid selection; for URA selection, add additionally 0.1% (w/v) casamino acids (Difco).
> 2% (w/v) Agar.

B. Instrumentation

The electrical parameters described below were defined using a BioRad gene pulser with pulse controller using 0.2-cm-gap disposable cuvettes (Becker and Guarente, 1991). A slight modification of the protocol (see Section IV below) has been developed for the BRL cell porator with 0.15-cm microelectroporation chambers (Lorow-Murray and Jessee, 1991). Both of these devices deliver exponential pulses. For square-wave generators, we direct the reader to Meilhoc *et al.* (1990).

III. Procedures

1. Inoculate 500 ml YPD broth in a 2-l flask. Grow with vigorous shaking at 30°C to $OD_{600nm} = 1.3–1.5$ (approximately 1×10^8 cells/ml).
2. Harvest the culture by centrifugation and resuspend in 100 ml YPD broth. Add 2.0 ml sterile 1 *M* HEPES, pH 8.0 (20 m*M* final concentration) and then add 2.5 ml sterile 1 *M* dithiothreitol (DTT; 25 m*M* final concentration) while swirling gently. Incubate 15 min at 30°C with gentle shaking.
3. Bring to 500 ml with Milli-Q H$_2$O.
4. Concentrate the cells approximately 1000-fold with several centrifugations, resuspending the successive pellets as follows:

1st pellet:	500 ml Milli-Q H$_2$0
2nd pellet:	250 ml Milli-Q H$_2$0
3rd pellet:	20 ml 1 *M* sorbitol
4th pellet:	0.5 ml 1 *M* sorbitol

   Resuspension should be vigorous. The rotor, the speed and exact duration of centrifuge spins are not critical. All solutions should be ice-cold. Final volume of resuspended yeast should be about 1.0–1.5 ml.
5. Mix 40 μl of yeast suspension per transformation with ≤100 ng DNA (in ≤5 μl) and transfer to a cold 0.2-cm sterile electroporation cuvette. The DNA should be in a low-ionic-strength buffer such as TE (10 m*M* Tris-HCl, 1 m*M* EDTA, pH 8.0). Incubation time can be varied for convenience.
6. Pulse at 1.5 kV, 25 μF, 200 Ω. ($E_0 = 7.5$ kV/cm; τ varies around 4.6 ms.)
7. Immediately add 1 ml cold 1 *M* sorbitol to the cuvette and recover the yeast, with gentle mixing, to a culture tube.
8. Spread aliquots of the transformation directly on selective plates containing 1 *M* sorbitol. A period of outgrowth is not required.

IV. Modifications

A. *Schizosaccharomyces pombe*

This protocol can be used for *S. pombe* by omitting DTT treatment and increasing the initial voltage to 2.0 kV ($E_0 = 10$ kV/cm; H. Prentice, personal communication; Olesen *et al.*, 1991). Note that the growth media and selection plates described above must be altered to reflect the nutritional preferences of fission yeast [see Moreno *et al.* (1991) for details]; be certain to retain 1 M sorbitol in the selection plates.

B. Life Technologies/BRL Cell-Porator

We have not tested other electroporation devices, but Lorow-Murray and Jessee (1991) report the following adaptation of our protocol for the device manufactured by Life Technologies/BRL. Prepare cells for electroporation as described above with the omission of the DTT pretreatment. Mix 20 μl of the concentrated yeast suspension with 2 ng plasmid, and transfer to a chilled, 0.15-cm-gap, disposable microelectroporation chamber. Electroporate at 400 V, 10 μF, using the low resistance setting. Recover 10 μl to a sterile microcentrifuge tube containing 0.5 ml ice-cold 1 M sorbitol. Mix gently. Plate aliquots directly on selection plates. The transformation efficiencies should be similar to those that we report with the BioRad Gene Pulser. Note, however, that Lorow-Murray and Jessee (1991) observe no improvement in efficiency by incorporating 1 M sorbitol in the selection plates.

C. Freezing

Lorow-Murray and Jessee report (1991) that yeast prepared without DTT can be frozen in a dry ice/ethanol bath and stored at $-70°C$ for subsequent transformation, with a 5- to 100-fold loss in efficiency.

Acknowledgments

This work was supported by grants from the National Institutes of Health (to L. Guarente) and by a postdoctoral fellowship from the American Cancer Society (to D. M. Becker). D.M. Becker is currently a Special Fellow of the Leukemia Society of America.

References

Becker, D. M. and Guarente, L. (1991). Transformation of yeast by electroporation. *Methods Enzymol.* 194, 182–187.

Delorme, E. (1989). Transformation of *Saccharomyces cerevisiae* by electroporation. *Appl. Environ. Microbiol.* **55**, 2242–2246.

Fikes, J. D., Becker, D. M., Winston, F., and Guarente, L. (1990). Striking conservation of TFIID in *Schizosaccharomyces pombe* and *Saccharomyces cerevisiae*. *Nature (Lond.)* **346**, 291–294.

Hill, D. E. (1989). Integrative transformation of yeast using electroporation. *Nucleic Acids Res.* **17**, 8011.

Hinnen, A., Hicks, J. B., and Fink, G. R. (1978). Transformation of yeast. *Proc. Natl. Acad. Sci. USA* **75**, 1929–1933.

Hood, M. T., and Stachow, P., (1990). Transformation of *Schizosaccharomyces pombe* and *Saccharomyces cerevisiae*. *Nucleic Acids Res.* **18**, 688.

Ito, H., Fukada, Y., Murata, K., and Kimura, A. (1983). Transformation of intact yeast cells treated with alkali cations. *J. Bacteriol.* **153**, 163–168.

Lorow-Murray, and Jessee, (1991). High efficiency transformation of *Saccharomyces cervesiae* by electroporation BRL/Life Technologies. *Focus* **13**, 65–68.

Meilhoc, E., Masson, J.-M., and Teissie, J. (1990). High efficiency transformation of intact yeast cells by electrified pulses. *Bio/Technology* **8**, 223–227.

Moreno, S., Klar, A., and Nurse, P. (1991). Molecular genetic analysis of fission yeast *Schizosaccharomyces pombe*. *In* "Methods in Enzymology," (C. Guthrie and G. R. Fink, eds.) vol. 194, pp. 795–823. Academic Press, San Diego, California.

Olesen, J. T., Fikes, J. D., and Guarente, L. (1991). The *Schizosaccharomyces pombe* homolog of *Saccharomyces cerevisiae* HAP2 reveals selective and stringent conservation of the small essential core protein domain. *Mol. Cell. Biol.,* **11**, 611–619.

Rech, E. L., Dobson, M. J., Davey, M. R., and Mulligan, B. J. (1990). Introduction of a yeast artificial chromosome vector into *Saccharomyces cerevisiae* cells by electroporation. *Nucleic Acids Res.* **18**, 1313.

Weaver, J. C., Harrison, G. I., Bliss, J. G., Mourant, J. R., and Powell, K. T. (1988). Electroporation: High frequency occurrence of a transient high-permeability state in erythrocytes and intact yeast. *FEBS Lett.* **229**, 30.

Appendix A
Yeast Electrotransformation by the Method of Delorme (1989)[a]

1. Culture *S. cerevisiae* cells overnight in YPD.
2. Dilute cultures 1 : 100 in fresh YPD.
3. Shake at 30°C to an A_{600nm} of 0.35–0.75.
4. Harvest by centrifugation and resuspend in YPD at A_{600nm} equal to 10–20.
5. To 200 μl of freshly prepared chilled cells in a 0.4 cm cuvette, add 0.1 μg of purified transforming DNA.
6. Pulse at 2.25 kV/cm with a 25-μF capacitance setting on a BioRad Gene Pulser.
7. Spread electroporated cells directly on SDCAA plates (0.67% yeast nitrogen base without amino acids, 2% glucose, 2% agar, amino acids as necessary for strain auxotrophies, with 0.5% casamino acids for URA selection).
8. Incubate 2 days at 30°C.

[a]Notes: The efficiency expected is about $1.0–4.5 \times 10^4$ colonies/μg DNA depending on the strain and DNA used.

32

Protocols of Electroporation and Electrofusion for Producing Human Hybridomas

Martin I. Mally, Michael E. McKnight, and Mark C Glassy

Sci-Clone, Inc., San Diego, California 92126

I. Introduction

A. Immortalization of Human Lymphocytes

Murine monoclonal antibodies (mAbs) have become very important tools in the research, diagnosis, and therapy of human diseases. However, immunogenicity of the murine mAbs has hampered their usefulness as long term immunotherapeutic agents, primarily due to the human anti-mouse antibody (HAMA) response. Therefore, the need has arisen for the generation of human mAbs to bypass the HAMA response during treatment. In addition, human mAbs probably recognize a different

Guide to Electroporation and Electrofusion
507

repertoire of antigens when compared to murine mAbs, rendering them more clinically useful. A major dilemma facing researchers today is the inability to efficiently produce stable, immunoglobulin-secreting human hybridomas as readily as their murine counterparts. To date, very few competent human fusion partners are available, and the most widely used fusion protocols result in very low fusion frequencies (typically 1 hybrid per 10^5–10^6 cells). More efficient means of producing stable human hybridomas have recently become available that should expedite progress in the field. Electroporation, or the electrically induced transfer of genes into cells, and electrofusion, whereby the application of transient electric fields induces somatic cell hybridization, offer many advantages over current protocols for immortalizing immunocompetent B cells, and will be discussed here.

Immortalization of B lymphocytes has generally been accomplished by either (1) fusion with a myeloma or lymphoblastoid cell line using polyethylene glycol (PEG) as the chemical fusogen (Kohler and Milstein, 1975), or (2) infection with the B-tropic Epstein-Barr virus (EBV) prior to fusion (Rosen *et al.*, 1977; Zurawski *et al.*, 1978). Both methodologies have serious drawbacks. As mentioned, with few exceptions, there presently is a lack of quality human fusion partners (i.e., able to produce genetically stable hybrids), the fusion frequency is low, and the stability of the hybridomas is short-lived. EBV infection relies on the presence of EBV receptors (the C3b receptor predominantly found on immature B lymphocytes), and the transformed cells generally secrete immunoglobulins (Ig) at low levels, which decline rapidly with time (Rosen *et al.*, 1977; Zurawski *et al.*, 1978). Electroporation circumvents the need for a fusion partner by supplying DNA sequences that code for immortalization functions. The chromosomal instability inherent when two cells are fused together is also avoided during electroporation, as only a relatively small amount of DNA is introduced into the cells. Electrofusion has the advantages that (1) it is a physical effect (as is electroporation), acting independently of cellular, genetic or biochemical predispositions; (2) the use of PEG as a fusogen can be avoided, as it is cytotoxic to certain human cell lines; and (3) fewer cells are required, an important consideration when rare, antigen-specific human B lymphocytes are involved. Both electroporation and electrofusion are less labor-intensive than traditional PEG-based fusion protocols, and have been used successfully on a variety of cell types, including those from plants, bacteria, yeast, and mammalian cells.

B. Principles of Electroporation

Several methods exist in which exogenous genetic materials are introduced permanently into cells. These include gene transfer, or transfection, by calcium phosphate-DNA coprecipitation (Graham and Van der Eb, 1973; Wigler *et al.*, 1977; Oi *et al.*, 1983), DEAE-dextran neutralization (McCutchan and Pagano, 1968; Farber *et al.,* 1975), the use of viral vectors (Mulligan *et al.,* 1979; Miller *et al.,* 1983), liposomes (Fraley *et al.,* 1980; Schaefer-Ridder *et al.,* 1982; Felgner *et al.,* 1987), or protoplast fusion (Schaffner, 1980; Sandri-Golden *et al.,* 1981; Rassoulzadegan *et al.,* 1982; Oi *et al.,* 1983), and the direct microinjection of DNA into the nucleus of recipient cells (Anderson *et al.,* 1980; Capecchi, 1980). Briefly exposing cells to a high-voltage direct current (DC) electric pulse temporarily and reversibly permeabilizes the cellular membrane (Zimmermann, 1982; Wong and Neumann, 1982), resulting in membrane pores, and has been used successfully to facilitate cellular uptake of DNA (Neumann *et al.,* 1982; Potter *et al.,* 1984; Toneguzzo *et al.,* 1986; Chu *et al.,* 1987; Knutson and Yee, 1987). This technique, termed electroporation, is rapidly becoming the method of choice for transfecting lymphocytes, as they are quite refractory to most other gene transfer techniques (Potter *et al.,* 1984). Electroporation of cultured cells in the presence of exogenous DNA results in the random, and in certain cases stable, integration of that DNA into the genome of a select number of cells. The resultant electrodomas, which express the newly incorporated genes, are then identified by their altered phenotype (i.e., drug resistance), similar to the method by which hybridomas are selected in HAT media. Similarly, electroporation of activated B cells in the presence of DNA sequences encoding immortalization functions (see Section III,A) can generate hybridoma-like cells, that is, immortalized, Ig-secreting lymphocytes, without the need for a fusion partner and its associated drawbacks. In this manner, a pure population of cells with the desired characteristics is obtained.

Optimization of a number of electrical and biological parameters must initially be accomplished for each cell type in order to achieve high-efficiency electroporation (Chu *et al.,* 1987; Hama-Inaba *et al.,* 1987; Knutson and Yee, 1987; Mally, 1990; Mally and Glassy, 1990). Although optimization may be lengthy, once these conditions are experimentally determined, these procedures are rapid and straightforward. Included among the electrical and physical variables that must be examined are the electric field strength, pulse width, number of pulses, and type of electroporation chamber. The biological parameters include the choice of electroporation medium, cell type, physiologic state of the cells, DNA conformation and concentration, the temperature at which electroporation is performed, and drug sensitivities and plating densities of the cells after electroporation. An appropriate selection system must also be determined.

C. Principles of Electrofusion

Electrofusion uses precise, transient electric fields to induce somatic cell hybridization (Zimmermann, 1982; Zimmermann and Vienken, 1982). In preparation for fusion, cells are initially exposed to an alternating electric field (AC) to cause them to align into a formation typically identified as a string of pearls." This process is termed dielectrophoresis (Zimmermann, 1982; Hofmann and Evans, 1986). The current is then switched to a direct electrical field (DC) to begin the dielectric breakdown of the cell membranes, resulting in transmembrane pores. The pores of adjacent cells form small channels between the two cells, which eventually broaden, causing the cells to fuse. When the direct electrical field is removed, the remaining pores in the membrane of the heterokaryon cell close, yielding an intact hybrid cell. As with electroporation, a number of variables must be optimized in order to maximize electrofusion efficiency (Glassy and Hofmann, 1985; Hofmann and Evans, 1986; Pratt *et al.*, 1987; Glassy and Pratt, 1989). The fusion can be performed at room temperature or 4°C. The cells can be treated with facilitators such as pronase, which enzymatically cleaves cell surface glycoproteins, allowing for more direct membrane–membrane interaction, and may potentially yield more hybrids. The type, molarity, and pH of the fusion medium can be altered. Optimal cell concentrations vary between cell types, but the ratio of isolated lymphocytes to fusion partners is generally between 2 : 1 and 4 : 1. Cell type is also important to achieving high electrofusion efficiency. It has been demonstrated (Glassy *et al.*, 1983) that β lymphoma cells (previously immortalized by malignant transformation) fuse the best, followed by lymph-node lymphocytes (*in vivo* activated lymphocytes), with peripheral blood lymphocytes (predominantly in the G_0 phase of the cell cycle) fusing poorly. The physical variables include the voltage used to align, porate, and fuse the cells; the configuration of the fusion chamber used to house the cells; the pulse width and number of pulses; and the number of repetitions of the fusion sequence. It is also critical to have an appropriate selection system to allow the investigator to sort fused cells from nonfused cells (i.e., HAT selection).

II. Materials

A. Reagents

1. RPMI 1640 culture medium (M.A. Bioproducts)
2. Fetal bovine serum (M.A. Bioproducts)
3. 100× L-glutamine, 200 mm (GIBCO)
4. 100× Penicillin–streptomycin, 5000U/ml (GIBCO)
5. 0.4% Trypan blue (GIBCO)
6. Geneticin or G418 (GIBCO)
7. Mycophenolic acid (Sigma)
8. 50× HAT (Sigma) (5 × 10^{-3} M hypoxanthine, 2 × 10^{-5} M aminopterin, 8 × 10^{-4} M thymidine)
9. Xanthine (Sigma)
10. Guanine (Sigma)
11. Mannitol (Sigma)
12. Purified plasmid DNA containing the genes of interest

It should be noted that the basic molecular biology techniques, such as the growth and transformation of bacterial cells, plasmid isolation and purification, and restriction endonuclease digestion of DNA, will not be detailed here, as they are beyond the intended scope of this chapter. It is therefore recommended that the reader perform these techniques as presented in a methods manual, such as "Molecular Cloning: A Laboratory Manual," by J. Sambrook, E. F. Fritsch, and T. Maniatis.

B. Instrumentation

Several electroporation and electrofusion instruments are commercially available, with a wide range of price and sophistication. In our laboratory, we have used for electroporation both the Cell-Porator (Bethesda Research Laboratories) and the Gene Pulser (Bio-Rad Laboratories) systems with equal success. Presterilized electroporation chambers with 0.4-cm interelectrode gaps were used. Both instruments use capacitor discharges to generate electrical pulses that decay exponentially. The operator can adjust both the voltage and capacitance settings on each instrument, yet only the Gene Pulser displays the pulse width (as well as the voltage and capacitance settings) after the pulse has been delivered, allowing a more detailed analysis of the actual pulse. The Cell-Porator sample chamber can accommodate up to four samples at a time, enabling sequential electroporations to be performed rapidly (the Gene Pulser sample chamber holds only a single sample), and can maintain sample temperature (i.e., water or ice can be placed in the water-tight compartment). The Cell-Porator is more user-friendly (and less expensive), but the Gene Pulser allows the investigator to more accurately examine the parameters of the decay waveform. More expensive systems also available include the geneZapper from International Biotechnologies, Inc. (IBI), the Advanced Gene Transfer System from Baekon, Inc., and the BTX 100 system from Biotechnologies and Experimental Research, Inc. (BTX).

We have used the BTX Electro Cell Manipulation System 401A for our electrofusion experiments. Fusions were performed with glass slides, as well as 1-ml and 50-ml chambers. The use, description, and design ideology of the 401A electrofusion apparatus have all been previously described (Hofmann and Evans, 1986; Pratt *et al.*, 1987). Although other electrofusion units are commercially available, we have not had any practical experience with them.

III. Procedures

A. Electroporation

To obtain the highest electroporation efficiencies, actively proliferating B lymphocytes should be used, as replicating cells appear to have an increased ability to take up and incorporate exogenous DNA when compared to quiescent cells (Potter *et al.*, 1984; Chu *et al.*, 1987; Stopper *et al.*, 1987). Therefore, electroporation should be performed on previously activated (either *in vivo* or *in vitro*) B lymphocytes. Regional draining lymph nodes, tonsils, and spleen cells are excellent sources of *in vivo* activated B cells. *In vitro* activation of B cells can be accomplished by stimulation with polyclonal activators such as the mitogens phytohemaglutinin and bacterial lipopolysaccharides, Ig-specific antibodies, or with specific antigen in the case of *in vitro* immunization (Koda and Glassy, 1990).

Several genes have been identified that confer on primary cells the ability to grow indefinitely in culture by increasing the proliferative potential of the cells and by facilitating their *in vitro* establishment. When transfected and expressed in primary cells, these so-called "immortalizing" genes allowed the cells to overcome cellular senescence without causing a tumorigenic conversion of the cells (Houweling *et al.*, 1980; Ruley, 1983; Land *et al.*, 1983; Rassoulzadegan *et al.*, 1983; Jenkins *et al.*, 1984; Eliyahu *et al.*, 1984). It is these genes that appear promising for the generation of immortalized, Ig-secreting lymphocytes via electroporation.

A series of simple, yet critical, preliminary experiments must be performed in order to determine optimal culture selection and electroporation conditions. A brief description of these experiments follows.

1. *Determination of Optimal Drug Concentration*

Since mammalian cells differ in their sensitivities to geneticin (G418) and myco-phenolic acid (an explanation for choosing these drugs will become evident later), the drug concentrations required for selection must be determined for each cell type. A simple method to accomplish this is to culture activated cells in microtiter plates in the presence of varying amounts of the drugs. Drug concentrations approximately 25–50% higher than the minimal concentration required to kill 100% of the cells within 7–14 days should be chosen for selection. It should be noted that these drugs are most effective against replicating cells.

Recombinant electrodomas are routinely selected by their ability to grow in culture medium supplemented with drugs that are cytotoxic to mammalian cells. The two most commonly used drug markers in transfection experiments are bacterial genes, which are nonfunctional in mammalian cells and have no mammalian coun-terparts, and can thus be introduced into any cell, precluding the need for genetically mutant cells (i.e., HAT-sensitive fusion partners). These dominant selectable genes code for neomycin resistance (neo) and for the enzyme xanthine phosphoribosyl-transferase (gpt). Placement of these genes under the control of eukaryotic promoter elements allows their expression in mammalian cells, rendering cells resistant to the drugs (Mulligan and Berg, 1980, 1981; Colbere-Garapin *et al.*, 1981; Southern and Berg, 1982). Mammalian cells are normally resistant to neomycin, yet are sensitive to one of its derivatives, so-called G418 or geneticin (Davies and Jimenez, 1980;

Chen and Okayama, 1987). This aminoglycoside is toxic to most eukaryotic cells by interfering with the functioning of the 80S ribosomes and the subsequent inhibition of protein synthesis. However, electrodomas expressing the neo gene survive when cultured in the presence of G418. Mycophenolic acid inhibits *de novo* purine nucleotide synthesis via inhibition of inosine monophosphate (IMP) dehydrogenase (Franklin and Cook, 1969), preventing formation of xanthine monophosphate and therefore of guanosine monophosphate (GMP). Although mycophenolic acid alone prevents extensive growth of most cells, the addition of aminopterin, which blocks *de novo* synthesis of IMP, the first purine nucleotide intermediate, ensures complete inhibition of *de novo* synthesis of all purines. Thus, only electrodomas expressing the transfected gpt gene are able to synthesize GMP from xanthine, a process that occurs very inefficiently in normal cells.

For neomycin selection, geneticin (G418) is dissolved in 100 mM HEPES (N-2-hydroxyethylpiperazine-N'-2-ethane sulfonic acid), pH 7.3, to approximately 40 mg/ml (actual drug concentration; note that G418 represents approximately 40–50% of the total weight of the drug preparation), the solution is sterile filtered (0.2 μm), and aliquots are stored at $-20°C$. For gpt selection, mycophenolic acid is dissolved to a concentration of 50 mg/ml in absolute ethanol and aliquots are stored at $-20°C$.

a. Aliquot 1×10^4 activated B cells/well (see below for the appropriate culture medium) in a 96-well microtiter plate.

b. Add G418 to a final concentration of 0–1200 μg/ml, or mycophenolic acid to 0–100 μg/ml. Mix components in each well. Be sure to use 5–10 wells per drug concentration to accurately assess the effects of the drugs.

c. Incubate in a 37°C humidified atmosphere with 5% CO_2 for 7–14 days, replacing media and drugs as needed.

d. Determine cell viability. Typically, effective drug concentrations in the range of 400–800 μg/ml for G418 and 1–25 μg/ml for mycophenolic acid are used for mammalian cells.

Complete culture medium for G418 selection consists of RPMI 1640 supplemented with 10% fetal bovine serum, 2 mM L-glutamine, and 50 units/ml of a penicillin–streptomycin solution. However, for gpt selection, the media requirements are different. It is necessary to use dialyzed fetal bovine serum and a medium in which guanine is absent (i.e., RPMI 1640, Dulbecco's modified Eagle's medium, Ham's F12 medium), as guanine overcomes the inhibitory effect on mycophenolic acid by allowing the cells to use the purine salvage pathway. Activated B cells should be plated in medium consisting of RPMI 1640 supplemented with 10% dialyzed fetal bovine serum, 250 μg/ml xanthine, 15 μg/ml hypoxanthine (or 25 μg/ml adenine), 10 μg/ml thymidine, 2 μg/ml aminopterin, 2 mM L-glutamine, and 50 units/ml of a penicillin–streptomycin solution.

2. Determination of Optimal Field Strength

The optimal electric field strength must be independently and accurately determined for the activated B cells. In general, maximum transfection efficiency occurs when approximately 20–60% of the cells exposed to the electric pulse remain viable (Boggs *et al.*, 1986; Knutson and Yee, 1987). Thus, a quick, though perhaps inaccurate, method of estimating the appropriate voltage is to determine the voltage at which approximately 50% cell viability occurs (Chu *et al.*, 1987; Knutson and Yee, 1987). However, the method of choice is to electroporate cells in the presence of varying concentrations of a marker plasmid (which had previously been linearized by restriction endonuclease digestion), culture the electrodomas in selective medium, and determine the transfection efficiency by simply counting the number of proliferating colonies.

a. After washing activated B cells twice in serum-free culture medium, resuspend cells to concentrations of 1×10^7 cells/ml and 2×10^7 cells/ml in serum-free media.

b. Place 5×10^6 cells in electroporation chambers and incubate at room temperature for 10 min.

c. Deliver field strengths varying between 500 and 1000 V/cm (a typical range for lymphocytes) with capacitance settings between 750 and 1000 μF. Varying the number of pulses can also be performed at this step.

d. After delivery of the pulse(s), continue incubating the cells at room temperature for an additional 10 min.

e. Dilute cells 1 : 10 in complete culture medium and incubate in a 37°C humidified atmosphere with 5% CO_2.

f. Determine cell viability after 12–14 h in culture.

When drug selection is applied to the electrodomas as the means of optimizing the field strength, include the following steps in the above procedure: (1) prior to electroporation, to the cells in the electroporation chambers, add linearized plasmid DNA (containing the drug marker gene of interest) to a final concentration of 1–20 μg/ml, mix gently by inversion, and incubate at room temperature for 10 min; (2) 48 h after electroporation, replace culture medium with the appropriate selective medium containing the optimal drug concentrations as previously determined; and (3) determine transfection efficiency after 10–14 days in culture by counting the number of antibiotic-resistant colonies.

3. Generation of Ig-Secreting Electrodomas

Following optimization of the various electrical and biological parameters, the generation of immortalized, Ig-secreting lymphocytes by electroporation can begin.

However, unless precautions are taken to obtain a "pure" population of B cells (e.g., sheep erythrocyte rosetting, cell panning, complement-mediated lysis), the investigator will undoubtedly electroporate and immortalize some unwanted cell types. This problem can be overcome by specifically targeting the immortalizing genes to B lymphocytes, using Ig-specific enhancer elements. Enhancers are DNA sequences that belong to the set of eukaryotic regulatory elements that increase the transcriptional efficiency of genes, independent of their position and orientation (Khoury and Gruss, 1983; Atchison and Perry, 1986; Ptashne, 1988). In the specific case of the Ig genes, the enhancers function only in lymphoid cells, thus operating in a tissue-specific manner. Therefore, placing Ig-specific enhancers adjacent to sequences coding for immortalizing genes would yield molecular constructs capable of specifically immortalizing B lymphocytes. In fact, such constructs have been used for the production of transgenic mice, which proceeded to develop tumors representing different stages of B-cell maturation (Adams et al., 1985; Langdon et al., 1986; Alexander et al., 1987 Schmidt et al., 1988). However, transgenic mice carrying similar constructs yet lacking the Ig enhancer failed to develop tumors (Adams et al., 1985), indicating the efficacy and specificity of the constructs containing the enhancer elements. Therefore, molecular constructs containing sequences coding for immortalization functions under control of Ig-specific enhancers appear attractive for the generation of immortalized, Ig-secreting lymphocytes even in the presence of other cell types, as only the B cells will be targeted.

It will be necessary to perform electroporation in the presence of a plasmid containing a drug-selective marker with one containing the immortalizing gene in order to select for successfully transfected electrodomas. It has been reported that the simultaneous transfection of two separate plasmids by electroporation yielded cotransfection efficiencies ranging from 23 to 100%, which is higher than would be expected from the random uptake of two plasmids (Miller and Temin, 1983; Boggs et al., 1986; Toneguzzo et al., 1986). Therefore, electrodomas that grow in selective culture media have a high probability of also expressing the gene(s) of interest, thus simplifying the molecular constructs the investigator needs.

Prior to a detailed description of an electroporation protocol for generating immortalized, Ig-secreting lymphocytes, several important points need to be considered. Activated lymphocytes should be used, and all manipulations performed at room temperature in order to increase electroporation efficiency. If the electrodomas can be grown in serum-free medium, then characterization of the secreted Ig will be simplified. A low plating density of cells in selective medium minimizes cross-contamination between electrodomas and simplifies the cloning of a single electrodoma, and, in turn, the purification of the antibody.

a. Wash activated B cells twice in serum-free culture medium and resuspend to 1–2 × 10^7 cells/ml in serum-free media.
b. Aliquot 3–5 × 10^6 viable cells in electroporation chambers.

c. Add sterile, linearized plasmid DNA to the cell suspensions, invert to mix, and incubate at room temperature for 10 min. For best results, the plasmids should be linearized at restriction sites distant from the marker or immortalizing genes. Use the previously determined optimal amounts of DNA in a molar ratio of 5 : 1 or 10 : 1 of the plasmid containing the immortalizing gene to the plasmid containing the gene conferring drug resistance, respectively.

d. Invert to mix, and electroporate at the previously optimized voltage and capacitance settings, at room temperature.

e. Incubate at room temperature for 10 min.

f. Dilute cells 1 : 10 in complete culture medium (i.e., RPMI 1640 supplemented with 10% fetal bovine serum and 2 mM L-glutamine), and incubate for 48 h in a 37°C humidified atmosphere with 5% CO_2.

g. Harvest cells and dilute to 1×10^5 viable cells/ml in selective culture medium.

h. Aliquot 200 μl/well into microtiter plates, and continue incubation for an additional 2–3 weeks until drug-resistant colonies appear, replacing media and drug every 2–4 days to maintain selective pressure.

i. Supernatants from the wells are then screened for the presence of Ig using one of the commonly used detection methods: radioimmunoassay (RIA), enzyme-linked immunosorbent assay (ELISA), or fluorescence immunoassay (FIA). Typically, a nonisotopic method such as ELISA or FIA is preferred (Gaffar and Glassy, 1988). The details for these techniques will not be presented here, as they can be found in many laboratory manuals. Those colonies secreting the highest levels of Ig are expanded in culture, and the antibodies are characterized biochemically and immunologically.

B. Electrofusion

As with electroporation, maximal electrofusion efficiencies depend on the experimental optimization of a number of variables. High fusion efficiencies are best obtained by using B lymphocytes with high variabilities that had previously been activated either *in vivo* or *in vitro*. These cells, as well as the fusion partner, must retain their high variabilities when placed in the electrofusion medium of choice. Typically, an iso-osmolar solution of 0.3 M mannitol is used as the fusion medium, although solutions with varying electrolyte composition have been used successfully. It has been demonstrated that the addition of 0.2–0.5 mM calcium acetate to the fusion medium also facilitates mammalian cell fusion. Therefore, incubation of the B lymphocytes and fusion partner in the fusion medium for several hours, followed by determination of cell viability, will allow the investigator to decide on an appropriate medium to be used during fusion. Also, other fusion facilitators, such as enzymatic treatment with pronase, may increase the fusion yields.

The system most commonly used to select for electrodomas following electrofusion is drug resistance in HAT medium. The fusion partner is generally sensitive to the components in HAT medium, allowing only successfully fused cells to survive under selective pressure.

1. Cell Preparation

Normal growth medium contains a large variety of ions and charged macromolecules with high conductivity that will significantly interfere with the electrical pulses generated by the electrofusion instrumentation and will need to be removed. This is most easily accomplished by extensively washing the chosen cell types in some iso-osmolar solution with low conductivity. A 0.3 M solution of mannitol has been our solution of choice, though other low-conductivity compounds may work as well or better, depending on the source and type of human cells used. Both fusion partners will need to be treated in the same manner.

The physiological state of each cell type is also critical and cells that are in mid-log growth (between 6×10^5 and 8×10^5 cells/ml) are optimal.

a. Cells to be used for fusion, both immunized lymphocytes and fusion partner, should be mixed together at the appropriate ratio (typically, 2 : 1, or 4 : 1), washed three times (500 $\times$ g for 10 min for each wash), and resuspended in 0.3 M mannitol, pH 7.2. Cell concentrations are usually 1×10^6 cells/ml, and optimal concentrations will need to be investigated for each type used.

b. The fusion partner is drug selectable and will die in HAT medium.

2. Determining Optimal Electrofusion Parameters

The membrane compositions of different mammalian cells vary significantly such that each cell type you see will need to be optimized for electrofusion. Under the same experimental parameters, one cell type may completely lyse whereas the other cell type may undergo dielectric breakdown and successfully fuse. There may be as much as a 50–100% difference in the applied field strengths (kV/cm) of the two fusion partners, and this will need to be carefully investigated to obtain the highest cell variabilities and fusion yields.

During a typical fusion sequence, the cells are subjected to a brief AC pulse to cause a dielectric breakdown at the plasma membrane with subsequent pore formation, followed by a DC pulse to cause membrane fusion and heterokaryon formation. The conditions of the AC pulse for electroporation described above are similar to those required here for electrofusion.

The fusion sequence, at least with the BTX 401A instrument, is composed of three steps: alignment, membrane poration, and the fusion event itself.

a. Cells suspended in 0.3 M mannitol are placed in the appropriate fusion chamber (see below).

b. A fusion sequence consisting of 20 s of 580 V/cm (maximum field strength) of AC 800-kHz, sinusoidal wave shape followed by a 3-s, 250-V/cm increase in AC amplitude to 830 V/cm was used. These dielectrophoretic conditions were followed immediately by two DC pulses of 7.69 kV/cm (maximum field strength), each having a duration of 20 μs. All of these parameters are keyed into the electrofusion unit.

c. After completing the electrofusion sequence, the cells were allowed to sit in the fusion chamber for 10 min before removal. The cells were then washed and resuspended in RPMI 1640 medium as described above and seeded at 1×10^5 cells per well in 96-well microtiter plates.

d. The following day, the wells containing cells were re-fed with complete RPMI 1640 medium containing 10% fetal bovine serum, plus HAT components. The cells should be re-fed with fresh medium approximately twice a week until hybrids appear, usually within 2–4 weeks postfusion.

3. Selection of a Fusion Chamber

Most commercial electrofusion instruments have several fusion chambers available, depending upon the demands of the investigator. Optimization of electrofusion parameters was performed on a microslide (20 μl volume) that was equipped with two parallel wires positioned 0.5 mm apart. Since the wires were round, a divergent field was induced that elicits dielectrophoresis in cells. It should be noted that the field geometry is quite different in the microslide electrodes compared to the coaxial chambers.

The majority of our experience has been with the BTX coaxial fusion chamber. This chamber has a divergent field geometry (Hofmann and Evans, 1986), which is necessary for a maximum dielectrophoretic effect. However, a divergent field does create an inhomogeneous field concentration when the DC pulse is applied. Therefore, the field strength across the chamber varies with the radial position. This causes the actual transmembrane voltage potential applied to each cell to depend on the cell's location in the chamber. Excessive inhomogeneous field geometry can have deleterious effects on fusion efficiency.

The major deciding factor in choosing which fusion chamber to use involved the volume and number of cells available. Relatively few cells ($<10^6$) should use a small volume chamber, whereas large cell numbers ($>10^7$) should use a large volume chamber. The BTX 401A unit has available a 50-ml chamber for large volumes, such as the number of cells in an entire mouse spleen.

4. Hybridoma Generation

For optimal fusion results, the cell partners are mixed in the appropriate fusion medium and the fusion sequence is keyed into the instrument of choice. After the cells are added to the fusion chamber, the investigator must decide how many reiterations or "doses" of the fusion sequence need to be applied to the cells. Typically, two "doses" are performed. After removing the cells from the fusion chamber, they are plated in 96-well microtiter plates and processed as in typical PEG-mediated fusions. After electrodoma (hybridoma) colonies become visible to the naked eye, screening for both Ig production and target antigen recognition is performed, usually by ELISA, fluorescence, or RIA (Gaffar and Glassy, 1988).

IV. Summary

Briefly exposing eukaryotic cells to high voltage electric pulses results in the transient breakdown of localized areas in the cellular membrane and subsequent pore formation (Knight, 1981), thereby increasing the permeability of the cells to exogenous macromolecules (electroporation), as well as leading to the fusion between cellular membranes (electrofusion). Both techniques offer advantages over the classical methods of producing hybridomas, including efficiency of hybridoma production, and provide the investigator alternative approaches for the generation of immortalized, Ig-secreting lymphocytes. In conclusion, the techniques discussed throughout this book have the potential to advance the field of human hybridoma production, resulting in novel human antibodies, which in turn will aid in our understanding of the human immune system, allowing the development of better diagnostic tools and clinically useful reagents.

References

Adams, J. M., Harris, A. H., Pinkert, C. A., Corcoran, L. M., Alexander, W. S., Cory, S., Palmiter, R. D., and Brinster, R. L. (1985). The c-myc oncogene driven by immunoglobulin enhancers induced lymphoid malignancy in transgenic mice. *Nature (Lond.)* **318**, 533–538.

Alexander, W. S., Schrader, J. W., and Adams, J. M. (1987). Expression of the c-myc oncogene under control of an immunoglobulin enhancer in Eu-myc transgenic mice. *Mol. Cell. Biol.* 7, 1436–1444.

Anderson, W. F., Killos, L., Sanders-Haigh, L., Kretschmer, P. J., and Diacumakos, E. G. (1980). Replication and expression of thymidine kinase and human globin genes microinjected into mouse fibroblasts. *Proc. Natl. Acad. Sci. USA* 77, 5399–5403.

Atchison, M. L., and Perry, R. P. (1986). Tandem kappa immunoglobulin promoters are equally active in the presence of the kappa enhancer: Implications for models of enhancer function. *Cell* 46, 253–262.

Boggs, S. S., Gregg, R. G., Borenstein, N., and Smithies, O. (1986). Efficient transformation and frequent single-site, single-copy insertion of DNA can be obtained in mouse erythroleukemia cells transformed by electroporation. *Exp. Hematol.* 14, 988–994.

Capecchi, M. R. (1980). High efficiency transformation by direct microinjection of DNA into cultured mammalian cells. *Cell* 22, 479–488.

Chen, C., and Okayama, H. (1987). High-efficiency transformation of mammalian cells by plasmid DNA. *Mol. Cell. Biol.* 7, 2745–2752.

Chu, G., Hayakawa, H., and Berg, P. (1987). Electroporation for the efficient transfection of mammalian cells with DNA. *Nucleic Acids Res.* 15, 1311–1326.

Colbere-Garapin, F., Horodniceanu, F., Kourilsky, P., and Garapin, A-C. (1981). A new dominant hybrid selective marker for higher eukaryotic cells. *J. Mol. Biol.* 150, 1–14.

Davies, J., and Jimenez, A. (1980). A new selective agent for eukaryotic cloning vectors. *Am. J. Trop. Med. Hyg.* 29, 1089–1092.

Eliyahu, D., Raz, A., Gruss, P., Givol, D., and Oren, M. (1984). Participation of p53 cellular tumour antigen in transformation of normal embryonic cells. *Nature (Lond.)* 312, 646–649.

Farber, F. E., Melnick, J. L., and Butel, J. S. (1975). Optimal conditions for uptake of exogenous DNA by Chinese hamster lung cells deficient in hypoxanthine guanine phosphoribosyltransferase. *Biochim. Biophys. Acta* 390, 298–311.

Felgner, P. L., Gadek, T. R., Holm, M., Roman, R., Chan, H. W., Wenz, M., Northrop, J. P., Ringold, G. M., and Danielson, M. (1987). Lipofection: A highly efficient, lipid-mediated DNA-transfection procedure. *Proc. Natl. Acad. Sci. USA* 84, 7413–7417.

Fraley, R., Subramani, S., Berg, P., and Papahadjopoulos, D. (1980). Introduction of liposome-encapsulated SV40 DNA into cells. *J. Biol. Chem.* 255, 10431–10435.

Franklin, T. J., and Cook, J. M. (1969). The inhibition of nucleic acid synthesis by mycophenolic acid. *Biochem. J.* 113, 515–524.

Gaffar, S. A., and Glassy, M. C. (1988). Applications of human monoclonal antibodies in non-isotopic immunoassays. *In* "Reviews on Immunoassay Technology" (S. B. Pal, ed.), Vol. I, pp. 123–145. Macmillan Press, Basingstoke, United Kingdom.

Glassy, M. C., and Hofmann, G. A. (1985). Optimization of electrocell fusion parameters in generating human-human hybrids with UC 729-6. *Hybridoma* 4, 61.

Glassy, M. C., and Pratt, M. (1989). Generation of human hybridomas by electrofusion. *In* "Electroporation and Electrofusion in Cell Biology" (E. Neumann, A. E. Sowers, and C. A. Jordan, ed.), pp. 271–282. Plenum Press, New York.

Glassy, M. C., Handley, H. H., Hagiwara, H. H., and Royston, I. (1983). UC 729-6, a human lymphoblastoid B cell line useful for generating antibody secreting human-human hybridomas. *Proc. Natl. Acad. Sci. USA* 80, 6327–6331.

Graham, F. L., and Van der Eb, A. J. (1973). A new technique for the assay of infectivity of human adenovirus 5 DNA. *Virology* 52, 456–467.

Hama-Inaba, H., Takahashi, M., Kasai, M., Shiomi, T., Ito, A., Hanoaka, F., and Sato, K. (1987). Optimum conditions for electric pulse-mediated gene transfer to mammalian cells in suspension. *Cell Structure Function* 12, 173–180.

Hofmann, G. A., and Evans, G. A. (1986). Electronic genetics—Physical and biological aspects of cellular electromanipulation. *IEEE Eng. Med. Biol.* 5, 6–25.

Houweling, A., Van den Elsen, P. J., and Van der Eb, A. J. (1980). Partial transformation of primary rat cells by the leftmost 4.5% fragment of adenovirus 5 DNA. *Virology* 105, 537–550.

Jenkins, J. R., Rudge, K., and Currie, G. A. (1984). Cellular immortalization by a cDNA clone encoding the transformation-associated phosphoprotein p53. *Nature (Lond.)* 312, 651–654.

Khoury, G., and Gruss, P. (1983). Enhancer elements. *Cell* 33, 313–314.

Knight, D. E. (1981). Rendering cells permeable by exposure to electric fields. *Tech. Cell. Physiol.* 113, 1–20.

Knutson, J. C., and Yee, D. (1987). Electroporation: Parameters affecting transfer of DNA into mammalian cells. *Anal. Biochem.* 164, 44–52.

Koda, K., and Glassy, M. C. (1990). In vitro immunization for the production of human monoclonal antibody. *Hum. Antibod. Hybridomas* 1, 15–22.

Kohler, G., and Milstein, C. (1975). Continuous cultures of fused cells secreting antibody of predefined specificity. *Nature (Lond.)* 256, 495–497.

Land, H., Parada, L. F., and Weinberg, R. A. (1983). Tumorigenic conversion of primary embryo fibroblasts requires at least two cooperating oncogenes. *Nature (Lond.)* 304, 596–602.

Langdon, W. Y., Harris, A. W., Cory, S., and Adams, J. M. (1986). The c-myc oncogene perturbs B lymphocyte development in Eu-myc transgenic mice. *Cell* 47, 11–18.

Mally, M. I. (1990). Genetic engineering of human lymphocytes for the production of monoclonal antibodies. *Hum. Antibod. Hybridomas* 1, 27–33.

Mally, M. I., and Glassy, M. C. (1990). Generating immortalized immunoglobulin-secreting human lymphocytes by recombinant DNA technology. *In* "Electromanipulation in Hybridoma Technology: A Laboratory Manual" (C. A. K. Borrebaeck and I. Hagen, ed.), pp. 71–88. Stockton Press, New York.

McCutchan, J. H., and Pagano, J. S. (1968). Enhancement of the infectivity of simian virus 40 deoxyribonucleic acid with diethylaminoethyl-dextran. *J. Natl. Cancer Inst.* 41, 351–357.

Miller, A. D., Jolly, D. J., Friedmann, T., and Verma, I. M. (1983). A transmissable retrovirus expressing human hypoxanthine phosphoribosyltransferase (HPRT): Gene transfer into cells obtained from human deficient in HPRT. *Proc. Natl. Acad. Sci. USA* 80, 4709–4713.

Miller, C. K., and Temin, H. M. (1983). High-efficiency ligation and recombination of DNA fragments by vertebrate cells. *Science* 220, 606–609.

Mulligan, R. C., and Berg, P. (1980). Expression of a bacterial gene in mammalian cells. *Science* 209, 1422–1427.

Mulligan, R. C., and Berg, P. (1981). Selection for animal cells that express the *Escherichia coli* gene coding for xanthine-guanine phosphoribosyltransferase. *Proc. Natl. Acad. Sci. USA* 78, 2072–2076.

Mulligan, R. C., Howard, B. H., and Berg, P. (1979). Synthesis of rabbit β-globin in cultured monkey kidney cells following infection with a SV40 β-globin recombinant genome. *Nature (Lond.)* 277, 108–114.

Neumann, E., Schaefer-Ridder, M., Wang, Y., and Hofschneider, P. H. (1982). Gene transfer into mouse myeloma cells by electroporation in high electric fields. *EMBO J.* 1, 841–845.

Oi, V. T., Morrison, S. L., Herzenberg, L. A., and Berg, P. (1983). Immunoglobulin gene expression in transformed lymphoid cells. *Proc. Natl. Acad. Sci. USA.* 80, 825–829.

Potter, H., Weir, L., and Leder, P. (1984). Enhancer-dependent expression of human κ immunoglobulin genes introduced into mouse pre-B lymphocyte by electroporation. *Proc. Natl. Acad. Sci. USA.* 81, 7161–7165.

Pratt, M., Mikhalev, A., and Glassy, M. C. (1987). The generation of Ig-secreting UC 729-6 derived human hybridomas by electrofusion. *Hybridoma* 6, 469–477.

Ptashne, M. (1988). How eukaryotic transcriptional activators work. *Nature (Lond.)* **335**, 683–689

Rassoulzadegan, M., Binetruy, B., and Cuzin, F. (1982). High frequency of gene transfer after fusion between bacteria and eukaryotic cells. *Nature (Lond.)* **295**, 257–259.

Rassoulzadegan, M., Naghashfar, Z., Cowie, A., Carr, A., Grisoni, M., Kamen, R., and Cuzin, F. (1983). Expression of the large T protein of polyoma virus promotes the establishment in culture of "normal" rodent fibroblast cell lines. *Proc. Natl. Acad. Sci. USA.* **80**, 4354–4358.

Rosen, A., Gergely, P., Jondal, M., Klein, G., and Britton, S. (1977). Polyclonal Ig production after Epstein-Barr virus infection of human lymphocytes in vitro. *Nature (Lond.)* **267**, 52–54.

Ruley, H. E. (1983). Adenovirus early region 1A enables viral and cellular transforming genes to transform primary cells in culture. *Nature (Lond.)* **304**, 602–606.

Sambrook, J., Fritsch, E. F., and Maniatis, T. (1989). "Molecular Cloning: A Laboratory Manual." Cold Spring Harbor Laboratory, Cold Spring Harbor, NY.

Sandri-Goldin, R. M., Goldin, A. L., Levine, M., and Glorioso, J. C. (1981). High-frequency transfer of cloned herpes simplex virus type 1 sequences to mammalian cells by protoplast fusion. *Mol. Cell. Biol.* **1**, 743–752.

Schaefer-Ridder, M., Wang, Y., and Hofschneider, P. H. (1982). Liposomes are gene carriers: Efficient transformation of mouse L cells by thymidine kinase gene. *Science* **215**, 166–168.

Schaffner, W. (1980). Direct transfer of cloned genes from bacteria to mammalian cells. *Proc. Natl. Acad. Sci. USA.* **77**, 2163–2167.

Schmidt, E. V., Pattengale, P. K., Weir, L., and Leder, P. (1988). Transgenic mice bearing the human c-myc gene activated by an immunoglobulin enhancer: A pre-B-cell lymphoma model. *Proc. Natl. Acad. Sci. USA.* **85**, 6047–6051.

Southern, P. J., and Berg, P. (1982). Transformation of mammalian cells to antibiotic resistance with a bacterial gene under control of the SV40 early region promoter. *J. Mol. Appl. Genet.* **1**, 327–341.

Stopper, H., Jones, H., and Zimmermann, U. (1987). Large scale transfection of mouse L-cells by electropermeabilization. *Biochim. Biophys. Acta* **900**, 38–44.

Toneguzzo, F., Hayday, A. C., and Keating, A. (1986). Electric field-mediated DNA transfer: Transient and stable gene expression in human and mouse lymphoid cells. *Mol. Cell. Biol.* **6**, 703–706.

Wigler, M., Silverstein, S. Lee, L-S., Pellicer, A., Cheng, Y-C., and Axel, R. (1977). Transfer of purified herpes virus thymidine kinase gene to cultured mouse cells. *Cell* **11**, 223–232.

Wong, T-K., and Neumann, E. (1982). Electric field mediated gene transfer. *Biochem. Biophys. Res. Commun.* **107**, 584–587.

Zimmermann, U. (1982). Electric field-mediated fusion and related electrical phenomena. *Biochim. Biophys. Acta* **694**, 227–277.

Zimmermann, U., and Vienken, J. (1982). Electric field-induced cell-to-cell fusion. *J. Memb. Biol.* **67**, 165–182.

Zurawski, V. R. Jr., Haber, E., and Black, P. M. (1978). Production of antibody to tetanus toxoid by continuous human lymphoblastoid cell lines. *Science* **199**, 1439–1441.

33

Human Hybridoma Formation by Hypo-Osmolar Electrofusion

S. M. Maseehur Rehman,[1] Susan Perkins,[1] Ulrich Zimmermann,[2] and Steven K. H. Foung[1]

[1]Department of Pathology, Stanford University School of Medicine
Stanford, California 94305

[2]Institute of Biotechnology, University of Wurzburg, Germany

I. Introduction

The ability to produce monoclonal antibodies and their applications have contributed greatly to virtually all fields of biomedical research. Most of the antibodies produced, however, have been of rodent origin, with three distinct limitations in their applications to human diseases. First, in some antigenic systems (e.g., major histocompatibility complex antigens), the majority of antigens produced when human cells are injected into a rodent are generally species-specific rather than specific for polymorphic determinants. This suggests an antibody response to different epitopes on complex antigens in humans and mice. This difference has also been observed with bacterial and viral antigens, such as *Mycobacterium tuberculosis* and rabies virus. Human monoclonal antibodies (HMAbs) identify distinct epitopes of *M. tuberculosis* or rabies virus antigens not previously recognized by murine antibodies [1, 2]. Another example is Rh antigens on human red cells, which are associated with hemolytic diseases of the newborn. In spite of many attempts, no murine monoclonal antibodies have been produced to Rh antigens. Second, a number of antibody-mediated effector functions are dependent on the Fc portion of Ig. For example, complement fixation and Fc binding to monocytes, macrophages, and natural killer cells are greater with human IgG_1 and IgG_3 than the IgG subclasses of rodent immunoglobulins [3, 4]. These properties are potentially important in identification of viral epitopes and responsible for eliciting antibody-dependent cell-mediated cytotoxicity (ADCC) or enhancement of viral entry into susceptible cells via Fc binding to Fc receptors [5, 6]. Last, rodent antibodies will elucidate a human anti-species antibody response when used in therapy. This can lead to neutralization of the desired effects of the antibodies and the induction of serum sickness [7]. Because of these effects, much effort has been devoted to "humanize" murine antibodies for therapy.

Technologies to generate HMAbs by their ability to isolate individual antibodies from large numbers of antibodies provide a useful tool to define the specificity of individual antibodies and to determine their functional and clinical significance. Human monoclonal antibody technology using a combination of B-cell selection and *in vitro* activation permits the isolation of virtually any antibody within the human B-cell repertoire. Recombinant DNA approaches may represent an alternative and efficient way to generate and select for immunoglobulins of a desire specificity. By generating random recombination of heavy and light chain genes, a greater diversity of antibodies than is represented on the repertoire of intact B cells is possible with these approaches [8]. To evaluate antibodies that are involved in human disease, however, the monoclonal antibody approach is the one that will reflect the antibodies that are encoded by human B cells *in vivo*. Therefore, the techniques described in this chapter present the ideal approach toward an evaluation of the role of antibodies in the pathogenesis or resolution of human disease.

Two approaches commonly used to immortalize human B cells are transformation by Epstein-Barr virus (EBV) and fusion of human B cells with appropriate murine, murine–human, or human lymphoblastoid cell lines [9, 10]. While these approaches have been successful, a persistent problem is the rarity of antigen-specific B cells in peripheral blood and lymphoid tissues. In contrast to rodents, deliberate immunization of humans with many antigens of interest is not practical. The problem is overcome partly by expanding the B-cell pool with EBV activation. Another solution is to develop methods capable of immortalizing relatively small numbers of antigen-specific B cells. This has been achieved using electric field-induced cell fusion or electrofusion [11, 12]. The placement of cells in an alternating electrical field leads to close membrane contact; subsequent delivery of a high-intensity direct-current pulse results in pore formation between cells, leading to cytoplasmic bridges and cell fusion. In comparison to polyethylene glycol-facilitated cell fusion, the efficiency in hybridoma formation is greater using electrofusion techniques [13]. Hybridoma formation efficiency of 10^{-3} has been achieved by electrofusion with a mouse–human heteromyeloma using as low as 10^6 input EBV-activated B cells. This is substantially higher than an efficiency of 10^{-5} to 10^{-4} associated with polyethylene glycol-induced cell fusions, which also require a higher number of input B cells.

Further enhancement of hybridoma formation efficiency has recently been achieved by electrofusion under hypo-osmolar conditions, allowing even a smaller number of EBV-activated B cells to be fused [14–16]. Microscopically, the cell surface area of membrane contact between two cells in an alternating electric field is markedly greater under these conditions. Another possible factor contributing to the increased efficiency is nuclear swelling, which may facilitate nuclear fusion in the hybridoma formation process. In a series of fusions between a mouse–human heteromyeloma and $1–3 \times 10^5$ EBV-activated B cells, hybridoma formation efficiencies of 10^{-5} to 10^{-4} were achieved under iso-osmolar conditions while efficiencies of 10^{-4} to 10^{-3} were achieved in hypo-osmolar conditions [16]. Under hypo-osmolar conditions, a tenfold decrease in the required number of input EBV-activated B cells is associated with a hybridoma formation efficiency equal to or greater than that achieved using a higher input cell number under iso-osmolar conditions. A critical factor in the development of hypo-osmolar fusion techniques is the duration of exposure of cells to these conditions. In our studies, prolonged exposure greater than 20–30 min led to a poor hybridoma formation efficiency. Other parameters that appear to affect hybridoma yield include the ratio of B cells to fusion partner, washing procedure, postfusion incubation time, and the elimination of molecules not normally toxic to intact cells (such as pH indicator) from the growth medium for the first 24 h after cell fusion.

The method in this chapter should serve only as a guide for investigators to develop protocols that will be successful with their unique conditions and reagents (e.g., fusion partners). In our experience, direct application of any method is difficult, and it should be used as a framework for other investigators in the field.

II. Materials

A. Reagents

1. Complete Growth Medium (10% FCS/IMDM)

> Iscove's modified Dulbecco's medium (IMDM) (Gibco)
> Fetal calf serum (FCS) (without antibiotics)
> L-Glutamine (100×, 29.2 mg/ml, Gibco)
> FM (100×): 0.45 g sodium pyruvate, 0.1 g bovine insulin, and 1.32 g *cis*-oxalacetic acid (Sigma). Dissolve in 100 ml double-distilled water at room

temperature with a stir bar (approximately 1 h); sterile filter; freeze in aliquots
at $-20°C$; freeze–thaw one time only.

2-Mercaptoethanol (2-ME, 1000 × stock, Biorad): In fume hood dilute 0.5 ml
of 2-ME into 6.6 ml double-distilled water; add 5 ml of this dilution to 95
ml double-distilled water; sterile filter; freeze in aliquots at $-20°C$; thaw
once, with subsequent storage at $4°C$.

2. HT/IMDM (To plate cells on the day of fusion only, the medium should contain
no pH indicator.)

Complete growth medium with 15% heat-inactivated FCS.

HT (100 ×): 0.776 g thymidine and 0.2772 g hypoxanthine (Sigma). Dissolve
in 200 ml double-distilled water at $70°C$, sterile filter; store in aliquots at
$-20°C$.

3. HAT/Ouabain 10^{-6} M

HT/IMDM

Aminopterin (1000 ×): 0.0176 g aminopterin (Sigma). Dissolve in 5 ml 1 N
NaOH; add 40 ml double-distilled water, adjust pH to 7.0–7.3; bring
volume to 50 ml; store in aliquots at $-20°C$.

Ouabain (100 ×): 0.0584 g ouabain (Sigma). Dissolve in 100 ml IMDM to
yield (1000 ×) stock; freeze in aliquots at $-20°C$; dilute (1 : 1000) just
prior to use. Ouabain is light sensitive, so medium should be covered with
aluminum foil.

4. Fusion Media: iso-osmolar (300L3) and hypo-osmolar (75L3):

Sorbitol (Merck 7759), iso-osmolar (300L3), 25.5 g (280mM); hypo-osmolar
(75L3), 6.38 g (70 mM)

Mg^{2+} acetate (Merck 5819), 53.6 mg (0.5 mM)

Ca^{2+} acetate (Merck 9325), 7.9 mg (0.1 mM)

Pure bovine serum albumin (Serva 11930), 0.5 g (1 mg/ml)

Dissolve dry ingredients in double distilled water; bring to 500 ml total volume,
sterile filter.

B. Instrumentation

The Z1000 Zimmermann cell fusion system was used to develop this methodology
(GCA/Precision Science Group, Chicago, IL). This system is, however, no longer
available. Another system successfully adapted for this electrofusion technique has

been developed by the Institute of Biotechnology, University of Wurzburg, manufactured and distributed by Biomed, Theres, FRG.

C. Miscellaneous Equipment

1. Helical fusion chambers: An electrode assembly of two platinum wire 200 μm in diameter wound in a helix with 200 μm spacing inside a receptacle to hold the cells (GCA).
2. Sterile hood.
3. Incubator for cells: 37°C; 6–$6\frac{1}{2}$% CO_2.
4. Microscope.
5. Room-temperature centrifuge.
6. Sterile plastic serologic pipets (Corning).
7. Pipetman: P20 (1–20 μl volume), P200 (20–200 μl volume), P1000 (200–1000 μl volume), Rainin; tips, Robbins Scientific.
8. Sterile Pasteur pipets (VWR).
9. Sterile tubes (Corning): 12 × 75 culture tubes and 15-ml conical centrifuge tubes.
10. Racks to hold 12 × 75 tubes and 15-ml conical tubes. Styrofoam racks that are sold containing the 15-ml conical tubes work well to support the helical chambers, and help cushion the cell inside from rough movements.
11. Sterile 96-well flat-bottom microtiter trays (Lindbro).
12. 0.02% Trypan blue (Gibco).

D. Cells

1. K_6H_6/B5. This heteromyeloma [17] is maintained in a horizontal flask in 10% FCS/IMDM and expanded by splitting a relatively dense cell concentration (approximately 5 × 10^5/ml) with 1–4 times the volume of fresh medium every 3 days. Change flasks once a week. Cells are fused the day after feeding.
2. EBV activation. The B cells are washed in 10% FCS/IMDM and resuspended in 30% FCS/IMDM containing 5–35% (v/v) supernatant from the marmoset line B95-8 as a source of EBV. Cells are plated in U-bottom 96-well microtiter trays (Lindbro) at 10^4 cells/well and fed 30% FCS/IMDM until the medium is spent enough to be assayed for activity.

III. Procedure

This procedure is for helical chamber fusions with K_6H_6/B5 and EBV-activated B cells in hypo-osmolar (75 L3) fusion medium.

When EBV-activating human B cells for the purpose of producing secreting hybrids, cells are activated at 10^4 B cells/well, and the wells contain approximately $1–3 \times 10^5$ EBV-activated B cells by the time the supernatant is spent enough to test for antibody activity. This fusion procedure was developed to fuse one well of EBV-activated cells immediately after the detection of antibody, without having to expand to increase cell number.

Because the fusions are performed with a very small number of cells and the EBV are sensitive to the hypo-osmolar medium, attention to detail is required to minimize cell loss. Timing is critical, so all supplies, equipment, and reagents must be prepared ahead of time and be available for immediate use. Adhere to the time limits outlined in the procedure. Because timing is so critical only a limited number of fusions can be performed at one time. Minimize the length of time the cells spend in suboptimal conditions, such as cell pellets at the end of a wash. Resuspend cells gently and avoid air bubbles. Since many cells will adhere to glass, use plastic serologic pipets or tips to manipulate the cells.

The K_6H_6/B5 and the EBV are pooled, washed once in iso-osmolar medium, and then fused in helical fusion chambers in hypo-osmolar medium at low DC voltage. The fused cells are washed out of the chambers and plated in medium containing no pH indicator. All steps are performed at room temperature using sterile technique.

A. Preparation

1. Sterilize helical fusion chambers.
2. Turn on and program the power supply:

> Duty cycle: 100%
> Alignment voltage limit: 50
> Preset alignment: 30 s
> Alignment frequency: 1000 kHz
> Alignment voltage: Set knob at 0. It will be turned up to 6 V just before fusion.
> DC fusion voltage: 20 V
> Fusion pulse duration: 15 μs
> Alignment off time between pulses: 10 ms
> Number of fusion pulses: 3
> Time between fusion pulses: 1.0 s

3. Because the number of EBV-activated B cells available is very small and it is undesirable to lose a significant portion of them with a cell count, an estimation of cell number is made in one of two ways:

    a. A cell count from the well of cells to be fused using 10 μl of suspended cells in 1 ml trypan blue:

      i. Estimate the volume of medium contained in the well by measuring the volume contained in a neighboring well of no interest.

      ii. Gently mix the cells to be fused with a P200 prior to removing 10 μl with a P20 for the cell count.

    b. Count cells contained in a different well of the same stimulation that visually has about the same cell density, using 50 μl of cells in 450 μl trypan blue:

      i. Measure the volume in the well of cells to be counted.

      ii. Mix cells to count gently with a P200 prior to taking 50-μl sample.

4. After calculating the estimated number of EBV-activated cells, calculate the number of $K_6H_6/B5$ needed for the fusion (2 $K_6H_6/B5$: 1 EBV) and the final volume of HT/IMDM containing no pH indicator needed to plate it. Cells are plated in microtiter trays at $5 \times 10^3 - 10^4$ cells/well in a volume of approximately 120 μl/well.

5. Label 96-well flat-bottom microtiter trays for the fusions and plate IMDM containing 100 μ/ml penicillin and 100 μg/ml streptomycin (Gibco) in the outer wells to prevent evaporation of the medium containing the fused cells. Three wells should be designated for EBV controls, and three wells for $K_6H_6/B5$ controls. Place the trays in the 5% CO_2 incubator until they are needed to plate cells.

6. Place calculated volume of HT/IMDM without pH indicator needed to plate the fusion in a conical centrifuge tube.

7. Label and fill any other tubes that will assist in keeping the work flow smooth once the fusion process has begun.

B. Fusion Procedure

1. Harvest $K_6H_6/B5$ to a conical centrifuge tube. Mix cells well. Add 50 μl cell suspension to 450 μl trypan blue for cell count. Screw cap tightly on conical tube.

2. Count cells. Calculate the volume of cell suspension needed to fuse to the well of EBV activated cells at 2 $K_6H_6/B5$: 1 EBV ratio.

3. Add the calculated volume of $K_6H_6/B5$ to a 15-ml conical tube for the fusion. Add approximately 0.4 ml to a second 15-ml conical tube to use a control.

4. Use EBV-activated cells from a different well of the same stimulation as an EBV control, preferably from a well containing cells similar in size, state of activation, and secreting a small amount of the antibody. Mix the cells in the well and transfer them to a third 15-ml conical tube.

5. Use a P200 to harvest the well of EBV-activated cells for fusion and add the

cells to the conical containing $K_6H_6/B5$ for fusion. Rinse the well several times with 10% FCS/IMDM.

6. Centrifuge the three conical tubes at 250–300 × g for 10 min. Aspirate the supernatant. Set the controls to one side. Screw caps on tightly.

7. Wash the cells for fusion by resuspending them in 300 L3 with a P1000 to 10^6 cells/ml. Underlayer the cell suspension with an equal volume of 300 L3 with the P1000. Using the same pipet tip for underlayering will prevent some cell loss. Centrifuge at 250–300 × g for 10 min.

8. While the fusions are washing, dilute and plate the control cells: Add 1 ml HT/IMDM without pH indicator to each control tube. Resuspend the cells and plate them at 2 drops/well with a 5-ml plastic serologic pipet (3 wells of each population). Discard the remaining control cells.

9. Aspirate the supernatant from the washed cells for fusion.

10. One fusion at a time:

   a. Unwrap an autoclaved chamber and place the parts in a styrofoam rack: the electrode assembly with the helix up and electrodes to connect the cable down, and the receptacle with the open end up.

   b. Resuspend the cell pellet with 170 µl 75L3 with a P200 and transfer all of the cell suspension from the conical tube into the very bottom of the receptacle. Lift the receptacle out of the racks while you add the cells, so that you can see that they end up on the bottom rather than along one side. Avoid creating air bubbles.

   c. Hold the receptacle in one hand and lift the electrode assembly out of the rack with the other. Invert the electrode assembly over the receptacle and insert it slowly inside the receptacle with a twisting motion, as if there were screw threads. Once the helix has touched the cell suspension, be sure to keep moving it downward while you twist to prevent air bubbles from forming.

   d. Once the electrode assembly is firmly seated in the receptacle, wrap parafilm around their interface and return the chamber to the rack with the receptacle down and the electrodes sticking up. If performing more than one fusion, all chambers should be filled before the actual fusion process begins. Place them in a styrofoam rack with sufficient space between the chambers. It is important that they not be moved once the fusion process has begun.

11. Take the rack containing the chamber to the power supply. The cells must be fused within 5–15 min of being placed in the hypo-osmolar fusion medium.

   a. Plug the connector cables leading from the power supply into the electrodes. Make sure they are firmly attached.

   b. Push the alignment button on.

c. Quickly turn the alignment voltage adjust knob until the alignment voltage meter reads 6 V.

d. After 30 s has elapsed on the process timer, push the DC cell fusion pulse trigger. With 30 s preset alignment, the alignment voltage will gradually decrease to zero over the next 30 s.

e. After the alignment has shut off, disconnect the cables.

12. At 10–15 min after the cells have been fused, wash them out of the chamber with the HT/IMDM containing no pH indicator premeasured for plating.

a. Remove the helical assembly from the receptacle with a twisting motion. Replace the receptacle in the rack, but continue to hold the helical assembly with the electrodes used to connect the cable up. There are cells in both the receptacle and the helix that need to be recovered.

b. With a P1000 set at 500 µl wash the cells out of the helix with HT/IMDM (no indicator) into the receptacle. Hold the helical assembly at an angle over the receptacle so the liquid will run down the helix and drop into the receptacle. The tip of the pipetman should travel in parallel with the helical electrodes up and down the side facing you, and the helix should then be rotated in order to reach all surfaces. Use the same pipet tip and do two 500-µl washes in this manner—cover the helix thoroughly with the first 500-µl wash and then quickly with the second wash as a rinse. If more than one fusion is being performed, all cells should be washed out of the helixes with plating medium into the receptacles before performing the next steps.

13. Gently pipet up and down in the receptacle with a 1-ml plastic serologic pipet and transfer the cells back into the conical tube containing the premeasured HT/IMDM without pH indicator. Rinse the receptacle with the same medium.

14. Plate in the prepared 96-well microtiter trays at two drops/well with a 5-ml plastic serologic pipet.

C. Hybrid Feeding Schedule

Day 1: Cells fused and plated in HT/IMDM with no pH indicator.
Day 2: Cells fed an equal volume of HAT/ouabain 10^{-6} M (2 drops/well with 5 ml serologic pipet).
Day 6: Cells fed HAT/ouabain 10^{-6} M (one-third to one-half of media removed; fed 1 drop/well with 10 ml serologic pipet.
Day 10: Cells fed HAT/ouabain 10^{-6} M.
Day 14: If all cells in the control wells are dead, cells should now be fed with HT.

At this point, some hybrids are visible macroscopically. Continue feeding twice weekly until supernatants are sufficiently spent from hybrid growth for assay.

Acknowledgments

The authors wish to thank Judy Campbell, who typed this manuscript. This study was supported in part by grants HL33811, AI22557, AI26031, and DA06596 from the National Institutes of Health to Steven K. H. Foung and grants of the Deutsche Forschungsegmeinschaft (SFB 176) and of the Federal Ministry of Research and Technology, FRG (DFVLR 01QV354), to Ulrich Zimmermann.

References

1. Dietzschold, B., Gore, M., Casali, P., Ueki, Y., Rupprecht, C. E., Notkins, A. L., and Koprowskik, H. (1990). Biological characterization of human monoclonal antibodies to human monoclonal antibodies. *J. Virol.* **64**, 3087.
2. Wallis, R. S., Alde, S. L. M., Havlir, D. V., Amir-Tahmasiek, M. H., Daniel, T. M., and Ellner, J. J. (1989). Identification of antigens of *Mycobacterium tuberculosis* using human monoclonal antibodies. *J. Clin. Invest.* **84**, 214.
3. Steplewski, Z., Sun, L., Shearman, C., *et al.* (1988). Biological activity of human-mouse IgG_1 hr, IgG_2 hr, IgG_3 and IgG_4 chimeric monoclonal antibodies with antitumor specificity. *Proc. Natl. Acad. Sci. USA* **85**, 4852.
4. Lubeck, M. D., Steplewski, Z., Baglia, F., Klein, M. H., Dorrington, K. J., and Koprowski, H. (1985). The interaction of murine IgG subclass proteins with human monocyte Fc receptors. *J. Immunol.* **135**, 1299.
5. Inkeless, J. (1989). Function and heterogeneity of human Fc receptors for immunoglobulin G. *J. Clin. Invest.* **83**, 355.
6. Takeda, A., Tuazon, C., and Ennis, F. (1988). Antibody enhanced infection by HIV-1 via Fc receptor-mediated entry. *Science* **242**, 580.
7. Miller, R., Maloney, D., McKillop, J., and Levy, R. (1981). In vivo effects of murine hybridoma monoclonal antibody in a patient with T-cell leukemia. *Blood* **58**, 78.
8. Huse, W. D., Sastry, L., Iverson S. A., Kang, A. S., Alting-Mees, M., Burton, D. R., Benkovic S. J., and Lerner, R. A. (1989). Generation of a large combinatorial library of the immunoglobulin repertoire in phage lambda. *Science* **246**, 1275.
9. Engleman, E. G., Foung, S. K. H., Larrick, J. A., and Raubitschek, A., eds. (1985). *"Human Hybridomas and Monoclonal Antibodies."* Plenum Press, New York.
10. James, K., and Bell, G. T. (1987). Human monoclonal antibody production. *J. Immunol. Methods* **100**, 5.
11. Zimmermann, U. (1986). Electrical breakdown, electropermeability and electrofusion. *Rev. Physiol. Biochem. Pharmacol.* **105**, 175.
12. Zimmermann, U. Electrofusion of cells. *In* "Methods of Hybridoma Formation" (A. H. Bartal and Y. Hirshant, eds.), Humana Press, Clifton, NJ, p. 97.
13. Foung, S. K. H., and Perkins, S. (1989). Electric field-induced cell fusion and human monoclonal antibodies. *J. Immunol. Methods.* **116**, 117.

14. Schmitt, J. J., and Zimmermann, U. (1989). Enhanced hybridoma production by electrofusion in strongly hypo-osmolar solutions. *Biochem. Biophys. Acta* **983,** 42.

15. Zimmermann, U., Gessner, P., Schnettler, R., Perkins, S., and Foung, S. K. H. (1990). Efficient hybridization of mouse-human cell lines by means of hypo-osmolar electrofusion. *J. Immunol. Methods* (in press).

16. Foung, S. K. H., Perkins, S., Kafadar, K., Gessner, P., and Zimmermann, U. (1990). Development of microfusion techniques to generate human hybridomas. *J. Immunol. Methods* (in press).

17. Carroll, W. L., Thielemans, K., Dilley, J., and Levy, R. (1986). Mouse X human heterohybridomas as fusion partners with human B cell tumors. *J. Immunol. Methods* **89,** 61.

34

Electrically Induced Fusion and Activation in Nuclear Transplant Embryos

James M. Robl, Philippe Collas, Rafael Fissore, and John Dobrinsky

Paige Laboratory, University of Massachusetts, Amherst, Massachusetts 01003

I. Introduction

Cloning of mammalian embryos by nuclear transplantation involves the fusion of a cell from a donor embryo containing 8–64 cells to an enucleated, mature oocyte. During normal fertilization the sperm not only contributes genetic material to the oocyte, it also "activates" the oocyte to continue its progression through the cell cycle and initiate cleavage. In nuclear transplantation the oocyte must be activated artificially. Following fusion and activation the transplanted nucleus undergoes a

Guide to Electroporation and Electrofusion

complete morphological remodeling, which includes a dramatic increase in size (Stice and Robl, 1988). This morphological change in the nucleus is believed to be associated with a functional reprogramming and results in a nucleus that is functionally and morphologically equivalent to a pronucleus and capable of supporting development to term (Stice and Robl, 1988).

Because the objective of cloning is to multiply embryos, efficiency is of utmost importance. Some of the earliest work in nuclear transplantation was done using an invasive microsurgical technique (Illmensee and Hoppe, 1981). A nucleus obtained by rupturing a cell and devoid of surrounding plasma membrane was aspirated into a micropipet. The pipet was then inserted through the plasma membrane and into the cytoplasm of a recipient cell. Because the plasma membranes of both the donor and recipient cells were ruptured, survival rates were very low. McGrath and Solter (1983) devised a nondisruptive method of transplanting nuclei that involved cell fusion. With this method enucleation was accomplished by pulling the genetic material into the pipet in a plasma membrane-bounded vesicle. The vesicle was pinched off from the recipient cell without rupturing the plasma membrane. A donor nucleus could be obtained using a similar procedure, then placed adjacent to the recipient cell. This early work on nuclear transplantation was done using mouse embryos and Sendai virus was the fusigenic agent. Subsequent work on nuclear transplantation indicated that Sendai virus was not highly fusigenic in other species (Robl and First, 1985; Willadsen, 1986; Robl *et al.*, 1987). Furthermore, other fusigenic agents such as polyethylene glycol or liposomes did not prove useful with this system (J. M. Robl, unpublished observations). Electrically induced fusion, however, has proved highly effective in all species tested (Table 1).

A fundamental aspect of cloning is the reprogramming of the donor nucleus as a result of activation of the recipient oocyte. Activation of the oocyte can be accomplished by various chemical or physical treatments including alcohols, anesthetics, calcium injection, calcium ionophore, calcium-free medium, protein synthesis inhibitors, hyaluronidase, phorbol ester, cold shock, heat shock, and electrical shock (Whittingham, 1980; Cutherbertson and Cobbold, 1985). Electrical shock, although not necessarily more effective than other treatments, is convenient and easier to manipulate. Because an electrical pulse can be used both to fuse and to activate nuclear transplant embryos, electrical fusion and activation are essentially the only methods being used in cloning work.

Cloning is now being investigated in many laboratories throughout the world, both industrial and academic. In the future it will be important both as a basic research tool and for the multiplication of genetically superior farm animals. Its primary limitation is its efficiency. Although progress in refining electrical fusion and activation techniques has been impressive, more work needs to be done. This work should focus not only on improving efficiency but also on understanding the basic processes mediating electrical fusion and activation. The purpose of this chapter

Table 1
Parameters for Electrical Fusion in Mammalian Embryos

Species	Cell types fused	Field strength (kV/cm)	Pulse duration (μs)	Number of pulses	Alignment	Maximum fusion (%)	Authors
Sheep	16-cell/oocyte	0.75	100	3	AC	~90	Willadsen (1986)
	16-cell/oocyte	1.25	80	1	AC	90	Smith and Wilmut (1989)
Cow	4-8-cell/zygote	1.0	40	1	AC	78	Robl et al. (1987)
	2-8-cell/oocyte	1.1	30	1	Manual	75	Prather et al. (1987)
	32-64-cell/oocyte	0.75	50	3	AC	~70	Bondioli et al. (1990)
Pig	2-8-cell/oocyte	1.2	30	1	AC	~80	Prather et al. (1989)
	2-cell/2-cell	0.7-0.79	70	2	AC	75	Clement-Sengewald (1988)
Rabbit	2-cell/2-cell	0.70-0.79	50	1	AC	56	Clement-Sengewald (1988)
	2-cell/2-cell	2.5	35	1	Manual	90	Ozil and Modlinski (1986)
	8-cell/oocyte	3.2	60	1	Manual	84	Stice and Robl (1988)
	8-cell/oocyte	3.6	60	3	Manual	94	Collas and Robl (1990)
Mouse	2-cell/2-cell	0.60-0.89	70	1	AC	85	Clement-Sengewald (1988)
	2-cell/2-cell	1.0	60	1	AC	80	Smith (1989)
	2-cell/2-cell	1.0	250	2	Manual	88	Kubiak and Tarkowski (1985)
Rat	PN-8-cell/zygote	1.5	200	2	AC	81-92	Kono et al. (1988)

is to describe the method of cloning by nuclear transplantation with special emphasis and discussion on electrical fusion and activation.

II. Materials

A. Oocytes and Embryos

Recipient oocytes and donor embryos are collected from the oviducts of mature mixed-breed female rabbits superovulated with six consecutive subcutaneous injections of 0.3 mg follicle-stimulating hormone (FSH; Burnes Biotech, Omaha, NE) given 12 h apart and 75 IU human chorionic gonadotropin (hCG; Sigma, St. Louis, MO) given 12 h after the last FSH injection.

B. Media and Reagents

Oocytes and embryos are flushed, manipulated, and cultured in an Earle's balanced salt solution (EBSS) supplemented with 10% fetal calf serum (FCS). EBSS has a bicarbonate buffer; therefore it is always layered with oil and pre-equilibrated in a 5% CO_2 incubator for at least 12 h before use. Following equilibration the pH will be maintained for at least an hour outside the incubator. The manipulation and postfusion medium is EBSS/10% FCS with 7.5 µg/ml cytochalasin B (CCB; Sigma). The zona pellucida is removed from donor embryos in Dulbecco's phosphate-buffered saline (DPBS) acidified with HCl to pH 2.5. Blastomeres are separated in calcium- and magnesium-free DPBS with 0.25% trypsin (ZDPBS; Sigma). The fusion medium (FM) is 0.30 M mannitol containing 100 µM $CaCl_2$ and 100 µM $MgCl_2$.

C. Micromanipulation Equipment

Micromanipulation is done on a Nikon Diaphot microscope with Narishige hydraulic micromanipulators.

D. Electrical Pulse Generator

Electrical pulses are given using a BTX Electro Cell Manipulator 200 (BTX Inc., San Diego, CA) and monitored with a BTX Optimizor–Graphic Pulse Analyzer.

The chamber for pulsing embryos consists of two 0.5-mm stainless steel wire electrodes mounted 0.5 mm apart on a glass microscope slide.

III. Procedures

A. Micromanipulation

1. General

Methods for manufacturing glass micropipets and setting up the micromanipulators have been discussed in detail elsewhere (Robl, 1988) and will not be covered here. The procedure described below is for rabbit embryos; however, the general procedures are similar for any species.

2. Donor Blastomere Isolation

Donor embryos at the 32-cell stage are flushed from the oviduct at approximately 60 h following mating and injection with hCG. At this time the rabbit embryo has a thick mucin coat over the zona pellucida (other species do not have a mucin layer). To isolate the embryo the zona pellucida is dissolved by incubation in acidified DPBS for approximately 30 s. The mucin coat is then torn and the embryo is expelled using two finely drawn glass needles. The zona-free embryo is then placed in ZDPBS for 5–15 min and the blastomeres are separated by repeated aspiration of the embryo into a 50-μm fire-polished pipet.

3. Recipient Oocyte Preparation

Recipient oocytes are recovered 14–15 h after hCG administration. The cumulus cells are removed using 1 mg/ml bovine testis hyaluronidase (Sigma), and corona cells are removed by gentle pipetting with a small-bore pipet.

4. Nuclear Transplantation

Oocytes and isolated blastomeres are transferred to 50-μl drop of EBSS/10% FCS with CCB under oil in a 100 × 15 mm plastic petri dish and incubated for 15 min before manipulation. The recipient oocyte is first enucleated (Fig. 1A, B, and C). Enucleation entails removing the first polar body and a portion of membrane bounded cytoplasm using a 30-μm beveled and sharpened pipet. Enucleation is assessed by visualization of the metaphase chromosomes in the pipet (Fig. 1C). A blastomere is then aspirated into the same pipet and placed into the perivitelline space and tightly apposed to the enucleated oocyte (Fig. 1D, E, and F).

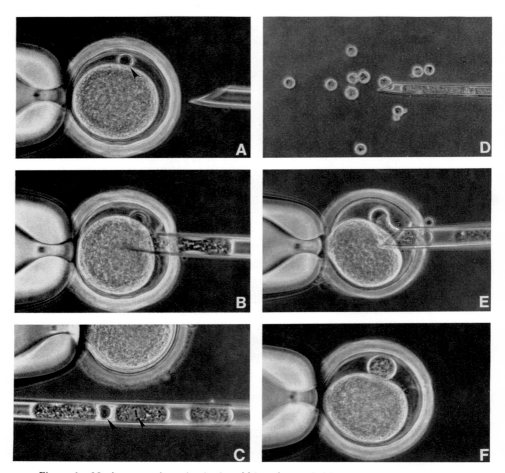

Figure 1   Nuclear transplantation in the rabbit embryo. (A) Mature oocyte with single polar body (arrow). Chromosomes lie beneath the polar body. (B) Aspiration of chromosomes in a cytoplasmic vesicle. (C) Chromosomes (small arrow) and polar body (large arrow) in the enucleation pipet. (D) Aspiration of an isolated donor blastomere from an eight-cell embryo. (E) Insertion of an elongated donor blastomere into the perivitelline space of the recipient oocyte. (F) Enucleated oocyte and adjacent donor blastomere.

Figure 2   Apparatus for alignment and electrical fusion of nuclear transplant embryos. (A) Electrical pulsing instrument attached to a fusion chamber mounted on a dissecting microscope. (B) Fusion chamber consisting of two wire electrodes mounted on a glass slide. Glass slide is placed in a 100-mm plastic petri dish and overlaid with 30 ml of fusion medium. (C) Cells are manually aligned between the electrodes by fluid flow from a large bore pipet.

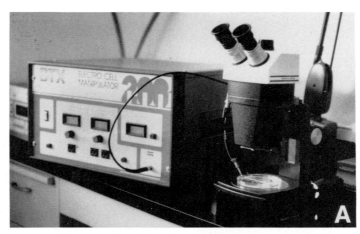

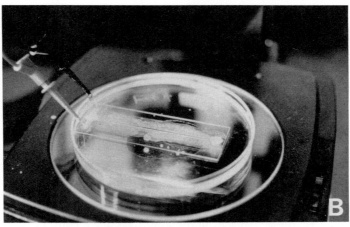

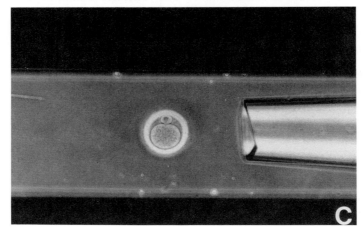

B. Electrical Pulsing

1. *Chamber Setup*

The fusion chamber is placed in a 100 × 15mm plastic petri dish and secured to the bottom with stopcock grease (Fig. 2A and B). The chamber is then overlaid with approximately 30 ml of FM. The petri dish with fusion chamber is secured to a dissecting microscope with a warming plate (Fryer Co., Carpentersville, IL). The FM is allowed to reach a temperature of 30°C before being used for fusion. A second dissecting microscope is used for handling embryos between pulses.

2. *Embryo Equilibration*

Because FM has a higher density than EBSS, embryos initially float to the surface. After being washed thoroughly they will remain on the bottom and can be transferred to the fusion chamber.

3. *Alignment*

Prior to pulsing, cells are manually aligned in the chamber so that the membranes to be fused are parallel to the electrodes (Fig. 2C). This is done using a glass embryo-handling pipet.

4. *Pulsing*

Embryos are given three square DC pulses of 3.6 kV/cm for 60 μs each, each pulse 30 min apart. Fusion generally occurs before the second pulse, and fused embryos are not aligned for the second and third pulse. All embryos remain in FM for the entire 1-h period.

5. *Monitoring Nuclear Transplant Embryos*

Embryos are monitored for fusion, lysis, fragmentation, presence of a swollen nucleus and cleavage. Fusion generally occurs within 30 min (Fig. 3) and is verified before the second and third pulses. Lysis is the irreversible breakdown of the plasma membrane. Embryos that undergo lysis appear granular and fill the entire zona pellucida. Lysis can take place in one cell or the other and can occur either before or up to 6 or 12 h after fusion. Fragmentation is a nonuniform cleavage. Many fragmented cells are anucleate. Fragmentation often occurs within 3 h of the pulse and is more common in aged oocytes. Approximately 18 h after the pulses, embryos may show only a swollen nucleus or appear as normal two- or four-cell embryos with uniform blastomeres. Embryos with swollen nuclei may or may not eventually

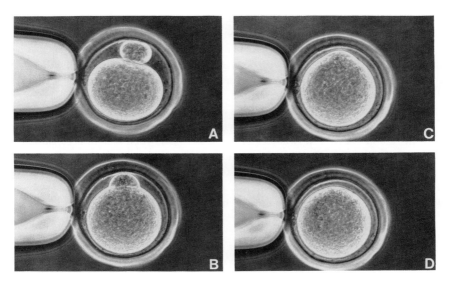

Figure 3 Fusion in nuclear transplant embryos. (A) Pulsed embryo prior to fusion. (B) Fusion between the oocyte and donor blastomere has just been initiated. The time required for initiation of fusion varies with the age of the oocyte from 10 to 25 min. (C) Rounding-off of donor blastomere following fusion. This process occurs within 2–3 min following the initiation of fusion (D) Completely fused and rounded-off cells.

divide but are probably not normal. Two-cell and four-cell embryos are each capable of giving rise to offspring (Collas and Robl, 1990).

bcFrom our current work (Collas and Robl, 1990) the results that might be expected using this procedure with 8- to 16-cell donors are the following: fusion, 78% (180/230); activation, 72% (130/180); development to the two- to four-cell stage, 85% (110/130); and development to term, 21% (23/110).

IV. Discussion

A. Fusion

1. Mechanism of Fusion

Although many models have been proposed as to the detailed mechanism of electrofusion, all involve pore formation. Evidence using erythrocyte membranes indicates that two classes of pores must be considered (Sowers and Lieber, 1986).

Large pores, reaching a radius of at least 8.4 nm, open, but immediately reclose within 100–200 ms. A second class of pores, with a radius of about 0.5 nm, stay open for an indefinite period of time. These small pores may or may not be due to incomplete reclosing of the large pores. Although it has been generally thought that the large pores are the primary mediators of fusion, evidence obtained by delaying cell contact for a 15-s interval after the pulse indicates otherwise (Sowers, 1989). In this study erythrocyte membranes were able to fuse when contact was initiated at a time when, presumably, only small pores were present. Our work with the rabbit oocyte supports this evidence. We have been able to detect uptake of fluoresceinated dextran (4000 Da) in only a small portion of oocytes pulsed at 3.6 kV/60 μs, which results in close to 100% fusion. Fusion is presumably mediated by pores smaller than would allow uptake of the 4000-Da fluorescein dextran, which would classify them as small pores. We have not yet investigated the presence of these small pores.

2. Alignment and Cell Contact

Alignment of the cells such that the membranes to be fused are perpendicular to the electrical field is necessary to achieve high rates of fusion. In one study with two-cell mouse embryos, 92% (22/24) fused when the cells were aligned prior to the fusion pulse and 68% (13/19) when the cells were randomly aligned (Robl *et al.*, 1987). Alignment can be done either manually or using an AC field. A retrospective analysis by Smith and Wilmut (1989) with sheep embryos indicated that AC alignment resulted in higher fusion than manual alignment. Our experience indicates that AC alignment is effective; however, problems can arise with embryos that are sticky or in which the blastomere is very small in comparison to the size of the oocyte. In the later case there is no distinct axis for alignment. Although both manual and AC alignment can be used, we have found no advantage to this over manual alignment alone.

The AC frequency, voltage, and duration have not been extensively investigated nor precisely defined. A frequency of 600–1000 kHz, a voltage of 5–6V, and a duration of 5–10 s is effective in most species. Because the alignment process is monitored under a microscope the voltage and duration can be adjusted to give a maximum effect for each set of embryos. Caution must be observed in applying the AC for longer than 20 s (Clement-Sengewald, 1988) and at higher than 10 V as this will cause a reduction in survival rate.

With other cells, AC is used to induce cell contact along with alignment. In nuclear transplantation the confined space of the zona pellucida ensures contact between the cells if they are sufficiently large. Physical contact is maintained in rabbit nuclear transplant embryos when a minimal amount of cytoplasm is removed for enucleation and a blastomere the size of a 32-cell stage or larger is used. When smaller blastomeres from embryos beyond the 32-cell stage are used, contact is maintained primarily by the blastomere sticking to the oocyte surface. In cases in which the blastomere does not stick, contact can be induced by an AC field; however, fusion rates are generally low in this situation because contact is lost at the cessation of the AC field. In general, fusion rates decline with smaller blastomeres (Prather et al., 1987).

3. Fusion Medium

Fusion is generally done in a 0.30 M mannitol solution containing 100 μM MgSO$_4$ and 50–100 μM CaCl$_2$ in water. A small amount of BSA (0.01 g/l) can be added to reduce the stickiness of the zona pellucida. This is important when using in vitro matured oocytes. The concentration of mannitol can be adjusted if the embryos are found to shrink or swell in the medium. Furthermore, sucrose or glucose can be substituted for mannitol.

The original reason for using a low-conductivity, sugar-based medium (conductivity of a mannitol fusion solution is approximately 34 times less than PBS) was to reduce heating caused by exposure during alignment to a continuous AC. Whether or not significant heating occurs is subject to question (Sowers, 1989). If AC alignment is not used in nuclear transplantation, the use of low-conductivity medium should not be important. However, the type of medium may also affect fusion and activation rate. Studies in the mouse (Kubiak and Tarkowski, 1985; Clement-Sengewald, 1988) indicate that high fusion rates can be achieved in ionic medium. However, a small study by Robl et al. (1987) with cow oocytes showed a dramatic reduction in fusion rate in ionic medium (1/9) as opposed to a low-conductivity medium (8/9). The difference in results between these studies may simply be that optimum pulse parameters may be different for the two types of medium, and this was not tested in the latter study. Little work has been done beyond these studies to investigate the importance of specific media components in fusion.

4. Field Strength and Pulse Duration

Reported pulse voltages and durations vary greatly between species and investigators (Table 1). Most success has been achieved with settings of about 1 kV/cm and 60 μs duration. One might expect that fusion rates could be increased by optimizing pulse voltage and duration to a point where lysis would cause a reduction in viability. This is generally true; however, the results of work by Smith (1989) and Clement-Sengewald (1988) indicate that fusion rate declines before dramatic increases in lysis occur. This indicates a possible transient destabilization of the membrane that is not conducive to fusion at a high voltage or long duration. Furthermore, there is a decline in viability, other than what is associated with lysis at high voltages or long duration.

5. Multiple Pulses

Multiple consecutive pulses (pulses given in immediate succession) are frequently used in electrofusion protocols. Knight (1981) has proposed that successive pulses would, at least initially, increase the number of pores because of slight shifts in the cells between pulses. It would be expected that successive pulses would therefore be beneficial for fusion. In several large studies with two-cell embryos Clement-Sengewald (1988) found that maximum fusion rates occurred with one pulse in the mouse, seven pulses in the rabbit, and two pulses in the pig. One explanation for the species variation may be that the rabbit is more resistant to electrical pulses and with a relatively low voltage multiple pulses may be of benefit by increasing the number of pores formed. In our studies with the rabbit we have used a much higher field strength than with other species to achieve high fusion rates (Stice and Robl, 1988; Collas and Robl, 1990). Increasing field strength may increase pore numbers just as multiple pulses would.

In our hands, consecutive pulses have not improved fusion efficiencies in the rabbit; however, fusion rates are improved when nonfused cells are repeatedly re-aligned and pulsed. With our current system (Collas and Robl, 1990), nuclear transplant embryos are given three pulses 30 min apart. Nonfused cells are realigned with each successive pulse. This procedure improved fusion rates from 79% (45/57) for a single pulse to 94% (67/71) for three pulses.

6. Disparity in Blastomere Size

Electropore formation is related to transmembrane potential. Transmembrane field strength induced by a specific applied field strength is directly related to the radius of the cells to be fused (Sowers, 1989). In other words, large cells require a lower field strength for pore formation than small cells. The critical field strength for an advanced donor cell theoretically should be up to eight times that for the oocyte recipient, considering differences in diameter. Although the recipient oocyte does lyse more frequently than the donor blastomere, the disparity in response between the two cell types to an electrical pulse is certainly not what is theoretically predicted. To date there is no evidence to indicate that the disparity in cell size affects fusion rate.

7. Oocyte Age

In the rabbit (P. Collas and J. M. Robl, unpublished observation) and the cow (K. Endo and J. M. Robl, unpublished observations), oocyte aging is associated with a decline in the rate of fusion and an increase in the time required to complete fusion. In the cow, 71% (39/55) of oocytes matured for 24 h fused with a single pulse whereas 40% (23/57) of oocytes matured for 42 h fused. In the rabbit, 100%

(31/31) 18 h post-hCG oocytes fused within 16 min of the pulse, whereas 83% (24/29) of 24 h post-hCG oocytes fused, and of those that fused, fusion required 40 min. Interestingly, fertilized embryos of the same age (24 h post-hCG) did not fuse at all (0/22) in response to a similar pulse. One difference between the fertilized embryo and the oocyte is that cortical granule exocytosis occurs as a result of fertilization. The plasma membrane may be altered by the fusion of the cortical granule membrane or by the action of the cortical granule contents. The reduction in fusion efficiency in response to oocyte age may be the result of a low rate of spontaneous cortical granule exocytosis.

8. Chamber Geometry

Essentially nothing has been reported on direct comparisons of chamber type on fusion or alignment. Most investigators use wire electrode chambers with gaps of 200–1000 μm.

B. Activation

1. Mechanism of Electrically Induced Oocyte Activation

Sperm-induced oocyte activation is thought to be mediated by transient increases in intracellular calcium. Transient increases in calcium have been measured following fertilization in mouse (Cuthbertson and Cobbold, 1985) and hamster (Igusa and Miyazaki, 1986). Manipulation of intracellular calcium by injection of calcium (Fulton and Whittingham, 1978) can stimulate activation. Electrical pulses are also thought to cause activation by inducing an elevation of intracellular calcium (Whittingham, 1980). Our evidence in the rabbit (R. Fissore and J. M. Robl, unpublished observations) indicates that electrical pulses cause an immediate increase in intracellular calcium from a baseline of 56 nM to a peak of 160 nM. Calcium then declines back to baseline with a half-time of approximately 85 s. Interestingly, oocytes that activate show a response that is similar to the response of oocytes that do not activate. Also, the level of calcium response does not relate to activation. The source of this calcium is external and probably dependent on electrically induced pores because oocytes pulsed in calcium-free medium do not show an increase in calcium.

2. Activation Medium

Little is known of the specific requirements of the medium for activation. Because the calcium transient caused by electrical stimulation is the result of an influx of calcium it would then be expected that calcium would be required in the medium.

Ozil (1990) verified this in an experiment with rabbit oocytes in which none of 105 oocytes activated in medium without electrolytes whereas up to 100% activated with the addition of 10 μM $CaCl_2$. The concentration of calcium in the medium determines the peak elevation in intracellular calcium (R. Fissore and J. M. Robl, unpublished observations). Because the calcium peak is not related to activation, the concentration of calcium in the medium may not be critical for activation.

3. Field Strength and Pulse Duration

In a study by Collas *et al.* (1989) no precise field strength or pulse duration was found to be optimum for activation. Activation rate was much more influenced by age of the oocyte and number of pulses than pulse parameters. This may be explained by the observation that all oocytes respond to an electrical pulse with a transient increase in intracellular calcium. Some other factor within the oocyte, however, ultimately determines whether or not the oocyte will activate in response to this calcium transient. Furthermore, the level of calcium response, which would be related to pore number and ultimately to field strength and pulse duration, is not related to activation. It is therefore expected that manipulating pulse field strength and duration would have little effect on activation.

4. Multiple Pulses

The rate of activation of oocytes is greatly improved with the use of multiple pulses. In the mouse, activation efficiency increased linearly with up to six pulses, each 30 min apart. Furthermore, activation rate increased linearly with the interval between pulses from 5 to 60 min (Collas *et al.*, 1989). In neither experiment could the increase be attributed solely to aging of the oocyte. Ozil (1990) found that activation rates of essentially 100% can be achieved in rabbits by pulses at a 4-min interval using a logarhythmic decline in pulse duration. Multiple electrical pulses presumably cause repeated calcium transients, which induce activation in oocytes not capable of responding to a single pulse.

5. Oocyte Age

Age of the oocyte is the single most important factor influencing oocyte activation. In contrast to fusion rate, activation rate increases with aging of the oocyte. This has been shown with electrical stimulation in the mouse (Collas *et al.*, 1989), rabbit (Collas and Robl, 1990), and cow (Ware *et al.*, 1989) oocyte. Activation rates are close to zero using a single pulse on oocytes at the time of ovulation and increase over a period of 12–14 h to essentially 100%. This change is not due to a change in calcium response to the electrical pulse. The factor that is thought to be responsible for inhibiting activation in recently ovulated oocytes is cytostatic factor (reviewed by Masui and Shibuya, 1987), which has recently been identified as the proto-oncogene product c-mos (Sagata *et al.*, 1989). Interestingly, the degradation of c-mos is regulated by calcium (Schollmeyer, 1988; Watanabe *et al.*, 1989), which may explain the fact that successive pulses are beneficial for activation.

6. Chamber Geometry

We have investigated round and rectangular electrodes and chambers with electrode gaps from 200 to 1000 μm for activation of mouse oocytes (Collas *et al.*, 1989). Although we believe that essentially any chamber could be used effectively for activating oocytes, the optimum field strength for each type of chamber would vary considerably. For example, in comparing a chamber with 1-mm wire electrodes to a chamber with 1-mm rectangular electrodes, a very high lysis rate was observed at 3 kV/cm with wire electrodes but no lysis at 4 kV/cm with the rectangular electrodes. Furthermore, using chambers with wire electrodes, but of 500- or 200-μm gaps, instead of 1 mm, resulted in a lower rate of lysis.

V. Conclusion

Highly effective procedures for electrically fusing and activating mammalian oocytes are required for successful cloning of mammalian embryos. Fusion has been the least difficult problem to solve. Fusion rates of 80–90% can be achieved when optimum pulse parameters are used and when nonfused embryos are repetitively realigned and repulsed. Activation, on the other hand, has been a more perplexing problem. The most effective methods, to date, of achieving high activation rates are to use aged oocytes or to give multiple electrical pulses. With either approach activation rates of close to 100% can be achieved.

Acknowledgments

We would like to thank Lisa Korpiewski for her assistance with preparation of the manuscript. We would also like to thank Dr. Kenji Endo and Teresa Conway for their contributions in the laboratory. This work was supported in part by U.S. Department of Agriculture grant 88-37240-4106 to J. M. Robl.

References

Bondioli, K. R., Westusin, M. E., and Looney, C. R. (1990). Production of identical bovine offspring by nuclear transfer. *Theriogenology* **33**, 165–174.

Clement-Sengewald, A. (1988). Elektrofusion and Enukleation von Blastomeren bei Saugetierembryonen. Dissertation zur Erlangung der tiermedizinischen Doktorwurde, Tierarztlichen Fakultat der Ludwig-Maximilians-Universitat, Munchen, FRG.

Collas, P., and Robl, J. M. (1990). Factors affecting the efficiency of nuclear transplantation in the rabbit embryo. *Biol. Reprod.* 43, 877–884.

Collas, P., Balise, J. J., Hofman, G. A., and Robl, J. M. (1989). Electrical activation of mouse oocytes. *Theriogenology* 32, 835–844.

Cuthbertson, K. S. R., and Cobbold, P. H. (1985). Phorbol ester and sperm activate mouse oocytes by inducing sustained oscillations in cell Ca^{++}. *Nature* 316, 541–542.

Fulton, B. P., and Whittingham, D. G. (1978). Activation of mammalian oocytes by intracellular injection of calcium. *Nature* 273, 149–151.

Igusa, Y., and Miyazaki, S. (1986). Periodic increase of cytoplasmic free calcium in fertilized hamster eggs measured with calcium-sensitive electrodes. *J. Physiol. (Lond.)* 377, 193–205.

Illmensee, K., and Hoppe, P. C. (1981). Nuclear transplantation in *Mus musculus*: Developmental potential of nuclei from preimplantation embryos. *Cell* 23, 9–18.

Knight, D. E. (1981). Rendering cells permeable by exposure to electric fields. *Tech. Cell. Physiol.* P113, 1–20.

Kono, T., Shioda, Y., and Tsunoda, Y. (1988). Nuclear transplantation of rat embryos. *J. Exp. Zool.* 248, 303–305.

Kubiak, J. Z., and Tarkowski, A. K. (1985). Electrofusion of mouse blastomeres. *Exp. Cell. Res.* 157, 561–566.

Masui, Y., and Shibuya, E. K. (1987). Development of cytoplasmic activities that control chromosome cycles during maturation of amphibian oocytes. *In* "Molecular Regulation of Nuclear Events in Mitosis and Meiosis" (R. A. Schlegel, M. S. Halleck, and P. N. Rao, eds.), Academic Press, New York.

McGrath, J., and Solter, D. (1983). Nuclear transplantation in the mouse embryo by microsurgery and cell fusion. *Science* 220, 1300–1302.

Ozil, J.-P. (1990). The parthenogenetic development of rabbit oocytes after repetitive pulsatile electrical stimulation. *Development* 109, 117–127.

Ozil, J.-P., and Modlinski, J. A. (1986). Effects of electric field on fusion rate and survival of 2-cell rabbit embryos. *J. Embryol. Exp. Morphol.* 96, 211–228.

Prather, R. S., Barnes, F. L., Sims, M. L., Robl, J. M., Eyestone, W. H., and First, N. L. (1987). Nuclear transplantation in the bovine embryo: Assessment of donor nuclei and recipient oocyte. *Biol. Reprod.* 37, 859–866.

Robl, J. M. (1988). Principles of micromanipulation. *In* "Hands-on IVF, Cryopreservation, and Micromanipulation" (N. L. First and A. H. DeCherney, eds.), University of Wisconsin Biotechnology Center, Madison, Series No. 2.

Robl, J. M., and First, N. L. (1985). Manipulation of gametes and embryos in the pig. *J. Reprod. Fertil. Suppl.* 33, 101–114.

Robl, J. M., Prather, R., Barnes, F., Eyestone, W., Northey, D., Gilligan, B., and First, N. L. (1987). Nuclear transplantation in bovine embryos. *J. Anim. Sci.* 64, 642–647.

Sagata, N., Watanabe, N., Vande Woude, G. F., and Ikawa, Y. (1989). The c-mos proto-oncogene product is a cytostatic factor responsible for meiotic arrest in vertebrate eggs. *Nature* 342, 512–518.

Schollmeyer, J. E. (1988). Calpain II involvement in mitosis. *Science* 240, 911–913.

Smith, L. C. (1989). Studies on the early development of mammalian embryos by nuclear transplantation. Ph.D. thesis, University of Edinburgh, UK.

Smith, L. C., and Wilmut I. (1989). Influence of nuclear and cytoplasmic activity on the development *in vivo* of sheep embryos after nuclear transplantation. *Biol. Reprod.* **40**, 1027–1035.

Sowers, A. E. (1989). The mechanism of electroporation and electrofusion in erythrocyte membranes. *In* "Electroporation and Electrofusion in Cell Biology" (E. Neuman, A. E. Sowers, and C. A. Jordan, eds.), Plenum Press, New York.

Sowers, A. E., and Lieber, M. L. (1986). Electropores in individual erythrocyte ghosts: diameters, lifetimes, numbers, and locations. *FEBS Lett.* **205**, 179–184.

Stice, S. L., and Robl, J. M. (1988). Nuclear reprogramming in nuclear transplant rabbit embryos. *Biol. Reprod.* **39**, 657–664.

Ware, C. B., Barnes, F. L., Maiki-Laurila, M., and First, N. L. (1989). Age dependence of bovine oocyte activation. *Gamete Res.* **22**, 265–275.

Watanabe, N., Vande Woude, G. F., Ikawa, Y., and Sagata, N. (1989). Specific proteolysis of the c-mos proto-oncogene product by calpain on fertilization of *Xenopus* eggs. *Nature* **342**, 505–511.

Whittingham, D. G. (1980). Parthenogenesis in mammals. *Oxford Rev. Reprod. Biol.* **2**, 205–231.

Willadsen, S. M. (1986). Nuclear transplantation in sheep embryos. *Nature* **320**, 63–65.

Part IV

Instrumentation for Electroporation and Electrofusion

35

Pulse Generators for Electrofusion and Electroporation

Bruce M. Chassy,[1] James A. Saunders,[2] and Arthur E. Sowers[3]

[1]Department of Food Science, University of Illinois at Urbana-Champaign, Urbana, Illinois 61801

[2]Plant Sciences Institute, Beltsville Agricultural Research Center, United States Department of Agriculture, Beltsville, Maryland 20705

[3]Department of Biophysics, School of Medicine, University of Maryland at Baltimore, Baltimore, Maryland 21201

I. Introduction

The original scientific observations documenting electrofusion and electroporation were conducted with experimenter-designed pulse generators. As methods for the electrical manipulation of cells became established, commercial generators appeared on the market. Development of improved apparati continues in both private and public laboratories. However, existing commercial equipment can be used for a wide variety of applications. The first-time buyer of electrofusion or electroporation equipment is faced with a complex buying decision. This chapter gives a brief description of many of the available units and, to the extent that the information was available, gives technical specifications.

The generators summarized in this chapter fall into two broad categories of design: equipment used for electrofusion and equipment employed for electroporation. A few generators can be useful for both applications. In general, the electroporation power supplies are designed to deliver a single discharge, which transiently permeabilizes the cell membrane. Electrofusion generators provide the additional capability of delivering an AC cell prealignment signal. A variety of approaches to prealignment are represented in the commercially available generators; the end user must decide which is optimal for the application. In electroporation procedures, low voltages and long pulse lengths are frequently applied to cells and protoplasts; bacteria and yeasts require higher field strengths. The majority of the generators deliver either decaying exponential or square-wave pulses. The potential buyer needs to decide on the intended purpose, electrofusion or electroporation, and select a generator with the appropriate electrical parameters. Frequently, the literature can serve as a guide to what capabilities are needed in a generator. Specific examples can be found in the accompanying chapters. Voltage, effective field strength, time constant or capacitance, number of repetitions, and prealignment parameters are of primary concern. Other considerations are the provision for data observation and collection, temperature control, type of controls provided, versatility, customer service, quality of construction, electrical safety, warranty, and available sample chambers. The prices charged for commercial generators vary between about $1000 to upward of $30,000 (US) for a complete system; the wide range of costs represents the widely differing capability and complexity of the apparatus.

For those qualified to undertake construction of an electronic pulse generator, a homebuilt apparatus is an economical alternative. Simple capacitive discharge generators, which deliver an exponentially decaying waveform suitable for bacterial electroporation, can be assembled with a few hundred dollars in parts. More complex apparatus, some of which use generic high-power pulse generators not specifically designed for electrical manipulation of cells, can also be used to advantage. Homebuilt machines also allow electrical manipulations that cannot be obtained with commercial generators. Some investigators have reported extremely satisfactory results with power supplies of their own design or construction. The would be home builder should be extremely cautious in their construction and attention to safety factors. Generators produce dangerous high-voltage, high-current discharges that are potentially lethal.

II. Commercially Available Generators Designed for Electrofusion and Electroporation

A. Bioelectronics

Bioelectronics Corporation[1]
1852 Thunderbird Road
Troy, MI 48084
Tel. (313) 362-2727

Biolectronics Corporation manufactures generators designed for electrofusion, injection (the company's term for electroporation), and extraction (the removal of molecules from a cell). A unique waveform is described as an oscillating pulse with a sharp spike at the end. An optional electronically controlled centrifuge is used to bring cells together and to flatten cells prior to the electrofusion or electroporation pulse. The product line includes the model 10000 high-voltage electronic cell processor, the model 20000 electronic centrifuge accessory, and the model 30000 universal cell injector.

1. Model 10000

Type: Unique waveform with end spike
Volts output: 10–3500 V
Pulse duration: 10–40 μs in four steps
Dataout/display: Digital meters
Electrofusion: Yes
RF field: 0–400 V (rich in harmonics); 400 kHz, timer controlled
Sample chamber: 2-ml autoclavable sample chambers; multiple chambers, up to 50, can be loaded in the centrifuge for a series of experiments
Additional features: Model 20000 centrifuge offers up to 14,000 × g.

2. Model 30000

A stand-alone cell injector centrifuge with no RF alignment capabilities and built in 3000-V, 15-μs pulse generator.

B. Bio-Rad

Bio-Rad Laboratories
3300 Regatta Blvd.

---

[1]Mention of a trademark or proprietary product does not constitute a recommendation by the authors over other products which may also be suitable.

Richmond, CA 94804
Tel. (800) 424-6723
Telex 3720184
FAX (510) 741-1047

1. Gene Pulser

Microprocessor-controlled pulse generator for the electroporation of eukaryotic and prokaryotic cells. Delivers exponentially decaying pulse. Accessory units provide increased capacitance and resistive timing. Company offers sterile, individually packaged 1-, 2-, and 4-mm sample cuvettes and Electro-Competent *Escherichia coli* cells.

Type: Exponential decay

Volts output: 50–2500 V in 10-V increments

Time constant: From 5 μs to 1 s; four capacitors of 0.25, 1.0, 3.0, 25 μF are included; optional capacitance extender has four capacitors with 125, 250, 500, and 960 μF

Dataout/display: Digital readout of voltage delivered and time constant; oscilloscope output possible

Electrofusion: Yes, without dielectrophoresis

Maximum field strength: 25 kV/cm with 1-mm cuvette

Sample chamber: 1-, 2-, and 4-mm sterile disposable cuvettes with maximum working volumes of 80, 400, and 800 μl, respectively

Additional features: Capacitance extender for 125, 250, 500, and 960 μF; pulse controller contains parallel resistors of 0, 100, 200, 400, 600, 800, and 1000 Ω, which allow high-voltage pulses to be delivered. Also contains 20-Ω current-limiting resistor in series with sample.

C. Braun

B. Braun Diessel Biotech Gmbh.
Postfach 110
Carl-Braun Strasse 1
3508 Melsungen, Federal Republic of Germany
Tel. 49-566-171-3943
FAX 49-566-171-3702

1. Biojet MI

High-voltage power supply generating short pulses of defined duration or long pulses of nonregulated duration.

Type: Exponential

Volts output: 0–15,000V (27,000 V/cm for 5.5-mm electrode gap)

Time constant: 0–59 s with manual control; 1 min to 59 min 59 s with automatic operation

Dataout/display: Digital display

Electrofusion: No

Sample chamber: 0.1–6.0 ml, up to 60 ml on request, electrode gap 5.5 mm

Additional features: Microprocessor controlled; stores up to 20 different user-defined programs

2. Biojet CF 100

Cell electrofusion instrument offering a high-frequency range of up to 10 MHz plus a pulse generator for square-wave pulse of controlled duration. Can store up to 50 user-defined programs in battery-buffered RAM

Type: Square wave

Volts output: 0–400 V, duration 5–100 μs, at AC off time of 10–999.9 s

Number of pulses: 1–9 at pulse interval 0.1–999.9 s

Dataout/display: Parameters available on screen; conductivity check built in

Electrofusion: Yes

Alignment parameters: 0–15 V for frequencies below 5 MHz, 0–12 V for frequencies below 8 MHz, 0–8 V for frequencies above 8 MHz; three time settings allowed with 1-s steps, 0–1000 s, 0–500 s, 0–500 s

Sample chamber: Various

Additional features: Special alternating field procedure for aligning cells which show a large difference in size, and menu-guided operation; has ramped or constant post-alignment mode of 1–999 s

3. Biojet CF 50

All-purpose-built cell electrofusion instrument. Offers a frequency range of up to 2 Mhz plus a pulse generator for square wave pulses of up to 9.999 ms. Displays all relevant electrofusion parameters on one display. Electrofusion conditions are entered using the keyboard; voltages are set with two adjusting potentiometers. Operation is either manual or following preset values. Memorizes the last electro-fusion condition used.

Type: Square wave

Electrofusion: Yes

Volts output: 15–250 V

Time constant: 5–9999 sec

Number of pulses: 1–99 at pulse interval of 0.1–9.9 s

Dataout/display: All operating parameters displayed on one screen

Alignment parameters: 0.5–20 V of 1–999 μs duration (0.1–2 MHz); AC off time is fixed

Sample chamber: Various including microslide and helical chamber

Additional features: Protoplast electrofusion mode provides plant biologists (and others working with large cells) with an even finer voltage control than during normal operation; has ramped postalignment mode of 1–999 s

D. BTX

BTX
3742 Jewell Street
San Diego, CA 92109
Tel. (619) 270-0861, (800) 289-2465
FAX (619) 483-3817

1. Electro Cell Manipulator 600

Single-unit design with over 1000 pulse lengths, two voltage ranges, and dual-display monitoring. The ECM 600 is designed for use with all types of cells: prokaryotes and eukaryotes. Internal monitoring measures true peak voltage and pulse length.

Type: Exponential decay

Volts output: two voltage ranges, high-voltage mode 300–2500 V, low-voltage mode 50–500 V

Time constant: Capacitance range from 25 to 3175 μF in 25-μF increments and 10 internal timing resistance settings from 13 to 720 ohms gives 1270 pulse-length options

Dataout/display: Instantaneous monitoring—set voltage, set capacitance, peak discharge voltage, pulse length, mode; pulse settings and measured parameters readily visible at all times

Electrofusion: No

Maximum field strength: >40 kV/cm with 0.56-mm Flatpacks and optional Flat-pack Slider

Sample chamber: 11 designs, reusable or sterile disposable, 0.56-, 1- or 2-mm gap

Additional features: Short-circuit- and arc-proof without sample solution limitations; BTX Electronic Genetics Database Service

2. *Electro Cell Manipulator 200*

Pulse power generator with both AC and DC voltage output. Equipment is designed for versatile application to plant, bacterial, and cell electroporation and electrofusion including embryo cloning. Operation in either automatic or manual mode. A built-in conductance meter monitors chamber media both before and after the pulse. Operates with all BTX electrodes including microslides, which are parallel wires adhered to glass slides for direct viewing with a microscope.

Type: Square wave
Volts output: 0–75 V AC, 0–960 V DC
Time constant: 0–99 μs (1-μs increments), pulses 1–99 automatic, unlimited
    manual
Dataout/display: Volts, time constant, conductance
Electrofusion: Yes
Alignment parameters: 1 MHz RF, 75 V zero to peak, 0–99 s, special BTX sine
    wave, automatic and manual modes
Sample chamber: Microslide, 1.9- and 3.5-mm cuvette, unique Meander fusion
    chamber sputter-coated design with 0.05-, 0.1-, and 0.2-mm gaps on slide, and
    molded cuvette chambers in 1- and 2-mm gap sizes; for electroporation; 11
    designs, reusable or sterile disposable, 0.56, 1, or 2 mm gap
Additional features: BTX Electronic Genetics Database Service

3. *T 800 Electroporator*

The T800 has the same features as the ECM 200, but does not include the alignment capability or the conductance meter.

Type: Square wave
Volts output: 0–960 V DC
Time constant: 0–99 μs (1-μs increments), pulses 1–99 automatic, unlimited
    manual
Dataout/display: Pulse amplitude (V)
Electrofusion: No
Sample chamber: Reusable chambers of 1.9- and 3.5-mm gap and sterile disposable
    cuvettes of 1- and 2-mm gaps width
Additional features: BTX Electronic Genetics Database Service

4. *Graphic Monitoring—Optimizor—Graphic Pulse Analyzer*

Microprocessor-controlled storage pulse analyzer, combining an oscilloscope and computer. It connects in-line between the pulse generator and the chamber. Measures, calculates, and displays multiple parameters in the cell chamber after the pulse is delivered. Displays graphical representation of pulse shape. Included with BTX ECM200 and T800 Super Systems. Optional Power Plus external voltage booster for use with ECM 200 and T 800. Employs a 2.6-fold stepup transformer to boost signal.

E. EquiBio s.a.

EquiBio s.a.
Rue Dossin, 45, B-4000
Liege, Belgium
Tel. 32 41 520009
FAX 32 41 520019

1. Cellject C2

Type: Decaying exponential waveform
Volts output: 2.5 kV at 40 µF, 450 V from 150 to 2100 µF, voltage adjustable from 100 to 2500 V with 10 V resolution
Time constant: Capacitor choice 40, 150, 300, 450, 600, 900, 1200, 1500, 1800, 2100 µF
Pulse time controller: 74, 90, 132, 192, 282, 412, 600, 1320 Ω and infinite
Dataout/display: Printer output RS 232 (300–9600 baud), single display with parameter selection
Electrofusion: No
Maximum field strength: 12,500 V/cm (C4 = 6.125 kV/cm)
Sample chamber: Disposable sterile 2- and 4-mm cuvettes with a capacity of 110–800 µl
Additional features: Continuous microprocessor control; the Cellject Twin is based on a patented double pulse process in which the electroporation pulse is followed by a second long-duration, low-amplitude pulse said to improve electroporation frequency

F. Hoefer Scientific Instruments

Hoefer Scientific Instruments
654 Minnesota Street
P.O. Box 77387
San Francisco, CA 94107

Tel. (415) 282-2307, (800) 227-4750
Telex 470778
FAX (415) 821-1081

1. PG 200 Progenetor II

Modern design features internal power supply, variable capacitance, choice of platinum reusable electrodes or disposable cuvettes. Designed for electroporation of plant, bacterial and animal cells.

Type: Exponential decay or square wave
Volts output: 0–500 V
Discharge Interval: 10 μs to 0.1 s in 10-μs intervals, from 1 to 10 s in 10-ms intervals
Maximum field strength: 25,000 V/cm depending on electrode
Time constant: Capacitance 0–2700 μF in 30 steps
Capacitor Values: (in μF) 100, 220, 490, 760, 1200, can be set in any combination
Charge time: less than 1 s to charge 1200 μF to 300 V
Dataout/display: Digital voltage meter
Electrofusion: No
Sample chamber: Two ring and multiple-plate electrodes, disposable cuvettes
Additional features: Electroporation chamber interlock; PG200 will not go through charge/discharge cycle until the lid is closed; the electrode must also be inside the chamber; automatic discharge of excess charge on capacitors

G. International Biotechnologies, Inc.

International Biotechnologies, Inc.
Subsidiary of Eastman Kodak Company
25 Science Park
New Haven, CT 06511
Tel. (800) 243-2555
FAX (203) 786-5694

1. geneZAPPER

Microprocessor-controlled pulse generator for the electroporation of plant, animal and bacterial cells. Designed for reproducible pulse delivery.

Type: Exponential RC time constant
Volts output: Settings 50–2500 V DC, in 1-V increments

Time constant: There are 34 capacitance settings between 7 and 1550 μF in 50-μF increments.
Dataout/display: 7 Segment LED displays voltage, capacitance, and time constant; BNC output connector
Electrofusion: No
Maximum field strength: 12,500 V/cm
Sample chamber: 2-, 4-, and 10-mm Reusable stainless steel plate accepts standard disposable cuvettes, clip adapter accepts presterilized, disposable, 2- or 4-mm foil-lined cuvettes
Additional features: Safety features include electrode safety guard, two discharge buttons, automatic discharge internally; protocol reference database service
Wave Controller: Selectable resistance accessory expands the capabilities of the geneZapper 450/2500 Electroporation System for bacterial cells.

H. Jouan

Jouan, Inc.
125-D Route 7, East
P.O. Box 2716
Winchester, VA 22601
Tel. (703) 665-0863
Telex 904 348

1. Cellular Electropulsator PS 10

Type: Square wave
Volts output: 0–1500V
Time constant: 5 μs to 24 ms
Dataout/display: Time, frequency
Electrofusion: Yes
Maximum field strength: 10,000 V/cm
Sample chamber: 0.15-, 0.25-, and 0.4-cm electrodes, various cuvettes, and 150 μl to 60 ml/min flow through cell
Additional features: Digital adjustment of voltage, duration, and frequency

I. Life Technologies, Inc.

Life Technologies, Inc.
8717 Grovemont Circle
P.O. Box 9418

Gaithersburg, MD 20898
Tel. (880) 828-6686
FAX (800) 331-2286

1. BRL Cell Porator Electroporation System

Microprocessor-controlled exponential decaying pulse power supply designed for electroporation of all kinds of cells. Employs optional voltage-booster for bacterial electroporation. Features ice bath for sample cuvettes. Electrocompetent *Eschericia coli* are also available.

Type: Exponential decay
Volts output: Two voltage ranges, high-voltage mode 50–2500 V, low-voltage mode 10–400 V
Time constant: Capacitance in eight steps to deliver time constants for 0.1–200 ms; 3–150 ms with voltage booster
Dataout/display: Digital readout of voltage delivered; oscilloscope output included
Electrofusion: No
Maximum field strength: 2.66 kV/cm; 16.6 kV/cm with voltage booster
Sample chamber: 4- and 1.5-mm sterile disposable cuvettes with working volumes of 0.24–1 ml and 25 μl, respectively.
Additional features: Toll-free techline; all units are safety interlocked

J. Shimadzu Corporation

Shimadzu Corporation
1, Nishinokyo-Kuwabaracho, Nakagyo-Ku
Kyoto 604, Japan
Tel. 075-823-1351
Telex 05422-166 SHMDS J
FAX 075-822-2617

1. SSH-1

The SSH-1 (somatic cell hybridizer) is designed primarily as a convenient cell electrofusion instrument with automatic operation. The capability to deliver high-intensity DC pulses also makes it suitable for electroporation of cells.

Type: Square wave
Volts output: 40–700 V

Duration of pulses: 10–500 μs pulse width; 10 pulses at intervals from 0.1 to
 999 s
Dataout/display: Equipped with standard graphic printer for all parameters and
 graphs; direct readouts of parameters on status indicators
Electrofusion: Yes
Maximum field strength: 14,000 V/cm
Alignment parameters: 0–40 V AC, 0–999 s pre- and post-pulse, 0.25, 0.5, 1, or
 2 MHz RF
Sample chamber: Eight chambers; 10 μl to 1.6 ml capacity with parallel electrodes
 of 0.5–4 mm; suitable for microscopic observation
Additional features: Sample cell temperature control system optional; remote keypad
 available; microprocessor-controlled and keypad programable; 29 types of auto-
 clavable chambers and five types of disposable chambers available.

2. GTE-10

Microprocessor-controlled exponential decaying pulse power supply designed for
electroporation of all kinds of cells. Wide range of time constants and voltages.

Type: Exponential decay
Volts output: Two voltage ranges, high-voltage mode 50–2500 V, low-voltage
 mode 50–500 V
Time constant: Capacitance in steps of 1, 3, 10, 35 μF; Up to 2.5 s with external
 capacitance accessory
Dataout/display: Digital readout of pulse settings and measured parameters readily
 visible at all times
Electrofusion: No
Maximum field strength: Up to 54,000 V/cm with 0.5-mm chamber
Sample chamber: 29 Types of autoclavable chambers and five types of disposable
 chambers available.
Additional features: Optional temperature controlled cell; external capacitance enhancers of 50–200, 200–800, and 800–3200 μF and resistive time constant
 controller available; optional external controller and data out interface

III. Homemade and Commercial Pulse Generators

Growth in the use of electroporation and electrofusion during the 1980s was accompanied and assisted by the concurrent growth in the availability of commercial
instruments specifically designed for the biotechnology market. This section is
devoted to (1) a discussion of the design and use of homemade equipment and (2)
the existence of other commercially available equipment that may be useable for

electroporation and electrofusion beyond that designed for the biotechnology market. We refer here to pulse generators made for industrial and engineering purposes and not intended for the electroporation and electrofusion market. Although not intended in any way as an endorsement, we mention that: Avtech (Ogdensburg NY 13669), Cober Electronics, Inc. (Stamford, CT 06902), Instrument Research Company (Columbia, MD 21045), Grass Instrument Co. (Quincy, MA), and Velonix (Santa Clara, CA 95050) all have product lines of pulse generators having enough voltage and power outputs that should be useful in many electroporation or electrofusion applications. Other manufacturers and sources can be found by consulting specialized catalogs, buying guides, research and development telephone directories, and electronic buyers handbooks. The authors have utilized a surplus property division as a source of a power supply unit used in a previously completed project. It is likely that units recovered from surplus property sites will not be accompanied by operation manuals. Hence the usefulness of such units may depend on individual resourcefulness.

All of the devices that these sources offer are likely to be much heavier, physically, and primarily designed for generally heavier use than many of those devices designed for the biotechnology market. Biotechnology applications such as electroporation and electrofusion will have to be optimized by the researcher with little input from the commercial manufacturers, since the companies will not be able to supply matching chambers to a customer. Such chambers may have to be procured from another manufacturer or be homemade. When sufficient expertise is available, homemade equipment will be cheaper, more readily modified or repaired, and more appropriate for mechanistic studies. However, the time required to build one's own equipment may excede the investment expectations of many researchers.

Homemade devices for supplying a strong electrical pulse for electroporation and electrofusion have been described recently in considerable detail (Mischke et al., 1986; Miles and Hochmuth, 1987; Sowers, 1989; Speyer, 1990). The general construction of scientific apparatus may be found in Moore et al. (1983). It is advisable to have means for measuring and checking the electrical performance of these devices. The most practical and advisable device is a storage screen oscilloscope. As voltages that can be harmful and potentially lethal are involved, it is important to consider the risk/benefit ratio of using homemade devices. Homemade high-voltage pulse generators have been in use in the laboratory of the authors for nearly a decade without serious incident. Care in design and precautions in the use of such equipment make accidental shock exposure no more hazardous or probable than other accidental hazards in laboratory activities (fire, broken glass lacerations, other human error).

In all cases, the design of electroporation and electrofusion apparatus must involve the consideration of (1) the desired field strength, $E = V/s$, where E is in volts (V) per unit distance s (e.g., mm, cm), (2) the desired duration of the pulse (decay half-time in the case of exponentially decaying waveforms and pulse width in the case of square wave pulses), and (3) the pulse power that is either available from the generator or needed by the chamber. Most pulse generators will produce a waveform that is either qualitatively "square" or "exponentially decaying." Although some research projects have utilized either complex waveforms or complex pulse protocols, the attributes of these waveforms may be annecdotal and are without a rigorously established validity. An apparatus that generates only low power levels will never be useable with large-volume chambers because these chambers will simply "overload" the generator. The outcome of the converse possibility—meaning the use of a strong generator with a chamber requiring only low power levels—is not predictable without technical information that is often not available from manufacturers of equipment for the biotechnology market but may be available from manufacturers of equipment for the industrial market. An example is the capacitor discharge (exponentially decaying) units. A large chamber will require a large capacitor. However, a small chamber may require a faster decay half-time than can be delivered by the lowest amount of capacitance available. Compensation for this may be possible by placing an external resistor in parallel with the chamber. The functioning of square wave generators can be based on one of many approaches which are usually complex. In some cases the fidelity of the square-waveform pulse may depend on technical details beyond the scope of this chapter.

Electrofusion requires, in addition to what is needed for electroporation, a means of bringing the membranes of cells into close contact. While many approaches can be used for this purpose, one of the most convenient methods involves the use of dielectrophoresis (Pohl, 1978). Dielectrophoresis is the name of a phenomenon in which cells in suspension become relocated (or "aligned") into positions such that they appear as "pearl chains" as a low-strength alternating electric field is generated within the suspension. While the theory is complex, the practical result is easy and convenient to obtain under conditions useful for membrane fusion applications. Misunderstandings of what is required for dielectrophoresis are common, and the reader is referred to other chapters within this volume for further information (see Chapter 8, Sowers). The fact that both a low-strength alternating field required for dielectrophoresis and a high-strength direct-current pulse must be delivered to the chamber in electrofusion protocols suggests that both sources can be connected to the chamber. However, a means must be provided to prevent the high-strength direct-current fusion pulse from damaging the source of the low-strength alternating current. The simplest way to satisfy this requirement is with a mechanical switch or relay (Mischke et al., 1986). A more elegant way to satisfy this requirement is with an electrical network. An example is given in Sowers (1989). Many other possibilities are obvious if simple electrical circuit theory is used. The interested reader is referred to Malmstadt et al. (1963, 1981) for further information.

References

Malmstadt, H. V., Enke, C. G., and Toren, E. C. (1963). "Electronics for Scientists." Benjamin, New York.

Malmstadt, H. V., Enke, C. G., and Crouch, S. R. (1981). "Electronics and Instrumentation for Scientists." Benjamin/Cummings, Menlo Park, NJ.

Miles, D. M., and Hochmuth, R. M. (1987). Micromanipulation and elastic response of electrically fused red cells. *In* "Cell Fusion" (A. E. Sowers, ed.), Plenum Press, New York, pp. 441–456.

Mischke, B. S., Saunders, J. A., and Owens, L. D. (1986). A versatile low-cost apparatus for cell electrofusion and electro-physiological treatments. *J. Biochem. Biophys. Methods* **13**, 65–75.

Moore, J. H., Davis, C. C., and Coplan, M. A. (1983). "Building Scientific Apparatus." Addison-Wesley, London.

Pohl, H. A. (1978). "Dielectrophoresis." Cambridge University Press, Cambridge.

Sowers, A. E. (1989). The mechanism of electroporation and electrofusion in erythrocyte membranes. *In* "Electroporation and Electrofusion in Cell Biology" (E. Neumann, A. E. Sowers, and C. A. Jordan eds.), Plenum Press, New York, pp. 229–256.

Speyer, J. F. (1990). A simple and effective electroporation apparatus. *Biotechniques* **8**, 28–30.

Index